The Ocean Surface

Wave Breaking, Turbulent Mixing and Radio Probing

The Ocean Surface

Wave Breaking, Turbulent Mixing and Radio Probing

Edited by

Y. TOBA
Department of Geophysics, Tohoku University, Japan

and

H. MITSUYASU
Research Institute for Applied Mechanics, Kyushu University, Japan

D. Reidel Publishing Company

A MEMBER OF THE KLUWER ACADEMIC PUBLISHERS GROUP

Dordrecht / Boston / Lancaster

Library of Congress Cataloging in Publication Data

The Ocean surface.

    Proceedings of the Symposium on Wave Breaking, Turbulent Mixing and Radio
Probing of the Ocean Surface, held at Tonoku University, Sendai, Japan on 19–25 July,
1984, sponsored by the IOC/SCOR Committee on Climate Changes and the Ocean and
the ICSU/WMO Joint Scientific Committee for the World Climate Research Programme.
      1.   Ocean waves–Congresses.   2.   Oceanic mixing–Congresses.   3.   Oceanography–
Remote sensing–Congresses.   4.   Turbulence–Congresses.   5.   Ocean-atmosphere inter-
action–Congresses.   I.   Toba, Y. (Yoshiaki), 1931–   .   II.   Mitsuyasu, Hisashi,
1929–   .   III.   IOC/SCOR Committee on Climate Changes and the Ocean.   IV.
WMO/ICSU Joint Scientific Committee.   V.   Symposium on Wave Breaking, Turbulent
Mixing and Radio Probing of the Ocean Surface (1984: Tohoku University)
GC206.O24   1985   551.47   85-2479
ISBN 90-277-2021-5

Published by D. Reidel Publishing Company
P.O. Box 17, 3300 AA Dordrecht, Holland

Sold and distributed in the U.S.A. and Canada
by Kluwer Academic Publishers,
190 Old Derby Street, Hingham, MA 02043, U.S.A.

In all other countries, sold and distributed
by Kluwer Academic Publishers Group,
P.O. Box 322, 3300 AH Dordrecht, Holland

Printed in The Netherlands

CONTENTS

Wave dynamics, wave statistics and wave modeling

Wave models

Wave dynamics and microwave probing

Radar, SAR, SLAR, scatterometry

Scatterometry and altimetry case studies

Remote sensor development

Mixed layer models for climate study

PREFACE

We are now entering an age when even the most delicate of climate changes may have significant effects on world economics, and on the very habitability of some areas of the world. This sensitivity can be related to a rapid increase in world population and possible changes in the environment brought about by man's extensive utilization of natural resources. This makes understanding of the geophysics of climate change one of the most important scientific problems of the end of this century.

The oceans are a central part of the climate system, both because they cover such a large percentage of the globe and because of their large thermal capacity and inertia. The oceans and the atmosphere interact directly at their interface, so that an understanding of the physical processes which occur in the upper boundary layer of the ocean, from the sea surface to the oceanic pycnocline, will be crucial to our understanding of climate variability.

The most conspicuous physical process at the sea surface is the generation and growth of wind waves. Progress in understanding the hydrodynamics involved has not, however, been rapid; both theory and experiment have been plagued by complications and nonlinearities. The problem of the initial generation of wavelets from calm water was finally solved satisfactorily only about five years ago. Attention is now focussed on the considerable progress made in the last few years in elucidating various characteristics of the waves themselves, such as instabilities of steep waves and breaking phenomena, and the details of the forcing of existing waves by the coupled air flow above them.

The possibility of making large-scale measurements of ocean waves from satellites has simultaneously opened a new epoch. The urgency of understanding the basic physics involved in the remote sensing of wind waves is emphasized by the swift approach of the "satellite measurement age" of the 1990's. The central unsolved problem of the day is the explanation of the relation between the fine structure on the sea surface and the observed electromagnetic scattering from it.

An IUCRM *Symposium on Wave Dynamics and Radio Probing of the Ocean Surface* was held at Miami Beach in 1981 as one of the activities of the IUGG/URSI Inter-Union Commission on Radio Meteorology. The Symposium reviewed recent progress in wave dynamics and prediction and relations with electromagnetic probing of the ocean surface.

The present *Symposium on Wave Breaking, Turbulent Mixing and Radio Probing of the Ocean Surface*, held at Tohoku University, Sendai, Japan on 19-25 July, 1984, was designed to serve as a sequel to the first IUCRM Symposium, with special emphasis this time on wave breaking as a key process connecting wave dynamics, the oceanic mixed layer and electromagnetic probing of the ocean surface. It was sponsored by the IOC/SCOR Committee on Climatic Changes and the Ocean (CCCO) and the ICSU/WMO Joint Scientific Committee for the World Climate Research Programme (JSC); the principal financial support came from the Japanese Ministry of Education, Science and Culture. This book is the Proceedings of the Symposium.

The Symposium was held on the campus of the Faculty of Science of Tohoku University, located on a hill and surrounded with green trees, overlooking the city of Sendai, itself known as the "Capital of Trees". About 180 scientists from 15 countries took part. The order in which the papers were given has been rearranged for the book into eleven groups, and includes both oral and poster papers. Special invited reviews are to be found at the beginning of four of the groups: Nonlinear Wave Dynamics; Wave Dynamics, Statistics and Wave Modelling; Wave Dynamics and Microwave Probing; and Mixed-Layer Models for Climate Study. These reviews were contributed by M.S. Longuet-Higgins, O.M. Phillips, G.R. Valenzuela and J.D. Woods. They run the gamut from comprehensive reviews of particular subjects to highly original contributions to the general subject of the meeting. These, combined with the other 73 papers, provide an up-to-date picture of the state of progress in this field of research.

The members of the Local Organizing Committee at Tohoku University were Y. Toba (Chairman), W. Brutsaert, H. Kamiyama, J. Kondo, H. Oya, N. Shuto, G. Takeda (Dean of the Faculty of Science), and M. Tanaka. The members of the International Advisory Committee, who were also very active in the organization of the meeting and should also be credited with its success, were H. Mitsuyasu (Chairman), T. Asai, M. Coantic, F. Dobson, Y. Furuhama, K. Hasselmann, N. Iwata, I.S.F. Jones, M.S. Longuet-Higgins, A.S. Monin, O.M. Phillips, R.W. Stewart, Y. Sugimori, T. Teramoto, G.R. Valenzuela, and J.D. Woods. Both Committees made up the Program Committee.

We express our special thanks to Prof. Asai, Dr. Dobson, Dr. H. Günther, Prof. Iwata, Dr. Jones, Prof. Longuet-Higgins, Prof. Phillips, Prof. Sugimori, Prof. Teramoto, Dr. Valenzuela and Prof. Woods; they served as session chairmen and also as referees for the papers. Prof. Woods, Chairman of the Liaison Committee of the CCCO and the JSC, and Prof. Nakao Ishida, President of Tohoku University, gave the keynote addresses. We express our sincere thanks to them for their kind cooperation.

Praise is also due to all members of the Physical Oceanography Laboratory of Tohoku University, including Mr. K. Hanawa, Dr. H. Kawamura, Miss Y. Inohana, Mr. A. Kubokawa, Mr. H. Mitsudera, and Mr. Y.

Yano for their devoted efforts in the local arrangements for the
Symposium.

    Lastly, we express our deep appreciation to the following
scientific organizations, which co-sponsored the Symposium: the
International Union of Geodesy and Geophysics (IUGG), the International
Union of Radio Science (URSI), the Scientific Committee on Oceanic
Research (SCOR), the International Association for the Physical Sciences
of the Ocean (IAPSO), the American Meteorological Society (AMS), the
Oceanographical Society of Japan, the Meteorological Society of Japan,
the Marine Meteorological Society, la Société franco-japonaise
d'océanographie, the Institute of Electronics and Communication
Engineers of Japan, the Society of Naval Architects of Japan, the Japan
Institute of Navigation, the Japan Society of Fluid Mechanics, the Flow
Visualization Society of Japan, the Remote Sensing Society of Japan, the
Japan Society for Natural Disaster Science and the Society of Airborne &
Satellite Physical & Fisheries Oceanography; and to the following
organizations which provided financial support of one kind or another in
addition to the support of the Ministry of Education, Science and
Culture: Asia Air Survey, Co., Ltd., Electrical Communication
Laboratories, NTT, Giken Kogyo Co., Ltd., Hidaka Foundation for the
Promotion of Oceanic Research, IBM Japan, Ltd., the International
Association for the Physical Sciences of the Ocean (IAPSO), Japan
Weather Association, Japex Geoscience Institute, Inc., Kajima
Foundation, Kokusai Denshin Denwa Co., Ltd.(KDD), Kokusai Kogyo Co.,
Ltd., Mitsubishi Electric Corporation, Miyagi Prefecture, National
Science Foundation, USA, Nippon Electric Co., Ltd., Remote Sensing
Technology Center of Japan, Sendai City, Shimadzu Science Foundation,
the 77 Bank, Ltd., Tohoku Department Store Association, Tohoku Electric
Power Co., Inc., Torey Science Foundation, Toshiba Corporation, and
Tsurumi-Seiki Kosakusho Co., Ltd.

     Sendai and Fukuoka
     25 December 1984

                              Yoshiaki Toba and Hisashi Mitsuyasu

A NEW WAY TO CALCULATE STEEP GRAVITY WAVES

M.S. Longuet-Higgins
Department of Applied Mathematics and Theoretical
Physics, University of Cambridge,
Silver Street, Cambridge CB3 9EW, England.

ABSTRACT. A simple and efficient way to calculate steep gravity waves
is described, which avoids the use of power series expansions or integral equations. The method exploits certain relations between the coefficients in Stokes's expansion which were discovered by the author in
1978.
     The method yields naturally the critical wave steepnesses for bi-
furcation of regular waves into non-uniform steady waves. Moreover,
truncation of the series after only two terms yields a simple model for
Class 2 bifurcation.
     The analysis can be used to discuss the stability of steep gravity
waves and to derive new integral relations. Particularly relevant to
breaking waves are some new relations for the angular momentum. The
level of action y_a for a limiting wave can also be expressed in terms
of the Fourier coefficients.

1.   INTRODUCTION

It is very desirable to be able to understand progressive nonlinear
gravity waves, and to calculate them efficiently, for two reasons:
     (1)   Such waves easily become unstable and may themselves break.
     (2)   They influence the behaviour of shorter waves which ride on
the backs of the longer waves, and may determine whether the shorter
waves will break.
     Modern methods of calculating steep surface waves may be said to
have begun with the work of Schwartz (1974) who discovered that the
first Fourier coefficient a_1 in Stokes's series was not a monotonically
increasing function of the wave steepness ak. Hence a_1 could not be
used as an expansion parameter over the whole range of waves. The same
was also found to be true of the wave speed c (Longuet-Higgins 1975)
and of many integral properties such as the density I of horizontal
momentum, and the kinetic and potential energies T and V. Thus I and
(T+V) both take maximum values together at ak = 0.429, while c and
(T-V) take maximum values together at ak = 0.436. An asymptotic theory,
showing that these are only the first of an infinite number of maxima

1

Y. Toba and H. Mitsuyasu (eds.), The Ocean Surface, 1–15.

and minima attained before the limiting steepness, was given by Longuet-Higgins and Fox (1978).

Later, bifurcations of the regular series of uniform waves into series of non-uniform, but steady, waves was discovered by Chen and Saffman (1980). The critical wave steepness for bifurcations of Class 2 and Class 3 apparently were close together, as shown in Figure 1.

Precise numerical calculations of steep waves have generally employed power series expansions with the wave height 2ak, for example, as a small parameter, and followed if necessary by Padé summation. Such methods require the use of a large number of coefficients, which must first be calculated. Other authors (Chen and Saffman) have used integral equations for the surface elevation, or other variable, but this does not in itself provide much insight into or understanding of the results.

In this paper we shall describe a new method of calculation, which is based on a system of quadratic relations between the Fourier coefficients a_n. These relations were discovered and proved by the author (Longuet-Higgins 1978a) but they have only recently been exploited (see Papers I, II and III). In this method the use of the complicated power series is avoided. Here we shall give a summary, showing not only that the method is efficient and accurate, but it also gives some physical insight into the occurrence of the bifurcations. In addition it provides a powerful analytical tool for solving other problems such as the nature of the normal-mode instabilities and the relationships between various integral properties of surface gravity waves.

2.   STEADY, PROGRESSIVE WAVES

Let us take rectangular coordinates (X,Y) in a frame of reference moving with the phase-speed c, the X-axis being horizontal and the Y-axis vertically upwards. The origin (0,0) may be chosen so that the mean surface level is $Y = -c^2/2g$ (the constant in Bernoulli's equation then vanishes). We choose units so that $g = 1$ and so that the period in the X-direction is 2π; thus the wavelength is a submultiple of 2π. After Stokes (1880) we express (X,Y) in terms of the velocity potential ϕ and streamfunction ψ by the Fourier series

$$(Y - iX) + (\psi - i\phi) = \tfrac{1}{2}a_o + \sum_{n=1}^{\infty} a_n e^{in(\phi+i\psi)/c} . \qquad (2.1)$$

For regular, symmetric waves the coefficients a_n are all real. The free surface is taken as the streamline $\psi = 0$.

Then it was shown in Paper I that the a_n satisfy the quadratic relations

$$a_0 + a_1 a_1 + 2a_2 a_2 + 3a_3 a_3 + \ldots = -c^2$$

$$a_1 + a_0 a_1 + 2a_1 a_2 + 3a_2 a_3 + \ldots = 0$$

$$a_2 + a_1 a_1 + 2a_0 a_2 + 3a_1 a_3 + \ldots = 0 \qquad (2.2)$$

$$a_3 + a_2 a_1 + 2a_1 a_2 + 3a_0 a_3 + \ldots = 0$$

. . . .

These can also be expressed in the form

$$F_n \equiv \frac{\partial F}{\partial a_n} = 0 \ , \quad n = 0,1,2,\ldots \qquad , \qquad (2.3)$$

where F is a certain cubic expression in the a_i (see Section 5). Now
the differential system derived from (2.2) is

$$dF_n \equiv \sum_m \frac{\partial F_n}{\partial a_m} \, d \, a_m = 0 \qquad (2.4)$$

in which it is clear from (2.3) that the coefficients $\partial F_n/\partial a_m$ are
symmetric in the suffixes n and m. In fact equations (2.4) may be
written as

$$\underset{\sim}{C} \times (d\underset{\sim}{a})^T = -dc^2 \, (1,0,0,\ldots)^T \qquad (2.5)$$

where $\underset{\sim}{a} = (\tfrac{1}{2}a_0, a_1, a_2, \ldots)$ and $\underset{\sim}{C}$ is the symmetric matrix

$$\underset{\sim}{C} = \begin{pmatrix} (1+1) & (a_1 + a_1) & (2a_2 + 2a_2) & \cdots \\ (a_1 + a_1) & (1+a_0 + 2a_2) & (2a_1 + 3a_2) & \cdots \\ (2a_2 + 2a_2) & (2a_1 + 3a_3) & (1+2a_0 + 4a_4) & \cdots \\ \vdots & \vdots & \vdots & \end{pmatrix} \qquad (2.6)$$

the elements all being simple linear combinations of the a_n.
    If $\mu = \mu(\underset{\sim}{a})$ denotes any parameter, say the wave amplitude

$$ak = a_1 + a_3 + a_5 + \ldots \qquad (2.7)$$

which varies monotonically throughout the range considered, then

$$d\mu = \sum_m \frac{\partial \mu}{\partial a_m} \, d \, a_m \ . \qquad (2.8)$$

    To calculate solutions to equation (2.2) numerically, it is conven-
ient to replace the first of equations (2.4) (the only one involving c^2)
by equation (2.8) and then use Newton's method of approximation. Thus
if $\underset{\sim}{a}^{(1)} = (\tfrac{1}{2}a_0^{(1)}, a_1^{(1)}, a_2^{(1)}\ldots)$ is any first approximation, we estimate
the next approximation $\underset{\sim}{a}^{(2)} = \underset{\sim}{a}^{(1)} + \delta\underset{\sim}{a}$ by solving

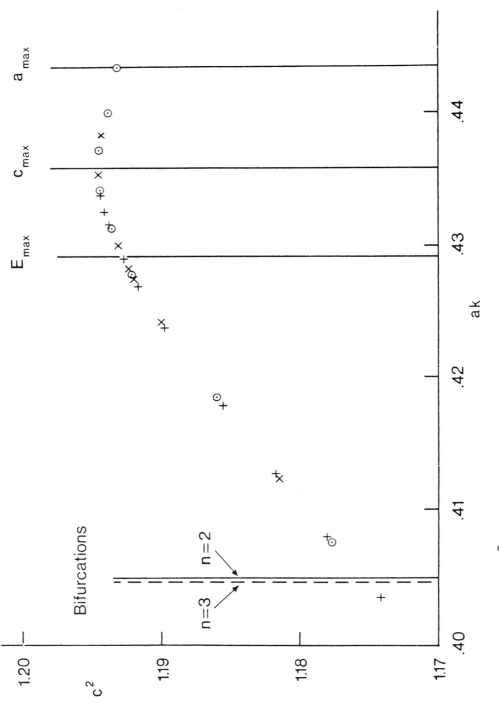

Figure 1.   Graph of c^2 vs. ak for steep gravity waves.   Circular plots denote previous determination by power series expansions.   Crosses denote present method.

$$
\begin{pmatrix}
0 & 1 & 0 & \cdots \\
(a_1+a_1) & (1+a_0+2a_2) & (2a_1+3a_3) & \cdots \\
(2a_2+2a_2) & (2a_1+3a_3) & (1+2a_0+4a_4) & \cdots \\
\vdots & \vdots & \vdots &
\end{pmatrix}
\times
\begin{pmatrix}
\tfrac{1}{2}da_0 \\
da_1 \\
da_2 \\
\vdots
\end{pmatrix}
=
\begin{pmatrix}
\mu-\mu^{(1)} \\
-F^{(1)} \\
-F^{(2)} \\
\vdots
\end{pmatrix}
$$

$$(2.9)$$

where μ is the prescribed value of the parameter and $\mu^{(1)}$, $F_n^{(1)}$ are the values of μ and F_n when $a = a^{(1)}$.

The above procedure is simple to programme, and converges very rapidly — two or three iterations will generally give six-figure accuracy. The value of the phase-speed c may then be calculated from the first equations (2.2). As seen in Figure 1, the results agree precisely with other, more elaborate, methods using power series expansions.

Instead of ak we might employ other parameters μ, for example $\mu = c$ or $\mu = a_0$. However c is not monotonic over the whole range. The condition that c have a turning point (dc = 0) is simply that

$$| \underset{\sim}{c} | = 0 .$$

$$(2.10)$$

From this condition it may easily be found that c^2 has a first maximum when ak = 0.436, in agreement with previous estimates (see Longuet-Higgins 1975; Longuet-Higgins and Fox 1978).

An alternative parameter which also is monotonic over the whole range of ak which tends to 1 for limiting waves is the "head"

$$Q = 1 - \tfrac{1}{2}q^2_{crest} .$$

$$(2.11)$$

In terms of the coefficients a_n we have by Bernoulli's equation

$$Q = 1 + Y_{\phi=0} = 1 + \tfrac{1}{2}a_0 + a_1 + a_2 + \cdots$$

$$(2.12)$$

so that all the terms $\partial\mu/\partial a_m$ in the top row of (2.9) are now unity. Calculations with this parameter also converge rapidly.

A third parameter that is sometimes convenient is the lowest coefficient a_0. However a_0 does have turning points where $da_0 = 0$. These are given by

$$| \underset{\sim}{c}' | = 0$$

$$(2.13)$$

where a prime denotes the matrix derived by omitting the first row and first column of the matrix. In Paper II it is found that the first stationary value of a_0 (a minimum) occurs when ak = 0.434, that is at a wave steepness intermediate between those for E_{max} and c_{max}.

Lastly, it was shown in II that the maximum energy, or E_{max}, corresponds to the vanishing of the determinant

$$\begin{vmatrix} (3c^2+a_0) & 4c^2 & 0 & 0 & \cdots \\ 1 & (1+1) & (a_1+a_1) & (2a_2+2a_2) & \cdots \\ 0 & (a_1+a_1) & (1+a_0+2a_2) & (2a_1+3a_3) & \cdots \\ 0 & (2a_2+2a_2) & (2a_1+3a_3) & (1+2a_0+4a_4) & \cdots \\ \vdots & \vdots & \vdots & \vdots & \end{vmatrix} \qquad (2.14)$$

and hence

$$| \underset{\sim}{c} | \, / \, | \underset{\sim}{c'} | = 4c^2/(3c^2+a_0) \quad . \qquad (2.15)$$

From this relation the corresponding value of ak was calculated as
ak = 0.4291, in agreement with previous estimates (Longuet-Higgins 1975).

3.   BIFURCATIONS

        As mentioned in the Introduction, a Class 2 system of waves, having
a horizontal period equal to two wavelengths, was found by Chen and
Saffman (1980) to have a bifurcation point, at which the regular, or
uniform, series of waves can branch into a series of non-uniform waves;
every alternative wave being higher than the preceding (or following) wave.

        For their calculation Chen and Saffman used an integral equation.
However, the analytic condition for bifurcation can be express-d much
more lucidly in terms of the Fourier coefficients a_n defined above.   In
fact, as shown in II, in the regular Class 2 system, all the <u>odd</u> Fourier
coefficients a_{2j+1} will vanish, and at the bifurcation point the odd in-
crements da_{2j+1} must satisfy the matrix equation

$$\begin{pmatrix} (1+a_0+2a_2) & (3a_2+4a_4) & (5a_4+6a_6) & \cdots \\ (3a_2+4a_4) & (1+3a_0+6a_4) & (5a_2+8a_8) & \cdots \\ (5a_4+a_6) & (5a_2+8a_3) & (1+5a_0+10a_5) & \cdots \\ \vdots & \vdots & \vdots & \end{pmatrix} \times \begin{pmatrix} da_1 \\ da_3 \\ da_5 \\ \vdots \end{pmatrix} = \begin{pmatrix} 0 \\ 0 \\ 0 \\ \vdots \end{pmatrix} \quad (3.1)$$

If the above matrix is denoted by $\underset{\sim}{C_2}$, then the condition for bifurcation
is simply that

$$| \underset{\sim}{C_2} | = 0 \quad . \qquad (3.2)$$

The elements a_{2i} of (3.1) are related to those of the corresponding coefficients a_j in the Class 1 system by $a_{2j} = \frac{1}{2}ja_j$. Hence the condition
(3.2) is easily computed, and bifurcation is found to occur at
ak = 0.40496 in agreement with the value found by Chen and Saffman.
        A similar criterion for Class 3 bifurcation (Paper II, Section 7)

leads to the value ak = 0.40469 for the critical steepness. This again
agrees with Chen and Saffman's value, to at least three significant
decimals.

The above results not only serve to confirm the numerical efficiency of the method. In addition it has the advantage of giving insight
into the physical reasons for bifurcation. For if in equations (2.4) we
retain only the terms in a_0, a_1 and a_2, we get the very simple quadratic
system

$$a_1 + a_0 a_1 + 2a_1 a_2 = 0$$

$$a_2 + a_1 a_1 + 2a_0 a_2 = 0 \qquad \qquad (3.3)$$

These equations may easily be solved for a_1 and a_2 in terms of a_0, with
the result shown in Figure 2. Regular Class 1 solutions, having
$a_2 \ll a_1$, lie in the branch AA'. Regular Class 2 solutions, with $a_1 = 0$,
lie on the a_2-axis. The nonuniform Class 2 solutions lie on the branch
BB'. The matrix in (3.1) consists of only one term, so the criterion
(3.2) reduces to

$$1 + a_0 + 2a_2 = 0 \quad . \qquad \qquad (3.4)$$

This and equations (3.3) are satisfied at the bifurcation point
$a_0 = -\frac{1}{2}$, $a_1 = 0$, $a_2 = -\frac{1}{4}$. The critical wave steepness is

$$ak = \mid 2a_2 \mid = 0.5 \qquad \qquad (3.5)$$

which differs from the more accurate value 0.405 by less than 20%. This
clearly indicates that the bifurcation is a consequence of a quadratic
interaction between the fundamental harmonic a_2 and its first subharmonic a_1.

A similarly simplified analysis can be given for the Class 3 bifurcation, involving only the three harmonics a_1, a_2, a_3, with a_0 as parameter (see Paper II, Section 7).

Can anything be said about Class 1 bifurcation? Consider first the
symmetric case (2.1) in which the coefficients a_n and their differen-
tials da_n are all real. From (2.4) it is clear that, given a point on
the regular branch of Class 1 waves and a small perturbation dc^2, then
provided $\mid \underset{\sim}{C} \mid \neq 0$ a unique solution da will exist, namely

$$da_n = (\Delta_n / \mid \underset{\sim}{C} \mid) \, dc^2 \qquad \qquad (3.6)$$

where Δ_n is the determinant derived from $\underset{\sim}{C}$ by replacing its (n+1)th
column by $(1,0,0,\ldots)^T$. Hence there are no symmetric bifurcations except possibly at stationary values of c^2.

To extend the argument to asymmetric perturbations we may write

$$a_n = p_n + iq_n, \qquad da_n = dp_n + iq_n \qquad \qquad (3.7)$$

p_n and q_n being real. The system of equations for
$\underset{\sim}{dp} = (\frac{1}{2}dp_0, \, dp_1, \, dp_2, \ldots)$ and $\underset{\sim}{dq} = (dq_1, \, dq_2, \ldots)$ takes the form

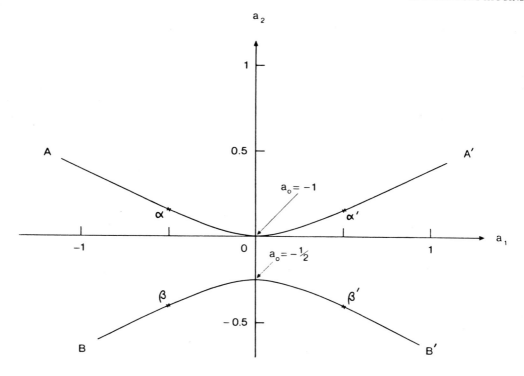

Figure 2. Class 2 bifurcation in the simplified model described by
equations (3.3). The a_2-axis represents regular Class 2
waves, BB′ the bifurcated system. AA′ represents regular
Class 1 waves.

$$\begin{pmatrix} \underset{\sim}{P} & \underset{\sim}{Q} \\ \underset{\sim}{Q}^T & \underset{\sim}{R} \end{pmatrix} \times \begin{pmatrix} \underset{\sim}{dp}^T \\ \underset{\sim}{dq}^T \end{pmatrix} = -dc^2 (1,0,0,\ldots)^T \qquad (3.8)$$

where $\underset{\sim}{P}$ is the same function of $\underset{\sim}{p}$ as $\underset{\sim}{C}$ is of $\underset{\sim}{a}$; $\underset{\sim}{R}$ is similar to $\underset{\sim}{P}$ but with the sign of the second term in each element of the matrix reversed; and $\underset{\sim}{Q}$ is a function of the q_n only (see Paper III, Section 8). When a_n is real (i.e. $\underset{\sim}{q} = \underset{\sim}{0}$), the system (3.8) splits into two independent systems:

$$\underset{\sim}{C} \times \underset{\sim}{dp}^T = -dc^2 (1,0,0\ldots)^T, \qquad \underset{\sim}{D} \times \underset{\sim}{dq}^T = 0 , \qquad (3.9)$$

where $\underset{\sim}{D}$ is obtained from $\underset{\sim}{R}$ by writing $p_n = a_n$. The first system in (3.9) is the same as (2.5) for increments of the regular waves. It has unique solutions dp except possibly when $dc = 0$. The second system always has solutions $\widetilde{dq}_n = np_n$ representing a pure phase-shift, and in fact we find $|\underset{\sim}{D}| \equiv 0$. However if we replace one equation of the second system by the condition $dq_1 = 0$, fixing the phase, the ratios $dq_2 : dq_3 : \ldots$ are determined uniquely. But the determinant of the modified system is now the principal minor $|\underset{\sim}{D'}|$, and the non-zero solutions exist only when $|\underset{\sim}{D'}| = 0$. Finally, it was shown numerically in II that $|\underset{\sim}{D'}|$ does not vanish over the range of interest ($ak < .436$) and particularly not when $E = E_{max}$ ($ak = 0.429$). Hence there are no other bifurcations, symmetric or asymmetric, in this range, apart from the pure phase shifts.

This finding confirms some rough numerical work reported by Chen and Saffman (1980) which led them to believe there were no Class 1 bifurcations in steep gravity waves. However, it leaves open the possibility of a bifurcation, symmetric or otherwise, when $dc = 0$.

4.   INSTABILITIES OF STEEP WAVES

An interesting problem, related to the phenomena of bifurcation, is to determine accurately the stability of steep gravity waves. Here we consider only the two-dimensional instabilities, but their behaviour may throw light on the more general case.

The first accurate calculations were made in Longuet-Higgins (1978b), by expressing the cartesian coordinates (x,y) in the form

$$\left. \begin{aligned} x &= X(\phi,\psi) + \varepsilon\xi(\phi,\psi)e^{-i\sigma t} , \\ y &= Y(\phi,\psi) + \varepsilon\eta(\phi,\psi)e^{-i\sigma t} . \end{aligned} \right\} \qquad (4.1)$$

Here ϕ, ψ and the time t are taken as independent variables; X and Y represent the unperturbed, finite-amplitude wave, seen in a frame of reference moving with the phase speed, and $\varepsilon(\xi,\eta)e^{-i\sigma t}$ represents a small perturbation of order ε, varying harmonically with the time. The free surface, which is generally not a streamline, is represented by

$$\psi = \varepsilon f(\phi)e^{-i\sigma t} . \qquad (4.2)$$

X and Y are considered known. The set of functions ξ, η, f and the corresponding normal-mode frequencies σ are to be determined.

If we adopt the Fourier expansion (2.1) for X and Y together with

$$\left.\begin{array}{c} (\eta - i\xi) = \sum\limits_{n=0}^{\infty} \; (\alpha_n + i\beta_n) e^{\; in(\phi + i\psi)/c} \\[3mm] f = \sum\limits_{n=1}^{\infty} \; (\gamma_n + i\delta_n) e^{\; in(\phi + i\psi)/c} \end{array}\right\} \qquad (4.3)$$

then substitution of these expansions into the appropriate boundary conditions (see Longuet-Higgins 1978a, Section 2) leads to a set of linear equations for the coefficients α_n, β_n, γ_n, δ_n having the form

$$\underset{\sim}{A} \times \underset{\sim}{\omega}^T = 0 \qquad (4.4)$$

where

$$\underset{\sim}{\omega} = (\alpha_0, \alpha_1, \ldots \; ; \; \beta_0, \beta_1, \ldots \; ; \; \gamma_1, \gamma_2, \ldots, \; ; \; \delta_1, \delta_2, \ldots) \qquad (4.5)$$

and $\underset{\sim}{A}$ is a matrix whose elements are functions of the coefficients a_n, and c. By truncating the series after a given wavenumber N and then successively increasing N, an approximation was obtained which converged satisfactorily for wave steepnesses ak less than about 0.42.

In the range $0 < ak < 0.42$ the frequency σ_m of each "superharmonic" normal mode was a decreasing function of ak. A linear extrapolation beyond ak = 0.42 suggests that for the lowest non-trivial mode m = 2, σ_2^2 would pass through zero, and the mode become unstable, at around ak = 0.436, where the phase-speed c was a maximum. This could be explained on physical grounds.

Subsequently Tanaka (1983) made more elaborate calculations, involving a stretching of coordinates near the wave crest, which enabled him to carry the computation to higher values of ak. His calculations, while essentially agreeing with those of Longuet-Higgins (1978a) up to ak = 0.42, suggested that σ_2^2 changed sign at a slightly lower value of ak, namely ak = 0.429, close to the energy maximum E_{max}. No explanation was found, however.

Now the method of the present paper has enabled us to study the problem analytically (see Paper III). In the first place we are interested particularly in the limit $\sigma \to 0$. In that limit (4.2) is independent of t, that is a constant γ_0 which we have chosen to be zero. The equations for the coefficients α_n and β_n reduce to the simpler system

$$\begin{pmatrix} \underset{\sim}{M} & \underset{\sim}{0} \\[3mm] \underset{\sim}{0} & \underset{\sim}{N} \end{pmatrix} \times \begin{pmatrix} \underset{\sim}{\alpha}^T \\[3mm] \underset{\sim}{\beta}^T \end{pmatrix} = \begin{pmatrix} \underset{\sim}{0} \\[3mm] \underset{\sim}{0} \end{pmatrix} \qquad (4.6)$$

in which case each element of $\underset{\sim}{M}$ and $\underset{\sim}{N}$ is a quadratic expression in the Fourier coefficients a_n. However, by elementary algebra it may be shown that

$$\underset{\sim}{M} = \underset{\sim}{B} \times \underset{\sim}{C}, \qquad \underset{\sim}{N} = \underset{\sim}{B} \times \underset{\sim}{D} \qquad (4.7)$$

where $\underset{\sim}{C}$ and $\underset{\sim}{D}$ are the two matrices defined in (2.9) and Section 3, and
where

$$\underset{\sim}{B} = \begin{pmatrix} b_0 & b_1 & b_2 & b_3 & \cdots \\ 0 & b_0 & b_1 & b_2 & \cdots \\ 0 & 0 & b_0 & b_1 & \cdots \\ 0 & 0 & 0 & b_0 & \cdots \\ \vdots & \vdots & \vdots & \vdots & \end{pmatrix} \qquad (4.8)$$

with

$$b_0 = 1, \qquad b_n = n\,a_n, \qquad n = 1,2,\ldots \quad . \qquad (4.9)$$

Now the system (4.6) splits into two;

$$\underset{\sim}{M} \times \underset{\sim}{\alpha}^T = \underset{\sim}{0} \qquad\qquad\qquad \underset{\sim}{N} \times \underset{\sim}{\beta}^T = \underset{\sim}{0} \qquad (4.10)$$

defining the symmetric and antisymmetric parts of the perturbation,
respectively. But the matrix $\underset{\sim}{B}$ is non-singular, since $|\underset{\sim}{B}| = 1$. There-
fore equations (4.10) are simply equivalent to the equation (3.9) for
the symmetric and anitsymmetric parts of a Class 1 bifurcation.

Our conclusion is that if a normal-mode perturbation exists at zero
limiting frequency, then in the limit as $\sigma \to 0$ it must take the form of
a pure phase-shift, except possibly when dc = 0.

In his paper (1983) Tanaka reported that at the normal mode (n=2)
became unstable at around ak = 0.429, and concluded that a bifurcation
must exist at this point. However, we can now see that if the limiting
form of the normal mode is a pure phase-shift, this conclusion is not
necessarily correct; the only bifurcation may be the phase-shift which
can occur at any value of ak.

5.   NEW INTEGRAL RELATIONS

We shall here summarise some further results that have been proved
with the aid of the Fourier coefficients (see Paper I).

In a stationary frame of reference, in which the particle velocity
at infinite depth is zero and the origin of y is in the mean surface
level, we may define the average densities of momentum I, and kinetic
potential energies T and V by

$$\left.\begin{aligned} I &= \overline{\int_{-\infty}^{y_s} u \, dy} \\ T &= \overline{\int_{-\infty}^{y_s} \tfrac{1}{2}(u^2 + v^2) \, dy} \\ V &= \overline{\int_{0}^{y_s} g y \, dy} \end{aligned}\right\} \qquad (5.1)$$

where $y = y_s$ at the surface and an overbar denotes the average value over a complete period (these definitions are valid also for non-uniform steady waves of Class m). It is known already that

$$2T = cI \qquad (5.2)$$

(Levi-Cività 1925) and that for waves of fixed length but variable amplitude

$$dE = c \, d I \qquad (5.3)$$

where $E = T + V$.

Now if we define

$$K = \tfrac{1}{2} \sum_{n=1}^{\infty} n \, a_n^2 \qquad (5.4)$$

where the a_n are defined by (2.1), it may be shown (see Paper I) that

$$I = c K \qquad (5.5)$$

and so from (5.2)

$$2T = c^2 K . \qquad (5.6)$$

By the first of equations (2.2), K is related to a_0 by

$$c^2 + a_0 = -2K . \qquad (5.7)$$

If we define further

$$J = \tfrac{1}{2} \sum_{n=1}^{\infty} a_n^2 \qquad (5.8)$$

it may also be shown (Paper I, Section 6) that

$$6V = J + 2c^2 K + K^2 \qquad (5.9)$$

and so

$$\left.\begin{aligned} 6E &= J + 5c^2 K + K^2 \\ 6L &= -J + c^2 K - K^2 \end{aligned}\right\} \qquad (5.10)$$

where $L = T - V$.

Lastly the function F introduced in Section 2 can be written explicitly as

$$F = (J + a_0 K + \alpha) + \tfrac{1}{4}(a_0 + c^2)^2 \qquad (5.11)$$

where J and K are defined by (5.8) and (5.4) and

$$\alpha \equiv \quad a_1(a_1a_2 + a_2a_3 + a_3a_4 + \ldots)$$
$$+ 2a_2(a_1a_3 + a_2a_4 + a_3a_5 + \ldots)$$
$$+ 3a_3(a_1a_4 + a_2a_5 + a_4a_6 + \ldots)$$
$$+ \ldots \quad . \tag{5.12}$$

The relations (2.3) may be verified directly by inspection. From (5.11) it follows that

$$\frac{\partial F}{\partial c^2} = \tfrac{1}{2}(a_0 + c^2) = -K \tag{5.13}$$

by (5.7), hence

$$dF \equiv \frac{\partial F}{\partial c^2} dc^2 + \sum_n \frac{\partial F}{\partial a_n} da_n = -K \, dc^2 \quad . \tag{5.14}$$

But from (5.3) the Lagrangian $L = T - V$ satisfies

$$dL = d(2T - E) = I \, dc = \tfrac{1}{2} K \, dc^2 \tag{5.15}$$

and therefore follows by integration that

$$F = -2L \quad . \tag{5.16}$$

6.   ANGULAR MOMENTUM

An integral quantity that is particularly relevant to breaking waves is the angular momentum density (see Longuet-Higgins 1980). The Eulerian-mean angular momentum density $\overline{A}_E$ may be defined as

$$\overline{A}_E = \int_{-\infty}^{y_s} (yu - xv) \, dy \quad . \tag{6.1}$$

In Paper I it is proved that a very simple relation exists between $\overline{A}_E$ and the Lagrangian density L, namely

$$\overline{A}_E = 2cL/g \quad . \tag{6.2}$$

From this and equations (5.2) and (5.3) it follows that

$$d\overline{A}_E = (3T + 2V) dc/g \tag{6.3}$$

so that $\overline{A}_E$, like L, is stationary at the same values of the wave amplitude as in the phase-speed c.

Lastly we note a result concerning the "level of action" y_a of the wave train. This was defined (Longuet-Higgins 1980) as the elevation above the surface level of points about which the Lagrangian-mean angular momentum $\bar{A}_L$ is exactly zero. In that paper it was shown that for waves of limiting steepness y_a is almost exactly equal to the crest height y_{max}. This helps to explain how a breaking wave can lose some horizontal mass and momentum by ejecting a whitecap near the wave crest without destroying the shape of the wave as a whole.

However, this coincidence between y_a and y_{max} was in the first place a numerical one; it was not known whether the coincidence was exact.

In Paper I the analysis has been carried further. In fact by expressing y_a in terms of the Fourier coefficients a_n, and by noting that for limiting waves

$$y_{max} = c^2 = -(K + \tfrac{1}{2}a_0) \tag{6.4}$$

it is possible to show that the equality $y_a = y_{max}$ for limiting waves would imply a certain identity between the coefficients a_n, which is apparently not satisfied. Hence the conjectured equality between y_a and y_{max} is not exact.

7.   CONCLUSIONS

The method of analysis by Fourier coefficients provides a quick and accurate way of calculating nonlinear gravity waves, without the use of power-series expansions. It yields very directly the critical values of the wave steepness corresponding to certain types of bifurcation, and by truncation of the Fourier series it provides a simple physical model for Class 2 bifurcations. As an analytical tool the Fourier series can be used to investigate some outstanding problems concerning the behaviour of normal-mode instabilities. And it has been used to settle a question concerning the level of action of waves of limiting amplitude, with implications for wave breaking.

The method has as far been applied to two-dimensional waves in deep water. It could certainly be extended to finite depth and possibly to three-dimensional waves also, so as to give insight into the important three-dimensional instabilities calculated by McLean (1982).

REFERENCES

Chen, B. and Saffman, P.G.   1980 Numerical evidence for the existence of new types of gravity waves of permanent form on deep water. Studies in Appl. Math. 62, 1-21.

Lamb, H.   1932 Hydrodynamics, 6th ed. Cambridge Univ. Press 738 pp.

Longuet-Higgins, M.S. 1975 Integral properties of periodic gravity
    waves of finite amplitude.
    Prof. R. Soc. Lond. A. 342, 157-174.

Longuet-Higgins, M.S. 1978a Some new relations between Stokes's
    coefficients in the theory of gravity waves.
    J. Inst. Maths. Applics. 22, 261-273.

Longuet-Higgins, M.S. 1978b The instabilities of gravity waves of
    finite amplitude in deep water. I. Superharmonics.
    Proc. R. Soc. Lond. A. 360, 471-488.

Longuet-Higgins, M.S. 1980 Spin and angular momentum in gravity waves.
    J. Fluid Mech. 97, 1-25.

Longuet-Higgins, M.S. 1984a (Paper I) New integral relations for
    gravity waves of finite amplitude.
    J. Fluid Mech. 149, 205-215.

Longuet-Higgins, M.S. 1984b (Paper III) On the stability of steep
    gravity waves.
    Proc. R. Soc. Lond. A 396, 269-280.

Longuet-Higgins, M.S. 1985 (Paper II) Bifurcation in gravity waves.
    J. Fluid Mech. (in press).

Longuet-Higgins, M.S. and Fox, M.J.H. 1978 Theory of the almost-
    highest wave. Part 2. Matching and analytic extension.
    J. Fluid Mech. 85, 769-786.

McLean, J.W. 1982 Instabilities of finite-amplitude water waves.
    J. Fluid Mech. 114, 315-330.

Schwartz, L.W. 1974 Computer extension and analytic continuation of
    Stokes' expansion for gravity waves.
    J. Fluid Mech. 62, 553-578.

Stokes, G.G. 1880 Supplement to a paper on the theory of
    oscillatory waves.
    Mathematical and Physical Papers 1, 225-228, Cambridge Univ. Press.

Tanaka, M. 1983 The stability of steep gravity waves.
    J. Phys. Soc. Japan 52, 3047-3055.

Williams, J.M. 1981 Limiting gravity waves in water of finite depth.
    Phil. Trans. R. Soc. Lond. A 302, 139-188.

NONLINEAR WATER WAVE THEORY VIA PRESSURE FORMULATION

W. H. Hui and G. Tenti
Department of Applied Mathematics
University of Waterloo
Waterloo, Ontario, Canada   N2L 3G1

ABSTRACT.   It is well-recognized that the main difficulties of water
wave theory in the classical formulation are due to the surface boundary
conditions which are nonlinear and have to be satisfied at an unknown
boundary.   A new formulation is given in which the pressure is regarded
as an independent variable and the continuity equation replaced by two
stream functions.   It is shown that in the cases of three-dimensional
steady wave motion and two-dimensional unsteady wave motion, the free
surface boundary conditions become linear and need be satisfied at a
fixed boundary.   The governing equations become more complex.   However,
they are amenable to symbolic computation in conjunction with a singular
perturbation method.   In particular, both the standing wave solution of
Penney and Price and the Stokes wave solution are reproduced analytically
to very high orders.   Furthermore, the study of the time evolution of a
sinusoidal wave train reveals the almost periodic behaviour of the waves,
for which the correlation function and the energy spectrum are in good
agreement with some hitherto unexplained observations of waves in Lake
Ontario.

1.   INTRODUCTION

We consider the classical problem of surface gravity waves on a large
body of water, for which the inviscid and imcompressible fluid model is
assumed to apply.   For the special case of irrotational flow, and in a
Cartesian system where (x,z) is the horizontal plane and y is oriented
upwardly, this leads to the well-known problem:

$$\nabla^2 \phi = 0 \ , \tag{1}$$

$$\eta_t + \eta_x \phi_x + \eta_z \phi_z - \phi_y = 0 \ , \tag{2}$$

$$\phi_t + \frac{1}{2} (\nabla\phi)^2 + g\eta = C \ , \tag{3}$$

on $y = \eta(x,z,t)$

$$\phi_n = 0 \ , \quad \text{on the solid boundary.} \tag{4}$$

17

Y. Toba and H. Mitsuyasu (eds.), The Ocean Surface, 17–24.
© *1985 by D. Reidel Publishing Company.*

Here $\Phi(x,y,z,t)$ is the potential, $y = \eta(x,z,t)$ the equation of the free surface, g the acceleration of gravity, C an arbitrary constant, and subscripts denote partial differentiation.

The first attempt at finding a solution of (1)–(4) goes back to the pioneering work of Stokes (1847, 1880), who limited himself to looking for a solution in the form of a progressive wave of permanent shape. Then, of course, the flow appears stationary in a frame translating at the speed c of the wave. Thus Stokes was able to reformulate the problem in terms of the complex variable $z = x + iy$ and the complex potential $w = \Phi + i\psi$, where ψ is the streamfunction and all quantities are made dimensionless by means of the wave number k and the phase speed c in the standard way. Moreover, Stokes realized the convenience of regarding z as a function of w , for then the problem is rendered a fixed boundary problem that he suggested be solved by a Fourier series expansion of the type (for infinite depth, say)

$$x = -\Phi - \sum_{n=1}^{\infty} \frac{a^{(n)}}{n} e^{-n\psi} \sin n\Phi , \tag{5}$$

$$y = -\psi + \sum_{n=1}^{\infty} \frac{a^{(n)}}{n} e^{-n\psi} \cos n\Phi \tag{6}$$

which gives, at $\psi = 0$, the wave profile in parametric form. From a computational point of view, it is clear that the difficulty resides in the constraints imposed by the free surface boundary condition on the Fourier coefficients $a^{(n)}$, which are determined by an infinite system of nonlinear equations. A systematic way of proceeding is to write these coefficients as a perturbation series in a small parameter ϵ as

$$a^{(n)} = \sum_{k=0}^{\infty} \alpha_k^{(n)} \epsilon^{n+2k} , \quad n = 1,2,\ldots . \tag{7}$$

Then, as it turns out, the coefficients $\alpha_k^{(n)}$ are determined numerically by solving a set of nonlinear equations (Schwartz, 1974; Cokelet, 1977; Longuet–Higgins, 1978).

Stoke's Fourier expansion method has been followed for more than a century by most researchers in this area, and it seems to have escaped everybody's attention that the problem can be considerably simplified if a perturbation expansion is performed first. Then, as outlined in Section 2, the coefficients at each order can be explicitly determined in terms of those of the preceeding orders and, in fact, a recursion formula can be derived either by hand or using modern symbolic computation systems, thus eliminating computer-generated noise and opening up the possibility of examining the convergence properties of the Stokes wave in detail.

The Stokes procedure is unfortunately limited to steady, two-dimensional, irrotational waves. For unsteady waves, the free surface is no longer a stream surface and the theory is still in a much less

developed state. Notable efforts in this direction are the standing
waves analysis of Penney and Price (1952), which was tested
experimentally by Taylor (1953), and the work of Longuet-Higgins and
Cokelet (1976, 1978) who developed a numerical method for calculating
the deformation of the free surface in any irrotational motion which
is periodic in the horizontal coordinate. In Section 3 we outline a
new analytical approach to the problem of unsteady waves.

2. STEADY WAVES

As mentioned above, a more convenient way of obtaining a parametric
representation of the Stokes wave is to look directly for a perturbation solution

$$x = -\phi - \sum_{n=1}^{\infty} \epsilon^n x^{(n)} \tag{8}$$

$$y = -\psi + \sum_{n=1}^{\infty} \epsilon^n y^{(n)} \tag{9}$$

where

$$x^{(n)} = \sum_{k=1}^{n} \beta_k^{(n)} e^{-k\psi} \sin k\phi \tag{10}$$

$$(n = 1, 2, \ldots)$$

$$y^{(n)} = \sum_{k=1}^{n} \beta_k^{(n)} e^{-k\psi} \cos k\phi \tag{11}$$

and the perturbation parameter ϵ may be chosen to be the amplitude of
the first harmonic or the wave height. The advantage of this formulation is that the free surface boundary condition then yields a set of
linear equations for the determination of the coefficients $\beta_k^{(n)}$ of
the form

$$\sum_{k=1}^{n} (k - 1) \beta_k^{(n)} \cos k\phi = \sum_{i} \cdots \sum_{j} B_{i\ldots j} \cos(i + \cdots -j)\phi , \tag{12}$$

where the quantities $B_{i\ldots j}$ involve the β-coefficients only up to
order $n - 1$. Then Eq. (9) can be written in the form (at $\psi = 0$)
$y = \sum_{n} B_n(\epsilon) \cos n\phi$, with a similar expression for x , where each of
the Fourier coefficients is given by an infinite series in ϵ .

As an example, with ϵ interpreted as a coefficient of the first
harmonic, i.e. $B_1(\epsilon) = \epsilon$, we show below the explicit form of $B_2(\epsilon)$,
$B_{21}(\epsilon)$ and of the dispersion relation calculated from Eq. (12) to
order 21 with the MAPLE symbolic computation system running on a
VAX 11/780 computer:

$$B_2(\epsilon) = \epsilon^2 + \frac{1}{2}\epsilon^4 + \frac{29}{12}\epsilon^6 + \frac{1123}{72}\epsilon^8 + \frac{502\ 247}{4\ 320}\epsilon^{10}$$

$$+ \frac{244\ 787\ 899}{259\ 200}\epsilon^{12} + \frac{884\ 130\ 455\ 111}{108\ 864\ 000}\epsilon^{14}$$

$$+ \frac{3\ 325\ 337\ 418\ 580\ 279}{45\ 722\ 880\ 000}\epsilon^{16}$$

$$+ \frac{12\ 891\ 044\ 455\ 831\ 800\ 281}{19\ 203\ 609\ 600\ 000}\epsilon^{18}$$

$$+ \frac{25\ 578\ 862\ 562\ 531\ 003\ 535\ 667}{4\ 032\ 758\ 016\ 000\ 000}\epsilon^{20} + \ldots \tag{13}$$

$$B_{21}(\epsilon) = \frac{41\ 209\ 797\ 661\ 291\ 758\ 429}{7\ 567\ 605\ 760\ 000}\epsilon^{21} + \ldots \tag{14}$$

$$\frac{c^2 k}{g} = 1 + \epsilon^2 + \frac{7}{2}\epsilon^4 + \frac{229}{12}\epsilon^6 + \frac{6175}{48}\epsilon^8$$

$$+ \frac{8\ 451\ 493}{8\ 640}\epsilon^{10} + \frac{4\ 162\ 161\ 883}{518\ 400}\epsilon^{12}$$

$$+ \frac{13\ 441\ 768\ 667}{193\ 536}\epsilon^{14} + \frac{57\ 077\ 417\ 875\ 339\ 637}{91\ 445\ 760\ 000}\epsilon^{16}$$

$$+ \frac{110\ 875\ 985\ 690\ 364\ 678\ 853}{19\ 203\ 609\ 600\ 000}\epsilon^{18}$$

$$+ \frac{83\ 926\ 522\ 731\ 752\ 447\ 156\ 327}{1\ 536\ 288\ 768\ 000\ 000}\epsilon^{20} + \ldots \tag{15}$$

Of course, the coefficients $\beta_k^{(n)}$ are related to the $\alpha_k^{(n)}$ of Eq. (7) by $\beta_k^{(n)} = \frac{1}{k}\alpha_{n-k}^{(k)}$. But the point is that the $\alpha_k^{(n)}$'s can only be obtained by solving a nonlinear system of equations. In contrast, Eq. (12) gives directly a recursion formula for the coefficients $\beta_k^{(n)}$, from which, among other things, the convergence properties of the Stokes wave may be studied without the trouble of computer-generated noise. The details of this and of the symbolic computation will be reported elsewhere, along with the generalization to the finite depth case.

3. UNSTEADY WAVES.

For unsteady flow, the free surface is unfortunately no longer a stream surface. However, it is a surface of constant pressure, and we take advantage of this fact to reformulate the theory as follows. We replace the continuity equation by two stream functions $\psi_j(x_m,t)$,

$j = 1,2$; $m = 1,2,3$, so that the velocity components are given by

$$v_\ell = \epsilon_{\ell mn} \frac{\partial \psi_1}{\partial x_m} \frac{\partial \psi_2}{\partial x_n} \quad , \quad \ell = 1,2,3 \, .$$ Next we regard the pressure as an

independent variable and write

$$\left. \begin{aligned} \psi_j &= \psi_j(x,z,p,t), \\[2mm] y &= y(x,z,p,t) \, , \end{aligned} \right\} \tag{16}$$

where the reference system is the one defined in the introduction. Specialization to two-dimensional, irrotational and infinite depth flow then leads to a set of governing equations of the form

$$\left. \begin{aligned} \psi_{xx} + \psi_{pp} &= N_1(y,\psi) \, , \\[2mm] y_x &= N_2(y,\psi) \, , \\[2mm] y_p &= N_3(y,\psi) \, , \end{aligned} \right\} \tag{17}$$

with boundary conditions

$$\left. \begin{aligned} y_t + \psi_x &= 0 \quad \text{at} \quad p = 0 \, , \\[2mm] y &\to \frac{-p}{\rho g} \quad \text{and} \quad \psi \to 0 \quad \text{as} \quad p \to +\infty \, , \end{aligned} \right\} \tag{18}$$

and initial conditions

$$\left. \begin{aligned} y(x,0,0) &= \epsilon f(x) \\[2mm] \psi(x,0,0) &= \epsilon g(x) \, , \end{aligned} \right\} \tag{19}$$

where the functions f and g can be specified as desired. The non-linear functions N_1, N_2 and N_3 appearing on the right-hand side of (17) are rather complicated. However, when a solution is sought in terms of an ϵ-perturbation expansion, they turn out to depend only on the known values of y and ψ of the previous orders. Thus the entire problem becomes a linear one, making it again possible to obtain high order solutions by means of symbolic computation. Notice also that the wave profile is obtained explicitly as $y(x,o,t)$, and that both the progressive (Stokes) and the standing (Penny and Price) wave solutions

can be simply obtained as particular cases by choosing appropriate
initial conditions.

The present theory, however, is much more general. As an example,
consider the case of the evolution of a progressive train of initially
sinusoidal waves, i.e. let f(x) = g(x) = cos x in dimensionless
notation. Then it is easily found that the solution $y(x,p,t) = \epsilon y^{(1)}$ + $\epsilon^2 y^{(2)}$ + ... takes the form

$$y^{(1)} = e^{-p}\cos(x - t) , \tag{19}$$

$$y^{(2)} = \frac{1}{2} e^{-2p}[\cos 2(x - t) - \cos\sqrt{2}t \cos 2x - \frac{1}{\sqrt{2}} \sin\sqrt{2}t \sin 2x], \tag{20}$$

with higher order terms involving $\cos\sqrt{n}t$ and $\sin\sqrt{n}t$.

4. CONCLUDING REMARKS.

This brief outline of some new developments in the theory of gravity
waves on large bodies of water has focussed on two main points. First,
for steady, two-dimensional, and irrotational waves we have shown that
the perturbation method is more convenient than the Fourier method,
originally suggested by Stokes (1880) and followed by many subsequent
investigators, as the latter leads to nonlinear equations while the
former yields a linear problem particularly suitable for symbolic
computation. It should be noted, in this regard, that this only gives
a parametric representation of the Stokes wave, while for a calculation
of the explicit wave profile the method suggested by Hui and Tenti
(1982) is preferable.

Second, we have presented a general theory for unsteady waves in
which the evolution of any intially prescribed shape can be followed
to any desired order. It recovers the progressive and standing wave

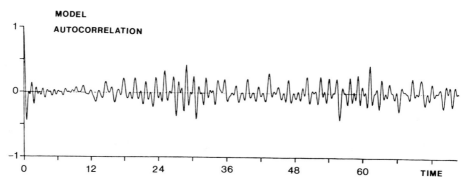

Fig. 1.   Postulated wave-height correlation function based on the
present formulation. The wave shape for this is illustra-
tive example is assumed by $y = \frac{1}{19} \cos t \sum_{n=2}^{20} \cos \sqrt{n}t$ and the
correlation function $C(\tau) = \frac{1}{2(N-1)} \sum_{n=2}^{N} [\cos(\sqrt{n}-1)\tau + \cos(\sqrt{n}+1)\tau]$.

AUTOCORRELATION

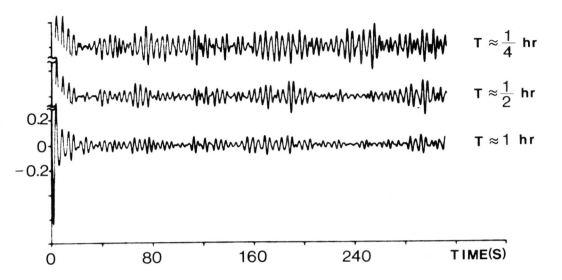

Fig. 2.   Wave-height auto-correlation function as a function of time
            lag for various averaging times   T; occasion 29405 (0500 UT,
            J.D. 294, 1976). Waves were recorded by a fixed capacit-
            ance wave gauge in water of 12 m depth in Lake Ontario.

solutions as special cases, and shows that in general the evolution
leads to almost-periodic behaviour. This has some important implications for the analysis of wave spectra, and in particular:
     (a) The autocorrelation function exhibits the same almost-
periodic behavior (Fig. 1), and therefore never dies out with increasing time lag.
     (b) The frequency spectrum indicates some local concentration of
energy near $\omega = \sqrt{2}\ \omega_{peak}$.
Both of these predictions seems to be confirmed by the wave measurements
in Lake Ontario (Donelan et al., 1984) as seen in Fig. 2 and 3.

NORMALIZED

SPECTRUM

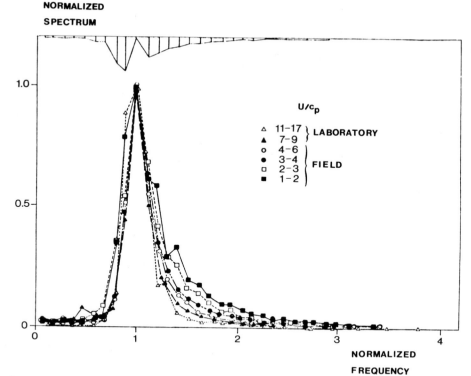

Fig. 3.   Normalized frequency spectra grouped into class by U/c$_p$.
The vertical bars at the top of the figure are an estimate
of 90% confidence limits based on the standard error of the
mean.

REFERENCES

Cokelet, E. D., *Phil. Trans. Roy. Soc.* A286, 183-230 (1977).

Donelan, M. A., Hamilton, J., and Hui, W. H., *Phil. Trans. Roy. Soc.*
(to appear; 1984).

Hamilton, J. Hui, W. H. and Donelan, M. A. J. Geophys. Res.
84, 4875-4884 (1979)

Hui, W. H., and Tenti, G., *J. Appl. Math. Phys.* (ZAMP), 33, 569-589
(1982).

Longuet-Higgins, M. S., *J. Inst. Maths. Applcs.*, 22, 261-273 (1978).

Longuet-Higgins, M. S., and Cokelet, E. D., *Proc. Roy. Soc. Lond.* A350,
1-26 (1976) and A364, 1-28 (1978).

Penney, W. G., and Price, A. T., *Phil. Trans.* A244, 254-284 (1952)

Schwartz, L. W., *J. Fluid Mech.* 62, 553-578 (1974).

Stokes, G. G., *Trans. Camb. Phil. Soc.* 8, 441-455 (1847) and *Math. Phys.
Papers*, Vol. 1, 314-326 (1880).

Taylor, Sir G., *Proc. Roy. Soc. Lond.* A218, 44-59 (1953).

PARTICLE TRAJECTORIES IN NONLINEAR CAPILLARY WAVES

S. J. Hogan
Mathematical Institute
University of Oxford
St.Giles
Oxford
England

Abstract.   The particle trajectories of nonlinear capillary waves on
water of infinite depth are calculated. For the steepest wave, a
particle moves through nearly eight wavelengths in one orbit at an
average drift velocity of almost 90% of the phase speed of the wave.
These results are presented in exact analytic form.
      The effect of finite depth is also included and results can be
given in terms of elliptic integrals and functions.
      Numerical results for the trajectories of particles in
capillary-gravity waves on water of infinite depth show that increased
surface tension leads to an increase in the horizontal distance
travelled by a particle and in the magnitude of the surface velocity.
      These results have implications for parasitic capillaries on
steep waves and for observations of the wind-drift current.

1.   INTRODUCTION

For steady deep water waves of small amplitude, the particle
trajectories are circles whose radius decreases exponentially with
depth. But, for gravity waves at least, the trajectories are open
in waves of finite amplitude leading to a mean horizontal drift, or
mass transport, known as the Stokes (1847) drift. Recently Longuet-
Higgins (1979) and Srokosz (1981) have extended these results to include
the highest gravity wave, as a prelude to the calculation of
trajectories in breaking waves.
      Capillary waves are often present in wind wave fields but their
effect on the motion of particles has not been considered until now.
The wave profiles are known to be different from gravity waves. It
turns out that the trajectories for nonlinear waves are radically
different with some particles being transported several wavelengths at
high average speeds in the course of one orbit.
      The bulk of this contribution comprises a summary of the main
results of two recent papers by the author (Hogan 1984a, b) to where
the interested reader is referred for details of the calculation.
Section 2 deals with nonlinear capillary waves (Hogan 1984a), and

Y. Toba and H. Mitsuyasu (eds.), The Ocean Surface, 25–30.

section 3 with nonlinear capillary-gravity waves (Hogan 1984b).
Section 4 is a discussion of these results.

2.   TRAJECTORIES IN NONLINEAR CAPILLARY WAVES

(a)   Infinite depth

For two-dimensional steady irrotational incompressible periodic
inviscid nonlinear wave motion at the surface of an infinitely deep
fluid with surface tension as the only restoring force, the wave
profile has been given exactly by Crapper (1957). With respect to
Cartesian axes moving with the phase speed   c   we have for the profile

$$z = \frac{\chi}{c} + \frac{4i}{k} \left\{ 1 - \frac{1}{1 + Ae^{ik\chi/c}} \right\}$$

(2.1)

where z = x + iy (x horizontally to the left, y vertically downwards),
χ is the complex potential, k = $2\pi/\lambda$ is the wave number, λ is the
wavelength. A is related to the wave steepness. It lies in the
range $0 \leq A \leq 0.45467$ for physically realistic solutions. In
addition

$$c^2 = \frac{S}{\rho k} \frac{(1-A^2)}{(1+A^2)}$$

(2.2)

where S is the surface tension and ρ is the density of the fluid.
In addition capillary wave streamlines are free surfaces for lower
amplitude waves so in this section, without loss of generality, we
consider the highest wave and its streamlines only and set
A = 0.45467.
     From this solution we find that the total time   T   taken to
complete one orbit is given by

$$\frac{cT}{\lambda} = \frac{1 + 13B^2 + 19B^4 - B^6}{(1-B^2)^3}$$

(2.3)

where B = A exp(-kψ/c) and ψ is the streamfunction. The time
averaged drift velocity U is given by

$$\frac{U}{c} = \frac{16B^2 (1 + B^2)}{1 + 13B^2 + 19B^4 - B^6}$$

(2.4)

and the total distance travelled by a particle in one orbit is given
by [X], where

$$\frac{[X]}{\lambda} = \frac{16B^2 (1 + B^2)}{(1-B^2)^3}$$

(2.5)

For the highest wave we find cT/λ = 8.99556, U/c = 0.88883 and

$[X]/\lambda$ = 7.99556.   These are considerably larger than corresponding
results for the highest gravity wave.
It is also possible to derive details of the orbits relative
to a frame of reference fixed at great depths.   These are given in
Figure 1, where X = x - ct and Y = y.   In Figure 2 we plot the drift
velocity ratio U/c as a function of the mean displacement of fluid
particles from the surface $(\overline{Y}_o - \overline{Y}_c)/\lambda$.
Full details of these calculations are given in Hogan (1984a).

(b)   Finite depth

As shown by Taylor (1959) and Kinnersley (1976) there are two
generalisations to equation (2.1) when the depth is finite,
corresponding to symmetric and antisymmetric wave forms.   As shown
in Hogan (1984c), finite depth leads to an increase in trajectory
length and particle velocity for symmetric waves and to a decrease in
both quantities for antisymmetric waves.   The description of this
problem requires elliptic integrals and functions.

3.   TRAJECTORIES IN NONLINEAR GRAVITY-CAPILLARY WAVES

In this case identical assumptions are invoked as in section 2(a) but
now gravity is also included as an additional restoring force.   No
exact solution is known to this problem so the solution was   computed
as detailed in Hogan (1984b).   The relevant parameter is

$$\kappa = \frac{Sk^2}{\rho g}$$

                                                                      (3.1)

where   g   is the acceleration due to gravity.
For small values of κ, that is long wavelengths, the particle
trajectories are very similar to those of pure gravity waves ($\kappa = 0$).
On the other hand, gravity need only be small for the results of
section 2(a) to be changed considerably.   Thus in Figure 3 we show
trajectories of particles in a wave with $\kappa = 0.1$ at its largest
steepness (h = half the crest-to-trough height).   In Figure 4 we show
the effect of increasing gravity on the drift velocity ratio U/c.
The case $\kappa = \infty$ corresponds to Figure 2.
Several families of gravity-capillary waves can exist at each
value of κ.   Perhaps the most striking example is at $\kappa = 0.5$, the
first Wilton ripple.   In Figure 5 we present particle trajectories of
members of two families at $\kappa = 0.5$ each with the same value of h = 0.20.
In Figure 6 we give the drift velocity proifle for the two waves.

4.   DISCUSSION

Particles at the surface of nonlinear capillary waves can travel very
large distances relative to the wavelength.   Gravity reduces this
effect but nevertheless the actual distance travelled can be the same

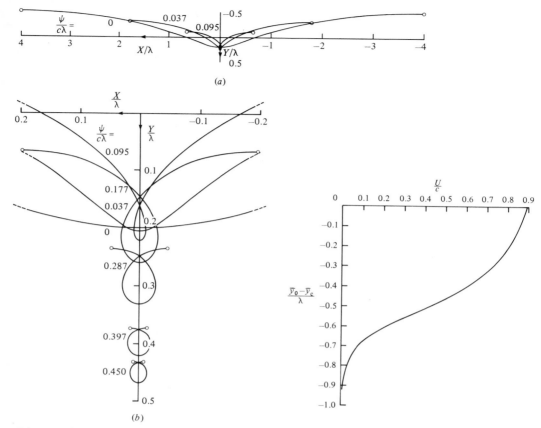

(a)

(b)

Figure 1 (a) Particle trajectories for capillary waves along stream-lines $\psi/c\lambda$ = 0, 0.037 and 0.095. (b) Full trajectories for $\psi/c\lambda$ = 0.177, 0.287, 0.397, 0.450 together with part trajectories from Figure 1(a).

Figure 2 Drift velocity ratio U/c as a function of the mean displacement of fluid particles $(\overline{Y}_o - \overline{Y}_c)/\lambda$ for capillary waves.

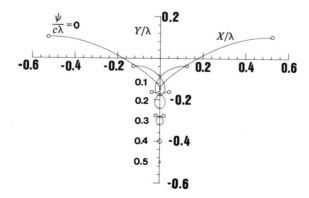

Figure 3
Particle trajectories for κ 1.0, h = 0.8069 along streamlines $\psi/c\lambda$ = 0, 0.1, 0.2, 0.3, 0.4, 0.5.

order of magnitude as for a longer wave. In addition the surface drift
velocity can be very large for short waves as well as penetrating deep
into the fluid.

    Given that capillary waves are often present in wind-wave fields,
these results may have some bearing on observations of the wind-drift
current. In addition the presence of parasitic capillaries near the
crest of steep gravity may result in particles being moved at a
different rate than had previously been expected.

    The trajectories of particles in different families of gravity-
capillary waves can be strikingly different.

ACKNOWLEGEMENT

The author gratefully acknowledges support form King's College
Cambridge in the form of a Junior Research Fellowship.

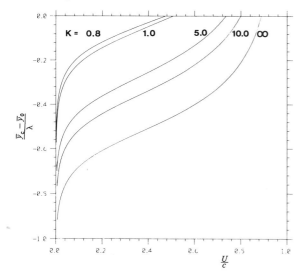

Figure 4   Drift velocity
ratio U/c for highest waves
with $\kappa = 0.8$, 1.0, 5.0, 10.0 and
infinite as a function of
$(\bar{\bar{Y}}_o - \bar{\bar{Y}}_c)/\lambda$.

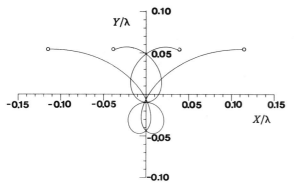

Figure 5 Free surface particle
trajectories for both Wilton
ripples at $\kappa = 0.5$, h = 0.20
drawn with coincident
crests and troughs.

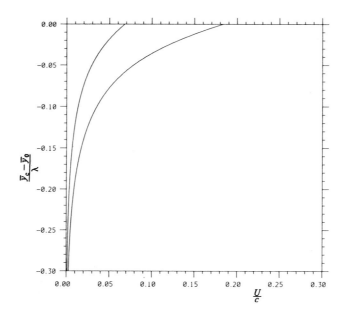

Figure 6   Drift velocity ratios for both Wilton ripples at
                  $\kappa = 0.5$, $h = 0.20$ as a function of $(\bar{Y}_o - \bar{Y}_c)/\lambda$.

REFERENCES

Crapper, G.D. 1957 An exact solution for progressive capillary waves
                    of arbitary amplitude. J. Fluid Mech. 2, 532-540
Hogan, S.J.   1984a Particle trajectories in nonlinear capillary waves
                    J. Fluid Mech. 143, 242-252.
Hogan, S.J.   1984b Particle trajectories in nonlinear gravity-
                    capillary waves. J. Fluid Mech. (to appear).
Hogan, S.J.   1984c Particle trajectories in nonlinear capillary waves
                    on sheets of fluid. In preparation,
Kinnersley, W. 1976 Exact large amplitude capillary waves on sheets
                    of fluid. J. Fluid Mech. 77. 229-241.
Longuet-Higgins, M.S. 1979 The trajectories of particles in steep,
                    symmetric gravity waves. J. Fluid Mech. 94,
                    497-517.
Srokosz, M. 1981 A note on particle trajectories in the highest wave.
                    J Fluid Mech. 111, 491-495.
Stokes, G.G. 1847 On the theory of oscillatory waves. Trans. Cam.
                    Phil. Soc. 8, 441-455.
Taylor, G.I., 1959 The dynamics of thin sheets of fluid II. Waves
                    on fluid sheets. Proc. Roy. Soc. Lond. A 253.
                    296-312.

WAVE BREAKING AND NONLINEAR INSTABILITY COUPLING

M.-Y. Su and A.W. Green
Naval Ocean Research and Development Activity
NSTL, MS 395 29   U.S.A

ABSTRACT. Experimental results are presented that show evidence of strong coupling between two different types of instabilities for finite-amplitude surface gravity waves in deep water. A consequence of this coupling is the three-dimensional, crescent-shaped breaking waves of wave trains and wave packets with the initial wave steepness $(a_o k_o)$ as low as 0.12. A second consequence is to provide a new mechanism for observed energy dissipation and directional energy spreading during evolution of waves when about $0.14 \leq a_o k_o \leq 0.18$, corresponding to the most commonly observed wave steepness during the rapid growth stages of wind-generated ocean waves.

1.   INTRODUCTION

Wave breaking may be the most dramatic surface phenomenon in open seas and along shallow coasts, which has long gripped people's imagination. Its practical importance to marine operations, gas and aerosol exchange at the air-sea interface, energy tranfer and dissipation among ocean waves, and other concerns has sitimulated a continuous study of this phenomenon during the past century (Cokelet, 1977). There is only a meager understanding of dynamic mechanisms responsible for this common phenomenon although some progress has been made in the past decade.

Lighthill (1965) first recognized that gravity waves of finite-amplitude (so-called Stokes waves) are unstable subject to two-dimensional perturbations. Benjamin and Feir (1967) elucidated the nature of this instability in terms of two side-band components nonlinearly coupled with the unperturbed primary waves. The growing instability results in modulation of the envelope of a wave train. Many investigations follow and led to the discovery of a new type of two-dimensional instability with a much more rapid growth rate for the wave steepness $ak \geq 0.41$ very near the Stokes limit of $ak = 0.443$ (Longuet-Higgins, 1978).

Another surprising property of steep waves which are intrinsically three-dimensional in nature becomes apparent shortly afterward; the

31

existence of three-dimensional instability and symmetric bifurcations
for gravity waves with $a_o k_o > 0.25$ has been independently discovered
experimentally (Melville, 1982; Su, 1982 and Su, et. al., 1982) and
theoretically (McLean, 1982; Mieron, Saffman and Yuen, 1982). The
combination of the new type of instability and bifurcation leads to the
formation of crescent-shaped breaking waves that closely resemble those
commonly observed in open oceans.

In this paper we present experimental evidence to show that the above
mentioned, two distinctively different types of wave instabiltiies, one
two-dimensional and another three-dimensional, can be nonlinearly
coupled. One significant result of this coupling is to cause the evolution of wave trains and packets to induce breaking with considerably
lower initial wave steepness (for $a_o k_o$ as low as 0.12) which lie within
the ranges found in real ocean waves (Su, 1984). This highly nonlinear
instability coupling may partially account for the frequent wave
breaking on deep oceans under strong wind forcing and thus bring us a
step closer to a full understanding of this long-standing geophysical
problem.

2. MAIN FEATURES OF WAVE INSTABILITIES

We shall follow the terminology of McLean, et. al. (1981) to call the
essentially two-dimensional, side-band modulational instability as Type
I instability, and to call the predominantly three-dimensional instability, Type II instability. To easily distinguish the description of the
coupling between Type I and Type II, we summarize the important features
of these two types of instabilites.

For Type I: (1) The two side-band perturbations are coupled with the
unperturbed primary waves to produce a modulation in the wave envelope,
(2) the most unstable mode is in the direction of the primary waves and
has a wave number about equal to $(a_o k_o)^{-1} k_o$, and (3) usually only the
most unstable mode manifests itself in the natural evolution of wave
trains and packets.

For Type II: (1) The most unstable mode has a wavenumber component in
the direction of primary waves equal to $1/2 \ k_0$, but it is always
three-dimensional; (2) its growth rate is smaller than that of Type I
for small $a_o k_o$ and becomes larger for $a_o k_o > 0.26$, and (3) its
two-dimensinal manifestations occur only for $a_o k_o > 0.41$, i.e., close to
the Stokes limit.

So far, Types I and II are described as if they are two independent
physical processes in the evolution of wave trains and packets. In
reality, they are co-exiting. For the special case of two-dimensional
wave evolution, Longuet-Higgins and Cokelet (1978) show, by the
time-stepping computation, that subharmonic instabilities of Type I lead
to a local steepening of the waves, which then induces instabilities of
Type II. Their computations further show that for large $a_o k_o \geq 0.25$,

the combined action of Type I and II lead to two-dimensional wave
breaking. In our experiments to be described below, the wave trains are
allowed to undergo both two- and three-dimensional modulations. We then
observed that the similar combined action can occur at much lower $a_o k_o$
compatible with the average ocean wave steepness in the rapid growth
stage.

3.   COUPLING BETWEEN TYPES I AND II

3.1.   Experimental Results

The experiments to be described here are conducted in a wave tank 167 m
long with a cross-section of 3.7 m by 3.7 m. The waves are generated by
a plunger-type wavemaker and are measured by the capacitive-type gauges
along the length of the tank. The range of wave steepness used is from
$a_o k_o$ = 0.09 to 0.20, with the primany wave frequency f_o = 1.23 Hz
remaining fixed. More details about the experimental set-up can be
found in Su, et al. (1982).

We shall first describe experimental results for continuous (uniform)
wave trains. The wave measurements are used to derive power spectra of
the surface displacements E $(a_o k_o; f)$, where f denotes the frequency,
and analog strip chart records of the surface displacement, $\eta(a_o k_o; t)$,
where t denotes the time. We are particularly interested in the growth
of the two side-bands of Type I, with f_1, and f_2 denoting the lower and
upper side-band frequency, respectively. Several typical examples of E
$(a_o k_o; f)$ and associated $\eta(a_o k_o; t)$ can be found in Su and Green (1984).
We found that the maximun modulation of the wave train due to Type I
instability corresponds to the stage when the maximum and equal growth
of the two side-band components occur and are approximately equal to
one-half of the spectral power at the primary frequeney f_0; i.e.,

$$\text{Max}[E(a_o k_o; f_1)] = \text{Max } [E(a_o k_o ; f_2)] = 1/2 \ E(a_o k_o; f_o) \qquad (1)$$

The average locations (exprssed in actual distance from the wavemaker x,
and in the corresponding dimensionless form x/λ_0, where λ_0 is the wave
length of the primary waves) of the above condition (1) with respect to
a range of $a_o k_o$ from 0.09 to 0.20 is given by curve (a) in Figure 1. It
is obvious that the x/λ_0 for a given $a_o k_o$ decreases with increasing
$a_o k_o$, since Type I has a larger growth rate for a larger wave steepness.
Visual obsevations during the wave measurements show that wave breaking
occurs at locations centered around the maximum wave modulation. Curves
(b) and (c) in Figure 1 bound the beginning and ending locations of the
observed wave breaking. The shapes of wave breaking are composed of
two-dimensional and threedimensional forms; the frequency of three-
dimensionl forms are found to be higher than the two-dimensional forms.
Futhermore, this tendency becomes even more dominant as $a_o k_o$ increases
from $a_o k_o \geq 0.14$.

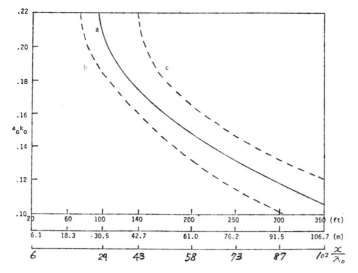

Figure 1. Stages of evolution of wave trains with respect to varying initial wave steepness; $a_o k_o$:
(a) Max $[E(a_o k_o; f_1)]$=Max $[E(a_o k_o; f_2)]$,
(b) Starting distance of observed wave breaking, and (c) Ending distance of observed wave breaking.

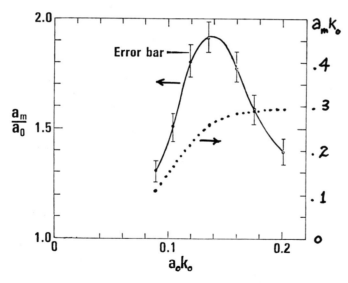

Figure 2. The ratio of the maximum wave amplitude a_m to the initial wave amplitude a_o for various initial wave steepness $a_o k_o$. Also shown is the variation of $a_m k_o$ vs. $a_o k_o$.

From the ensemble of $\eta(a_o k_o;t)$, we find that the ratio of the maximum
wave amplitude, a_m, to the initial wave amplitude, a_o, is not a
monotonic increasing function of $a_o k_o$, but has a maximum around $a_o k_o =$
0.14, as shown in Figure 2. It is noted that a_m/a_o increases very
rapidly from $a_o k_o = 0.09$ to 0.14, with the maximum of a_m/a_o reaching
1.9; the wave amplitude almost doubles its height from the original due
to Type I modulational instability. The decrease of a_m/a_o with $a_o k_o$
from 0.14 to 0.20 is equally considerable. Plotted also on Figure 2 is
the variation of $a_m k_m$ vs. $a_o k_o$. We note that $a_m k_m$ increase rapidly from
0.11 as $a_o k_o$ increase from 0.09, but the former level off at 0.29 for
$a_o k_o \gtrsim 0.16$.

We have conducted a series of experiments using wave packets of
different length (N), i.e., varying the number of waves in the initial
generation of wave packets with uniform amplitude. A typical example
with $a_o k_o = 0.16$ and N = 10 is shown in Figure 3 by a sequence of
photographs taken from a nearby tower. (This experiment is conducted in
a different outdoor wave tank, 150 m long with a cross-section of 3.7 m
wide, and 1 m deep. Photographs (a) through (f) show the six sequential
stages of the evolution of the same wave packet. The wavemaker is
visble in Figure 3(a) on the extreme left. Stage (c) corresponds to the
maximum modulation of the wave packet with a clear manifestation of two
rows of three-dimensional, crescent-shaped, spilling, breaking waves
atop the two-dimensional, long wave crest. In stages (d) and (e),
futher wave breaking with longer crest-wise dimensions are visible.
Finally, three separate envelope-solition like wave groups are formed in
stage (f).

For the growth rate of Type I and the stages of beginning and ending of
wave breaking of wave packets, the two control parameters $a_o k_o$ and N are
both important. For a fixed $a_o k_o$, the growth rate is higher and wave
breaking occurs sooner for smaller N. The exact characteristics of
starting and stopping the wavemaker also plays a role in providing
varying degrees of initial perturbations. It certainly deserves more
detailed experimental study and will not be discussed further.

3.2.   Interpretation

We suggest the following physical interpretation for the experrimental
results described here as the coupling between Type I and Type II
instabilities, whose main features have been described in §2. Under the
initial action of Type I instability, the wave trains (or packets) with
$a_o k_o \geq 0.12$ may undergo a considerable modulation in its envelope;
subsequently, a few of the waves in the middle of the maximum modulation
will have local wave steepness $ak > 0.20$, which will be high enough to
trigger the Type II instability. As $a_o k_o \geq 0.15$, these locally steeper
waves may reach $ak \geq 0.30$, at which steepness previous experiments
(Melville, 1982; Su, et. al., 1982) show extremely fast growth of Type
II, thus leading to violent three-dimensional wave breaking. This may
explain the leveling off of $a_m a_o$ with respect to $a_o k_o$ near $a_o k_o = 0.14$
in Figure 2 as the transfer of the two-dimensional wave energy to

Figure 3. (a-f) A typical sequence of six stages of
the evolution of a wave packet with a k_o = 0.16 and N=10
and f_o = 1.23 Hz showing three-dimensional wave breaking
in a channel of 150 m long and 3.7 m wide and 1 m deep.

three-dimensional wave form, plus dispersion into three-dimensional
higher wavenumber components and dissipation into turbulence is made.

4. DISCUSSIONS AND CONCLUSION

Possible relevance of experimental findings on Type I and II coupling to
deep-ocean wave breaking in growing seas is discussed. First, these
natural breaking waves are predominantly three-dimensional crescent-
shaped and are spilling rather than plunging forms.   Second, Donolan,
Longuet-Higgins and Turner (1972) have found, from direct observations,
that the average frequency of wave breaking is about equal to half of
that of underlying main waves. These authors attributed this feature to
the fact that the group velocity is equal to one-half of the phase
velocity.   Third, a recent statistical analysis on storm waves (Su,
1984) on "extreme wave groups", which are defined as wave groups where
each contains the maximum wave height in one 20-minute record, shows
that the extreme wave group consists of three zero-crossing waves in the
mean, whose average wave steepness is close to 0.16. The average wave
steepness of the highest waves is close to 0.20 and, hence, they are
very likely near the breaking stage. The three observations from the
open oceans, in all aspects, are remarkably in agreement with the
experimental observations of Types I and II coupling near $a_o k_o = 0.16$,
both qualitatively and quantitatively. As an alternative to the
explanation proposed by Donalean, et al. (1972) on the frequency of wave
breaking, we suggest that it is due to the strongest subharmonic
perturbation of the wavenumber equal to $1/2 \ k_0$ of Type II instability
that has the propagation velocity equal to the phase velocity of primary
waves (McLean, 1982; Mieron, Saffman and Yuen, 1982; Su, et al., 1982).

In summary, based on the above experimental results, statistical
analyses of field data and theoretical findings, we are led to suggest
that spilling wave breaking for large energy-containing waves near the
peak frequency can occur due to nonlinear coupling of Type I and Type II
instabilities for average wave steepness near 0.14 to 0.18.

5. REFERENCES

Benjamin, T.B. and J.E. Feir, J. Fluid Mech., 28, 417 (1967).
Cokelet, E.D., Nature, 267, 769 (1977).
Donelan, M., M.S. Longuet-Higgins and J.S. Turner, Nature, 239, 449
    (1972).
Lighthill, M.J., J. Inst. Math. Appl. 1, 269-306.
Longuet-Higgins, M.S., Proc. Roy. Soc. Lond., Ser A, 360-489 (1978).
Longuet-Higgins, M.S. and E.D. Cokelet, Proc. Roy. Soc. Lond., Ser A,
    364, 1 (1978).
McLean, J.W., J. Fluid Mech., 114, 315 (1982).
McLean, J.W., Y.C. Ma, D.U. Martin, P.G. Saffman and H.C. Yuen, Phys.
    Rev. Lett., 46, 817 (1981).
Melville, W.K., J. Fluid Mech., 115, 165 (1982).

Meiron, D.I., P.G. Saffman and H.C. Yuen, J. Fluid Mech. (1982).

Su, M.Y., J. Fluid Mech., 124, 73 (1982).

Su, M.Y., Phys. Fluid, 24, 2167 (1982).

Su, M.Y., Proc. OCEANS '84, Washington, D.C., 711 (Sept. 1984).

Su, M.Y., M. Bergin, P. Marler and R. Myrick, J. Fluid Mech., 124, 45
    (1982).

Su, M.Y. and A.W. Green, Submitted to Phys. Fluid, 27(11), 2595 (1984).

Yuen, H.C. and B.M. Lake, Adv. Appl. Mech., 22, 67 (1982).

STABILITY OF A RANDOM INHOMOGENEOUS FIELD OF WEAKLY NONLINEAR
SURFACE GRAVITY WAVES WITH APPLICATION TO THE JONSWAP STUDY

Peter A.E.M. Janssen
Department of Oceanography
Royal Netherlands Meteorological Institute
P.O. Box 201,
3730 AE De Bilt, Holland

ABSTRACT. The stability of a random inhomogeneous field of weakly
nonlinear surface gravity waves is studied and applied to the JONSWAP
study.

1. INTRODUCTION

Starting with the investigations of Phillips (1960) and Hasselmann
(1962, 1963) there has been much interest in the energy transfer due
to four-wave interactions in a nearly homogeneous random sea
(Hasselmann et al (JONSWAP) 1973; Watson & West 1975; Willebrand
1975). Longuet-Higgins (1976) derived the narrow band limit of
Hasselmann's equation by starting from the nonlinear Schrödinger
equation, describing the evolution of a narrow band, weakly nonlinear
wave train. All this nonlinear energy transfer occurs on a rather long
time scale since the rate of change of the action density is
proportional to n^3. Hence,$(\partial n/\partial t)/n = O(\varepsilon^4 \omega_o)$, where ε is the wave
steepness and ω_o a typical frequency of the wave field.
   A much faster energy transfer is possible in the presence of
spatial inhomogeneities. For an inhomogeneous sea, Alber (1978)
derived an equation describing the evolution of a random narrow band
wave train, using the Davey-Stewartson equations. Finally, starting
from the full equations of motion, Crawford, Saffman & Yuen (1980),
following Zakharov's (1968) approach, obtained a unified equation for
the evolution of a random field of deep-water waves which accounts for
both the effects of spatial inhomogeneity and the energy transfer
associated with a homogeneous sea. All these investigations on an
inhomogeneous random field of waves gave rise to a much faster energy
transfer $\left((\partial n/\partial t)/n = O(\varepsilon^2 \omega_o)\right)$.
   In this paper we wish to discuss nonlinear interactions in an
inhomogeneous wave field and we choose as our starting point the
nonlinear transport equation for the envelope spectrum, in the narrow
band approximation. The envelope spectrum is just the Fourier
transform of the autocorrelation function of the envelope of the wave

39

Y. Toba and H. Mitsuyasu (eds.), The Ocean Surface, 39–49.

train and contains all the information of the stochastic wave field we need. It should be emphasized that the assumption of an inhomogeneous wave field makes sense because Alber (1978) showed that a homogeneous spectrum is unstable to long-wavelength perturbations if the width of the spectrum is sufficiently small. For a Gaussian spectrum instability was found for $\sigma_\omega/\omega_0 < \varepsilon$, where ε is the width in frequency space, whereas Crawford et al (1980) found similar results for a Lorentzian shape of the spectrum. In the limit of vanishing bandwidth the deterministic results of Benjamin and Feir (1967) on the instability of a uniform wave train were rediscovered. In passing, we remark that this approach fills the gap between the deterministic evolution of wave train in the laboratory and the evolution of a narrow band, stochastic wave field in nature.

Here, we would like to concentrate on the physical interpretation of Alber's result that only spectra with a sufficiently large width are stable. This means that one would expect to find wave spectra in nature with a width σ_ω larger than $\varepsilon\omega_0$ as for smaller spectral width the random version of the Benjamin-Feir instability would occur resulting in a broadening of the spectral shape (Janssen, 1983). To test this conjecture we have applied Alber's result on the threshold for instability to the case of the Jonswap-spectrum, and we found that most of the cases of the Jonswap study were stable or just marginally stable.

The plan of this paper is as follows. In Sec. 2 we present the evolution equation of the envelope spectrum as obtained from the one-dimensional nonlinear Schrödinger equation and we briefly review the linear stability theory of a homogeneous spectrum of random, narrow-band wave trains. We also discuss some important differences between the energy transfer for a homogeneous field of surface gravity waves and a inhomogeneous field. Next, in Sec. 3, we discuss the conjecture that in case of B.F. instability the unstable sidebands will reshape the spectrum in such a way that the spectrum broadens and becomes stable again, thereby quenching the random version of the B.F. instability. We next apply in Sec. 4 the results on the threshold for stability to the Jonswap study and we conclude, with a summary of conclusions (Sec.4). For the mathematical details we would like to refer to Alber (1978) and Janssen (1983).

2. THE RANDOM VERSION OF THE BENJAMIN-FEIR INSTABILITY

In order to investigate the effect of inhomogeneities on the nonlinear energy transfer of weakly nonlinear water waves we study the nonlinear Schrödinger equation. It is well-known that this equation may be applied to the case of water waves with a narrowband spectrum and small wave steepness so that the surface elevation ζ is approximately given by

$$\zeta \simeq \text{Re}\left(A(x,t)\text{expi}(k_0 x - \omega_0 t)\right) \qquad (1)$$

Here ω_0 and k_0 are the angular frequency and the wave number of the carrier wave, which obey the deep-water dispersion relation $\omega_0 = (gk_0)^{\frac{1}{2}}$ (g is the acceleration of gravity), and $A(x,t)$ is the slowly varying complex envelope of the wave. The evolution of the envelope is determined by the following nonlinear Schrödinger equation:

$$i(\frac{\partial}{\partial t} + \omega_0' \frac{\partial}{\partial x})A + \frac{1}{2} \omega_0'' \frac{\partial^2}{\partial x^2}A - \frac{1}{2} \omega_0 k_0^2 |A|^2 A = 0, \tag{2}$$

where a prime denotes differentiation with respect to k_0. Transforming to a frame moving with the group velocity ω_0' and introducing dimensionless units $\tilde{t} = \frac{1}{2} \omega_0 t$, $\tilde{x} = 2 k_0 x$ and $\tilde{A} = k_0 A$, the equation for $\tilde{A}$ (which is for a uniform wave train just the wave steepness) reads

$$i\frac{\partial}{\partial t}A - \frac{\partial^2}{\partial x^2}A - |A|^2 A = 0 \tag{3}$$

where we have dropped the tilda. In a statistical description of waves one is interested in the time evolution of the two-point correlation function $\rho(x_1, x_2 t)$ defined as

$$\rho (x,r,t) \equiv < A (x_1,t) A^* (x_2,t) > \tag{4}$$

(* = complex conjugate) where the average coordinate $x = (x_1 + x_2)/2$, the separation coordinate $r = x_2 - x_1$ and the angle brackets denote an ensemble average. The inhomogeneity of the wave field is expressed by the fact that ρ is also a function of the average coordinate x. Assuming the quasi-Gaussian approximation, the transport equation for ρ is

$$i\frac{\partial}{\partial t}\rho - 2 \frac{\partial^2}{\partial x\partial r}\rho - 2 \rho\big(\rho(x + \frac{1}{2}r,o) - \rho(x - \frac{1}{2}r,o)\big) = 0 \tag{5}$$

From (5) one can derive for the envelope spectrum W, defined as

$$W(x,p) = \frac{1}{2\pi} \int dr e^{ipr} \rho(x,r), \tag{6}$$

the transport equation

$$\frac{\partial}{\partial t}W + 2p \frac{\partial}{\partial x}W + 4 \sin (\frac{1}{2} \frac{\partial^2}{\partial p\partial x'}) W (x,p) \rho (x',o)\Big|_{x' = x} = 0, \tag{7}$$

where

$$\sin (\frac{1}{2}\frac{\partial^2}{\partial p\partial x'}) = \frac{1}{2i} \sum_{\ell = o}^{\infty} \frac{(\frac{i}{2} \frac{\partial^2}{\partial p\partial x'})^{2\ell + 1}}{(2\ell + 1)!} \tag{8}$$

Equation (7) describes the evolution of an inhomogeneous ensemble of narrowband weakly nonlinear wavetrains. The timescale for (reversible) energy transfer owing to spatial inhomogeneities can easily be estimated from (8) with the result

$$\frac{1}{\rho}\frac{\partial\rho}{\partial t} = O(\rho) = O(\varepsilon^2) \qquad (9)$$

If one includes deviations from Gaussian statistics, which are generated because nonlinearity gives rise to correlation between the different components of the envelope spectrum, an irreversible energy transfer is found. Crawford et al (1980) have shown, however, that these irreversible changes (including the nonlinear energy transfer associated with a homogeneous sea; Hasselmann 1962) occur on the much longer time scale $\tau_{irrev} = O(\varepsilon^{-4})$

As a matter of fact, in the one-dimensional case there is no nonlinear energy transfer in a homogeneous wave field. The reason is that only those waves interact nonlinearly that satisfy the resonance conditions $\vec{k}_1 + \vec{k}_2 = \vec{k}_3 + \vec{k}_4$ and $\omega_1 + \omega_2 = \omega_3 + \omega_4$. In one dimension these conditions can only be met for the combinations $k_1 = k_3$, $k_2 = k_4$ or $k_1 = k_4$, $k_2 = k_3$ and then the rate of change of the action density $n_1 = n(k_1)$, given by

$$\frac{d}{dt} n_1 = \int d\vec{k}_2 d\vec{k}_3 d\vec{k}_4 \ T \left(n_3 n_4 (n_1 + n_2) - n_1 n_2 (n_3 + n_4)\right)$$
$$\delta (\vec{k}_1 + \vec{k}_2 - \vec{k}_3 - \vec{k}_4) \ \delta (\omega_1 + \omega_2 - \omega_3 - \omega_4),$$

vanishes identically. Our argument is independent of the form of the transfer function T hence it also applies for narrow-band waves for which T is a constant. This is an important distinction between the homogeneous theory and the inhomogeneous theory of nonlinear interactions as we will see in a moment that in one dimension there certainly is energy transfer in an inhomogeneous wave field, namely owing to the random version of the Benjamin-Feir instability. Here, we would like to concentrate on the energy transfer owing to spatial inhomogeneities as we feel that this transfer might play an important rôle in controlling the shape of the wave spectrum. In this section we support this conjecture by means of a stability analysis of a homogeneous spectrum which shows that there is stability provided the width of the spectrum is sufficiently broad. In the next section we test this conjecture against data from the Jonswap study.

To see whether a homogeneous spectrum W_o (p) is a stable solution of Eq. (7) one proceeds in the usual fashion by perturbing W_o (p) slightly according to

$$W = W_o(p) + W_1(x_1 p), \ W_1 \ll W_o \qquad (10)$$

and one considers perturbations of the type

$$W_1 = \hat{W}_1(p) \ \text{expi} \ (kx - \omega t) \qquad (11)$$

After some analysis one arrives at a dispersionrelation for ω which for a Lorentz spectrum

$$W_o(p) = \frac{\langle A_o^2 \rangle \sigma}{\pi(p^2 + \sigma^2)}, \tag{12}$$

with σ the width of the spectrum and $\langle A_o^2 \rangle$ the mean-square wave steepness, is given by

$$\omega = 2k \left(-i\sigma \pm (\tfrac{1}{4}k^2 - \langle A_o^2 \rangle)^{\frac{1}{2}}\right) \tag{13}$$

Clearly we have instability for Im $(\omega) > 0$, i.e. provided $\sigma^2 + k^2/4 < \langle A_o^2 \rangle$ (Fig.1). In the limit of vanishing bandwidth σ the

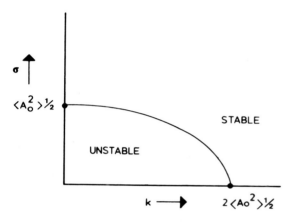

Fig. 1.   The region of a instability of the σ-k plane.

growth rate (13) reduces to the result of Benjamin & Feir (1967) for a deterministic   wavetrain   if   one   makes   the   identification $2\langle A_o^2 \rangle \Rightarrow A_o^2$. We also remark that finite bandwidth σ gives a reduction of the growth rate and that when

$$\sigma \gg \langle A_o^2 \rangle^{\frac{1}{2}} \tag{14}$$

the instability disappears. This criterion for stability admits a simple physical interpretation. It tells us that the growth rate of the Benjamin-Feir instability vanishes as the correlation length scale of the random wave field (ca. $1/\sigma$) is reduced to the order of the characteristic length scale for modulational instability of the wave system (ca. $2\pi/k_{max}$, where $k_{max} = 2^{\frac{1}{2}} \langle A_o^2 \rangle^{\frac{1}{2}}$ corresponds to maximum growth for $\tau \to 0$). Thus decorrelation of the phases of the wave envelope leads to stabilization of the wave train.

    We have reviewed the one-dimensional theory only. Extension to two dimensions is rather straightforward but not needed for present purposes as the same criterion for stability (14) is found (Alber 1978).

3. LONG-TIME BEHAVIOR OF THE RANDOM VERSION OF THE BENJAMIN-FEIR INSTABILITY

In the previous section we have seen that a random wave train is stable if its spectral width is sufficiently large. An interesting question is what happens if condition (14) is violated. Then, initially, the unstable sidebands will grow with a growth rate given by Eq. (13). It is, however, clear that the amplitude of the sidebands may become so large that nonlinear effects, such as the generation of second harmonics and modification of the equilibrium W_o (p), become important thereby considerably modifying the linear growth of the unstable sidebands. The long-time behavior of a slightly unstable sideband was determined by Janssen (1983). He found that near the threshold for instability (see Fig. 1) the evolution equation of the amplitude of the slightly unstable sideband is given by a Duffing equation with complex coefficients:

$$A\frac{\partial^2}{\partial t^2} \Gamma + B\frac{\partial}{\partial t} \Gamma + C\Gamma + D|\Gamma|^2\Gamma = 0 \tag{15}$$

where A, B, C and D are integrals over the spectrum W_o (p). The time behavior of Γ for some special cases is given in Fig. 2. The most

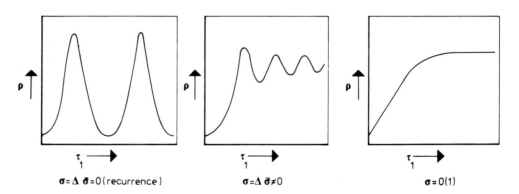

$$\sigma=\Delta \ \bar{\sigma}=0\,(\text{recurrence})\qquad\qquad \sigma=\Delta \ \bar{\sigma}\neq0\qquad\qquad\qquad \sigma=0(1)$$

Fig. 2. Time evolution of Γ (Eq. 15) for several cases.

interesting aspect to note is that Γ grows initially in agreement with linear theory whereas at later times the instability is quenched because of a broadening of the spectrum W_o (p) To be more specific, if one only takes into account the effect of the modification of the equilibrium then the unstable sideband grows until such a level that the width of the spectrum W_o (p) just satisfies condition (14) (with the equality sign). However, if also the generation of second harmonics is considered then the width σ will become larger (Janssen, 1983). In practice this means that it is not at all unlikely to find spectra with a width exceeding $\langle A_o^2\rangle^{\frac{1}{2}}$

4. APPLICATION TO THE JONSWAP STUDY

Let us apply the considerations on the stability of a homogeneous spectrum to the observations of the Jonswap study (1973). During this experiment, the generation of waves by wind blowing orthogonally from a straight shore was studied and a uniform good fit to nearly all the observed spectra was attained by the function

$$E(x) = \frac{\alpha g^2}{\omega_p^4} (x + 1)^{-5} \exp\{-\frac{5}{4} (x + 1)^{-4}\} \; \gamma^{\exp\{\frac{x^2}{2\sigma^2}\}},$$

$$\sigma = \{ \begin{array}{ll} \sigma_a, & x < 0 \\ \sigma_b, & x > 0 \end{array} \tag{16}$$

where $x = (\omega - \omega_p)/\omega_p$, ω_p is the peak frequency, α is Phillips' constant, γ is the peak enhancement and $\sigma_{a,b}$ is related to the spectral width. In order to apply the theory of par. 2 we assume that $\sigma_a = \sigma_b$. The narrowband approximation to the Jonswap spectra can be obtained by Taylor expansion around $x = 0$,

$$E(x) \simeq E(o) - \tfrac{1}{2} x^2 \frac{\partial^2}{\partial x^2} E(o) \simeq E_o / (1 + \tfrac{1}{2} \frac{E_o{''}}{E_o} x^2)$$

and the latter equality holds for small enough x. Evaluation of the curvature of the Jonswap spectrum at $x = o$ then gives

$$E(x) \simeq \frac{E_o}{1 + (\frac{x}{\sigma_J})^2} \tag{17}$$

where $E_o = \alpha \, g^2 \, \gamma \, e^{-5/4}/\omega_p^4$ and $\sigma_J^2 = 2\sigma^2/(20\sigma^2 + \ln \gamma)$. In Fig. 3 we

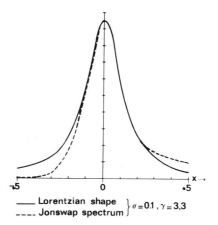

_____ Lorentzian shape ⎫
- - - - Jonswap spectrum ⎬ $\sigma = 0.1, \gamma = 3.3$

Fig. 3. Lorentzian fit to the JONSWAP spectrum.

have compared the Jonswap shape (16) with the narrow band approximation (17) and we have found good agreement up to $x \simeq 2\sigma$. Finally, the significant wave steepness spectrum $S = k_p^2 E(x)$ is then given by ($\omega_p = (g k_p)^{\frac{1}{2}}$)

$$S(x) = \frac{S_o}{1 + (\frac{x}{\sigma_J})^2}, \quad S_o = \alpha \gamma e^{-5/4} \tag{18}$$

Comparing the Lorentz spectrum (12) with (18) one sees at once that $\langle A_o^2 \rangle$ should be replaced by $\sigma_J \pi S_o$ and σ by σ_J in order to obtain the stability criterion for the narrow band approximation of the Jonswap spectrum (cf. (14)). Thus, the spectrum (17) is stable provided

$$\sigma_J > \pi S_o \tag{19}$$

Using the expression for σ_J this amounts to the following condition for σ,

$$\sigma^2 > \tfrac{1}{2}\pi^2 S_o^2 \frac{\ln \gamma}{1 - 10 \pi^2 S_o^2}, \quad S_o = \alpha \gamma e^{-5/4} \tag{20}$$

In order to compare (20) with observations we used $\gamma = 3.3$ which is, according to Jonswap, the mean of the observations of the peak enhancement for small fetch, whereas for α we used the fetch law

$$\alpha = 0.076 \ X^{-0.22}. \tag{21}$$

In Fig. 4 we have plotted the Jonswap observations for σ_b as a

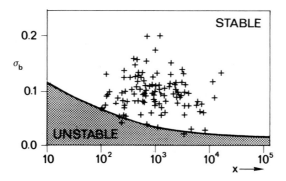

Fig. 4. The JONSWAP observations of the spectral width as a function of fetch. Most of the spectra are stable or marginally stable.

function of dimensionless fetch gx/U_{10}^2 and we have also shown the "width" σ for which the Jonswap spectrum is just marginally stable according to the linear theory of par. 2 (i.e. (20) with the equality sign). It is seen that most of the spectra are stable or marginally stable. We believe this is evidence for our conjecture that one is likely to find spectra with a "width" σ exceeding $\{\frac{1}{2}\pi^2 s_0^2 \ln \gamma/(1-10 \pi^2 s_0^2)\}^{\frac{1}{2}}$, as spectra with a smaller width would be unstable. For the narrowband approximation of the Jonswap spectrum the maximum growth rate of this sideband instability would be

$$\text{Im} (\omega) = \pi \sigma_J S_o \omega_o \qquad (22)$$

where we used (13) in dimensional form with $\langle A_o^2 \rangle = \pi \sigma_J S_o$. In order to get an estimate for the growth rate we use $\gamma = 3.3$, $\sigma_J = 0.1$ and $\alpha = 0.01$ with the result that

$$\text{Im} (\omega) = 2 \; 10^{-3} \omega_o \qquad (23)$$

This means that an unstable spectrum would change on a time scale of the order 500 $\omega_o^{-1} \simeq 250$ s for a typical peak frequency observed during Jonswap. Hence, an unstable spectrum would broaden very rapidly and since the spectra were obtained from 30 minutes time series one should expect to observe stable spectra only.

We note that other effects relevant for the energy balance of wind waves have a much longer time scale. Consider e.g. the increase of wave energy due to wind. Using the bight-of-Abaco parametrization (Snyder et al, 1981) the time scale τ_W for the increase of energy by wind is given by

$$\frac{1}{\tau_W} = 0.2 \; \frac{\rho_a}{\rho_w} \; \omega \; (\frac{U_5}{c} - 1) \qquad (24)$$

where ρ_a/ρ_w is the ratio of air density to water density, c is the phase speed of the waves and U_5 the wind speed at 5 meter height. For the already mentioned conditions during Jonswap (with $U_5 = 10$ m/s) one finds $\tau_W \simeq 4000 \; \omega_o^{-1}$ at the peak of the spectrum, which is much longer than the time scale for inhomogeneous nonlinear interactions. In the present analysis it is therefore justified to neglect the effect of wind. Again, this supports our conjecture that nonlinear interactions of a inhomogeneous wave field are important for controlling the shape of wind sea spectra under generating conditions.

5. SUMMARY OF CONCLUSIONS

We have reviewed the theory of a random inhomogeneous, nonstationary field of weakly nonlinear surface gravity waves. A linear stability analysis of a homogeneous spectrum shows that there is instability provided the spectral width is sufficiently small. A threshold for instability is therefore present. Near this threshold one can show that the unstable sidebands will reshape the spectrum in such a way that the spectral width becomes larger thereby quenching the random version of the Benjamin-Feir Instability. In the so-called quasilinear approximation (i.e. if one only takes into account the effect of reshaping) the width of the spectrum becomes so large that there is just marginal linear stability. In the full nonlinear theory the spectral width will exceed the threshold in order to quench the instability. Thus the threshold for linear stability is not a hard one as it is not unlikely that spectra with larger spectral width will occur.

We have applied the theory to the observations of the Jonswap study by investigating the threshold for unstability for the narrowband approximation to the Jonswap spectrum. We have found that most of the spectra in the Jonswap study were stable or just marginally stable. This provides evidence to the conjecture that the nonlinear interactions of a inhomogeneous field of surface gravity waves play an important rôle in controlling the shape of the wave spectrum.

In this paper we have only studied the nonlinear interactions of the waves on the fast time scale$_4$ ($\sim \epsilon^2$ t) . The effect of inhomogeneity on the longer time scale ϵ t remains to be investigated (cf. Crawford et al. 1980).

Acknowledgement:
The author would like to thank Evert Bouws for a useful discussion on the Jonswap study.

REFERENCES

Alber, I.E. 1978. The effects of randomness on the stability of two-dimensional surface wavetrains. Proc. R. Soc. Lond. A363, 545.

Benjamin, T.B. & Feir, J.E. 1967. The disintegration of wave trains on deep water. Part 1. Theory. J. Fluid Mech. 27, 417 – 430.

Crawford, D.R., Saffman, P.G. & Yuen, H.C. 1980. Evolution of a Wave random inhomogeneous field of nonlinear deep-water gravity waves. Wave Motion 2, 1 – 16.

Hasselmann, K. 1962. On the non-linear energy transfer in a gravity-wave spectrum. Part 1. General theory. J. Fluid Mech. 12, 481.

Hasselmann, K. 1963. On the non-linear energy transfer in a gravity-wave spectrum. Part 2. Conservation theorems, wave-particle analogy, irreversibility. J. Fluid. Mech. 15, 273.

Hasselmann, K. , Barnett, T.P., Bouws, E., Carlson, H., Cartwright, D.E., Enke, K., Ewing, J.A., Gienapp, H., Hasselmann, D.E., Krusemann, P., Meerburg, A., Müller, P., Olbers, D.J., Richter, K., Sell, W. & Walden, H. 1973. Measurements of wind-wave growth and swell decay during the Joint North Sea Wave Project (JONSWAP). Deutsche Hydrogr. Z. Suppl. A(80), no.12.

Janssen, P.A.E.M. 1983. Long-time behaviour of a random inhomogeneous field of weakly nonlinear surface gravity waves. J. Fluid Mech. 133, 113 – 132.

Longuet-Higgings, M.S. 1976. On the nonliner transfer of energy in the peak of a gravity-wave spectrum: a simplified model. Proc. R. Lond. A347, 311 – 328.

Phillips, O.M. 1960. On the dynamics of unsteady gravity waves of finite amplitude. Part 1. J. Fluid Mech. 9, 193 – 217.

Snyder R.L., Dobson, F.W., Elliot, J.A. and Long, R.B. 1981. Array
    measurements of atmospheric pressure fluctuations above surface
    gravity waves. J. Fluid Mech. 102, 1.
Watson, K.M. & West, B.J. 1975. A transport equation description of
    nonlinear ocean surface wave interactions. J. Fluid Mech. 70, 815
    − 826.
Willebrand, J. 1975. Transport in a nonlinear and inhomogeneous
    random gravity wave field. J. Fluid Mech. 70, 113 − 126.
Zakharov, V.E. 1968. Stability of periodic waves of finite amplitude
    on the surface of a deep fluid. Zh. Prikl. Mekh. Tekh. Fiz. 9, 86
    − 94 (English translation in J. Appl. Mech. Tech. Phys. 9, 190 −
    194.)

INITIAL INSTABILITY AND LONG-TIME EVOLUTION OF STOKES WAVES

L. Shemer[1] and M. Stiassnie[2]
1 Faculty of Engineering, Tel-Aviv University,
  Tel Aviv 69978, Israel
2 Dept. of Civil Engineering, Technion, Haifa 32000, Israel

ABSTRACT. The modified Zakharov equation is used to assess the long-time evolution of a system composed of a Stokes wave and two initially small disturbances.
   The most important result is that a kind of Fermi-Pasta-Ulam recurrence phenomenon (which has already been reported for class I instabilities), exists also for class II instabilities.

1. INTRODUCTION

In a recent study (Stiassnie & Shemer, 1984) we derived a modified version of the Zakharov integral equation for surface gravity waves. This version includes higher order, class II, nonlinear interaction as well as the more familiar class I interaction. A linear stability analysis of the new equation was used to study some short-time aspects of class I and class II instabilities of a Stokes wave, yielding result in agreement with those of McLean (1982). It is our opinion that the present knowledge of the long-time evolution of class I is limited and of class II is almost nil.

Class I instability:
Wave flume experiments by Lake et al (1977) have shown how the disturbances grew in time, reached a maximum and then subsided. Furthermore, the experiments showed how the unsteady wave train became, at some stage of its evolution, nearly uniform again. Yuen & Lake (1982) used a numerical solution of the Zakharov equation to show that the evolution may be recurring (Fermi-Pasta-Ulam recurrence) or chaotic, depending on the choice of modes included in the calculation. Stiassnie & Kroszynski (1982) used the nonlinear Schrödinger equation to study analytically the evolution of a three-wave system, composed of a carrier and two initially small 'side-band' disturbances. Their recurrence period (given by a simple formula) is in good agreement with the numerical results.

Y. Toba and H. Mitsuyasu (eds.), The Ocean Surface, 51–57.
© 1985 by D. Reidel Publishing Company.

Class II instability:
To our knowledge, the only information available is that of the
experiments by Su and Su et al. (1982). They have found that an initial
two-dimensional wave train of large steepness evolved into a series
of three-dimensional crescentic spilling breakers (class II), and was
followed by a transition to a two-dimensional moduled wave train (class
I). One can only speculate that the growth of the crescentic waves and
their disappearance are one cycle of a recurring phenomenon. Note that
any theoretical study of this process had to await the derivation of
the modified Zakharov equation.
     In the present paper we attempt to assess the long-time evolution
of three-wave systems composed of a Stokes wave (also called carrier)
and two most unstable, initially small disturbances.
     The long-time evolution of class II as well as class I instabili-
ties is considered for infinitely deep water. The theory is presented in
par. 2 and the results in par. 3.

2. THEORY

The smallest number of wave trains required to enable significant nonlinear interaction is three for class I as well as class II. In order
that significant interactions will occur, these three waves have to form
a nearly resonating 'quartet' for class I and a nearly resonating
'quintet' for class II. To form a 'quartet' or a 'quintet' out of three
waves, one can 'count' one of the waves, say-a, twice for class I and
three times for class II.
     Linear stability analysis is enabled by assuming that the initial
amplitudes of the two disturbances are much smaller than the amplitude
of the carrier wave.
     The wave numbers of the carrier (denoted by subscript a) and the
disturbances (b and c) are
$$\underline{k}_a = k_o(1,0); \quad \underline{k}_b = k_o(1+p,q); \quad \underline{k}_c = (J-p,-q) \qquad (2.1)$$
where:

$$J = \{ \begin{matrix} 1, \text{ for class I} \\ 2, \text{ for class II} \end{matrix} \qquad (2.2)$$

The regions of instability in the (p,q)-plane and the most-unstable
disturbances (having the maximum growth rate) are discussed in Stiassnie
and Shemer (1984).
     The free surface elevation for the three-wave system is given by
$$\eta = \sum_{j=a,b,c} a_j \cos(\underline{k}_j \cdot \underline{x} - \int_o^t \Omega_j dt + \theta_j) \qquad (2.3)$$
where $(x_1, x_2) = \underline{x}$ are the horizontal coordinates, t is the time and θ_j
are the initial phase shifts. The wavenumbers $\underline{k}_j$ are given in $(2.1)^j$
and the 'Stokes-corrected' frequencies Ω_j are given by:

$$\Omega_a = \omega_a + T_{aaaa}|R_a|^2 + 2T_{abab}|R_b|^2 + 2T_{acac}|R_c|^2 \qquad (2.4a)$$
$$\Omega_b = \omega_b + 2T_{baba}|R_a|^2 + T_{bbbb}|R_b|^2 + 2T_{bcbc}|R_c|^2 \qquad (2.4b)$$
$$\Omega_c = \omega_c + 2T_{caca}|R_a|^2 + 2T_{cbcb}|R_b|^2 + T_{cccc}|R_c|^2 \qquad (2.4c)$$

where ω_j is related to $\underline{k}_j$ by the linear dispersion relation

$$\omega_j = (g|\underline{k}_j|)^{\frac{1}{2}} \text{ and } R_j = \pi(\frac{2g}{\omega_j})^{\frac{1}{2}} a_j e^{i\theta_j} \qquad (2.5)$$

The governing equations for R_j are a discretized form of the Zakharov and modified Zakharov eqs. for class I and class II, respectively, given by:

$$\frac{dR_a}{dt} = -2iS_a^{(J)} \cdot (R_a^*)^J R_b R_c \text{expi}(\int_0^t \Omega_J dt) \qquad (2.6a)$$

$$\frac{dR_b}{dt} = -iS_b^{(J)} \cdot R_c^*(R_a)^{J+1} \text{expi}(-\int_0^t \Omega_J dt) \qquad (2.6b)$$

$$\frac{dR_c}{dt} = -iS_c^{(J)} \cdot R_b^*(R_a)^{J+1} \text{expi}(-\int_0^t \Omega_J dt) \qquad (2.6c)$$

where $\Omega_J = (J+1)\Omega_a -\Omega_b -\Omega_c$ $\qquad (2.7)$
and

	$S_a^{(J)}$	$S_b^{(J)}$	$S_c^{(J)}$
J=1	T_{aabc}	T_{bcaa}	T_{cbaa}
J=2	$\frac{1}{2}(U_{aaabc}^{(3)} +U_{aaacb}^{(3)})$	$U_{bcaaa}^{(2)}$	$U_{cbaaa}^{(2)}$

the * denotes the complex conjugate, and the interaction coefficients $T_{....}$, $U_{.....}$ are given in Stiassnie & Shemer (1984). Note that the present R_j is related to B_j of Stiassnie & Shemer through:

$$R_j = B_j \text{ expi}(\int_0^t (\Omega_j -\omega_j) dt) \qquad (2.8)$$

Applying the operation $R_j^* \cdot Eq(2.6j)+R_j \cdot Eq(2.6j)*$ on each of the Equations (2.6j) $j = a,b,c$ yields

$$\frac{d}{dt} |R_a|^2 = 4S_a^{(J)} \text{Im}\{(R_a^*)^{J+1} R_b R_c \text{expi}(\int_0^t \Omega_J dt)\} \qquad (2.9a)$$

$$\frac{d}{dt} |R_b|^2 =-2S_b^{(J)} \text{Im}\{(R_a^*)^{J+1} R_b R_c \text{expi}(\int_0^t \Omega_J dt)\} \qquad (2.9b)$$

$$\frac{d}{dt} |R_c|^2 =-2S_c^{(J)} \text{Im}\{(R_a^*)^{J+1} R_b R_c \text{expi}(\int_0^t \Omega_J dt)\} \qquad (2.9c)$$

A new real function Z is defined, so that

$$\frac{dZ}{dt} = \text{Im}\{(R_a^*)^{J+1} R_b R_c \text{expi}(\int_0^t \Omega_J dt)\} \qquad (2.10)$$

Substitution of (2.10) in (2.9) and integration yield:

$$|R_a|^2 = 4S_a^{(J)} Z + |r_a|^2 \qquad (2.11a)$$

$$|R_b|^2 =-2S_b^{(J)} Z + |r_b|^2 \qquad (2.11b)$$

$$|R_c|^2 =-2S_c^{(J)} Z + |r_c|^2 \qquad (2.11c)$$

where $r_j=R_j(t=0)$ are the initial values. Using (2.6) one can show that

$$\frac{d}{dt} \text{Re}\{(R_a^*)^{J+1} R_b R_c \text{expi}(\int_0^t \Omega_J dt)\} = -\Omega_J \frac{dZ}{dt}, \qquad (2.12)$$

which, after integration, gives

$$\text{Re}\{(R_a^*)^{J+1}R_b R_c \exp i(\int_0^t \Omega_J dt)\} = -\int_0^Z \Omega_J dZ + \text{Re}\{(r_a^*)^{J+1}r_b r_c\} \quad (2.13)$$

From (2.10) and (2.13) we obtain

$$(\frac{dZ}{dt})^2 = |R_a|^{2(J+1)}|R_b|^2|R_c|^2 - [-\int_0^Z \Omega_J dZ + \text{Re}\{(r_a^*)^{J+1}r_b r_c\}]^2 \quad (2.14)$$

The r.h.s of (2.14), after substitution of (2.7), (2.4) and (2.11) is a known polynomial in Z of order (J+3), denoted by $P_{J+3}(Z)$.

The solution of (2.14) is

$$t = \int_0^Z dZ/ \sqrt{P_{J+3}(Z)} \quad (2.15)$$

where Z is allowed to vary between two neighboring roots of the polynomial: Z=ZL and Z=ZR where ZL < 0, and ZR > 0.

From (2.15) it is clear that Z is periodic in time and that the recurrence period T is given by

$$T = 2\int_{ZL}^{ZR} dZ/ \sqrt{P_{J+3}(Z)} \quad (2.16)$$

For class I we write $P_4(Z) = \sum_{\ell=0}^{4} a_\ell Z^{(4-\ell)}$. When $a_o > 0$ then the 4 roots are $Z_4 > Z_3 > 0 > Z_2 > Z_1$; giving Z_3=ZR and Z_2=ZL, and (2.15) has the explicit solution

$$Z = \frac{Z_4 (Z_3 - Z_2) \text{sn}^2 (u, \kappa) - Z_3 (Z_4 - Z_2)}{(Z_3 - Z_2) \text{sn}^2 (u, \kappa) - (Z_4 - Z_2)} \quad (2.17a)$$

where sn is the Jacobian elliptic function of argument u and modulus κ :

$$u = \text{sn}^{-1}(\beta, \kappa) - a_o^{\frac{1}{2}} t/\gamma \quad (2.17b)$$

$$\beta = \sqrt{(Z_4 - Z_2)Z_3} / \sqrt{(Z_3 - Z_2)Z_4} \quad (2.17c)$$

$$\gamma = 2/\sqrt{(Z_4 - Z_2)(Z_3 - Z_1)} \quad (2.17d)$$

$$\kappa = \sqrt{(Z_3 - Z_2)(Z_4 - Z_1)} / \sqrt{(Z_4 - Z_2)(Z_3 - Z_1)} \quad (2.17e)$$

The recurrence period for this case is given by

$$T = \frac{2}{a_o^{\frac{1}{2}}} K(\kappa) \quad (2.18)$$

where K is a complete elliptic integral. Expressions similar to (2.17) and (2.18) exist for $a_o < 0$. For class II, where the polynomial is of order five, we cannot express the solution in terms of tabulated functions, and we integrate (2.15) and (2.16) numerically. Once Z is found, we use (2.11) to obtain $|R_j|$, (2.5) to obtain a_j and (2.3) to obtain η, (note that $\int_0^t \Omega_j(t)dt = \int_0^Z \Omega_j(Z) \sqrt{P_{J+3}(Z)} dZ$).

3. RESULTS

3.1 Initial instability

For infinitely deep water, the most unstable disturbances have the following wave-numbers:

Class I: $\underline{k}_b = k_o(1+P_I,0)$, $\underline{k}_c = k_o(1-P_I,0)$

Class II: $\underline{k}_b = k_o(1.5, q_{II})$, $\underline{k}_c = k_o(1.5, -q_{II})$

Thus, the class I evolving wave field is two-dimensional, whereas the class II wave field is three-dimensional but symmetric.

The values of P_I and q_{II} as functions of the initial steepness of the Stokes wave (h/λ, h and λ are the wave height and wave-length, respectively) are given by the solid lines and full symbols in Fig. 1. The lines, for class I and class II were derived from the Zakharov equation and the modified Zakharov equation, respectively. The dots and squares are from McLean (1982), obtained by a numerical stability analysis of an exact finite amplitude Stokes wave. The 'dashed' lines and hollow symbols on the same figure, give the growth-rate of the most unstable disturbances. ($Im\sigma$ in McLean, 1982). Here again, the lines are our results and the dots (for class I) and squares (for class II) are those of McLean. The agreement between the two sets of results is good for waves of small to moderate wave steepness, and less impressive for very steep waves. The most important result of this analysis is that for $h/\lambda > 0.1$, (from McLean, or $h/\lambda > 0.11$ from our calculation), the growth rate of the class II instability overtakes that of class I.

3.2 The recurrence period

The nondimensional recurrence period $\omega_a T$ as a function of the initial linear carrier steepness $a_o k_o$ ($a_o = a_a(t=0)$), is shown in Fig. 2 for three cases: (i) class I, $\theta = 0$, (ii) class I, $\theta = \pi/2$; (iii) class II, $\theta = \pi/2$. The phase-shift difference θ is given by $\theta_b + \theta_c - (J+1)\theta_a$ at t=0. For all three cases we chose the relative amplitude of the initial disturbance $\varepsilon_1 = a_b(t=0)/a_o = a_c(t=0)/a_o$ to be 0.1. Generally speaking, the recurrence period depends on three parameters: The carrier steepness $a_o k_o$, the relative amplitude of the initial disturbance ε_1, and the phase-shift difference θ. For class I Stiassnie & Kroszynsky (1982) obtained:

$$\omega_a T = \begin{cases} 2(a_o k_o)^{-2}[0.98 - 2\ln(\varepsilon_1) - \ln|\cos\theta|], & \theta \neq \pi/2 \\ 2(a_o k_o)^{-2}[1.67 - 4\ln(\varepsilon_1)], & \theta = \pi/2 \end{cases} \tag{3.1}$$

Eq. 3.1 is represented in Fig. 2 by the two lower dashed straight lines. These results, obtained from the nonlinear Schrödinger equation, are in fair agreement with the present class I claculations. The recurrence period for class II, ($\varepsilon_1 = 0.1$ and $\theta = 90°$) is given by the upper solid curve in Fig. 2. The dashed line below this curve has the slope 1:3, representing a relationship of the form $\omega_a T \alpha (a_o k_o)^{-3}$. The dependence of class II T on ε_1 and θ was found to be qualitatively similar to that of class I. Namely, the periods for $\varepsilon_1 = 0.01$ were found to be 1.65 to 2 times greater than those for $\varepsilon_1 = 0.1$, ($\ln(0.01)/\ln(0.1) = 2$); the largest period is obtained for $\theta = 90°$; and the smallest for $\theta = 0°$.

To obtain a better physical feeling, note that the recurrence periods for $k_o a_o = 0.36$ ($\theta = 90°$, $\varepsilon_1 = 0.1$) which are about equal for the two classes are 38 times the carrier period.

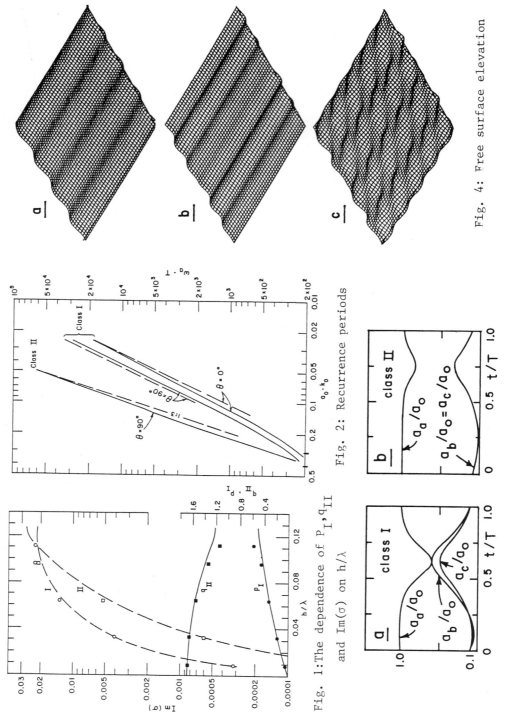

Fig. 1: The dependence of P_I, q_{II} and $\mathrm{Im}(\sigma)$ on h/λ

Fig. 2: Recurrence periods

Fig. 3: Periodic amplitude evolution

Fig. 4: Free surface elevation

3.3 Amplitude evolution

The evolution of the three amplitudes a_a/a_o, a_b/a_o, and a_c/a_o as a function of time t/T, (for $k_o a_o = 0.29$, $\varepsilon_1 = 0.1$, $\theta = 90°$) for class I and class II, is given in Fig. 3a and 3b respectively. Figures for the same $k_o a_o$ but different ε_1 are almost identical; and those with different $a_o k_o$, are rather similar. Both evolution processes have two distinct regions: (i) A region in which the disturbances stay smaller than their initial value. In this region the disturbances reach a minimum(at Z=ZR), which corresponds to an almost uniform wave train. (ii) A region in which the disturbances grow beyond their initial value. Here the disturbances reach a maximum (at Z=ZL),which corresponds to the most disturbed wave field. The order of appearance of these two regions depends on the sign of $\sin\theta$, see   2.9. Fig. 3 is typical for cases with $\sin\theta > 0$; For $\sin\theta < 0$ the order of the above- mentioned two regions is interchanged, so that the disturbances grow at the initial stage. Note that for class II $a_b = a_c$ throughout the evolution.

3.4 The free surface

Figure 4 shows the free surface elevation of $k_o a_o = 0.36$, $\varepsilon_1 = 0.1$ and $\theta = 90°$. The almost undisturbed wave-train is shown in Fig. 4a; the class I most modulated situation is given in Fig. 4b, and the crescentic shaped, class II waves in Fig. 4c.

REFERENCES

Lake, B.M. , Yuen, H.C. , Rungaldier, H. , and Ferguson, W.E. , 1977
     Nonlinear deep water waves: Theory and experiments 2. Evolution
     of a continuous wave train. J. Fluid Mech. 83, 49-74.
McLean, J.W. , 1982, Instabilities of finite-amplitude water waves.
     J. Fluid Mech. 114, 315-330.
Stiassnie,M. , and Kroszynski, U.I. , 1982, Long-time evolution of an un-
     stable water-wave train.J. Fluid Mech. 116 207-225.
Stiassnie, M. , and Shemer, L. , 1984, On modifications of the Zakharov
     equation for surface gravity waves. J. Fluid Mech. 143 47-67.
Su, M.-Y. , 1982, Three-dimensional deep-water waves. Part I. Experi-
     mental measurement of skew and symmetric wave patterns.
     J. Fluid Mech. 124, 73-108.
Su, M.-Y. , Bergin, M., Marler, P. , and Myrick, R. , 1982, Experiments
     on nonlinear instabilities and evolution of steep gravity-wave
     trains. J. Fluid Mech. 124, 45-72.
Yuen, H.C. , and Lake, B.M. , 1982, Nonlinear dynamics of deep-water
     gravity waves. Adv. Appl. Mech. 22, 67-229.

ON NON-LINEAR WATER WAVE GROUPS AND THE INDUCED MEAN FLOW

H. Tomita
Ship Research Institute
Shinkawa 6-chome Mitaka, Tokyo 181, Japan

ABSTRACT. The non-linear modulation of water wave groups was investigated and the interaction equations with induced flows are obtained.
Some applications are given to confirm the theoretical results.

1. INTRODUCTION

On the ocean surface, wind-generated waves consist of many components,
each of which has a different wave length and is propagated in different
directions. Such a confused ocean surface is usually handled from a
statistical point of view. However, if waves are propagated out of the
wind blowing region, they loose part of their stochastic property.
They might appear as wave groups by the effect of self modulation.
In this case, a standard method would be to follow the well known
Schrödinger equation-based modulation theory which is considered to be
a technique for treating the problem of modulational characteristics
of non-linear dispersive waves. However, it is inadequate to describe
waves whose steepness is as large as 0.15.
    In this paper, we show how to remove this constraint by including
the role of the mean flow in our considerations.

2. DERIVATION OF THE EQUATIONS

The governing equations of water waves for an arbitrary uniform depth
are as follows,
(1) Continuity equation

$$\frac{\partial^2 \phi}{\partial x^2} + \frac{\partial^2 \phi}{\partial y^2} = 0, \qquad -h < z < \zeta \tag{1}$$

(2) Boundary conditions

$$\frac{\partial \phi}{\partial t} + g\zeta + \frac{1}{2}\{\phi_x^2 + \phi_z^2\} = 0, \quad \frac{\partial \zeta}{\partial t} - \frac{\partial \phi}{\partial z} + \frac{\partial \phi}{\partial x}\frac{\partial \zeta}{\partial x} = 0, \quad z = \zeta \tag{2}$$

here, conventional notations are used. And $\partial \phi/\partial z = 0$ at $z = -h$.

59

Y. Toba and H. Mitsuyasu (eds.), The Ocean Surface, 59–64.
© 1985 by D. Reidel Publishing Company.

In order to solve these equations, the velocity potential and surface elevation are expanded into powers of ε (smallness parameter) such that

$$\Phi = \varepsilon\Phi^{(1)} + \varepsilon^2\Phi^{(2)} + \ldots, \qquad \zeta = \varepsilon\zeta^{(1)} + \varepsilon^2\zeta^{(2)} + \ldots \quad .$$

At the same time, Φ and ζ can be divided into two parts, $\Phi = \psi + \phi$, $\zeta = \xi + \eta$ so that $\Phi^{(i)} = \psi^{(i)} + \phi^{(i)}$, $\zeta^{(i)} = \xi^{(i)} + \eta^{(i)}$. In these expressions, ϕ and η are the periodic terms, and ψ and ξ denote the aperiodic currents and mean surface variations. In this analysis, we concentrate on the quasi-monochromatic solution so that we can adopt the following elementary solutions which include powers up to the first order in the periodic terms and up to second order in the aperiodic terms. Elementary solutions are

$$\phi^{(1)} = \frac{-i\omega \, \cosh(w+\delta)}{\kappa \, \sinh\delta}\alpha e^{i\theta} + c.c, \quad \psi^{(1)} = \beta, \quad \psi^{(2)} = \gamma,$$

$$\eta^{(1)} = \alpha e^{i\theta} + c.c, \quad \xi^{(1)} = 0, \quad \xi^{(2)} = \lambda$$

where κ, ω are constants which obey $\omega^2 = \kappa g\sigma$, $\sigma = \tanh\delta$, $\delta = \kappa h$. $\theta = \kappa x - \omega t$ is the phase angle of the wave and $w = \kappa z$. If we neglect the nonlinearity of the problem, we can obtain a solution by setting α, $\beta = $const. If the nonlinearity is to be taken into account and the perturbation method is used, α, β, γ and λ are not constants, but must be allowed to vary slowly with respect to space and time in order to remove the secular nature of the solution. By the aid of the Krylov-Bogoliubov-Mitropolski technique we expand these parameters into powers of ε

$$\frac{\partial\alpha}{\partial t} = \sum_j\varepsilon^j A^{t(j)}, \quad \frac{\partial\alpha}{\partial x} = \sum_j\varepsilon^j A^{x(j)}, \quad \frac{\partial\beta}{\partial t} = \sum_j\varepsilon^j B^{t(j)}, \quad \ldots\ldots, \text{ etc.}$$

The differential operaters can be split as follows:

$$\frac{\partial}{\partial t} = \omega\frac{\partial}{\partial\tau} - \omega\frac{\partial}{\partial\theta} + \varepsilon(A^{t(1)}\frac{\partial}{\partial\alpha} + A^{t(1)}\frac{\partial}{\partial\alpha} + B^{t(1)}\frac{\partial}{\partial\beta} + C^{t(1)}\frac{\partial}{\partial\gamma}) + O(\varepsilon^2), \text{ etc.}$$

Applying these relations to (1), we obtain, for the periodic terms

$$\frac{\partial^2\phi^{2}}{\partial\theta^2} + \frac{\partial^2\phi^{2}}{\partial w^2} = -\frac{2\omega}{\kappa^2}\frac{\cosh(w + \delta)}{\sinh\delta} A^{x(1)} e^{i\theta} + c.c \tag{3-1}$$

$$-\omega\frac{\partial\phi^{(2)}}{\partial\theta} + gn^{(2)} = \frac{i\omega}{\kappa\sigma} A^{t(1)} e^{i\theta} + \omega^2\frac{3\sigma^2-1}{2\sigma^2}\alpha^2 e^{2i\theta} + c.c, \tag{3-2}$$

$$\omega\frac{\partial\eta^{(2)}}{\partial\theta} + \kappa\frac{\partial\phi^{(2)}}{\partial w} = A^{t(1)} e^{i\theta} + \frac{2i\kappa\omega}{\sigma}\alpha^2 e^{2i\theta} + c.c$$

and for the aperiodic terms

$$B^{z\,(1)} = 0, \quad g\lambda = \omega^2(1 - 1/\sigma^2)|\alpha|^2 - B^{t\,(1)}. \tag{4}$$

If we choose the harmonic components as solutions of the equations (3),

$$\phi^{(2)} = f_1 \frac{(w+\delta)\,\sinh(w+\delta)}{\sinh\delta} e^{i\theta} + f_2 \frac{\cosh 2(w+\delta)}{\sinh 2\delta} e^{2i\theta} + C.C,$$

$$\eta^{(2)} = e_1 e^{i\theta} + e_2 e^{2i\theta} + C.C$$

we obtain immediately from (3-1) $f_1 = -\omega A^{x\,(1)}/\kappa^2$ and from (3-2) as the coefficients of $\exp(i\theta)$

$$-i\omega f_1\delta + ge_1 = i\omega A^{t\,(1)}/\kappa\sigma, \quad i\omega e_1 + \kappa f_1(1+\delta/\sigma) = A^{t\,(1)}$$

Eliminating e_1, f_1 from this simultaneous equations gives a relation between $A^{t\,(1)}$ and $A^{x\,(1)}$, $2gA^{t\,(1)} + \omega g(1+\delta/\sigma - \sigma\delta)/\kappa \cdot A^{x(1)} = 0$. This yields

$$A^{t\,(1)} + C_g A^{x\,(1)} = 0, \tag{5}$$

where $C_g = \omega/2\kappa(1+\delta/\sigma-\sigma\delta)$ is the group velocity of the wave train. In a similar fassion, from the coefficients of $\exp(2i\theta)$, we obtain

$$f_2 = -3i\omega(1-\sigma^2)\alpha^2/2\sigma^3, \quad ge_2 = \omega^3(3-\sigma^2)\alpha^2/2\sigma^4. \tag{6}$$

We can elaborate the same procedure one step further. The counterparts of (3-2) and (5) in this order are then

$$\omega \frac{\partial\psi^{(3)}}{\partial\tau} + B^{t\,(2)} + C^{t\,(1)} + g\xi^{(3)} = \frac{i\omega}{\kappa}^3 \frac{1-\sigma^2}{\sigma^2}(A^{x\,(1)}\overline{\alpha} - \overline{A^{x\,(1)}}\alpha),$$

$$\tag{7}$$

$$\omega \frac{\partial\xi^{(3)}}{\partial\tau} + L^{t\,(1)} - B^{z\,(2)} - C^{z\,(1)} - \kappa \frac{\partial\psi^{(3)}}{\partial w} = -\frac{2\omega}{\sigma}(A^{x\,(1)}\overline{\alpha} + \overline{A^{x\,(1)}}\alpha),$$

and

$$i(A^{t\,(2)} + C_g A^{x\,(2)}) + \mu A^{xx\,(2)} + \nu|\alpha|^2\alpha + \frac{\kappa\omega}{2}(\sigma - \frac{1}{\sigma})\lambda\alpha - \kappa B^{x\,(1)}\alpha = 0, \tag{8}$$

where

$$\mu = \frac{-g}{8\kappa\omega\sigma}[\{\sigma-\delta(1-\sigma^2)\}^2 + 4\delta^2\sigma^2(1-\sigma^2)], \quad \nu = \frac{-g\kappa^3}{4\omega\sigma^3}(9-10\sigma^2+9\sigma^4).$$

By summing (4) and (7), we obtain up to the 3-rd order

$$\frac{\partial\xi}{\partial t} - \frac{\partial\psi}{\partial z} = \frac{-2\omega}{\sigma} \frac{\partial|\alpha|^2}{\partial x}, \quad \frac{\partial\psi}{\partial t} + g\xi = \frac{\omega^2(\sigma^2-1)}{\sigma^2}[|\alpha|^2 - \frac{i}{\omega}(\frac{\partial\overline{\alpha}}{\partial t}\alpha - \frac{\partial\alpha}{\partial t}\overline{\alpha})]. \tag{9}$$

This represents the mean flow response of the inhomogeneity of the wave train in the case of arbitrary depths. (5) and (9) also yield 3-rd order modulation equation which is a modification of non-linear Schrödinger equation by the effect of mean flow

$$i(\frac{\partial \alpha}{\partial t} + C_g \frac{\partial \alpha}{\partial x}) + \mu \frac{\partial^2 \alpha}{\partial x^2} + \nu |\alpha|^2 \alpha + \{\frac{\kappa \omega}{2}(\sigma - \frac{1}{\sigma})\xi - \kappa \frac{\partial \psi}{\partial x}\} \alpha = 0 \qquad (10)$$

together with the auxiliary condition that ψ satisfies the Laplace equation.

3. APPLICATIONS

a) Mean flow response. In the first place, we now treat the well known problem of calculating a 2-nd order flow by means of waves that have a narrow band amplitude spectrum. First order waves are composed of infinitely many components, that is

$$\zeta^{(1)} = \sum_n 2|\alpha_n|\cos(\theta_n + \gamma_n), \quad \phi^{(1)} = \sum_n \frac{2\omega_n}{\kappa_n} \frac{\cosh\kappa_n(z+h)}{\sinh\kappa_n h} |\alpha_n| \sin(\theta_n + \gamma_n). \quad (11)$$

If we adopt the normal perturbation technique retaining only the terms of difference of phase angle $\Delta_{nm} = \theta_n - \theta_m + \gamma_n - \gamma_m$ to the 2-nd order, we obtain the equation for the slowly varying potential and surface elevation

$$\frac{\partial \phi^{(2)}}{\partial t} + g\zeta^{(2)} = \sum_{n,m} 2|\alpha_n||\alpha_m|\omega_n^2 \cos \Delta_{nm}$$

$$- \sum_{n,m} \frac{\omega_n \omega_m}{\sinh\kappa_n h} \frac{\cosh(\kappa_n + \kappa_m)h}{\sinh\kappa_m h} |\alpha_n||\alpha_m| \cos \Delta_{nm}. \qquad (12-1)$$

Considering that ω_n, ω_m's are concentrated around the dominant frequency ω, this expression can be simplified to $\omega^2(1-1/\sigma^2) \sum_{n,m} |\alpha_n||\alpha_m| \cos \Delta_{nm}$. Similaly,

$$\frac{\partial \zeta^{(2)}}{\partial t} - \frac{\partial \phi^{(2)}}{\partial z} = \frac{\omega}{\sigma} \sum_{n,m} 2|\alpha_n||\alpha_m|\Delta\kappa_{nm} \sin \Delta_{nm}. \qquad (12-2)$$

On the other hand, if we use the expression of varying amplitude, ie $|\alpha|^2 = \sum_{n,m} |\alpha_n||\alpha_m|\cos \Delta_{nm}$, (12) leads to

$$\frac{\partial \phi^{(2)}}{\partial t} + g\zeta^{(2)} = \frac{\omega^2(\sigma^2-1)}{\sigma^2}|\alpha|^2, \quad \frac{\partial \zeta^{(2)}}{\partial t} - \frac{\partial \phi^{(2)}}{\partial z} = -\frac{2\omega}{\sigma} \frac{\partial |\alpha|^2}{\partial x} \qquad (13)$$

The results coincide with those of (9), except for the last term of the second equation which yields higher order corrections to the mean surface elevation. (13) is also equivalent to the results obtained by Longuet-Higgins and Stewart (1962).

b) Instability of the wave train. In equations (9), (10) below, we introduce non-dimensional variables τ, v, w, Y, Ψ, A. For simplicity, we restrict ourselves to the case of large depth (large compared with the wave length but not necessarily compared with its group length). The equations simplify to

$$2i(\frac{\partial A}{\partial \tau} + \frac{1}{2}\frac{\partial A}{\partial v}) - \frac{1}{4}\frac{\partial^2 A}{\partial v^2} = A(|A|^2 + \frac{\partial \Psi}{\partial v})$$

$$\frac{\partial^2 \Psi}{\partial \tau^2} + \frac{\partial \Psi}{\partial w} = \frac{\partial}{\partial v}|A|^2, \quad \frac{\partial^2 \Psi}{\partial v^2} + \frac{\partial^2 \Psi}{\partial w^2} = 0$$

(14)

If we choose as a basic flow $A=A_0\exp(-0.5iA_0^2\tau)$, $\Psi=0$, ie, a steady wave train (STOKES WAVE), we can analyse the stability of small perturbed motions which are assumed to be $\hat{A}=C_1\exp[i(Kv-\Omega\tau)]$, $\hat{\Psi}=C_3\exp[i(Kv-\Omega\tau)]$. By the standard method, they lead to the dispersion relation

$$\{ (K-2\Omega)^2 + \frac{1}{4} K^2(2A_0^2 - \frac{1}{4} K^2)\}(\Omega^2-K) + \frac{1}{2}K^4A_0^2 = 0.$$

(15)

If we do not take the effect of mean flow into account, the last term in (15) can be omitted, and clearly the two terms in the bracket are individually zero. In this case, it is easy to find the marginal frequency of instability $K-2\Omega=0$, $2A_0^2-K^2/4=0$, which implies $K=2\sqrt{2}A_0$, $\Omega_M=\sqrt{2}A_0$, ie, the BENJAMIN-FEIR instability (1962).

    In a next approximation, we include the last term, but replace Ω^2-K by $-K$ in (15). The calculation is also easy. In this case, the marginal frequency of instability becomes

$$K = 2\sqrt{2}A_0 - 4A_0^2, \quad \Omega_M = \sqrt{2}A_0 - 2A_0^2.$$

(16)

The results are shown in Figure 1 together with the results of Benjamin (1967) and the numerical result of Longuet-Higgins (1978). Finally, if we include the entire terms of (15), it can be shown that to a very good approximation equation (16) results.

    The precise discussion will be reported elsewhere.

4.   CONCLUSION

The non-linear Schrödinger equation was modified by considering the effect of mean flow induced by the wave modulation itself.

    The effect of radiation stress was reexamined in the wave groups for the arbitrary depth. The stability characteristics could be derived for the case when the depth is not large compared with the group length.

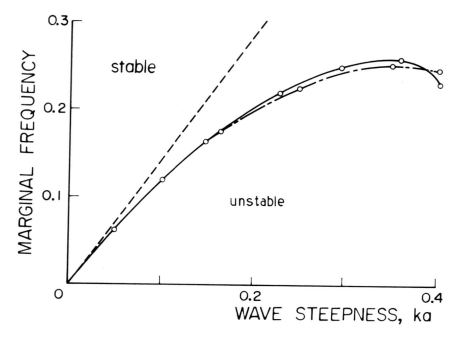

FIGURE 1.   MARGINAL FREQUENCY

```
------- nlS equation
---·---·-- Dysthe (1979) and the present result
——————— Longuet-Higgins
```

REFERENCES

Benjamin, T. B., 1967: Instability of periodic wavetrains in nonlinear
    dispersive systems. Proc. R. Soc. London, Ser. A 299, 59-75.
Dysthe, K. B., 1979: Note on a modification to the nonlinear Schrödinger
    equation for application to deep water waves. Proc. R. Soc. London,
    Ser. A 369, 105-114.
Longuet-Higgins, M. S., 1978: The instabilities of gravity waves of
    finite amplitude in deep water. II. Proc. R. Soc. London, Ser.
    A 360, 489-505.
Longuet-Higgins, M. S. and R. W. Stewart, 1962: Radiation stress and
    mass transport in gravity waves, with application to 'surf beats'.
    J. Fluid Mech, 13, 481-504.

NUMERICAL AND EXPERIMENTAL ANALYSIS OF NONLINEAR
DEFORMATION OF OCEAN WAVES ON 2-D AND 3-D SANDBARS

T. Hino [1], H. Miyata [2] and H. Kajitani [2]
1 Ship Research Institute, Mitaka Tokyo 181 Japan
2 Department of Naval Architecture, The University of Tokyo,
Bunkyo-ku, Tokyo 113 Japan

ABSTRACT.   The finite-difference method for the Navier-Stokes equations
is applied to the problems of nonlinear deformation of ocean waves on
sandbars and wave breaking in front of an advancing floating body.

1.   INTRODUCTION

A number of methods have been developed to analyse the various problems
concerning nonlinear behavior of ocean waves. The finite-difference
simulation method presented here is effective for the analysis of the
interactions of steep ocean waves with various sea topography or
artificial structures, because it exactly satisfies the nonlinear free
surface conditions and can cope with two or three dimensional body of
arbitrary shape.
    This method is a version of the TUMMAC (Tokyo University Modified
Marker-And-Cell) method [1), 2)] which is based on the MAC method [3)] and
employs the modified finite-difference schemes and boundary conditions.
    In this paper the outline of the solution method is presented and
calculated results of the regular waves over the 2-D or 3-D sandbar and
the wave breaking in front of an advancing floating body are discussed
in comparison with the experimental results.

2.   NUMERICAL METHOD

The basic equations are the Navier-Stokes equations and the continuity
equation for the incompressible fluid. In the 3-D case they are

$$\partial u/\partial t + \partial(u^2)/\partial x + \partial(u v)/\partial y + \partial(u w)/\partial z$$
$$= - \partial \phi/\partial x + \nu \cdot (\partial^2 u/\partial x^2 + \partial^2 u/\partial y^2 + \partial^2 u/\partial z^2) \qquad (1)$$

$$\partial v/\partial t + \partial(v u)/\partial x + \partial(v^2)/\partial y + \partial(v w)/\partial z$$
$$= - \partial \phi/\partial y + \nu \cdot (\partial^2 v/\partial x^2 + \partial^2 v/\partial y^2 + \partial^2 v/\partial z^2) \qquad (2)$$

Y. Toba and H. Mitsuyasu (eds.), The Ocean Surface, 65–70.
© 1985 by D. Reidel Publishing Company.

$$\partial w / \partial t + \partial (w u)/ \partial x + \partial (w v)/ \partial y + \partial (w^2)/ \partial z$$
$$= - \partial \phi / \partial z + \nu \cdot (\partial^2 w / \partial x^2 + \partial^2 w / \partial y^2 + \partial^2 w / \partial z^2) + g \quad (3)$$

$$\partial u / \partial x + \partial v / \partial y + \partial w / \partial z = 0 \quad (4)$$

where (u, v, w) are velocity components in (x, y, z) direction and ϕ, ν and g are pressure divided by the density of water, kinematic viscosity of water and gravitational acceleration, respectively.

The computational domain is divided into rectangular cells whose dimensions are (DX, DY, DZ) in (x, y, z) direction, respectively. A staggered mesh system is employed. Eqs.(1),(2) and (3) are expressed in the finite-differencing form using forward differencing in time, centered differencing in space for the diffusion and pressure terms and the combination of centered and donor cell differencing for the convection term. Thus eq.(1) is rewritten as below,

$$u_{i+1/2jk}^{n+1} = \xi_{i+1/2jk} - DT/DX \cdot (\phi_{i+1jk} - \phi_{ijk}) \quad (5)$$

where DT is time increment and ijk mean location of a cell (1/2 means location of a side of a cell) and n+1 means value at the (n+1)-th time step and values without superscripts are at the (n)-th time step.

$$\xi_{i+1/2jk} = u_{i+1/2jk} - DT \cdot (c o n v e c t i o n \quad t e r m s) +$$
$$+ DT \cdot (d i f f u s i o n \quad t e r m s) \quad (6)$$

In the same manner, eqs.(2) and (3) are rewritten as below,

$$v_{ij+1/2k}^{n+1} = \eta_{ij+1/2k} - DT/DY \cdot (\phi_{ij+1k} - \phi_{ijk}) \quad (7)$$

$$w_{ijk+1/2}^{n+1} = \zeta_{ijk+1/2} - DT/DZ \cdot (\phi_{ijk+1} - \phi_{ijk}) \quad (8)$$

Velocity values at the (n+1)-th time step can be calculated from velocity and pressure values at the (n)-th time step using eqs.(5),(7) and (8).

The next procedure is to calculate pressure from the newly obtained velocity field. Substituting eqs.(5),(7) and (8) into the centered difference expression of eq.(4) at the (n+1)-th time step, the Poisson equation for pressure is obtained and it is solved iteratively by the SOR method. The iterative equation of the SOR method for the Poisson equation is transformed as below,

$$\phi_{ijk}^{m+1} = \phi_{ijk}^{m} - 0.5 \cdot \omega \cdot \{ DT \cdot (1/DX^2 + 1/DY^2 + 1/DZ^2) \}^{-1} \cdot [D_{ijk}^{m+1}] \quad (9)$$

where m and ω are iteration number and relaxation factor, respectively, and D_{ijk} is divergence which is the LHS of the centered difference expression of eq.(4).

At every iteration step, the velocity values are always renewed using eqs.(5),(7) and (8) whenever the neighbouring pressure value is calculated. When the iteration is converged, the pressure at the (n)-th time step and the velocities at the (n+1)-th time step are obtained simultaneously.

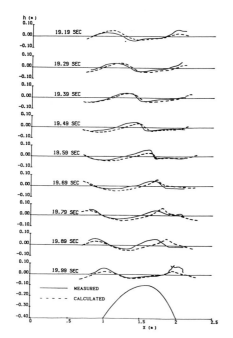

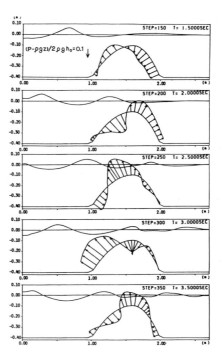

Fig.1   Measured and calculated
        wave profiles on a sandbar,
        $\lambda=1.0m$, $h_0=0.05m$.

Fig.2   Computed pressure
        distribution on a sandbar,
        $\lambda=1.0m$, $h_0=0.05m$.

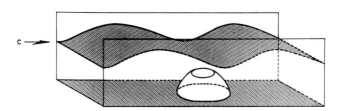

Fig.3   Sketch of a submerged mountainous obstacle in a channel.

Fig.4   Perspective view of measured wave configuration on a submerged
        mountainous obstacle, $\lambda=0.6m$, $h_0=0.02m$.

The basic solution algorithm is to integrate eqs.(1),(2) and (3) in time as a intial-value problem, and at each time step the Poisson equation for pressure is solved as a boundary-value problem.

The exact inviscid free surface condition is used. The kinematic condition is fulfilled by the Lagrangian movement of marker particles in the sandbar problems and by that of line segments in the breaking wave problem. The dynamic condition is fulfilled by using the atmospheric pressure in the solution procedure of the Poisson equation for pressure. The free-slip body boundary condition for arbitrary body configurations is applied to the sandbar problems and the no-slip body boundary condition to the breaking wave problem. Waves are generated by setting the velocity values on the inflow boundary according to the linear theory and the linear superposition is used to generate irregular waves.

3. NUMERICAL AND EXPERIMENTAL RESULTS

3.1 Wave Deformation over a 2-D Sandbar

The deformation of regular waves passing over a 2-D sandbar located on the flat bottom is analysed. The sandbar whose length and height are 1.0m and 0.3m, respectively, is composed of two parabolic curves. The water depth is 0.4m and the incident wave is a regular wave of the wavelength $\lambda=1.0$m and the amplitude $h_0=0.05$m. The period of the incident wave is about 0.806sec.   DX and DZ are 0.05m and 0.02m, respectively, and DT is 0.01sec.

The time evolution of the wave profile is shown in Fig.1.   The waves propagate from left to right and solid lines are experimental results while dashed lines are calculated ones. Both experimental and calculated results show that the forward face of the wave is steepened in its approach to the highest point of the sandbar. In the experiment the wave breaking occurs at the point slightly behind the highest point at which the computation shows a very steep wave crest.

The calculated pressure distribution on the surface of the sandbar is shown in Fig.2. The pressure pushing the body appears under the wave crest and the pressure pulling the body under the wave trough. The direction of the total drifting force is alternating between upwave and downwave.

3.2 Wave Deformation over a 3-D Sandbar

The deformation of 3-D regular waves over a 3-D axisymmetric sandbar located in a channel as seen in Fig.3 is analysed. The width and the water depth of a channel are 0.54m and 0.17m, respectively, and the sandbar of which diameter is 0.15m at its top and 0.306m at its bottom is located at the center of the channel. The incident wave is a regular wave of the wavelength $\lambda=0.6$m and the amplitude $h_0=0.02$m. The period is about 0.638sec. DX,DY and DZ are 0.03m, 0.03m and 0.01m, respectively, and DT nondimensionalized by the period of the incident wave is 0.01.

An example of the perspective views of the wave configurations

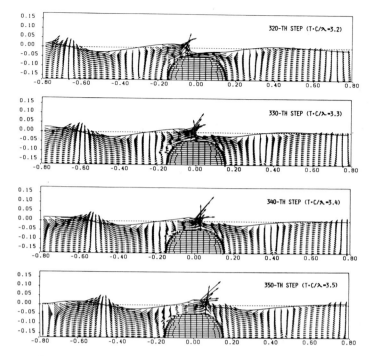

Fig.5   Computed velocity fields of waves on a submerged mountainous
        obstacle on the center-plane, $\lambda=0.6$m, $h_0=0.02$m.

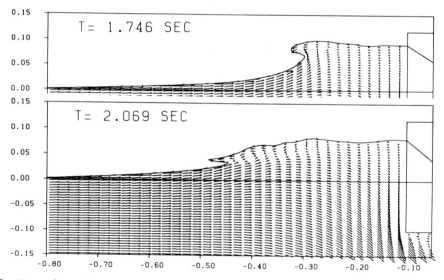

Fig.6   Wave breaking in front of an advancing 2-D floating body.

measured by the grid projection mehtod[4] is shown in Fig.4 The waves
propagate from up-right to down-left. The longitudinal position of the
center of the sandbar is at the fourteenth mesh line from left. A very
steep 3-D wave is formed when the wave crest passes near the center of
the sandbar. The wave breaking occurs immediately after this state.
    Fig.5 illustrates the calculated wave profiles and velocity vectors
on the centerplane. The forward face of the wave crest becomes steep
and the large upward velocities appear as the wave crest passes over
the sandbar.

3.3 Wave Breaking in front of an Advancing 2-D Floating Body

The present method applies to the analysis of the flow around a
floating 2-D body advancing steadily at uniform speed in deep water.
With some modifications the overturning of wave configurations and the
subsequent breaking process can be simulated by the new version. All
the cells are flagged in order to indicate where the free surface and
the body boundary exist. And the line segment on the free surface is
moved in the Lagrangian way by using the normally extrapolated
velocities.
    The length and draft of the floating body are 0.2m and 0.1m,
respectively. The Froude number based on the draft is 1.25. DX and DZ
are 0.01m and 0.005m, respectively, and DT is about 0.00135sec.
    The wave profiles and velocity vectors in front of the body are
shown in Fig.6. The sequence of the appearance of the steep wave,
overturning, breaking and the generation of vortex is clearly noted.

REFERENCES

1) Hino, T. et al.   'A Numerical Solution Method for Nonlinear Shallow
                     Water Waves (1st and 2nd Report)' J. Soc. Nav.
                     Arch. Japan, vol.153 and 154 (1983)
2) Aoki, K. et al.   'A Numerical Analysis of Nonlinear Waves Generated
                     by Ships of Arbitrary Waterline (1st and 2nd
                     Report)' J. Soc. Nav. Arch. Japan, vol.154 and
                     155 (1983, 1984)
3) Welch, J. E. et al. 'The MAC method --- A Computing Technique for
                     Solving Viscous, Incompressible, Transient Fluid-
                     Flow Problems Involving Free Surface' Los Alamos
                     Sci. Lab. LA-3425 (1965)
4) Kanai, M.   'Wave Analysis by Grid Projection Method'
                     J. Kansai Soc. Nav. Arch., Japan, vol.193 (1984)

ON THE GROWTH OF GRAVITY-CAPILLARY WAVES BY WIND [*]

Klaartje van Gastel, Peter A.E.M. Janssen
Gerbrand J.Komen
Royal Netherlands Meteorological Institute
P.O. Box 201,
3730 AE De Bilt, Holland

ABSTRACT. Generation and growth of gravity-capillary waves ($\lambda \sim 1$ cm) by wind is reconsidered. Our main objective is to establish the sensitivity of this process to strength and profile of wind and wind-induced shear current. To this end we give a quasi-analytical description of the process based on linear instability theory. We start from a coupled wind-current system and solve the equations by expansion in the small parameters ρ_a/ρ_w, $1/R_a$ and $1/R_w$. The solutions are valid for all friction velocities. We indicate where our work extends earlier results of Brooke-Benjamin (1959), Miles (1962), Valenzuela (1976) and Kawai (1979). We find that the phase velocity is mainly sensitive to the current profile and the growth rate to the wind profile. The growth rate of the initial wavelets, the first waves to be generated, scales with the friction velocity u_*, as u_*^3.

1. INTRODUCTION

Small-scale waves on the surface of the ocean can be of importance in various processes. For instance, they form a link in the imaging of large-scale phenomena like winds and currents from space (Schroeder et al 1982).

   This study is concerned with one aspect of these small-scale, or gravity-capillary, waves ($\lambda \sim 1$ cm), notably their growth by wind. Other studies on this subject are those of Miles (1962), based on analytical work of Brooke Benjamin (1959), and of Valenzuela (1976) and Kawai (1979), both numerical. All three use viscid versions of the linear instability theory as presented in its inviscid form by Miles in 1957. We have again taken up the subject to see whether an analytic, perturbative method leads to a good description of the growth of gravity-capillary waves.

---

[*] Extended Abstract. A full account is given by van Gastel, Janssen and Komen, 1984.

Y. Toba and H. Mitsuyasu (eds.), The Ocean Surface, 71–75.

2. THEORY

We also use linear instability theory. The outline of this theory is the following. Wind and current are presented as basic flows; the interface of air and water as a flat surface. The stability of this equilibrium to small-amplitude wave-like perturbations is studied. The perturbations of the flow appear as waves on the surface.

We took the following flows to model wind and current, called U_a and U_w resp.:

$$U_a = \frac{\xi u_*}{\varepsilon_a} + U_o \qquad\qquad 0 < \xi < \xi_1$$

$$U_a = U_1 + U_o + \frac{u_*}{\kappa}(\alpha - \tanh \tfrac{1}{2}\alpha) \qquad\qquad \xi > \xi_1 \qquad (1)$$

$$\sinh \alpha = \frac{2\kappa}{\varepsilon_a}(\xi - \xi_1)$$

$$\xi_1 = 5\varepsilon_a \qquad \kappa = 0.4 \qquad U_1 = 5u_*$$

$$U_w = U_o e^{\lambda \xi} \qquad\qquad \xi < 0 \qquad (2a)$$

$$\lambda = \frac{\delta U_o}{\varepsilon_a u_*} \qquad\qquad (2b)$$

Here ξ is the height coordinate made dimensionless by multiplying with the wavenumber (see eq. (3)), U_o is the value of the current at the surface, ξ_1 is the thickness of the viscous sublayer and u_* is the friction velocity in air. $\delta = \rho_a/\rho_w$; the ratio of the densities of air and water and ε_a is the inverse of a Reynolds-number: $\varepsilon_a = 1/R_a = \nu_a k/u_*$, where ν_a is the kinematical viscosity of air. For $\xi \gg \xi_1$ the windprofile is logarithmic. Eq. (2b) is a result of the continuity of shearing stress and couples the current to the wind. The small perturbations of these flows are related to a surface wave η of wavenumber k and phase velocity c:

$$\eta = \eta_o e^{ik(x-ct)} \qquad\qquad (3)$$

We solve the initial value problem, thus c (complex) as a function of k (real). We are interested in Re(c), the phase velocity, and 2k Im(c), the growth rate of the energy.

The governing equations for the perturbations are the Orr-Sommerfeld equation, continuity equations at the interface and the condition that the perturbations disappear towards infinity (Valenzuela, 1976; van Gastel et al, 1984). The Orr-Sommerfeld differential-equation can be deduced from the viscid Navier-Stokes equation with the use of a stream-function. The equations at the interface are the kinematical condition and continuity equations for tangential and normal velocity and shearing stress; the jump in the normal pressure must be counterbalanced by the surface tension.

To solve the equations we use the fact that δ, ε_a and ε_w $(= 1/R_w = \nu_w k/u_*)$ are small. This allows for the use of asymptotic methods. In water we solve the Orr–Sommerfeld equation analytically, exact to order ε_w. In air a critical height exists, i.e. a height where $U_a = c$. However, as the critical height is within the viscous sublayer this does not give rise to complications (Drazin & Reid, 1981). We find an analytical expression for the fast–varying solution. The other independent solution, which is a solution of the Rayleigh equation, is found numerically using the method of Janssen & Peeck (1983). The coefficients of the independent solutions in air and water and the complex phase velocity are calculated analyticallly from the continuity equations at the interface. Thus our study is near-analytical.

We find the following expansion for c:

$$c = c_o + c_1 + 0\ \left(\delta\varepsilon_a^{-2/3},\ \varepsilon_w^{1\frac{1}{2}}\right) \qquad * \qquad\qquad (4a)$$

$$\frac{c_1}{c_o} = 0\ \left(\delta\varepsilon_a^{-4/3}\right) \qquad\qquad\qquad\qquad (4c)$$

$$c_o = U_o + \sqrt{}\ \left(\frac{1}{\phi'_{iwo}}\ (\frac{g}{k} + \frac{Tk}{\rho_w}) + (\frac{\lambda U_o}{2\phi'_{iwo}})^2\right) - \frac{\lambda U_o}{2\phi'_{iwo}} \qquad (4c)$$

$$c_1 = u_* \cdot \frac{-i\varepsilon_w H + \delta\ (P - im\ Q)}{N} \qquad\qquad (4d)$$

Here ϕ_{iwo} is the solution of the Rayleigh equation in water, normalized to unity at the interface. A prime stands for differentiation to ξ, g is the gravitational constant and T is the surface tension. c_o is real; it is equal to the phase velocity of free gravity-capillary waves in the absence of a current plus the value of the current at the interface minus corrections due to shear in the current; $-i\varepsilon_w H/N$ is the viscous damping in water. H/N is real; in the absence of a current its value is 2. Deviations from this value are of the same order and due to shear of the current. P and Q are the amplitudes of the normal pressure and shear stress on the interface, respectively. The term $\delta\ (P - im\ Q)/N$ represents effect of the presence of air and wind on the surface and is proportional to δ. Both P and Q are complex. Full expressions for all of these quantities are given in Van Gastel et al (1984). P is an order $\varepsilon_a^{-2/3}$ larger than Q. As m = 0(1) this implies that the effect of the pressure on the growth of the waves is an order $\varepsilon_a^{-2/3}$ larger than the effect of the shear stress.

---

* Actually the order estimates depend on the Reynoldsnumber. Here the largest errors occuring are given. This happens for $0.007 < \varepsilon_a < 0.02$.

Miles (1962) found the same type of expression as (4). The main differences result from the fact that Miles assumed there was no current. This implies that the basic flows in his study do not fulfill the continuity equation for shearing stress at the surface. This in turn implies that formally there is no justification for Miles' results, in contrast to our case.

3. SUMMARY OF RESULTS

We have calculated phase velocities and growth rates as a function of wavenumber for several values of u_*. Plots of our results can be found in van Gastel et al (1984). For $u_* < 0.05$ m/s we find no growth. This value of the minimum wind speed capable of generating waves is in accordance with Miles (1962). The growth increases rapidly with increasing friction velocity. At fixed wave number there is no simple scaling law expressing this. However, the maximum of each curve is found to scale with u_*:

$$\beta_{max} \sim u_*^3 \qquad\qquad (5)$$

Kawai (1979) has shown that the wavenumber of this maximum coincides with the wavenumber of the initial wavelets, which are the first waves to be generated by the wind. We find that for 0.14 m/s $< u_* <$ 0.25 m/s this wavenumber is such that $R_a \simeq 36$.

We have also calulated the growth using a windprofile linear up to infinity, identical to the lin-log profile in the viscous sublayer. The growth is much larger than in the lin-log model, implying that the growth rate is very sensitive to the wind profile, even to that part which is above the critical height.

We studied the sensitivity of the growth to the current profile by comparing our results with those of others, notably Miles (1962), Valenzuela (1976) and Kawai (1979). Each study is based on a different current profile and identical wind profiles. The curves all lie near each other; differences are within 20% of the values. Empirical data of Larson & Wright fall within this same region but show an even larger spread. Thus the growth is relatively insensitive to the current profile.

Values for the (real) phase velocity that we find in our model are near those for free waves. This also follows from eq. (4b) and the fact that c_0 is independent of the flow in the air. Thus the phase velocity is independent of the profile of the wind. However, c does depend weakly on the strength of the wind, through eq. (2b). The sensitivity of the phase velocity to the current profile is stronger. We find variations of about 40%, depending on the particular profile chosen.

4. CONCLUSIONS

We have succesfully derived a quasi-analytical, perturbative description of growing gravity-capillary waves using a model in which the current is coupled to the wind. We find that the growth rate of the waves depends strongly on the wind speed; their phase velocity depends weakly on wind speed. The growth rate of the initial wavelets is proportional to u_*^3. The effect of the wind profile on the growth rate is large; on phase velocity negligible. The current profile mainly influences the phase velocity and hardly affects the growth rate. This knowledge of effects of wind and current on small-scale wave may be a help in understanding the images of the seas taken by microwave radar.

ACKNOWLEDGEMENTS

This work was partly supported by the Netherlands Organization for the Advancement of Pure Research (ZWO).

REFERENCES

Brooke Benjamin, T., 1959. Shearing flow over a wavy boundary, J. of Fluid Mech., 6, pp. 161-205.

Drazin, P.G. & Reid, W.H., 1982. Hydrodynamic Stability, Cambridge University Press, Cambridge.

Gastel, K. van, Janssen, P.A.E.M. & Komen, G.J., 1984. On Phase Velocity and Growth Rate of Wind-Induced Gravity-Capillary Waves. J. of Fluid Mech., to appear.

Janssen, P.A.E.M. & Peeck, H.H., 1984. On the quasilinear evolution of the coupled air-flow, water wave system. Submitted for Publication.

Kawai, S., 1979. Generation of Initial Wavelets by Instability of a Coupled Shear Flow and their Evolution to Wind Waves, J. of Fluid Mech., 93, pp. 661-703.

Larson, T.R. & Wright, J.W., 1975. Wind-generated gravity capillary waves: laboratory measurements of temporal growth rates using microwave backscatter, J. of Fluid Mech., 70, pp. 417-436.

Miles, J.W., 1957. On the Generation of Surface Waves by Shear Flows, J. of Fluid Mech., 3, pp. 185-204.

Miles, J.W., 1962. On the generation of surface waves by shear flows. Part 4, J. of Fluid Mech., 13, pp. 433-448.

Schroeder, L.C., Boggs, D.H., Dome, G., Halberstam, I.M., Jones, W.L., Pierson, W.J. & Wentz, F.J., 1982. The Relationship Between Wind Vector and Normalized Radar Cross Section Used to Derive SEASAT-A Satellite Scatterometer Winds, J. of Geophysical Research, 87, pp. 3318-3336.

Valenzuela, G.R., 1976. The Growth of Gravity-Capillary Waves in the Coupled Shear Flow, J. of Fluid Mech., 76, pp. 229-250.

NONLINEAR WAVES IN A DEVELOPING PROCESS

Y. L. Yuan[1], N. E. Huang[2] and C. C. Tung[3]
1 Institute of Oceanology, Academia Sinica, Qingdao, China
2 NASA/GSFC/GLAS, Goddard Space Flight Center, Greenbelt
   MD 20771
3 North Carolina State University, Raleigh, NC 27695

1. Introduction

As we know, a wind sea is a wave process with striking nonlinearity.
In the wave field there is an energy transport from higher frequencies
to lower ones and a sideband growth associated with instability of the
waves. On the other hand, when the wind blows over the sea, the air near
the surface will transmit energy to the water waves due to resonant and
coupling mechanisms. Wind-generated waves are, in fact, in a state of
continuous evolution with energy input.
  As for the wind-generated waves there are three kinds of coupling
mechanisms, that is, Miles' shear flow, Jefferys' defilade and Banner-
Melville's separation mechanisms. Even though a precisely mathematical
description of these mechanisms has not be obtained, they have things
in common, i.e., there is a phase shift between the atmospheric pressure
field and the wave field. This shift leads to a coupling energy transport
from air to water waves.
  If the phase shifts are different for various wave number components
the pressure field on the sea surface can be written as follows

$$p_a(\mathbf{r},t) = p_f(\mathbf{r},t) - \frac{1}{2\pi} \iint_k \beta(k)\frac{\partial \eta(k)}{\partial t}\exp(ikr)dk = p_f(\mathbf{r},t) - \frac{\partial Z}{\partial t} \tag{1}$$

where $p_f(\mathbf{r},t)$ denotes the free fluctuation part of the surface pressure,
$\beta(k)$ is a comprehensive coefficient, which can be estimated as

$$\beta = 0.4\frac{g\rho}{\sigma}\left(\frac{u_*}{c}\right)^2 \qquad \text{and} \qquad \gamma = \frac{|k|}{2\rho\omega}\beta = \bar{\gamma}\frac{\omega}{\sigma} \tag{2}$$

in which σ and c are the frequency and speed of the wave and $\omega = \sqrt{g|k|}$
is called zero-order dispersion relation.
  Thus the governing equations for the wave process can be written as

$$\nabla^2\phi + \frac{\partial^2\phi}{\partial z^2} = 0 \tag{3}$$

Y. Toba and H. Mitsuyasu (eds.), The Ocean Surface, 77–86.

$$\frac{\partial \phi}{\partial t} + \frac{1}{2}(|\nabla \phi|^2 + (\frac{\partial \phi}{\partial z})^2) + gz + \frac{p}{\rho} = 0 \qquad -\infty < z < \eta(\mathbf{r},t) \tag{4}$$

$$\frac{\partial \eta}{\partial t} = \sqrt{1 + |\nabla \eta|^2} \frac{\partial \phi}{\partial n} = \frac{\partial \phi}{\partial z} - \nabla \eta \nabla \phi \tag{5}$$

$$z = \eta(\mathbf{r},t)$$

$$\frac{\partial \phi}{\partial t} + \frac{1}{2}(|\nabla \phi|^2 + (\frac{\partial \phi}{\partial z})^2) + g\eta + \frac{p_a}{\rho} = 0 \tag{6}$$

$$|\nabla \phi|^2 + (\frac{\partial \phi}{\partial z})^2 \to 0 \qquad z \to -\infty \tag{7}$$

This is for irrotational motion of an incompressible ideal fluid with a free surface, whose undisturbed position is on the reference plane $(\mathbf{r},0)$ = $(x,y,0)$. ϕ is velocity potential.

2. Dispersion relation and time scale analysis

There are, at least, two small parameter in a developing nonlinear wave process: the typical wave slope ϵ and the specific energy input rate γ. Therefore, we may introduce the following multiple times

$$t_0 = t, \qquad t_1 = \epsilon t, \qquad t_2 = \epsilon^2 t \quad \text{and} \quad \tau = \gamma t \tag{8}$$

and, then, the corresponding expansions of the waves cut off to the third order in

$$\phi_0(k,t) = \epsilon \phi_{01}(k,t_0,t_1,t_2,\tau) + \epsilon^2 \phi_{02}(k,t_0,t_1,t_2,\tau) + \epsilon^3 \phi_{03}(k,t_0,t_1,t_2,\tau)$$

$$\eta(k,t) = \epsilon \eta_1(k,t_0,t_1,t_2,\tau) + \epsilon^2 \eta_2(k,t_0,t_1,t_2,\tau) + \epsilon^3 \eta_3(k,t_0,t_1,t_2,\tau)$$

Using a multiple scale Fourier transformation and with a series of complicated manipulations, we arrive at the following results:

$$(1) \quad \eta_1(k,\omega_0,\omega_1,\omega_2,\tau) = \exp(\omega_0 \tau)(A_1(k,\omega_1,\omega_2)\delta(\omega_0 - \sqrt{g|k|})$$

$$+ A_2(k,\omega_1,\omega_2)\delta(\omega_0 + \sqrt{g|k|})) \tag{9}$$

$$\phi_{01}(k,\omega_0,\omega_1,\omega_2,\tau) = -i(\omega_0/|k|)(1 + i\gamma)\eta_1(k,\omega_0,\omega_1,\omega_2,\tau) \tag{10}$$

The first order solution is entirely concentrated on the surface $\omega_0 = \pm\sqrt{g|k|}$, which is just the zero-order dispersion relation.

$$(2) \quad \omega_1 = 0 \tag{11}$$

$$\begin{Bmatrix} \eta_2(k,\omega,\tau) \\ \phi_{02}(k,\omega,\tau) \end{Bmatrix} = \int\!\!\int_{k_1 k_2} \int\!\!\int_{\omega'\omega''} \begin{Bmatrix} A(k,\omega_0,k_1,\omega_0',k_2,\omega_0'') \\ B(k,\omega_0,k_1,\omega_0',k_2,\omega_0'') \end{Bmatrix} \eta_1(k_1,\omega_0',\omega_2',\tau')$$

$$\eta_1(k_2,\omega_0'',\omega_2'',\tau'')\delta(k-k_1-k_2)\sum_{n=0}^{2}\delta(\omega_n-\omega_n'-\omega_n'')dk_1dk_2d\omega_n'd\omega_n'' \qquad (12)$$

$$(3) \quad \omega_2 = \omega_0 \iint_{k_1\omega_0'\omega_2'} (C_0(k,\omega_0',k_1,\omega_0') + i(2\gamma/\omega_0)C_1(k,\omega_0',k_1,\omega_0'))$$

$$|\eta_1(k_1,\omega_0',\omega_2',\tau')dk_1d\omega_0'd\omega_2'|^2 \qquad (13)$$

where the functions A, B, C_0 and C_1 can be found in a previous paper(1).
This result shows that in a developing process time scales proportional to first order in the wave slope do not exist. The energy input
rate has an essential effect on the dispersion relation. Even though the
dispersion relation does not change in form, a developing wave spectrum
has been introduced. The existence of the imaginary part of ω_2 due to
appearence of C_1 means that the waves are also growing at a rate of the
order $\gamma \in^2$. In the stationary case ($\gamma = 0$) our result can be written in
discrete form as obtained by Webber in 1977.
Using the "Wallops" spectrum (Huang et al, 1981) our computations
are in good agreement with the laboratory measurements of Ramamonjiarisoa (1976) in the energy containing frequency range. The deviation in
the high frequency range may be caused by wind drift in the surface
layer (see Fig. 1).

3. Hamilton equations and the main evolution equation

First of all, we introduce the following functional

$$F = \iint_r (\int_{-\infty}^{\eta} \frac{1}{2}(|\nabla\phi|^2 + (\frac{\partial\phi}{\partial z})^2)dz + \frac{1}{2}g\eta^2 + \frac{P_f}{\rho})dr \qquad (14)$$

We use the wave surface, $\eta(r,t)$, and the velocity potential on the surface, $\psi(r,t) = \phi(r, \eta(r,t),t)$, as canonical variables and then define
a function, G, as the variational derivative of a functional with respect to a canonical function, (.), if the following variational relation
holds

$$\delta_{(.)}F = \iint_{(r,k)} G(.)d(r,k) \quad \text{and denoting} \quad \frac{\delta F}{\delta(.)} = G. \qquad (15)$$

Through a series of tedious operations, it is easily proved that
all the solutions of the Stokes equation should satisfy the following
Hamilton equations

$$\frac{\partial\eta}{\partial t} = \frac{\delta F}{\delta\psi}, \quad \frac{\partial\psi}{\partial t} - \frac{1}{\rho}\frac{\partial z}{\partial t} = -\frac{\delta F}{\delta\eta}. \qquad (16)$$

Furthermore, in wave-number space we define a complex canonical
function

$$a(k,t) = \frac{f^{-1}(k)}{\sqrt{2}}(\eta(k,t) + if^2(k)\psi(k,t)) \qquad (17)$$

where $+(k) = \sqrt{|k|/\omega}$. According to the definitions (15) and (17) we can transform the Hamilton equations above to k-space and obtain

$$\frac{\partial a(k,t)}{\partial t} - i \left(\frac{\partial a(k,t)}{\partial t} + \frac{\partial a^*(-k,t)}{\partial t}\right) = -i \frac{\delta F}{\delta a^*(k,t)} . \qquad (18)$$

In fact, whether in the sea or in a wind tunnel tank, the waves always have a small typical slope, $\in$, so that the perturbation method can be used efficiently. To third order in $\in$ the solution of equations (3) and (7) can be written as

$$\phi(r,z,t) = \frac{1}{2\pi} \iint_k \left(\psi(k,t) - \frac{1}{2\pi} \iiint_{k_1 k_2} \psi(k_1,t)\eta(k_2,t)|k_1|\delta(k-k_1-k_2)dk_1 dk_2 \right.$$

$$+ \frac{1}{2(2\pi)^2} \iiiint_{k_1 k_2 k_3} \psi(k_1,t)\eta(k_2,t)\eta(k_3,t)|k_1|(|k-k_3|+|k-k_2|$$

$$\left. - |k_1|)\delta(k-k_1-k_2-k_3)dk_1 dk_2 dk_3 \right) \exp(|k|z)\exp(ikr)dk \quad (19)$$

Substituting this expression into the functional (14) and in consideration of the resonant conditions

$$k + k_1 = k_2 + k_3 \qquad \text{and} \qquad \omega(k) + \omega(k_1) = \omega(k_2) + \omega(k_3) \qquad (20)$$

the main part of the Hamilton equation (18) to third order should be

$$(1-i\gamma)\frac{\partial a(k)}{\partial t} - i\gamma\frac{\partial a^*(-k)}{\partial t} = -i\omega(k)a(k) + \frac{f(k)}{\sqrt{2}} P_f(k)$$

$$+\iint_{k_1 k_2} (a(k_1)a(k_2)V^{(-)}(k,k_1,k_2)\delta(k-k_1-k_2)+2a(k_1)a^*(k_2)V^{(-)}(k_1,k,k_2)$$

$$\delta(k-k_1+k_2)+a^*(k_1)a^*(k_2)V^{(+)}(k,k_1,k_2)\delta(k+k_1+k_2))dk_1 dk_2$$

$$+\iiint_{k_1 k_2 k_3} a^*(k_1)a(k_2)a(k_3)W(k,k_1,k_2,k_3)\delta(k+k_1-k_2-k_3)dk_1 dk_2 dk_3) \qquad (21)$$

in which the expressions of $V^{(-)}$, $V^{(+)}$, W and T in the following are also shown in a previous paper(5).

Since the first order time scale does not exist, as shown in section 1, we can introduce two kinds of time scale and a closure assumption as follows

$$t = t, \quad t = \in^2 t \quad \text{and} \quad a(k,t) = \in A(k,\tau,t) + \in^2 H(k,\tau,t) \qquad (22)$$

obtaining the first to third order equations

$$(1-i\gamma)\frac{\partial A(k)}{\partial t} - i\gamma\frac{\partial A^*(-k)}{\partial t} = -i\omega(k)A(k) - i\frac{f(k)}{\sqrt{2}\rho\in} P_f(k) \qquad (23)$$

$$(1-i\gamma)\frac{\partial H(k)}{\partial t} - i\gamma\frac{\partial H^*(-k)}{\partial t} = -i\omega(k)H(k) -i\iint_{k_1 k_2} (V^{(-)}_{0,1,2}A_1 A_2\delta(k-k_1-k_2)$$

$$+2V^{(-)}_{1,0,2}A_1 A_2^*\delta(k-k_1+k_2)+V^{(+)}_{0,1,2}A_1^* A_2^*\delta(k+k_1+k_2))dk_1 dk_2; \quad (24)$$

$$(1-i\gamma)\frac{\partial A(k)}{\partial \tau} - i\frac{\partial A^*(-k)}{\partial \tau} = -i\int\int_{k_1 k_2} (V^{(-)}_{0,1,2}(A_1 H_2 + A_2 H_1)\delta(k-k_1-k_2)$$

$$+2V^{(-)}_{1,0,2}(A_1 H_2^* + A_2^* H_1)\delta(k-k_1+k_2)+V^{(+)}_{0,1,2}(A_1^* H_2^* + A_2^* H_1^*)\delta(k+k_1+k_2))dk_1 dk_2$$

$$- i\int\int\int_{k_1 k_2 k_3} W_{0,1,2,3}A_1^* A_2 A_3\delta(k+k_1-k_2-k_3)dk_1 dk_2 dk_3 \quad (25)$$

The first order equation (23) is the well-known linear Phillips-Miles' equation describing the resonant and coupling mechanisms.

Considering only the coupling mechanism, the equations can be merged into the following equation

$$\frac{\partial \bar{a}(k,t)}{\partial t} - (\gamma - i\sqrt{1-\gamma^2})\omega(k)\bar{a}(k,t) = i\int\int\int_{k_1 k_2 k_3} T_{0,1,2,3}\bar{a}^*(k_1,t)\bar{a}(k_2,t)\bar{a}(k_3,t)$$

$$\delta(k+k_1-k_2-k_3)dk_1 dk_2 dk_3 \quad (26)$$

where $\bar{a}(k,t) = \epsilon A(k,\epsilon^2 t, t)$ is the main part of the waves. This is the developing evolution equation to third order in ϵ of the main part of the waves.

4. Generalized Schrödinger equation

If there is, indeed, a main wave-number, k_0, in the wave field, $\bar{a}(r,t)$, we can use the following expression to define the envelope, $B(r,t)$, of the waves as follows

$$\bar{a}(r,t) = B(r,t)\exp(i(k_0 r - \sqrt{1-\gamma_0^2}\omega_0 t)) \quad (27)$$

The image of the above in k-space should be

$$\bar{a}(k,t) = B(k-k_0,t)\exp(-i\sqrt{1-\gamma_0^2}\omega_0 t) \quad (28)$$

Substituting (28) into main governing equation, keeping all the terms of orders lower than $\gamma\epsilon^3$ or ϵ^4 and taking into account some delta operations in the narrow band case, we get

$$\frac{\partial B(\bar{k},t)}{\partial t} - \bar{\gamma}\omega_0 B(\bar{k},t) + i(1+i\bar{\gamma})(\omega(k)-\omega(k_0))B(\bar{k},t)$$

$$= -i\int\int\int_{\bar{k}_1 \bar{k}_2 \bar{k}_3} T(\bar{k}+k_0,\bar{k}_1+k_0,\bar{k}_2+k_0,\bar{k}_3+k_0)B^*(\bar{k}_1,t)B(\bar{k}_2,t)B(\bar{k}_3,t)$$

$$\delta(\bar{k}+\bar{k}_1-\bar{k}_2-\bar{k}_3)d\bar{k}_1 d\bar{k}_2 d\bar{k}_3 \quad (29)$$

The original image of equation (29) in physical space is

$$\frac{\partial B(r,t)}{\partial t} - \bar{\gamma}\omega_0 B(r,t) + (1+i\bar{\gamma})\left(\frac{\omega_0}{2k_0}\frac{\partial B(r,t)}{\partial x} + i\frac{\omega_0}{2k_0^2}\left(\frac{\partial^2 B(r,t)}{\partial x^2} - \frac{\partial^2 B(r,t)}{\partial y^2}\right)\right)$$

$$+ (1+\bar{\gamma}(1-\beta))k_0^3|B(r,t)|^2 B(r,t) = 0 \tag{30}$$

in which $\bar{k} = k-k_0$, $T_R(k_0) = k_0^3/4\pi^2$ and $\beta = (3-2\sqrt{2}) \ll 1$ $\tag{31}$

When $\bar{\gamma}$ approaches zero the equation will degenerate into the well-known Schrödinger equation, so we can refer to equation (30) as the generalized Schrödinger equation. The derivation of the equation shows that the Schrödinger equation is equivalent to the Stokes equation to third order in the narrow band case.

5. A Developing Nonlinear Single Wave

In consideration of the following discrete expression

$$\bar{a}(k,t) = \sum_j B_j(t)\exp(-i\sqrt{1-\gamma_j^2}\omega(k_j)t)\delta(k-k_j) \tag{32}$$

and the integral-limit process

$$\underset{\epsilon_1 \to 0}{L\ i\ m} \int\limits_{|k-k_j|<\epsilon_1} (...) \, dk \tag{33}$$

the main governing equation can be written in discrete form as follows

$$\frac{\partial B_j(t)}{\partial t} - \gamma_j\omega_j B_j(t) = -i\sum_{l,m,n} T(k_j,k_1,k_m,k_n)B_1^*(t)B_m(t)B_n(t)$$

$$\exp(i(\omega_j+\omega_1-\omega_m-\omega_n)t)\Delta(k_j+k_1-k_m-k_n) \tag{34}$$

in which $\Delta(k) = 1$, as $k = 0$ or $\Delta(k) = 0$, as $k \neq 0$.
Then the governing equation for the envelope of a single wave and its dispersion relation should be

$$\frac{\partial B(t)}{\partial t} - \bar{\gamma}\omega B(t) + iT_R|B(t)|^2 B(t) + \beta\bar{\gamma}T_R|B(t)|^2 B(t) = 0$$

$$\sigma(k) = \omega(k) + T_R(k)|B(t)|^2 \tag{35}$$

The solution of this equation can easily be obtained:

$$|B(t)|^2 = |B_0|^2\exp(2\bar{\gamma}\omega t)/(1+\frac{\beta T_R B_0}{\omega}^2(\exp(2\bar{\gamma}\omega t)-1)) \tag{36}$$

and

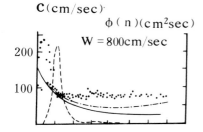

Fig.1: Comparison of theoretical
dispersion relation with the
experiment made by Ramamonjiarisoa.

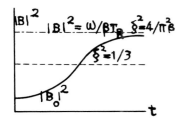

Fig.2: The illustration of the
developing process of a nonlinear
single wave.

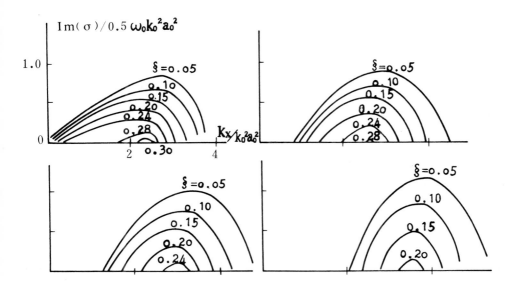

Fig.3: The computation of two-dimensional instability.

$$B(t) = B_0 \exp(\bar{\gamma}(\omega - \beta T_R \frac{\int_0^t |B|^2 dt}{t})t)\exp(-iT_R \int_0^t |B|^2 dt) \tag{37}$$

In the initial stage, since $T_R|B_0|^2/\omega = \pi^2 \xi^2/4 \ll 1$ and $\beta \ll 1$ the solution can be reduced to

$$|\dot{B}(t)|^2 = |B_0|^2 \exp(2\bar{\gamma}\omega t) \quad \text{and} \quad B = B_0 \exp(\bar{\gamma}\omega t)\exp(-iT_R|B_0|^2 t) \tag{38}$$

This means the wave envelope grows with a Miles' pattern and has a low frequency oscillation which is independent of the energy input. From equations (36) and (37) we find that with time the effect of dispersion increases so as to restrain the energy input, so that the wave approaches an equilibrium state $|B|^2 = \omega/\beta T_R$. In practice, however, this process will be stopped by wave breaking, because the wave slope is much larger than the breaking criterion ξ = wave hight/wave length/2 = $1/\sqrt{3}$. So the Miles' pattern is, in general, a good description of wave growth (see Fig. 2).

6. Three-Dimensional instability analysis

Now we assume that besides the main wave k_0 there are other two perturbation waves k_0+k and k_0-k, for which the following amplitude condition is satisfied:

$$|B_+|, |B_-| \ll |B_0| \tag{39}$$

Substituting these waves into the discrete equation (34) and keeping all the terms of orders up to $\gamma \epsilon^2$ or ϵ^3 we can easily obtain the linearized governing equation for the perturbation waves

$$\frac{\partial b_\pm}{\partial t} - (\bar{\gamma}\omega_\pm + i\xi)b_\pm = -ib_0^2 \exp(2\bar{\gamma}\omega t)(T_{1\pm}b_\mp^* - 2T_{2\pm}b_\pm) \tag{40}$$

where $b_\pm$ are equivalent amplitudes defined by $B_\pm = b_\pm \exp(-i(T_{R0}|b_0|^2 + \frac{1}{2}\Omega)t)$ and the expressions for Ω, $T_{1\pm}$, $T_{2\pm}$ and ξ can be found in our paper (6).

Based on the conclusions of Section 2 and the solution (38), $b_\pm$ should vary with two time scales

$$t_1 = 0.5\omega_0 k_0^2 a_0^2 t = t/T^{(1)} \quad \text{and} \quad t_2 = \bar{\gamma}\omega_0 t = t/T^{(2)} \tag{41}$$

and normally the relation, $\mu \equiv \bar{\gamma}\omega_0/(0.5\omega_0 k_0^2 a_0^2) \ll 1$, holds in nature. This means that the two time scales describing the variations of the wave envelope are well-separated. Introducing this small parameter, μ, and using the nondimensional times t_1 and t_2, we suppose

$$b_\pm(t_1, t_2) = \sum_j \mu^j b_\pm^{(j)} \exp(i\chi(t_1, t_2)) \tag{42}$$

It can be proved exactly that if the solution does exist, the wave phase will satisfy the following equations for the coefficients

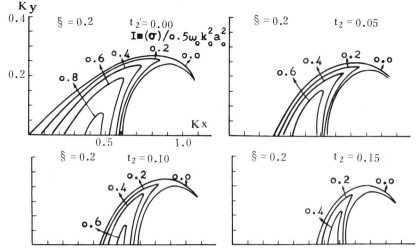

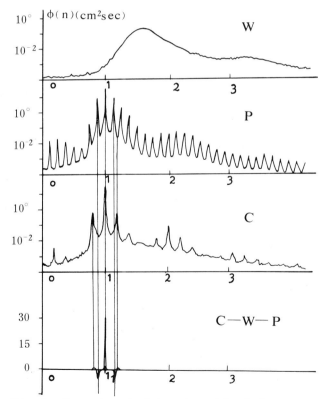

Fig.4: The computation of three-dimensional instability.

Fig.5: Experiment data: W: mid wind wave spectrum; P: paddled wave spectrum; C: paddled wind wave spectrum; C-W-P: the spectrum difference between C and W and P.

$$\left(\frac{\delta \chi_-^*}{\delta t_1} - \Omega + 2T_{2-}\exp(2t_2)\right)\left(\frac{\delta \chi_+}{\delta t_1} - \Omega + 2T_{2+}\exp(2t_2)\right) - T_{1+}T_{1-}\exp(2t_2) = 0 \qquad (43)$$

and

$$\frac{\delta \chi_+}{\delta t_2} + i\omega_+ = 0, \qquad \frac{\delta \chi_-}{\delta t_2} + i\omega_- = 0 \qquad (44)$$

Equation (44) suggests that every wave component will grow with the Miles' pattern and the sideband perturbation will evolve at a rate determined by equation (43) in time scale $T^{(1)}$.

In order to illustrate the evolution process we have made computations for various wave slopes and several time stages. The computation in Fig. 3a shows that in the initial stage ($t_2 = 0$) all the perturbations around the main wave are unstable when the main wave slope is less than 0.27. As the wave slope increases a stable area appears around the main wave number, and the lower and upper limits of the unstable range move away. On the other hand, as shown in Fig. 3b, c, d, the stable range can also reappear, and the unstable range also moves apart with time.

The computation of three-dimensional instability also produces the same conclusion (see Fig. 4).

Recently, an experiment was done by L. Bliven and N. Huang in the NASA Wallops Island wave tank. Fig 5a and b show the spectra of wind waves and patted waves, respectively. When the wind blows over the patted wave (see Fig. 5c) a striking frequency shift of the side-band waves appears and the developing growth determined by (44) keeps the main waves more distinct than that in the pure patted wave case.

References

1. Crawford, D. R., P. G. Saffman and H. C. Yuen (1980), Wave Motion 2, 1-16
2. Gu, D. F. and Y. L. Yuan (1984), Chinese Journal of Oceanology and Limnology (Submitted)
3. Riley, D. S., M. A. Donelai and W. H. Hui (1982), Boundary Layer Meterology 22, 209-225
4. Saffman, P. G. and H. C. Yuen (1982), Journal Fluid Mechanics 123, 459-476
5. Yuan Y. L., N. E. Huang and C. C. Tung (1984), Chinese Journal of Oceanology and Limnology, Vol. 2, No. 1
6. Yuan Y. L., N. E. Huang and C. C. Tung (1984), Chinese Journal of Oceanology and Limnology, Vol. 2, No. 2
7. Zakharov, V. Y. and M. M. Zaslavskiy (1982), Izvestiya, Atmospheric and Oceanic Physics 18, 747-753

THE EFFECT OF SHORT WAVES ON THE TRANSFER OF WIND MOMENTUM TO LONG
WATER WAVES

M. T. Landahl
Massachusetts Institute of Technology, Cambridge, MA 02139,
USA
and
The Royal Institute of Technology, 10044 Stockholm 70,
Sweden

ABSTRACT.   The transfer of wind momentum to long waves in the presence
of short waves is calculated using an inviscid third-order nonlinear
model. The long-wave steepness are assumed to be smaller than the
short wave one so that only the nonlinear effects due to finite short
wave amplitude is retained. An asymptotic solution valid for small
ratio of short to long wave length is developed. The theory yields a
nonlinear contribution to the long-wave growth rate proportional to
the square of the short-wave steepness. This is found to arise from
the modulation of the short wave Reynolds stresses by the long wave
which alters the phase relationship between wave slope and induced
pressure so as to increase the pressure component in phase with the
long-wave slope.

1. INTRODUCTION

The most commonly used model for the transfer of wind momentum to
water waves, Miles' (1957, 1962) linear instability model, has been
found to give reasonable agreement with observation for the short
waves, say for waves of wave lengths less than about 20 centimeters
(Larson and Wright, 1975, Valenzuela 1976, Plant and Wright, 1977) but
underestimates the momentum transfer for the longer waves.
        There are several different nonlinear mechanisms that may effect
wind wave growth. Resonant wave interaction (Hasselmann, 1962, 1963)
may transfer energy between wave numbers. Modulation of the turbulent
stresses in the air by the waves may alter the pressure component in
phase with the surface slope, an effect that has been investigated by
many authors, employing a variety of turbulence models (e.g. Manton
1972, Davis 1972, Townsend 1972, Gent and Taylor, 1976). These show
that such a mechanism could indeed be important.
        An interesting possibility for transfer of energy from the wind
to the longer gravity waves is through nonresonant interaction with
the short waves. This mechanism can act both in the water and in the

Y. Toba and H. Mitsuyasu (eds.), The Ocean Surface, 87–94.
© 1985 by D. Reidel Publishing Company.

air (Landahl et al. 1981). In their work the primary consideration was
the modulational momentum transfer through "leakage" through the water
to the long waves, but a preliminary analysis was also carried out of
the interaction in the air. It was discovered that modulation of the
short-wave induced Reynolds stresses could produce large phase shifts
between the long-wave induced pressure and the surface slope, thus
changing the direct transfer of momentum from the air to the long
wave. However, in the preliminary analysis a simplified model was
employed in which it was assumed that the thickness of the boundary
layer was much smaller than the wave length of the long wave. This
assumption is not justified in the case of a deep logarithmic boundary
layer. In his analysis of the linearized, inviscid problem, Miles
(1957) introduced an approximation valid for the deep boundary layer,
both in the long- and the short-wave limit. This can also be applied
for the problem of interaction between long and short waves.

2. FORMULATION OF THE MODEL

Consider interacting short and long two-dimensional (x,z) wind induced
water surface waves with the surface displacement defined by

$$z = \zeta^s(x,t) + \zeta^\ell(x,t) \tag{1}$$

where superscripts s and ℓ refer to the short and long waves,
respectively. The long-wave train is assumed to be uniform, whereas
the short waves may be modulated by the long waves. The long wave
length, $\lambda^\ell = 2\pi/k^\ell$, is assumed to be much greater than the short one,
$\lambda^s = 2\pi/k^s$. Thus, we introduce the scale ratio

$$\lambda^s/\lambda^\ell = k^\ell/k^s = \varepsilon \tag{2}$$

as a small parameter to be used in an asymptotic expansion of the
solution of the interaction problem. Additional small parameters in
the problem are the short and long wave steepnesses

$$\sigma_s = \sqrt{<(\partial\zeta^s/\partial x)^2>} \ ,$$

$$\sigma_\ell = \sqrt{<(\partial\zeta^\ell/\partial x)^2>} \ , \tag{3}$$

respectively, < > denoting (ensemble) average, and the density ratio
$s=\rho_a/\rho_w$ between air and water. For the purpose of the asymptotic
expansion we also assume that the short waves are steeper than the
long waves so, that

$$\sigma_s^2 = O(\sigma_\ell) = O(\varepsilon) \ . \tag{4}$$

The growth rate of the long waves will be determined to first order in
ε under the additional assumption that the density ratio s is of order
ε or less.

In the solution procedure adopted, the velocity and pressure fields are subdivided into overall mean, long wave, and short wave components as follows:

$$U_i = U_i^m + u_i^\ell + u_i^s \,, \tag{5}$$

and similarly for the pressure. This is accomplished through the application of two distinct averaging procedures. First, the overall average, denoted by $\langle \ \rangle$, of an ensemble of long and short waves is taken and subtracted. Then the long-wave average, denoted by an overbar ($\overline{}$) is taken and subtracted. The remainder represents the short-wave field (we will here ignore the turbulent field).

The mean wind velocity is taken to be parallel and of the form

$$U^m = U(z,t)\delta_{i1} \,, \tag{6}$$

δ_{ij} being the Kronecker delta.

The momentum transfer from the wind to the slowly growing water waves is given by the component of the induced pressure which is in phase with the interface slope. Following Miles (1957) we write the pressure as

$$p = \rho\, U_a^2 (\beta^s \zeta_x^s + \beta^\ell \zeta_x^\ell) \,. \tag{7}$$

where U_a is a reference air velocity. The pressure components in phase with the surface deflection are omitted for clarity. The modulational effects on the transfer coefficients may be expressed as

$$\beta^s = \beta_o^s + k^\ell \zeta^\ell \beta_1^s \,, \tag{8}$$

$$\beta^\ell = \beta_o^\ell + K_R \sigma_s^2 \,, \tag{9}$$

where subscript o denotes the linear values.

Since the wind-wave growth rate is proportional to the density ratio, s, and hence small, one may determine the aerodynamic coefficients by calculating the pressures induced on a quasi-steady and uniform wave train. Because only two-dimensional waves will be considered it is convenient to employ a stream function, Ψ, such that

$$u^\ell + u^s = \Psi_z, \quad w^\ell + w^s = -\psi_x \tag{10}$$

After substituting the stream function into the vorticity equation and subtracting the equation for the mean flow one obtains in the inviscid limit

$$[\partial/\partial t + (U + \Psi_z)\partial/\partial x - \Psi_x \partial\partial z]\nabla^2\psi - \Psi_x U_{zz}$$
$$= (\partial^2/\partial x\partial z)\langle\Psi_z^2\rangle - (\partial^2/\partial z^2)\langle\Psi_x \Psi_z\rangle \,. \tag{11}$$

We further set

$$\psi^S = \zeta^S \phi^S(z,t) \exp[ik^S(x - c^St)] , \tag{12}$$

$$\psi^\ell = \zeta^\ell \phi^\ell \exp[jk^\ell(x - c^\ell t] . \tag{13}$$

The long-wave and short-wave phases are distinguished by the use of j as the imaginary unit for the long waves and i for the short. One then obtains for the stream function amplitudes ϕ^S and ϕ^ℓ the following equations:

$$(U^\ell - c^S)[\phi^S_{zz} - (k^S)^2\phi^S] - U^\ell_{zz}\phi^S = 0 , \tag{14}$$

$$(U - c^\ell)[\phi^\ell_{zzi} - (k^\ell)^2\phi^\ell] - U''\phi^\ell = -jT^S_{zz} , \tag{15}$$

where $U^\ell = U + u^\ell$ and the short-wave induced Reynolds stresses are represented by

$$T^S = (\overline{\phi^S_x \phi^S_z} - \langle\phi^S_x\phi^S_z\rangle)/k^\ell\zeta^\ell . \tag{16}$$

They may be related directly to the aerodynamic coefficients $\beta^{S,\ell}$ by noticing that the momentum balance for a uniform wave train indicates that β is directly related to the wave induced Reynolds stresses through

$$\langle p^\ell\zeta^\ell_x \rangle = -\rho\langle u^\ell w^\ell\rangle ,$$

or

$$\beta^\ell = \langle\psi^\ell_x \psi^\ell_z \rangle/\sigma^2_\ell U^2_a , \tag{17}$$

and similarly for the short waves. Inside the matched layer $z = z_c$ (where $U(z_c) = c$) the Reynolds stress is constant, independent of z, and zero outside. Thus, it follows that

$$T^S = \beta_1 \sigma^2_s U^2_a[1 - H(z - z^S_c] , \tag{18}$$

where H is the Heaviside step function. From (17) it follows that

$$\beta^\ell = \text{Im}\{\phi^{\ell*}(0)\phi^\ell_z (0)\}/k^\ell , \tag{19}$$

star denoting complex conjugate. Following Miles' (1957) procedure we divide (15) by $(U - c^\ell)$, multiply it by $\phi^{\ell*}$, and integrate from 0 to ∞. This gives after integration by parts (treating T^S as a generalized function)

$$\text{Im}\{\phi^{\ell*}(0)\phi^{\ell*}_z(0)\phi^\ell_z (0)\} = -\pi|\phi^\ell_c|^2[U''/U']_{z^\ell_c}$$

$$+ (k^\ell)^2\beta^S_1\sigma^2_s U^2_a \text{Re}\{\int_{z^S_c}^\infty (U - c^\ell)\phi^{\ell*}dz\} , \tag{20}$$

divide (15) by $(U - c^{\ell})$, multiply it by $\phi^{\ell*}$, and integrate from 0 to ∞. This gives after integration by parts (treating T^S as a generalized function)

$$\text{Im}\{\phi^{\ell*}(0)\phi_z^{\ell*}(0)\phi_z^{\ell}(0)\} = -\pi \phi_c^{\ell}{}^2[U''/U']_{z_c^{\ell}}$$
$$+ (k^{\ell})^2 \beta_1^S \sigma_s^2 U_a^2 \text{Re}\{\int_{z_c^S}^{\infty} (U - c^{\ell})\phi^{\ell*}dz\} , \qquad (20)$$

where subscript c denotes values at the matched layer. From a first integral of (15) one finds

$$\phi_c^{\ell} = k^{\ell}I_1^{\ell} /(U')_{c^{\ell}} , \qquad (21)$$

where

$$I_1^{\ell} = \int_{z_c^{\ell}}^{\infty} (U - c^{\ell})\phi^{\ell}dz . \qquad (22)$$

Hence

$$\beta^{\ell} = -\pi k^{\ell} I^{\ell}{}^2[U''/(U')^3]_{z_c^{\ell}} + \beta_1^S \sigma_s^2 U_a^2 (c^{\ell} - c^S)^{-2} \text{Re}\{I_1^{\ell*}\} , \qquad (23)$$

where only the lowest order terms in long-wave wave number have been retained. To determine the integral I_1^{ℓ} we may use Miles' (1957) approximation

$$\phi^{\ell} = (c^{\ell} - U) \exp(-k^{\ell}z) , \qquad (24)$$

which satisfies the kinematic boundary condition at $z = 0$.
    To determine β_1^S we start from the modulated value

$$\beta^S = -\pi k^S |I^S|^2 [U_{zz}^{\ell} /(U_z^{\ell})^3]_{z_c^S} , \qquad (25)$$

where

$$I^S = \int_{z_c^{\ell}}^{\infty} (U^{\ell} - c^S)\phi^S dz , \qquad (26)$$

and expand in $k^{\ell}\zeta^{\ell}$ retaining only the lowest order terms. For ϕ^S we again follow Miles (1957) and set

$$\phi^S = (c^S - U^{\ell}) \exp[-k^S(z - \zeta^{\ell})] . \qquad (27)$$

After some calculations the following results emerge:

$$\beta_1^S/\beta_0^S = (c^{\ell} - c^S)\{[U'''/(U''U') - 3U''/(U')^2]_{z_c^S} + 4I_0^S/I_1^S\} - 1 , \qquad (28)$$

$$\beta^{\ell} = \beta_0^{\ell} + \beta_1^S \sigma_s^2 U_a^2 I_1^{\ell} /(c^{\ell} - c^S)^2 , \qquad (29)$$

where

$$I_p^{s,\ell} = k^{s,\ell} \int_{z_c^{s,\ell}}^{\infty} (U - c^{s,\ell})^{p+1} dz \quad . \tag{30}$$

β_o^s and β_o^ℓ are given by Miles' (1957) formula (c.f. (25))

3. APPLICATION TO A LOGARITHMIC WIND PROFILE

For a logarithmic wind profile

$$U = (u_*/\kappa)\ln(z/z_*) = U_a \ln(z/z_*) \quad , \tag{31}$$

where u_* and z_* are the friction velocity and the roughness height, respectively, $I_p^{s,\ell}/U^{p+1}$ are functions of kz_c, only. We may then express the results for the interaction coefficients in terms of the nondimensional integrals

$$\overline{I}_p^{s,\ell} = \int_{z_c^{s,\ell}}^{\infty} \left[\ln(Z/k^{s,\ell})\right]^{p+1} \exp(-Z)dZ \quad , \tag{32}$$

giving

$$\beta_o^{s,\ell} = \pi k^{s,\ell} (\overline{I}_1^{s,\ell})^2 \tag{33}$$

$$\beta_1^s = (c^\ell - c^\tau)\left[1 + 4\overline{I}_o^s/\overline{I}_1^s\right]/U_a - 1 \quad , \tag{34}$$

$$\beta_o^\ell = \pi k^\ell z_c^\ell (\overline{I}_1^\ell)^2 + \beta_1^s \sigma_s^2 U_a \overline{I}_1^{s\ell} /(c^\ell - c^s)^2 \quad . \tag{35}$$

The short waves cause a mean tangential stress on the air of

$$\langle\tau^s\rangle = \langle p^s\zeta_x^s\rangle = \rho\beta_o^s\sigma_s^2 U_a^2 \quad . \tag{36}$$

Assuming that the short waves provide the dominating contribution to the surface stress we therefore set

$$\rho\beta_o^s\sigma_s^2 U_a^2 = \rho u_*^2 \tag{37}$$

giving

$$\beta^\ell = \beta_o^\ell + (\beta_1^s/\beta_o^s)u_*^2\overline{I}_1^\ell /(c^\ell - c^s)^2 \quad . \tag{38}$$

The total long-wave growth rate thus depends only on the same parameters as in the linear case. It should be noted that in this simplified theory the effect of the short-wave Reynolds stresses on the mean velocity distribution has been neglected.

Numerical results are shown in Fig.1 for the parameters $z_*/\lambda^s = 0.001$, $u_* = 0.125 c^\ell$ for two fixed ratios of long to short wave lengths, $\lambda^\ell/\lambda^s = 10$ and 20. In order to avoid that the matched layer would be located close to the surface and in the viscous layer, the calculations were restricted to fairly large values of λ^s.

4. DISCUSSION

As seen from the figure, the nonlinear effect of the long-wave
modulated short-wave Reynolds stresses could be quite substantial,
giving an increase by a factor of two or more over Miles' (1957)
linear inviscid results. This increase is of the same order as that
obtained by Gent and Taylor (1976), who found that if the surface
roughness was allowed to vary along the wave, the energy input from
the wind to the wave could increase by a factor of up to three over
the linear value. However, in view of the approximations made in the
present theory, the numerical results should be only regarded as
suggestive as to the order of magnitude of the effects of the short
waves. In particular, Miles' (1957) approximation (16) is a fairly
coarse one, as found by Miles in his (1959) extension of the theory.
More accurate results would need direct numerical integration of the
governing equations, and some preliminary results have already been
obtained (see also Landahl et al., 1981).

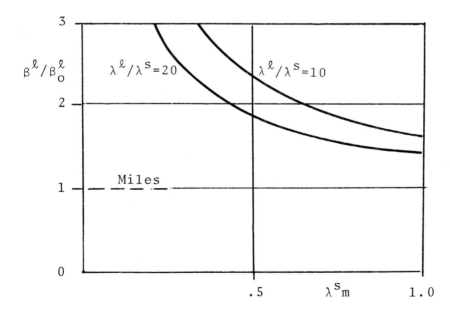

Fig. 1. Aerodynamic momentum transfer coefficient for long waves in the
presence of short (gravity) waves, $z_*/\lambda^s = 0.001$, $u_* = 0.125\ c^\ell$.

ACKNOWLEDGEMENT

This researh was supported in part by the National Science Foundation
under Grant CME-79-12132.

REFERENCES

Davis, R.E. 1972 On prediction of the turbulent flow over a wavy
boundary. J. Fluid Mech. 52, 287.
Gent, P.R. and Taylor, P.A. 1976 A numerical model for the air
flow above water waves. J. Fluid Mech. 77, 105.
Hasselman, K. 1962 On the non-linear energy transfer in a
gravity wave spectrum. Part 1. J. Fluid Mech. 12, 481.
Hasselmann, K. 1963 On the non-linear energy transfer in a
gravity wave spectrum. Part 2. J. Fluid Mech. 15, 273. Part 3.
Ibid. 15, 385.
Landahl,, M.T., J.A. Smith and S.E. Widnall 1981 The interaction
of long and short sind-generated waves. Paper presented at the IUCRM
Symposium on Wave Dynamics and Radio Probing of the Ocean Surface,
Miami, FA (to appear).
Larsson, T.R. and J. Wright 1975 Wind generated gravity-
capillary waves: laboratory measurements of temporal growth rates
using microwave backscatter. J. Fluid Mech. 70, 417.
Manton, M.J. 1972 On the generation of sea waves by a turbulent
wind. Boundary Layer Met. 2, 348.
Miles, J.W. 1957 On the generation of waves by shear flows. J.
Fluid Mech 3, 185.
Miles,J.W. 1959 On the generation of surface waves by shear
flows. Part 2. J. Fluid Mech. 6, 568.
Miles, J.W. 1962 On the generation of waves by shear flows. Part
4. J. Fluid Mech. 13, 433.
Plant, W.J. and J.W. Wright 1977 Growth and equilibrium of short
gravity waves in a wind wave tank. J. Fluid Mech. 82, 767.
Townsend, A.A. 1972 Flow in a deep turbulent boundary layer over
a surface distorted by water waves. J. Fluid Mech. 55, 719.
Valenzuela, G.R. 1976 The growth of gravity-capillary waves in a
coupled shear flow. J. Fluid Mech. 76, 229.

MEASUREMENTS OF ATMOSPHERIC PRESSURE OVER SURFACE GRAVITY WAVES DURING KonTur

D. Hasselmann [1], M. Dunckel [2], J. Bösenberg [2]
1.   Meteorologisches Institut
     Universität Hamburg
     Bundesstrasse 55
     D-2000 Hamburg 13
     FR Germany
2.   Max-Planck-Institut für Meteorologie
     Bundesstrasse 55
     D-2000 Hamburg 13
     FR Germany

ABSTRACT only.   As part of the KonTur experiment in the autumn of 1981 the atmospheric pressure over surface gravity waves was measured. The instrument carrier was a slim mast located in the North Sea, 27 km off the island of Sylt at station 8 of the JONSWAP array (Hasselmann et al., 1973). The pressure sensor was provided by R.L. Snyder and of the same type as used in previous experiments (Snyder, Dobson, Elliott, Long, 1981, (SDEL)).

Two resistance wires were used to measure wave height, η_1 and η_2. The pressure probe was mounted on a spar about 2 m away from the mast and at a height $z \approx 6$ m above mean water level, depending on the tide. The resistance wires were hung from two diametrically opposed spars and kept taut by a weight. One wire, η_2, was located directly under the pressure sensor, so that the distance between the wires was about 4 m. The pressure p was sampled at 10 Hz, but for the extended analysis the data were filtered to a 2 Hz series, to conform with the sampling rate of the resistance wires. The whole arangement was very similar to the P77 experiment (Hasselmann, D. et al., 1985), but in contrast to that experiment we experienced higher wind speeds: up to 12 m/s as measured with cup anemometers on the mast, and significant wave height H_s up to 2.1 m. The fairly rough weather conditions were the main reason why the pressure instrument had to be mounted so high above the mean water level. Altogether we collected 23 h of data during four days of operation, after which the instruments were lost and the spars twisted in a severe storm.

The data were analyzed mainly within the framework of SDEL. Thus we first obtained a set of cross-spectra at a bandwidth of 5/128 Hz and with 160 degrees of freedom. The $< \eta_1 \eta_2 >$ cross-spectra yield a rough directional wavespectrum and the $<p \eta_2>$ cross-spectra gave an estimate of the transfer-function between wave-induced atmospheric pressure and waveheight. The non-dimensional transfer function $\gamma = \alpha + i \beta$ is obtained as

$$\frac{\rho_a}{\rho} \rho g \gamma = <p\,\eta_2>/<\eta_2\,\eta_2> \qquad\qquad (1)$$

95

Y. Toba and H. Mitsuyasu (eds.), The Ocean Surface, 95–97.

here ρ_a, ρ and g denote density of air, density of water and acceleration of gravity and γ is a function of non-dimensional parameters of which we consider only $\lambda = k\,z$ and $\mu = (U_5/c)\cos\Theta$. Here k is wavenumber (modulus), U_5 is windspeed (at 5 m), c is the phase velocity and Θ is the angle between waves and wind. The vertical dependence was found to be essentially exponential in earlier work (SDEL, P77) and this was used to extrapolate to the surface. The mean value of μ at each frequency considered could then be inferred from the directional distribution, so that the dependence of α and β on μ could be analyzed. In this experiment μ ranged from -1 to +2; for negative values we have swell running against the wind, while values $0 < \mu < 1$ occur for swell running with the wind, and $\mu > 1$ is the range for windsea. In spite of the high windspeeds of $U_5 \approx 12$ m/s the data was limited to $\mu < 2$ because of the relatively large height $z \approx 6$ m of the pressure sensor above mean water level. If $\lambda = kz > 2.3$ (exp $(\lambda) > 10$) the error due to the extrapolation to the surface became too large to yield meaningful data. Thus the data for large k, thus small phase speed and high μ, could not be used. In order to estimate the dependence on μ, the value of μ was calculated for each spectral band by appropriately averaging over the directional wave spectrum. Currents were taken into account by transforming into a system with zero current. The currents were not measured but estimated to be tidal, with typical values of 0.3 m/s. From -1 to +3 the μ-axis was then divided into bins and the γ values belonging to each bin were properly weighted and averaged. The number of entries in each bin is rather variable, but four is a typical number.

Our results are in agreement with those of SDEL and P77 but higher than the results suggested by Hsiao and Shemdin (1983), while a direct comparison with the formula of Plant (1982) is not feasible, because our values of μ are too low for Plant's formulation to apply. In terms of the SDEL formulation we find

$$\beta = b\ (\mu - 1) \qquad \text{for } \mu > 1 \qquad\qquad b \approx 0.3 \pm 0.1$$

$$\beta = 0 \pm 0.05 \qquad \text{for } \mu < 1 \tag{2}$$

$$\alpha = -a\ (\mu - 1)^2 \qquad a \approx 0.9\ \mu < 1,\ a \approx 0.6 \text{ for } \mu > 1\ .$$

One difficulty with this formula, clearly recognized by SDEL, is the rather arbitrary height of 5m for fixing the velocity scale, which was selected for practical reasons. A scale based on the friction velocity u_* is more satisfactory and one might expect to see a difference between such a formulation and that of SDEL over our range of wind speeds U_5 in which the drag coefficient $C_D = (u_*/U_5)^2$ varies appreciably according, for instance, to Wu (1982). We have therefore tested various other formulations for α and β in which velocity was scaled with u_* using C_D as proposed by Wu, but were unable to detect any significant improvement of the fits. The interpretation of the statistics is difficult mainly because the errors due to the directional distribution are very difficult to estimate. In summary we can use (2) in applications, but if preferred a formulation like (2) with U_5 replaced by $b_*\ u_*$, where b_*^2 is a typical value of $1/C_D$, is also compatible with the data, as is a formulation in which U_5 is replaced by $U(k^{-1})$. Neither the swell range, $0 \leq \mu \leq 1$, nor the range for waves running against the wind, $\mu < 0$, shows any sign of generation or damping.

References

Snyder, R.L., Dobson, F.W., Elliott, J.A., and Long, R.B. (1981) "Array Measurements of atmospheric pressure fluctuations above surface gravity waves". J. Fluid Mech., 102, 1-59.

Hasselmann, K. et multi al. (1973) JONSWAP, Dtsch. Hydrog. Z., A. Suppl. No. 12

Wu, J. (1982) "Wind-Stress coefficient over sea surface from breeze to hurricanes". J. Geophys. Res., 87, C, 9704-9706.

Plant, W.J. (1982) "A relationship between wind stress and wave slope". J. Geophys. Res., 87, C, 1961-1967.

Hasselmann, D, Bösenberg, J., Dunckel, M., Richter, K., Grünewald, M., and Carlson, H. "Measurements of wave-induced pressure over surface gravity waves". in IUCRM Symposium on Wave Dynamics and Radio Probing of the Ocean Surface, Miami, 1981. Plenum Press (to have been published, 1985).

Hsiao, S.V., and Shemdin, O.H. (1983) "Measurements of wind velocity and pressure with a wave follower during MARSEN". J. Geophys. Res., 88, C, 9841-9849.

ATMOSPHERIC STABILITY EFFECTS ON THE GROWTH OF SURFACE GRAVITY WAVES

Peter A.E.M. Janssen and Gerbrand J. Komen
Department of Oceanography
Royal Netherlands Meteorological Institute
P.O. Box 201,
3730 AE De Bilt, The Netherlands

ABSTRACT

We study the effect of atmospheric stability on the growth of surface gravity waves. To that end we numerically solved the Taylor-Goldstein equation for wind profiles which deviate from a logarithmic one because stratification affects the turbulent momentum transport. Using Charnock's relation for the friction height z_o of the wind profile it is argued that the growth rate of the wave depends on the dimensionless phase velocity c/u_* (where u_* is the friction velocity) and a measure of the effect of atmospheric stability, namely the dimensionless Obukhov length gL/u^2_*, whereas it only depends weakly on gz_t/u^2_* (where z_t is the friction height of the temperature profile). We find that for given u_*, the growth rate as a function of c/u_* is larger for stable stratification ($L>o$) than for an unstable stratification ($L<o$). If one, on the other hand, considers the growth rate as a function of c/U_{10} (where U_{10} is the windspeed at 10 meters) the situation reverses and unstable growth is larger over most of the frequency range. We explain these features in a qualitative way.

1. INTRODUCTION

Miles' theory (1957, 1959) on the growth of surface gravity waves is in reasonable agreement with observations on wave growth. A comparison between theory and observations is made in fig. 1. The Snyder et al (1981) data were observed in the Bight of Abaco; the shaded area represents a fit by Plant (1982) based on laboratory and field data; the dotted curve is by Mitsuyasu and Honda (1983).

It is well known that atmospheric stability may influence wave growth, because density differences introduce additional forces. As a result the equation describing shear flow instability in a stratified fluid is the Taylor-Goldstein equation rather than the Rayleigh equation of Miles' theory. In addition the mean velocity profile will deviate from the usual logarithmic profile when stratification comes into play. To quantify these effects we extended Miles' calculation.

99

Y. Toba and H. Mitsuyasu (eds.), The Ocean Surface, 99–104.
© *1985 by D. Reidel Publishing Company.*

An extensive account of our work is given elsewhere (Janssen and Komen, 1984). Here, we will summarize the most important aspects. In

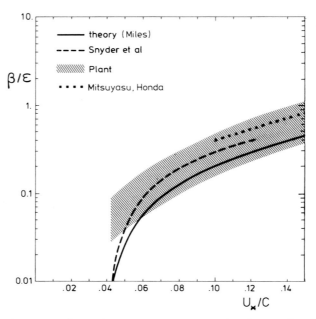

Fig.1.   A comparison between our numerical result for the dimensionless wave growth as a function of u_*/c (solid line) and the fits of Plant, Snyder et al and Mitsuyasu and Honda, for neutrally stable stratification.

section 2 we briefly indicate the treatment of the mean wind profiles; in section 3 we discuss shear-flow instability in the presence of density stratification; sections 4 and 5 give results and conclusions.

2. MEAN WIND PROFILES

Large and Pond (1982) have indicated how to treat mean wind profiles in turbulent shear flow in the presence of density stratification. In fact, they have shown that it is possible to derive the velocity and temperature profile as well as the usual boundary layer parameters when U_{10}, T_{10}, U_{sea} and T_{sea} are given. Here, U_{10} and T_{10} are velocity and temperature at a height of 10 m; U_{sea} and T_{sea} are the sea-surface current and temperature. We neglect, for simplicity, humidity effects, which are relatively small (Janssen and Komen, 1984). The profiles are found to be of the following form.

$$u(z) = u_{sea} + \Delta u(z/L; z_o/L) \tag{1}$$

$$T(z) = T_{sea} + \Delta T(z/L; z_t/L) \tag{2}$$

Here, z_o and z_t are the roughness heights for velocity and temperature, respectively. For z_o we take Charnock's relation; for z_t we take an empirical constant.

Important parameters are the friction velocity u_* and the Obukhov length L. To study stratification effects, we will consider the case of a wind speed of 10 m/s. Using Large and Pond (1982) we find that neutral stratification corresponds with a friction velocity of $u_* = 0.383$ m/s. When we consider the (rather extreme) case of a temperature difference of $\pm 10^o$ C, we find $u_* = 0.411$ m/s, L = - 30 for unstable stratification and $u_* = 0.325$ m/s, L = 45 m for stable stratification. Instability enhances momentum transfer.

It should be noted that Large and Pond (1982) deal with macro-profiles; they say nothing what happens to their profiles at heights of less than one wave amplitude. This is important, since in the ocean the shear-flow instabilities discussed here occur at or below one dominant-wave amplitude for $u_*/c > 0.1$. We will present results based on the profiles (1) and (2) for $u_*/c < 0.15$. One should be aware of the uncertainty for $u_*/c > 0.1$.

3. SHEAR FLOW INSTABILITY

Miles (1957) gave a relatively simple expression for the non-dimensional growth rate $\beta = (1/\sigma E) dE/dt$ in terms of the flow parameters:

$$\beta = \pi\varepsilon \frac{\sigma^2}{gk^2} |\chi_c|^2 \frac{W''_c}{|W'_c|} \tag{3}$$

with

$$(\frac{\partial^2}{\partial z^2} - k^2) \chi = \frac{W''}{W'} \chi \qquad \chi(o) = 1, \; \chi(\infty) = 0 \tag{4}$$

Here ε is the air-sea density ratio, σ is angular frequency, g is gravitational acceleration, k is wave number, $\chi(z)$ represents the dimensionless velocity of the wave induced air motion, satisfying the Rayleigh equation (4) and W = c - U (z), with c the phase velocity corresponding with frequency σ and U (z) the mean wind profile. The prime denotes differentiation with respect to z, and the subscript c indicates evaluation at the critical height: U (z_c) = c. For very short waves ($u_*/c > 0.15$) the critical height is in the viscous sublayer and the theory has to be modified. Here we consider $u_*/c < 0.15$ therefore.

It can be shown that stratification (in the so-called meteorological approximation) leads to the following generalization of the Rayleigh equation:

$$\left(\frac{\partial^2}{\partial z^2} - k^2\right)\chi = \left(\frac{W''}{W'} + \frac{g\rho_a'\rho_a}{W^2}\right)\chi \tag{5}$$

Here, ρ_a denotes the air density. The growth rate can be given in terms of χ as follows

$$\beta = \frac{\varepsilon}{2k}\mathcal{W}\left(\chi, \chi^*\right)_{z=0} \tag{6}$$

where the Wronskian $\mathcal{W} = -i\left(\chi'\chi^* - \chi\chi^{*\prime}\right)$. In neutrally stable stratification (6) can be shown to reduce to (3).

We solved (5) and (6) numerically. It turned out to be convenient to nondimensionalize all quantities with the help of g and u_*. A careful analysis showed that the growth rate depends on the dimensionless phase velocity $\tilde{c} = c/u_*$, the dimensionless roughness height $\tilde{z}_o$ (which is a constant, however), and further on $\hat{L} = Lg/u_*^2$ and $\tilde{z}_t = z_t g/u_*^2$. The dependence on $\tilde{z}_t$ was found to be very weak. Therefore in effect

$$\beta = \beta\left(c/u_*, Lg/u_*^2\right) \tag{7}$$

In the usual way the singularity due to the vanishing of W at the critical height, gives numerical complications. We avoided these by making a Frobenius expansion around the critical height.

4. RESULTS

Results are given in figure 2. Over most of the frequency range wave growth in unstable stratification is larger than in stable

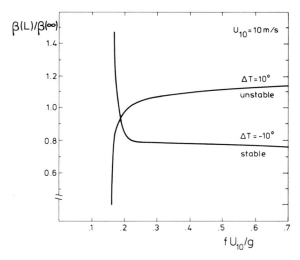

Fig. 2.   Numerical results for the modification of the wave growth due to atmospheric stability effects. The growth ratio is given as a function of fU_{10}/g.

stratification. The effects are of the order of 20% for the extreme temperature difference considered. For smaller temperature differences the effect will be smaller. A straightforward analysis of our results indicates first of all that eq. (7) can be approximated by (3) even in presence of stratification. Further we find that the main effect of the growth rates comes from the enhanced curvature of the wind profile: $\left(W_c''/|W_c'|\right)_{unstable} > \left(W_c''/|W_c'|\right)_{stable}$ for any value of U_{10}/c. However, for the lowest frequencies another effect comes into play. In those cases the critical height is fairly high, and it turns out that he dimensionless velocity amplitude of the air motion at the critical height is larger in stable stratification. This is because the strong curvature in unstable stratification damps the wave induced motion when one integrates (4) or (5) from the sea surface upwards.

It should be noted that if one transforms to c/u_* as an independent "(inverse) frequency" variable, one finds that stable growth exceeds unstable growth. This is perhaps contrary to intuition, but it can be understood by realizing that now for a fixed u_* $W_c''/|W_c'|$ hardly depends on stratification, whereas χ_c as above is larger for stable stratification. An important consequence is that for a given u_* the lowest values of u_*/c are excited in stable stratification. (See figure 3.).

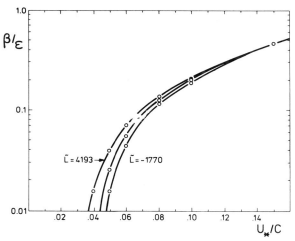

Fig. 3.   The dimensionless wave growth as a function of u_*/c for various cases of atmospheric stability.

5. CONCLUSIONS

The validity of our work is depends based on a few assumptions: the validity of Miles' theory, and extension of Large and Ponds profiles to the critical height. Under these assumptions we found that in the neutrally stable case wave growth is always determined by the parameter c/u_* if Charnock's relation for the friction height is valid. In the presence of thermal stratification, however, the growth rate becomes dependent on the dimensionless Obukhov length L. The largest effects occur at low frequencies.

It would be of interest to compute consequences for wave growth. For example, one might ask what is the maximum wave height and period that can be reached for a given wind speed, as a function of L. In stable situations we expect slow initial growth, but ultimately higher and longer waves would occur. The exact exploration of these questions requires the running of a spectral wave prediction model, and good knowledge of dissipative and nonlinear contribution to the evolution of the wave spectrum.

REFERENCES

Janssen, P.A.E.M.,   and G.J. Komen, 1985. Effect of atmospheric stability on the growth of surface gravity waves. To appear in Bound. Lay. Met.

Large, W.G. and Pond, S., 1982. Sensible and latent heat flux measurements over the ocean. J. Phys. Ocean. 12, 464 – 482.

Miles, J.W., 1957. On the generation of surface waves by shear flow. J. Fluid Mech. 3, 185 – 204.

Miles, J.W., 1959. On the generation of surface waves by shear flow, part 2. J. Fluid Mech. 6, 568 – 582.

Mitsuyasu, H. and T. Honda, 1982. Wind-induced growth of water waves. J. Fluid Mech. 123, 425.

Plant, W.J., 1982. A relationship between wind stress and wave slope. J. Geophys. Res. 87, 1961 – 1967.

Snyder, R.L., Dobson, F.W., Elliot, J.A. and Long, R.B., 1981. Array measurements of atmospheric pressure fluctuations above surface gravity waves. J. Fluid Mech. 102, 1 – 59.

NEW ASPECTS OF THE TURBULENT BOUNDARY LAYER OVER WIND WAVES

H. Kawamura and Y. Toba
Department of Geophysics
Tohoku University
Sendai 980 Japan

ABSTRACT. As in the boundary layer over a flat plate, ordered motions have been found in the air flow over wind waves, and their structure studied in detail. The ordered motions have a horizontal scale similar to the wavelength of the underlying wind waves. Possible interactions between the ordered motions and the wind waves are suggested.(This paper is an extended abstract. This study will be published in full elsewhere.)

1. INTRODUCTION

Kawamura et al.(1981) and Kawai(1982) have studied the air flow over wind waves, looking for characteristic events reoccuring randomly and hidden behind turbulence. They have employed the same kind of approach and techniques as recent studies on turbulent shear flow. In this study, three different experiments were carried out to investigate the whole turbulent boundary layer over wind waves. These are Ex.1: vorticity measurement in the vicinity of the water surface, Ex.2: measuring of two dimensional velocity in the lower part of the logarithmic layer and Ex.3: flow visualization combined with velocity measurement in the outer boundary layer. All the experiments were done in a wind-wave tunnel which is 0.15 m wide, 0.7 m high and 8 m long with a water depth of 0.52 m (See Kawamura et al., 1981). The wind speed was 5.75 m/s at the wind core at a fetch of 3.86 m, where Ex.3 was made. Ex.1 and Ex.2 were made at a fetch of 6.0 m. The experimental conditions were: friction velocity $u_* = 0.32$ m/s, boundary layer thickness $\delta = 6.1$ cm, significant wave height $H_{1/3} = 0.37$ cm for the fetch of 3.86 m; and $u_* = 0.29$ m/s, $\delta = 6.1$ cm and $H_{1/3} = 1.07$ cm for 6.0 m.

2. AIR FLOW SEPARATION OVER WIND WAVES (Ex.1)

In order to investigate the air flow structure near the water surface, hot-wires were mounted 0.15 cm apart vertically on a wave follower. The velocities u_1 and u_2 measured by the lower and the upper hot-wire, and the vorticity component $\Omega_y = u_1 - u_2$ at positions under the wind-wave crest

Y. Toba and H. Mitsuyasu (eds.), The Ocean Surface, 105–110.
© *1985 by D. Reidel Publishing Company.*

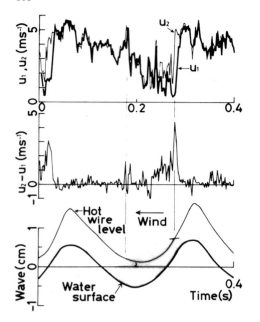

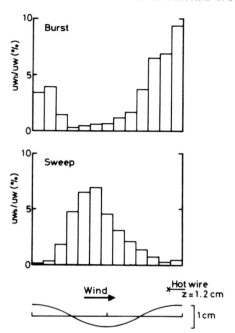

Figure 1. Typical example of the time series showing conspicuous peak of Ω_y . Expected separation bubble is indicated in the bottom figure.

Figure 3. Distribution of Reynolds stress production rate by bursts and sweeps at z=1.2 cm, relative to the wind wave phase.

• over leeward side
• over windward side

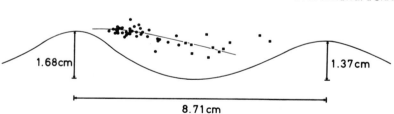

Figure 2. Detection of the high shear layer at the outer edge of the separation bubble.

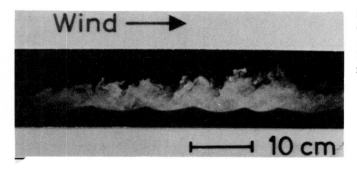

Figure 4. Visualization of large scale motion in the outer boundary layer with paraffin mist.

level, were obtained. In the time series of Ω_y conspicuous peaks
showing large positive values are observed: Figure 1 is a typical
example. The velocity signals show very slow air flow over the trough,
and the jumps of u_1 and u_2 correspond to the Ω_y peak. These variations in
the signals are well explained by the existence of a separation bubble
behind the wind-wave crest, as illustrated in Figure 1. Over the wind
waves two distinct air flow patterns are observed, with and without air
flow separation behind the crests.

The positions of the high shear layers at the edge of the separation
bubbles were detected by searching the Ω_y peaks in the record(Figure 2).
The separation bubble covers the entire trough. A representative velocity
difference across the shear layer (~ 0.15 cm wide) is 4 m/s, which gives
a Reynolds number of 400, as for a free shear flow. Since this value
indicates a very unstable flow condition from linear stability theory, we
expect strong turbulence will be generated in the shear flow. Evidence
of this unstable condition is the scattering of the detected positions of
the high shear layers over the windward side of the waves in Figure 2.

The separated shear flow reattaches on the windward side of the next
wave. It is known from experiments on the air flow over a solid wall
that the reattachment of the separated shear flow causes a considerable
pressure rise on the wall, balancing the turbulent shear stress generated
within it(Tani et al., 1961). It is thus suggested by analogy that a
significant pressure rise occurs on the windward side of the wind wave,
where the shear flow reattaches. The imbalance in the pressure
distribution caused by this reattachment can add energy to the wave
motion, and may be an important mechanism for wind wave growth.

3. BURSTING PHENOMENA OVER WIND WAVES (Ex.2)

The existence of bursting phenomena over wind waves, characterized by
intermittent blowing up of low-speed air masses(burst) and coming down of
high-speed air masses(sweep), were reported by Kawamura et al.(1981).
Both the burst and the sweep are distinguished by large negative peaks in
the time series of instantaneous Reynolds stress u(t) x w(t).

We measured the horizontal and vertical velocity fluctuations u(t)
and w(t) near the wind waves to obtain the contribution rate of the
bursting phenomena to Reynolds stress generation. A conditional sampling
and averaging technique was used, employing the u(t) x w(t) signal to
detect the burst and sweep events. The time t_c when the negative peak of
u(t) x w(t) exceeded a threshold level C, was determined, and the
Reynolds stress caused by the event was obtained from

$$uw_i = \int_{t_c - \tau_d/2}^{t_c + \tau_d/2} u(t) \times w(t) \, dt \, , \qquad i = \begin{cases} \text{burst, if } u(t) < 0, \\ \text{sweep, if } u(t) > 0. \end{cases}$$

After some trial analyses in order to choose C, we settled on a value ten
times the local mean Reynolds stress. The duration of the events τ_d was
chosen to be the period needed for one wind-wave length pattern to pass
through a point at the local mean velocity.

The results of the analysis show that, at z=1.2 cm, 73% (40% by the
burst and 33% by the sweep) of the local Reynolds stress is produced by

events whose period is 30% (17% and 13%) of the total period. Figure 3
shows the distribution of the Reynolds stress production rate by both the
events reletive to the wind wave phase. It is evident that production by
bursts occur over the windward side of the wind wave and by sweeps over
the leeward side.

4. ORDERED MOTION OVER WIND WAVES (Ex.3)

Our flow visualization used paraffin mist in order to observe large scale
motions in the outer boundary layer. The mist was continuously released
from a point source placed near the water surface at 2.8 m in fetch, and
its spreading manner was strobed with a strong slit light, photographed
at 3.8 m fetch. A typical result is shown in Figure 4, clearly
indicating the presence of large scale phenomena in the outer boundary
layer. The fluid marked with the mist (white area) near the water
surface is being swept into the outer part of the boundary layer, forming
a train of the characteristic shapes hereafter called "bulges". The
appearance of the bulges is similar to ordered motion visualized with
smoke in turbulent boundary layers over flat plates (e.g. Falco, 1977).
However, it is noteworthy that in the present case the bulges have a
horizontal length scale which corresponds to the wavelength of the wind
waves, seen in the photograph at the lower edge of the mist.
      To better understand the visualized large scale phenomena, we
combined hot-wire measurement with stroboscopic photographs of the
visualization. Before the experiment, it was verified that the
influence of the mist on the hot-wire measurement was essentially
negligible. Figure 5 shows successive pictures taken at 150 Hz with
simultaneous records of the velocity at z=7 cm and the surface displacement just under the hot-wire. We can follow each bulge's movement from
left to right, designated by "a" to "d" in the figure, passing the hot-
wire placed at the upstream end of the horizontal rod seen in the right
side of the photos. The variations relating to the passage of the bulges
are clearly distinguishable in the velocity records. Low velocity($u(t)<0$)
coincides with the time of the passage, usually coupled with upward
velocity($w(t)>0$) producing, as a results, large negative values of
$u(t)$ x $w(t)$. We conclude that the bulges with the peculiar velocity
variations are ordered motions associated with the wind waves.
      Figure 6 shows the evolution of the ordered motions from successive
pictures. Two types of evolution of the traveling downstream ordered
motion were observed. Type A: a short bulge in the inner layer grows in
height and becomes a taller bulge (Figure6 (a)). Type B: a taller bulge
does not change its height, but deforms its shape a little(Figure 6 (b)).
Type B can be considered to be the stage following type A.
      The taller bulge, which we call a fully developed(FD) ordered
motion, was studied by conditional sampling and averaging. Ten FD
ordered motions with the corresponding velocity fluctuations were sampled
from the photographs. Figure 7 shows the ensemble averaged velocity
distribution inside the FD ordered motion. Since the local Reynolds
stress at this height is -0.04 $m^2 s^2$, it is evident that the FD ordered
motions produce the larger part of the local Reynolds stress.

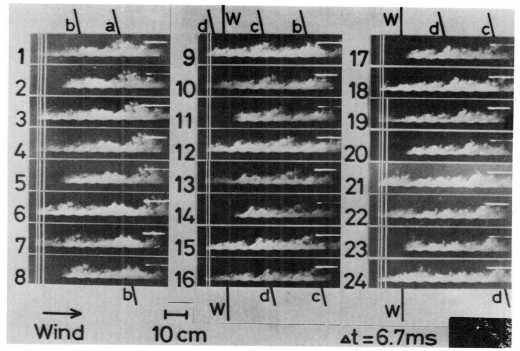

Figure 5. (a) A sequence of pictures of the paraffin mist
visualization. Lines "a"-"d" show the movement of the bulges and the
line "W" the movement of a wind wave.

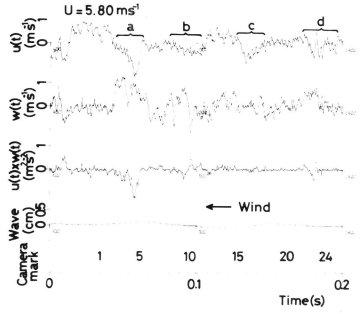

Figure 5. (b) Records
of the hot-wire at z=7
cm and the wave gauge
placed just under the
hot-wire, obtained
simultaneously with
the pictures in Figure
5 (a).

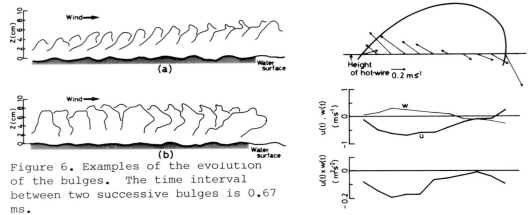

(a)

(b)

Figure 6. Examples of the evolution
of the bulges. The time interval
between two successive bulges is 0.67
ms.

Figure 7. Ensamble averaged u, w and u x w inside the fully developed
ordered motion at z=5 cm. The velocity vectors shown in the top figure
are referenced to an observer who moves at the mean speed of the bulges.

Considerable activity is noticeable on the rear part of the bulge; it is
caused by low horizontal velocity coupled with upward velocity.
Entrainment of the high speed air occurs at the streamwise interface.
These suggest interaction between the FD ordered motion in the outer
layer and the flow in the inner layer. In the top figure vectors of the
ensemble averaged velocity are referenced to an observer moving at the
averaged streamwise velocity of the FD. Large scale motion, slowly
rotating around the core, is seen to exist in the FD.
     We propose two mechanisms for the ordered motion, associated with
the separated flow, to explain the bursting phenomena. The formation
process of the ordered motion consists of the whole air mass composing
the separation bubble blowing up(called a "big burst"), spreading and
increasing its height(Figure 6(a)), becoming the FD. The maintenance
mechanism of the FD ordered motion is described as follows: when the FD
passes over the separation bubble, its original motion influences the
unstable high shear layer and sucks the energetic eddies from the
reattachment region("small burst"). The FD maintains its rotating motion
and turbulence in it by means of entrained eddies.

REFERENCES

Falco, R.E., 1977: 'Coherent motions in the outer region of turbulent
     boundary layers', Phys. Fluids, 20, S124-132.
Kawai, S., 1982: 'Structure of air flow separation over wind wave
     crests', Boundary-Layer Met., 23, 503-521.
Kawamura, H., K. Okuda, S. Kawai and Y. Toba, 1981: 'Structure of
     turbulent boundary layer over wind waves in a wind wave tunnel',
     Tohoku Geopys. Journ. (Sci. Rep. Tohoku Univ. Ser. 5), 28, 69-86.
Tani, I., M. Iuchi and H. Komoda, 1961: 'Experimental investigation of
     flow separation associated with a step or a groove', Aeronautical
     Research Institute, University of Tokyo, Report No.364, 119-136.

SPECTRAL CHARACTERISTICS OF BREAKING WAVES

O.M. Phillips
Department of Earth and Planetary Sciences
The Johns Hopkins University
Baltimore, Maryland 21218

ABSTRACT. In the equilibrium range of wind-generated waves, it is postulated that the processes of energy (or action) input from the wind, loss by wave breaking and net transfer by non-linear resonant wave interactions are of comparable importance throughout the range. Consideration of the action spectral density balance then indicates that the wave-number spectrum in this range is proportional to $(\cos \theta)^{1/2} \, u_* \, g^{-1/2} k^{-7/2}$, where θ is the angle between the wind and the wave-number k, and the frequency spectrum is of the form found empirically by Toba (1973), namely $u_* \, g \sigma^{-4}$. These forms have also been derived by Kitaigorodskii (1983) though on a quite different physical basis. The spectral rate of energy loss by wave breaking is found to be proportional to $(\cos \theta)^{3/2} \, u_*^3 \, k^{-2}$ and the spectral rate of momentum loss from the waves to $(\cos \theta)^{5/2} \, g^{-1/2} u_*^3 \, k^{-3/2}$. As the wave field develops with increasing fetch or duration, the total rate of energy input to the water turbulence by wave breaking increases as $\rho_a u_*^3 \times \ln(k_1/k_0)$ where k_1 and k_0 are the upper and lower wave-number limits to the range; the total momentum flux increases also but asymptotes to a fixed fraction of $\rho_a u_*^2$. The various constants of proportionality are found in terms of Toba's constant and a coefficient expressing the rate of energy input from wind to waves.

1. Introduction

The breaking of waves is a process that is ubiquitous over two-thirds of the surface of the globe. It is clearly responsible for part of the transfer of mechanical energy and of momentum from the atmosphere to ocean currents and turbulence, for the enhancement of heat transfer and especially the exchange of gases between the atmosphere and the ocean as well as augmenting substantially but locally the drag of the air on the water itself (Banner and Melville, 1976).

In the past few years a great deal of attention has been paid to the dynamics of breaking and the search for criteria under which waves might be expected to break. The remarkable and pioneering theory and

Y. Toba and H. Mitsuyasu (eds.), The Ocean Surface, 111–123.

numerical experiments of Longuet-Higgins and Cokelet (1976) have
traced the evolution of finite amplitude irrotational waves on deep
water, either as a result of their intrinsic instabilities or of
impulsive forcing to the point of wave breaking and just beyond.
Less fundamental have been attempts to find a single threshold
variable such as local vertical acceleration, or combination of such
variables, which determine the probability of breaking of an individual wave crest. This concept lay behind the original idea that led
to the simple $g^2 \sigma^{-5}$ saturation spectrum proposed a number of years
ago. The idea has been taken a great deal further in other directions
with interesting success in a series of three papers in 1983 written
by Snyder, Kennedy and Smith in various combinations. It does remain
difficult, though, to associate any single local variable with the
examples of breaking calculated by Longuet-Higgins and Cokelet; it
seems that the recent time history of the surface configuration is
more pertinent than a single local threshold variable.

        In this paper, a rather different approach is taken, more in
the spirit of Hasselmann (1974) in which the detailed configuration at
the point of incipient breaking is ignored - it disappears anyway as
soon as the wave breaks - while concentrating on the statistical consequences of the ensemble of breaking events at various points on the
sea surface. The initial goal is to use simple dynamical reasoning to
provide as reliable an estimate as possible for the average rate of
spectral energy loss resulting from breaking; in turn, this leads to
the form of the high frequency spectrum of gravity waves that was
inferred on empirical and dimensional grounds by Toba (1973) and also
to a series of simple expressions for quantities such as the spectral
distributions and total rates of the enrgy loss and momentum flux from
the waves of the equilibrium range by wave breaking. The account
given in this paper is necessarily brief; a more extended and detailed
discussion is expected to appear in the Journal of Fluid Mechanics.

2.   The Statistical Equilibrium of Short Waves

        The spectrum of a random distribution of surface waves can be
specified by

$$\psi(\underline{k}) = (2\pi)^{-2} \int \overline{\zeta(\underline{x}) \, \zeta(\underline{x}+\underline{r})} \; e^{-i\underline{k}\cdot\underline{r}} \; d\underline{r} ,$$

where ζ represents the local surface displacement and the integral is
over the entire separation ($\underline{r}$) plane. The dynamics of the field is,
however, more conveniently described (particularly when wave-current
interactions are involved) by the balance of action spectral density

$$N(\underline{k}) = (g/\sigma) \, \psi(\underline{k}) = (g/k)^{1/2} \, \psi(\underline{k}), \qquad (2.1)$$

where σ is the intrinsic frequency and the water density is divided out throughout. Following energy paths (see, for example, Phillips, 1980),

$$\frac{dN}{dt} = \frac{\partial N}{\partial t} + (\underset{\sim}{C} + \underset{\sim}{U}) \cdot \nabla N = - \nabla_{\underset{\sim}{k}} \cdot \underset{\sim}{T}(\underset{\sim}{k}) + S_w - D, \qquad (2.2)$$

where $\underset{\sim}{C}$ is the local group velocity. The various processes that modify the action spectral density following a wave group are represented on the right. $\underset{\sim}{T}(\underset{\sim}{k})$ represents the spectral flux of action resulting from resonant wave-wave interactions. These exchanges are conservative for gravity waves and the integral of this term over all wave-numbers vanishes. The rate of spectral input of wave action from the wind is expressed schematically by the term S_w and D represents the rate of loss by wave breaking and possibly the formation of parasitic capillaries at large gravity wave-numbers.

    For those components at wave-numbers large compared with that of the spectral peak, in a well-developed wave field under the continued action of the wind, the time scales of their growth are long compared with the internal time scales involved in wave-wave interactions, action input from the wind and loss by breaking, so that for these components the spectral balance reduces to

$$- \nabla_{\underset{\sim}{k}} \cdot \underset{\sim}{T}(\underset{\sim}{k}) + S_w - D = 0. \qquad (2.3)$$

In this equilibrium range, the detailed functional forms of each of these terms would be expected to depend on the nature of the spectrum $N(\underset{\sim}{k})$ in this range and it is of interest to enquire what spectral characteristics are associated with the possible balances among the three terms of (2.3).

    The spectral re-distribution of wave action has been the subject of pioneering investigations by Hasselmann (1962, 1968) and others; it can be represented as a "collision integral" over sets of four resonantly interacting gravity waves:

$$- \nabla_{\underset{\sim}{k}} \cdot \underset{\sim}{T}(\underset{\sim}{k}) = \iiint Q^2 \{ [N(\underset{\sim}{k}) + N(\underset{\sim}{k}_1)] N(\underset{\sim}{k}_2) N(\underset{\sim}{k}_3) -$$
$$- [N(\underset{\sim}{k}_2) + N(\underset{\sim}{k}_3)] N(\underset{\sim}{k}) N(\underset{\sim}{k}_1) \} \times \qquad (2.4)$$
$$\times \, \delta (\underset{\sim}{k} + \underset{\sim}{k}_1 - \underset{\sim}{k}_2 - \underset{\sim}{k}_3) \, \delta (\sigma + \sigma_1 - \sigma_2 - \sigma_3) \, d\underset{\sim}{k}_1 \, d\underset{\sim}{k}_2 \, d\underset{\sim}{k}_3$$

where the coupling coefficient Q is a complicated homogeneous function of the wave numbers $\underset{\sim}{k}, \ldots, \underset{\sim}{k}_3$ and is of order k^3 and δ represents

the Dirac delta function. Later work by Fox (1976) and Sell and
Hasselmann (1972) suggests that the interactions are primarily local
in the wave-number plane, so that the net action transfer to a given
wave-number interval is determined primarily by the action spectral
density in this vicinity. Near the spectral peak, of course, the flux
to neighboring wave-numbers is dominated by the peak itself, but in
the equilibrium range, the net flux to or from a wave-number band
should scale with the local value of N, i.e. $N(\underset{\sim}{k})$. Consequently,
since (2.4) is cubic in N and since $Q^2 \sim k^6$, the net spectral flux
divergence scales as

$$- \nabla_k \cdot T(\underset{\sim}{k}) \sim Q^2 N^3 k^4 / \sigma \sim N^3(\underset{\sim}{k}) k^{19/2} g^{-1/2}, \qquad (2.5)$$

as given by Kitaigorodskii (1983). This can be expressed equivalently
in terms of the dimensionless function, the "degree of saturation"

$$B(\underset{\sim}{k}) = g^{-1/2} k^{9/2} N(\underset{\sim}{k}) = k^4 \Psi(\underset{\sim}{k}),$$

defined by the author (1984) in terms of which (2.5) becomes

$$- \nabla_k \cdot T \sim g k^{-4} B^3(\underset{\sim}{k}). \qquad (2.6)$$

The rate of action (or energy) input from the wind has been
the subject of many theoretical and experimental investigations over
the past twenty years which have, if nothing else, demonstrated the
complexity and variety of the detailed processes involved. In order
to give a simple expression for S_w in (2.3) the best guide seems to be
provided by the analysis of careful experiments interpreted in the
light of only very general theoretical considerations. Plant (1982)
suggests from a survey of such measurements that

$$S_w \simeq 0.04 \cos\theta \; \sigma \, (u_*/c)^2 N(\underset{\sim}{k}), \qquad (2.7)$$

where θ is the angle between the wave-number $\underset{\sim}{k}$ and the wind, u_* is the friction velocity of the air flow over the water surface and $c = (g/k)^{1/2}$ the phase velocity of the component concerned. This form has been suggested by others as well; Mitsuyasu and Honda (1984) give a numerical coefficient of 0.05 and Gent and Taylor's (1976) calculation gives approximately 0.07. The form of (2.7) might also be justified on general dynamical grounds. The action and energy fluxes from wind to waves result from variations in surface stresses in phase with the orbital velocities at the surface; with stress variations of order $\rho_a u_*^2$ times the local wave slope, and orbital velocities also proportional to the slope, the net transfer rate must vary as $\rho_a u_*^2 N(k)$. For dimensional consistency, then $s_w \propto (\rho_a/\rho_w)\sigma (u_*/c)^2 N(k)$, which, apart from the numerical constant and the directional factor (less certain anyway) reduces to (2.7). In terms of the degree of saturation, this becomes

$$S_w = m \cos \theta \, g k^{-4} (u_*/c)^2 B(\underset{\sim}{k}), \qquad (2.8)$$

where $m = 0.04$, but may be rather larger.

The development of an expression for the rate of spectral action dissipation is more tentative. The author has argued (1984) that this will depend on the spectral level, represented by B (rather than the wind stress directly) since the occurrence of local breaking and the consequent energy loss is the result of a local excess of energy or action, however this excess is produced. It may, for example, arise from a local convergence in an underlying current which increases the local degree of saturation and consequently the intensity of breaking. In an active wind-generated wave field where wave-current interactions are negligible, the degree of saturation may be enhanced by the wind stress, but the extent to which wave breaking occurs still has as its primary causative property, the degree of saturation B. In the equilibrium range, B may be expected to vary only slowly with wave-number magnitude k, so that in spite of the localness in physical space of the dissipation process, the spectral rate of dissipation of wave action at a given wave-number $\underset{\sim}{k}$ in this range may be considered to be a function of B at that wave-number:

$$D(\underset{\sim}{k}) = g k^{-4} f(B(\underset{\sim}{k})). \qquad (2.9)$$

In summary, then, we have three physical processes that are pertinent to the equilibrium range in an active wind-generated sea, which balance among themselves and which scale as follow:

Spectral flux divergence $\quad gk^{-4}B^3(\underset{\sim}{k})$

Wind input $\quad m\cos\theta\ gk^{-4}(u_*/c)^2 B(\underset{\sim}{k})$ $\quad$ (2.10)

Dissipation $\quad gk^{-4}f(B(\underset{\sim}{k}))$

The form of the spectrum in this range depends upon the balances that may exist among these processes, and several alternatives may be visualized.

Kitaigorodskii (1983) has proposed the existence of a Kolmogoroff type of equilibrium range in wind-generated waves in which the energy input from the wind is assumed to occur primarily at the energy-containing scales with dissipation at much larger scales. This then postulates the existence of a range of wave-numbers over which the spectral flux divergence, wind input and dissipation are all negligible; the spectral energy flux ε_0 is constant over this range and the spectral form must be such as to accommodate this constant flux. On similarity grounds he gives for the (directionally averaged) energy spectrum

$$F(k) = \int_{-\pi}^{\pi} \psi(k,\theta)\,d\theta \sim \varepsilon_0^{1/3}\,g^{-1/2}\,k^{-7/2} \qquad (2.11)$$

and for the frequency spectrum

$$S(\sigma) \sim \varepsilon_0^{1/3}\,g\,\sigma^{-4}.$$

Arguing further that $\varepsilon_0 \propto (\rho_a/\rho_w)U^3$, where U is the mean wind speed, or, approximately that $\varepsilon_0 \propto u_*^3$, he obtains wave-number and frequency spectra of the forms $u_* g^{-1/2}k^{-7/2}$ and $u_* g\sigma^{-4}$ respectively for wave-

numbers and frequencies above those at which energy input from the
wind occurs and below those for which dissipation is regarded as
important.

The principal conceptual difficulty with Kitaigorodskii's
argument is the need to postulate that the energy input from the wind
is concentrated at wave-numbers close to those of the spectral peak.
To be sure, the air flow over the dominant waves may modify the rate
of energy input to smaller waves superimposed on the longer ones, but
it is difficult to see why it should be suppressed entirely. Indeed,
according to (2.7), the time scale for wind energy input is (for

$$N(\underline{k})/S_w \simeq 25 (c/u_*)^2 \sigma^{-1} = 25 g^2/u_*^2 \sigma^3,$$

which decreases rapidly as the frequency increases. Yet very careful
measurements of the frequency spectra of wind-generated waves by Toba
(1973) and more recently by Forristall (1981), Kahma (1981) and
Donelan et al.(1982) indicate strongly that over a considerable range
of frequencies higher than that of the spectral peak, the spectrum is
much better represented as $gu_* \sigma^{-4}$ than as the $g^2 \sigma^{-5}$ saturation form
proposed in 1958 by the author on much simpler dimensional grounds.
It seems that the matter demands re-consideration -- twenty-five years
is a pretty fair lifetime for a simple idea.

The basic point of this paper is to indicate how an equilibrium spectrum of the Toba type can be derived from a very different
assumption about the dynamical balances in the equilibrium range and,
as a bi-product, to infer a number of simple properties concerning the
statistics of the breaking events themselves. In contrast to the
hypothesis made by Kitaigorodskii, let us suppose that in the
equilibrium range of an active wind-generated sea, all of the three
processes represented in (2.10), namely the spectral flux divergence
resulting from wave-wave interactions, wind input and dissipation by
wave breaking, are comparable throughout the range. Since there are
no internal wave-number scales within the equilibrium range, the
ratios of the three terms must be constant over wave-numbers sufficiently far from the ends of the ranges:

$$B^3(\underline{k}) \propto m \cos\theta \, (u_*/c)^2 B(\underline{k}) \propto f(B(\underline{k})), \qquad (2.12)$$

whence it follows immediately that

$$B(\underline{k}) = \beta (\cos\theta)^{1/2}(u_*/c), \qquad (2.13)$$

and

$$f(B) \;=\; a\,B^{3}(\underset{\sim}{k}),\tag{2.14}$$

where β and a are numerical constants. (Note that, from the definition, $B(\underset{\sim}{k}) = B(-\underset{\sim}{k})$; in (2.13) $-\pi/2 < \theta < \pi/2$).

This then leads to a wave-number spectrum in the equilibrium range

$$\Psi(\underset{\sim}{k}) \;=\; k^{4}B(\underset{\sim}{k}) = \beta\,(\cos\theta)^{1/2}\,(u_{*}/c)\,k^{-4},$$

$$= \beta\,(\cos\theta)^{1/2}\,u_{*}\,g^{-1/2}\,k^{-7/2}\tag{2.15}$$

similar to that given by Kitaigorodskii on a quite different basis. The freqency spectrum can be found from (2.15), although care must be taken to restrict the range of frequencies to those below which the advection by the dominant waves (and the consequent Doppler shifting) becomes significant. The orbital speed of the dominant waves is approximately $2\,(\overline{\zeta^{2}})^{1/2}\,\sigma_{0}$, where σ_{0} is the frequency at the spectral peak, so that Doppler shifting becomes significant for components whose intrinsic phase velocity g/σ is not large compared to this. Accordingly, the frequency spectrum

$$\Phi(\sigma) \;=\; 2\int_{-\pi/2}^{\pi/2} k\,\Psi(\underset{\sim}{k})(\partial\sigma/\partial k)^{-1}\,d\theta\,\Bigg|_{k=\sigma^{2}/g}$$

$$= \alpha\,u_{*}\,g\,\sigma^{-4},\tag{2.16}$$

where $\alpha = 4\beta\int_{-\pi/2}^{\pi/2}(\cos\theta)^{1/2}d\theta = 9.4\beta$. This is the form found empirically by Toba (1973) from wind tunnel data and confirmed in field observations by Kawai, Okada and Toba (1977), Donelan et al. (1982) and others. The constant of proportionality measured by Toba in a wind-tunnel was approximately 0.02 and Donelan et al.'s field measurements are consistent with this, although Kawai, Okada and Toba's later field work gives a value of 0.06 ± 0.01. Kawai et al. give some explanation for the difference between this result and Toba's earlier estimate, though the reasons for the discrepancies may still not be well understood.

The expressions (2.13) and (2.14) allow us to estimate the spectral rates of dissipation of wave action, wave energy and wave momentum in the wind direction which are, respectively,

$$D(\underset{\sim}{k}) = g k^{-4} f(\beta) ,$$

$$= a\beta^{3} (\cos\theta)^{3/2} g k^{-4} (u_*/c)^{3} ,$$

$$= a\beta^{3} (\cos\theta)^{3/2} g^{-1/2} u_*^{3} k^{-5/2} ; \tag{2.17}$$

$$\varepsilon(\underset{\sim}{k}) = \sigma D(\underset{\sim}{k}) ,$$

$$= a\beta^{3} (\cos\theta)^{3/2} u_*^{3} k^{-2} ; \tag{2.18}$$

and

$$\tau(\underset{\sim}{k}) = (\varepsilon(\underset{\sim}{k})/c) \cos\theta ,$$

$$= a\beta^{3} (\cos\theta)^{5/2} g^{-1/2} u_*^{3} k^{-3/2} . \tag{2.19}$$

In the absence of more complete observational verification, not too much significance should be ascribed to the directional factors given in these expressions, but it is interesting to note that the directional distribution of the equilibrium range energy density that they indicate is quite broad.

The total fluxes of energy and momentum from the wind to the sea occur in three separate pathways: (a) directly by the mean shear stress on the water surface, (b) from wind to waves, resulting in wave growth and radiation from the generating area and (c) from wind to the waves of the equilibrium range, from which it is lost locally from the waves by breaking. The last of these can now be estimated from (2.18) and (2.19). If k_o represents the lowest wave-number associated with active wave breaking (which may be coincident with that of the spectral peak, but may be somewhat higher) and k_1 is the upper limit to this range, then the total rate of energy loss from the waves by breaking, or, equivalently, the rate of energy input to the surface layer turbulence in this way is

$$\varepsilon_o = 2 \int_{-\pi/2}^{\pi/2} \int_{k_o}^{k_1} \varepsilon(\underset{\sim}{k}) k \, dk \, d\theta ,$$

$$= 3.42 \, a\beta^{3} u_*^{3} \ln(k_1/k_o) , \tag{2.20}$$

in which the directionality factor in (2.18) is taken at face value.
This quantity is more usually expressed in terms of the air density;
restoring the density factors we have

$$\varepsilon_o = (3 \cdot 42 \, a\beta^3 \rho_w / \rho_a) \cdot \rho_a u_*^3 \, \ln(k_1 / k_o). \qquad (2.21)$$

The total momentum flux to the surface layer by wave breaking is
likewise

$$\tau_w = (5 \cdot 64 \, a\beta^3 \rho_w / \rho_a) \cdot \rho_a g^{-1/2} u_*^3 (k_1^{1/2} - k_o^{1/2}). \qquad (2.22)$$

According to Banner and Phillips (1974), freely travelling gravity
waves for which $c < u_*$ (or $k > g/u_*^2$ are strongly suppressed by the
wind drift induced by the direct shear stress at the water surface; if
$k_1 = g/u_*^2 \gg k_o$, then from (2.22),

$$\tau_w = (5 \cdot 64 \, a\beta^3 \rho_w / \rho_a) \cdot \rho_a u_*^2, \qquad (2.23)$$

which must, of course, be less than $\rho_a u_*^2$.
      Accordingly, as a wave field develops from, say, an initial
state of rest, the momentum flux to the surface layer by wave
breaking, initially zero, increases as the equilibrium range covers a
wider and wider interval of wave-numbers, approaching asymptotically a
fixed fraction of the total wind stress. The energy flux to the turbulence of the surface layer by wave breaking continues to increase,
albeit logarithmically. With $k_1 = g/u_*^2$ and $k_o = g/c_o$,

$$\varepsilon_o = (3 \cdot 42 \, a\beta^3 \rho_w / \rho_a) \ln(c_o / u_*) \rho_a u_*^3, \qquad (2.24)$$

and if the fetch and duration of the field are sufficient to generate
dominant waves moving at the wind speed,

$$\varepsilon_o \approx \left(3.42 \; a\beta^3 \rho_w / \rho_a \right) \ln \left(c_D^{-1} \right) \rho_a u_*^3 , \qquad (2.25)$$

where C_D is the drag coefficient.

3. Constraints on the Constants of Proportionality.

Among the interesting consequences of the analysis of the previous section are the relations it provides among various numerical
coefficients that have been inferred from independent sets of measurements, though none to high precision. The quantity m of (2.8)
expressing the rate of energy input from the wind is about 0.04 but
may be rather larger; Toba's constant α of (2.16) specifying the
spectral level of the frequency spectrum in the equilibrium range may
be bracketed by the value 0.02 and 0.06 found in different experiments. The constant involved in the wave-number spectrum in the
saturation range has not yet been measured directly, but the $(\cos \theta)^{1/2}$
directionality factor gives $\alpha = 9.36 \; \beta$; different but reasonable
directional distributions may give of up to fifty per cent or so in
the coefficient. One firm constraint that we have from (2.23) is that

$$5.46 \; a\beta^3 \rho_w / \rho_a < 1$$

or

$$a\beta^3 < 1.8 \times 10^{-4} . \qquad (3.1)$$

Now, in the action spectral density balance (2.12), f(B) certainly represents a loss and the wind input a gain; the calculations
of Sell and Hasselmann (1972), although not too reliable at these
large wave-numbers, indicate that the net spectral flux also represents a gain. Consequently, the rate of dissipation aB^3 must be
greater than or equal to the rate of wind input, so that

$$a\beta^3 \geq m \approx 0.04 \qquad (3.2)$$

If $\alpha = 0.02$ then $\beta = 2 \times 10^{-3}$ and $a\beta^3 > 8 \times 10^{-5}$, comfortably satisfying (3.1) and suggesting that fifty per cent or more of the total
wind stress is communicated to the surface layer by wave breaking. On
the other hand, if $\alpha = 0.06$, then $\beta = 7 \times 10^{-3}$ and $a\beta^3 > 2.8 \times 10^{-4}$,
which is inconsistent with (3.1). We conclude therefore that either a
value of 0.06 for Toba's constant, or a value of 0.04 for the wind-

wave coupling coefficient (or both) are too high. Nevertheless, even with somewhat smaller values one can also conclude that in a well-developed wind-wave field, (1) a substantial fraction of the total wind stress is communicated to the surface layer by wave breaking and (2) the energy flux as turbulence to the surface layer by wave breaking is a modest multiple ($\ln C_D^{-1} = 6.5$) of $\rho_a u_*^3$, and is certainly greater than the energy flux by the mean surface shear stress acting on the wind-induced mean drift.

    It is a pleasure to acknowledge the support of the Fluid Dynamics Branch of the Office of Naval Research under contract N00014-76-C-0184.

References

Banner, M.L. and O.M. Phillips, 1974: On the incipient breaking of small scale waves. J. Fluid Mech., 65, 647-56.

Banner, M.L. and W.K. Melville, 1976: On the separation of air flow over water waves. J. Fluid Mech., 77, 825-42.

Donelan, M.A., J. Hamilton and W.H. Hui, 1984: Directional spectra of wind-generated waves. Phil. Trans. Roy. Soc., A, xxx.

Forristall, Z., 1981: Measurements of a saturation range in ocean wave spectra. J. Geophys. Res., 86, 8075-84.

Fox, M.J.H., 1976: On the nonlinear transfer of energy in the peak of a gravity wave spectrum - II. Proc. Roy. Soc., A. 348, 467-83.

Gent, P.R. and P.A. Taylor, 1976: A numerical model of the air flow above water waves. J. Fluid Mech., 77, 205-28.

Hasselman, K., 1962: On the non-linear energy transfer in a gravity-wave spectrum. Part 1. General Theory. J. Fluid Mech., 12, 481-500.

Hasselmann, K., 1968: Weak interaction theory of ocean waves. Basic Developments in Fluid Dynamics, Vol. 2. M. Holt, Ed. Academic Press. 117-82.

Hasselmann, K., 1974: On the spectral dissipation of cean waves due to whitecapping. Boundary-Layer Meteorol., 6, 107-27.

Kahma, K.K., 1981: A study of the growth of the wave spectrum with fetch. J. Phys. Oceanogr., 11, 1503-15.

Kawai, S., K. Okada and Y. Toba, 1977: Field data support for three-seconds power law and gu $_*$ σ^{-4} spectral form for growing wind waves. J. Oceanogr. Soc. Japan, 33, 137-50.

Kennedy, R.M. and R.L. Snyder, 1983: On the formation of whitecaps by a threshold mechanism. Part 11: Monte Carlo Experiments. J. Phys. Oceanogr., 13, 1493-1504.

Longuet-Higgins, M.S. and E.D. Cokelet, 1976: The deformation of steep surface waves. Proc. Roy. Soc., A 350, 1-26.

Mitsuyasu, H. and T. Honda, 1984: The effects of surfactant on certain air-sea interaction phenomena. Wave Dynamics and Radio Probing of the Ocean Surface, Plenum Press, N.Y.

Phillips, O.M., 1980: The Dynamics of the Upper Ocean. Cambridge University Press, pp.336

Phillips, O.M., 1984: On the response of short ocean wave components
       at a fixed wave-number to ocean current variations. J. Phys.
       Oceanogr., 14, No 8.
Plant, W.J., 1982: A relationship between wind stress and wave slope.
       J. Geophys. Res., 87, 1961-67.
Sell, W. and K. Hasselmann, 1972: Computation of nonlinear energy
       transfer for JONSWAP and empirical wave spectra. Rep. Inst.
       Geophys., Univ. Hamburg.
Snyder, R.L. and R.M. Kennedy, 1983: On the formation of whitecaps by
       a threshold mechanism. Part 1: Basic Formalism. J. Phys.
       Oceanogr., 13, 1482-92.
Snyder, R.L., L. Smith and R.M. Kennedy, 1983: On the formation of
       whitecaps by a threshold mechanism. Part III: Field experiment
       and comparison with theory. J. Phys. Oceanogr., 13, 1505-18.
Toba, Y., 1973: Local balance in the air-sea boundary processes, III.
       On the spectrum of wind waves. J. Oceanogr. Soc. Japan, 29,
       209-20.

THE FULLY DEVELOPED WIND-SEA SPECTRUM AS A SOLUTION OF THE ENERGY
BALANCE EQUATION

G.J. Komen,[1] S. Hasselmann,[2] K. Hasselmann[2]
1) Royal Netherlands Meteorological Institute,
   P.O. Box 201, 3730 AE De Bilt, The Netherlands
2) Max-Planck-Institut für Meteorologie,
   Bundesstrasse 55, Hamburg, FRG

Extended abstract

     We consider the energy transfer equation for well developed ocean
waves under the influence of wind, and study the conditions for the
existence of an equilibrium solution in which wind input, wave-wave
interaction and dissipation balance each other. For the wind input we
take the parametrization proposed by Snyder et al (1981), which was
based on their measurements in the Bight of Abaco, and which agrees
with Miles' (1957, 1959) theory. The wave-wave interaction is computed
with an algorithm given by Hasselmann et al (1984). The dissipation is
less well-known, but we will make the general assumption that it is
quasi-linear in the wave spectrum with a factor coefficient depending
only on frequency and integral spectral parameters (cf. Hasselmann,
1974). Full details of this study are given elsewhere (Komen,
Hasselmann and Hasselmann, 1984). Here we summarize the main results.
In the first part of our study we investigated whether the assumption
that the equilibrium spectrum exists and is given by the Pierson-
Moskowitz spectrum with a standard type of angular distribution leads
to a reasonable dissipation function. We find that this is not the
case. Even if one balances the total rate of change for each frequency
(which is possible), a strong angular imbalance remains. This is
illustrated in fig. 1, in which the assumed asymptotic spectrum and
the corresponding source terms are given. The dissipation constant is
chosen such that at any given frequency the total rate of change
vanishes. As one can see there is no angular balance. Thus the assumed
source terms are not consistent with this type of asymptotic spectrum.
In the second part of the study we chose a different approach. We
assumed that the dissipation was given · and we performed numerical
experiments simulating fetch limited growth, to see under which
conditions a stationary solution can be reached. For the dissipation
we took Hasselmann's (1974) form with two unknown parameters. From our
analysis it follows that for a certain range of values of these
parameters a quasi-equilibrium solution results. We estimate the
relation between dissipation parameters and asymptotic growth rates.

125

Y. Toba and H. Mitsuyasu (eds.), The Ocean Surface, 125–128.
© 1985 by D. Reidel Publishing Company.

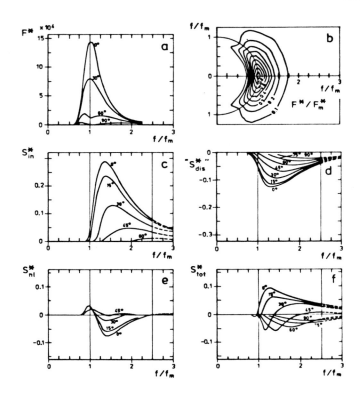

Fig. 1.   Two-dimensional spectrum (panel a, with polar isoline
           representation in panel b) and source functions for the
           Pierson-Moskowitz   spectrum   with   a   standard   angular
           distribution. The 2d dissipation source function "S_{dis}^*" is
           computed   by   requiring   a   1-dimensional   balance.   The   net
           source   function $S_{tot}^* = S_{in}^* + S_{nl}^* + S_{dis}^*$ is   significantly
           different from zero.

For equilibrium spectra, the input, dissipation and nonlinear transfer
source functions are all significant in the energy containing range of
the spectrum. The energy balance proposed by Zakharov and Filonenko
(1966) and Kitaigorodskii (1983), in which dissipation is assumed to
be significant only at high frequencies, yields a spectrum which grows
too rapidly and does not approach equilibrium. One of our equilibrium
solutions has a one-dimensional spectrum which lies close to the
Pierson-Moskowitz spectrum. The energy balance for this spectrum is
given in Figure 2. A striking feature is the angular distribution. In
the forward direction the spectrum peaks at a value higher than the
peak frequency of the one-dimensional spectrum. This is related to the
fact that the wind input has a strong maximum at a frequency well
above the one dimensional peak frequency. For more details we refer to
Komen, Hasselmann and Hasselmann (1984).

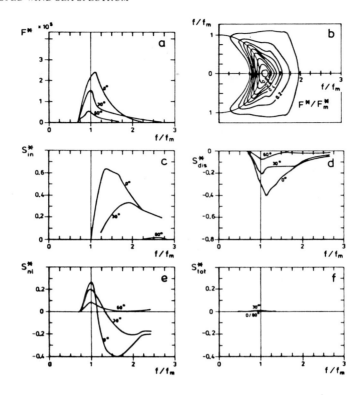

Fig. 2.   Two-dimensional spectrum (panels a and b) and source
functions for our best simulation run at large fetch. The
net source function (panel f) is two orders of magnitude
smaller than the individual source functions (panels c, d,
e). This may be contrasted with panel f of Fig. 1, for a
Pierson-Moskowitz spectrum with a prescribed spreading
function. The difference may be attributed to the
differences in the 2d spectral distributions, cf. panels a,
b, for the two cases.

REFERENCES

Hasselmann, K., 1974. On the spectral dissipation of ocean waves due
to white capping. Boundary-Layer Met. 6, 107 - 127.
Hasselmann, S., K. Hasselmann, J.H. Allender and T.P. Barnett, 1984.
Improved methods of computing and parametrizing the nonlinear
energy transfer in a gravity wave spectrum (submitted for
publication).
Komen, G.J., S. Hasselmann and K. Hasselmann, 1984. On the existence
of a fully developed wind-sea spectrum. Journ. Phys. Oceanogr.

Kitaigorodskii, S.A., 1983. On the theory of the equilibrium range in
    the spectrum of wind-generated gravity waves. J. Phys Oceanogr.
    13, 816 - 827.
Miles, J.W., 1957. On the generation of surface waves by shear flows,
    Part 1. J. Fluid Mech. 3, 185 - 204.
Miles, J.W., 1959. On the generation of surface waves by shear flows,
    Part 2. J. Fluid Mech. 6 568 - 582.
Snyder, R.L., F.W. Dobson, J.A. Elliot and R.B. Long, 1981. Array
    measurements of atmospheric pressure fluctuations above surface
    gravity waves. J. Fluid Mech. 102, 1 - 59.
Zakharov, V.Ye. and N.N. Filonenko, 1966. The energy spectrum for
    stochastic oscillations of a fluid surface. Dokl. Akad. Nauk
    SSSR, 170, No. 6, 1292 - 1295.

THE KINEMATICS OF SHORT WAVE MODULATION BY LONG WAVES

David E. Irvine
The Johns Hopkins University
Applied Physics Laboratory
Laurel, Maryland 20707, USA

ABSTRACT

The modulation of short wave spectra (defined as the variation of a given spectral component at either a fixed wavenumber or a fixed frequency) due to a long wave is studied
in the absence of overt wind effects. Simple action conservation is found to predict modulations dominated by effects
other than actual short wave amplitude modulations.

1.0 Introduction

The interaction of long and short gravity waves has been
studied primarily in the service of understanding some <u>other</u>
phenomenon. Interest in those other phenomena has been in
one case practical - the understanding of radar returns from
the ocean - and in the other, theoretical - the understanding
of wave growth. Only in the last fifteen years has the long
wave/short wave problem itself attracted sufficient attention to merit serious experimental investigation. An extensive review is given in the author's Ph.D. dissertation
(Irvine, 1983); this paper is a condensation of that thesis.

The next section outlines a simple linear theory to describe
the impact of long waves on both wavenumber and frequency
spectra of short waves. The governing equations for the two
cases are then contrasted. Finally, the implications are
discussed.

2.0 Linear Theory

The starting point in the analysis is the interaction of a
monochromatic short wave with a long wave. This is generalized to first a wavenumber spectrum of short waves and then
a frequency spectrum of short waves. The analysis owes much
both in spirit and detail to Hughes (1978).

Y. Toba and H. Mitsuyasu (eds.), The Ocean Surface, 129–134.
© *1985 by D. Reidel Publishing Company.*

The short wave is characterized by its wavenumber (k), intrinsic frequency (σ), apparent frequency (n), amplitude
with respect to the long wave (ξ), and energy density (E).
The only explicit impact of the wind is thru a constant
surface drift current (v_d). Finally, the presence of the
long wave is felt indirectly via its orbital velocity (v)
and vertical acceleration (Z_{tt}); the long wave amplitude is
Z. Altogether, this becomes

$$\frac{\partial}{\partial t}\left(\frac{E}{\sigma}\right) + \frac{\partial}{\partial x}\left([c_g + v + v_d]\frac{E}{\sigma}\right) = 0 \tag{1}$$

$$\frac{\partial k}{\partial t} + \frac{\partial n}{\partial x} = 0 \tag{2}$$

$$\sigma = \sqrt{g(x,t)k + \gamma k^3} \tag{3}$$

$$n = \sigma + k(v+v_d) \tag{4}$$

$$E = \frac{1}{2}\{\rho g(x,t)\overline{\xi^2} + \rho\gamma\overline{(\xi_x)^2}\} \tag{5}$$

$$g(x,t) = \overline{g} + Z_{tt}(x,t) \tag{6}$$

Apparent gravity (as sensed by the short wave) is given by
g(x,t); $\overline{g}$ denotes normal gravity. γ is surface tension, and
ρ is water density.

There are four assumptions underlying this analysis; two are
benign, two are severe. It is assumed that there is a substantial scale separation between long and short waves, and
that the long wave slope is small. These are the benign
assumptions. They allow one to compute v and Z easily, and
to replace the formally correct (but very cumbersome) linearization of the short waves about Z with a linearization
about the mean water level, using apparent gravity.

The severe assumptions are that the short wave system can be
linearized, and is one-dimensional in the direction of the
long wave propagation. The only truly defensible justification is the resulting simplicity and clarity, which hopefully lay a solid foundation for future analyses.

The generalization of the above to a wavenumber spectrum is
straightforward, and follows Hughes directly. The wavenumber spectrum for the short waves with respect to the long
wave surface (Z) is

$$<\xi^2(x,t)> = \int_k \psi(k;x,t)\,dk \tag{7}$$

where $<\xi^2>$ is a phase average with respect to long wave phase. Remember that this is a two scale problem; $\psi(k;x,t)$ varies rapidly with respect to k and "slowly" with respect to x and t. The phase averaged energy density and action density are given by

$$<E> = \int_k [\rho g(x,t) + \rho\gamma k^2]\psi(k;x,t)\,dk \tag{8}$$

$$<A> = \int_k \frac{[\rho g(x,t) + \rho\gamma k^2]\psi(k;x,t)\,dk}{\sigma} = \int_k \rho c\psi\,dk \tag{9}$$

Suppose the wavenumber variation to be given by

$$k = \kappa(k_o;x,t) \tag{10}$$

where k_o is the mean water level value. The action integral can now be taken over a fixed short wave wavenumber range:

$$<A> = \int_{k_o} \rho c(\kappa(k_o;x,t);x,t)\psi(\kappa(k_o;x,t);x,t)\frac{d\kappa}{dk_o}\,dk_o \tag{11}$$

The generalization of (1) is gotten by assuming each element of the integral is separately conserved:

$$\frac{\partial}{\partial t}\left(\rho c\psi\frac{d\kappa}{dk_o}\right) + \frac{\partial}{\partial x}\left((c_g+v+v_d)\,\rho c\psi\,\frac{d\kappa}{dk_o}\right) = 0 \tag{12}$$

The behavior of $(d\kappa/dk_o)$ is obtained from (2):

$$\frac{\partial}{\partial t}\left(\frac{d\kappa}{dk_o}\right) + \frac{\partial}{\partial x}\left(\frac{dn}{dk}\frac{d\kappa}{dk_o}\right) = 0 \tag{13}$$

where $\quad \dfrac{dn}{dk} = c_g + v+v_d \tag{14}$

Finally,

$$\frac{\partial (c\psi)}{\partial t} + (c_g + v + v_d) \frac{\partial}{\partial x} (c\psi) = 0 \tag{15}$$

A parallel derivation for a frequency spectrum is straight forward. First, the local spectrum is represented by

$$<\xi^2(x,t)> = \int_n \phi(n;x,t) \, dn \tag{16}$$

whence

$$<A> = \int_n \rho c\phi \, dn \tag{17}$$

Suppose the apparent frequency variation to be given by

$$n = (n_o;x,t) \tag{18}$$

(This simple statement is a direct result of the assumption that the short waves are one dimensional. For two dimensional waves, $n=v'(n_o,\theta_o;x,t)$, and where θ_o is the mean water level heading of the wave.)

Then

$$<A> = \int_{n_o} \rho c\phi \, \frac{dv}{dn_o} \, dn_o \tag{19}$$

Hence

$$\frac{\partial}{\partial t} \left(\rho c\phi \frac{dv}{dn_o} \right) + \frac{\partial}{\partial x} \left((c_g + v + v_d) \, \rho c\phi \, \frac{dv}{dn_o} \right) = 0 \tag{20}$$

The behavior of (dv/dn_o) is different from that of $(d\kappa/dk_o)$ in an important way:

$$\frac{\partial}{\partial t} \left[\frac{dk}{dn} \frac{dv}{dn_o} \right] + \frac{\partial}{\partial x} \left[\frac{dv}{dn_o} \right] = 0 \tag{21}$$

Thus

$$\frac{\partial}{\partial t} [c(c_g + v + v_d)\phi] + (c_g + v + v_d) \frac{\partial}{\partial x} (c(c_g + v + v_d)\phi] = 0 \tag{22}$$

This relation and its contrast to (15) constitute the principal result of this paper. A similar result is discussed
by Richter and Rosenthal (1981). The contrast lies in the
presence in (22) of the energy flux velocity, $c_g + v + v_d$; the
effect of this term is most easily seen in the case of a
steady long wave moving at speed C_{LW}. Equations (15) and
(22) become

$$\frac{\psi(k;\chi)}{\psi(k;o)} = \frac{c(k_o;o)}{c(k;\chi)} \cdot \frac{\psi(k_o;o)}{\psi(k;o)} = \frac{c(k_o;o)}{c(k;\chi)} \cdot \left(\frac{k_o}{k}\right)^{-p} \qquad (23)$$

$$\frac{\phi(n;\chi)}{\phi(n;o)} = \frac{c(k_o;o)}{c(k;\chi)} \cdot \left(\frac{c_g(k_o;o) + v_d}{c_g(k;\chi) + v + v_d}\right) \cdot \frac{\phi(n_o;o)}{\phi(n;o)} \qquad (24)$$

$$= \frac{c(k_o;o)}{c(k;\chi)} \cdot \left(\frac{c_g(k_o;o) + v_d}{c_g(k;\chi) + v + v_d}\right) \cdot \left(\frac{n_o}{n}\right)^{-q}$$

assuming MWL spectral shapes $\psi \sim k^{-p}$, $\phi \sim n^{-q}$. The long wave
phase is denoted by χ, where $\chi = x - C_{LW} t$.

3.   Discussion

For wavenumber spectra, $(k_o/k)^{-p}$ - the "spectral shape"
term - can easily account for 90% of the measured modulation,
even though k variations are not especially large.

Measurements of the modulation of short wave frequency spectra are often corrected for the effect of the "spectral
shape" term, $(n_o/n)^{-q}$, since variations in n are quite large
and would otherwise totally dominate the measurement. Less
widely recognized is the impact of the middle term, governed
by the variations in the energy flux velocity. When v and
c_g are of opposite sign, this term can become quite large.
This apparently dominated the measurements of Reece (1978).

The implications for the wind dependence of short wave spectral modulations are fairly clear. Variations of "spectral
shape" and v_d with wind speed may well dominate any direct
influence the wind may have on actual short wave amplitude
modulations. In fact, if the direct influence of the wind
is modelled as a relaxation process, the relative importance
of spectral shape and v_d is actually heightened. The reason
is simply that relaxation tends to reduce the short wave amplitude modulation, forcing any spectral modulation to be
almost totally due to something else.

4.   References

Hughes, B.A., 1978: The effect of internal waves on sur-
    face wind waves. 2. Theoretical analysis, J. Geophys.
    Res., 83, 455-465.

Irvine, D.E., 1983: The interaction of long waves and short
    waves in the presence of wind, Ph.D. Dissertation, The
    Johns Hopkins University.

Reece, A.M., 1978: Modulation of short waves by long waves,
    Boundary Layer Meteor., 13, 203-214.

Richter, K. and Rosenthal, 1981: Energy distribution of
    waves above 1 Hz on long wind waves, given at the IUCRM
    Symposium of Wave Dynamics and Radio Probing of the
    Ocean Surface, Miami, May 13-20.

THE EFFECT OF SURFACE CONTAMINATION ON THE DRIFT VELOCITY OF WATER WAVES

Kewal K. Puri and Bryan R. Pearce
University of Maine at Orono
Orono, Maine 04469

ABSTRACT. This paper presents a study on the effect of surface contamination on the damping of progressive waves and on the second order Eulerian drift. The results are shown to subsume those obtained by Phillips in the limiting case of an inextensible film. The quasi-steady states are examined.

Introduction. Longuet-Higgins (1953) developed a viscous model to analyze the drift of fluid particles, induced by a surface wave. As evidenced by the empirical studies of Bagnold (1947), Russell and Osorio (1957) and others, his results, since extended by Dore (1978, 1979) and others, marked a major improvement over those of Stokes (1947) for the corresponding case of the inviscid fluid. Longuet-Higgins' theory, however, does not admit viscous attenuation of the waves and is directed to an uncontaminated fluid. Liu and Davis (1977) admitted the former and Craik (1982) studied the effect of contamination following the assumption of Phillips (1977) that the surface adsorption forms an inextensible film.

Rather than following the 'film' theory, the focus in this paper is to assume the presence of a soluble or insoluble material on the free surface and to explore its effects on the drift velocities induced by a progressive wave on the free surface as well as on the quasi-steady solutions in the interior.

2. Formulation of the problem

A two dimensional wave motion is considered on the contaminated free surface of an incompressible fluid of depth d, density ρ and a small kinematic viscosity ν. The contamination consists of a monomolecular surfactant adsorption $\Gamma(x,\zeta,t)$ interacting with an underlying bulk concentration $c(x,z,t)$, both dimensionless and having equilibrium values, respectively, Γ_0 and 1. The origin is taken at the mean level, z-axis is directed vertically upward and x-axis is in the direction of the propogating wave,

135

Y. Toba and H. Mitsuyasu (eds.), The Ocean Surface, 135–143.
© 1985 by D. Reidel Publishing Company.

$$\zeta = Re[\alpha \exp\{i(x-t)\}] \qquad\qquad (2.1)$$

where the lengths and the time are referred to the scales defined by the inverse wave number, k^{-1} and the inverse frequency ω^{-1}. These reference scales are also used to non-dimensionalize all other variables that appear in the sequel. The equations of motion in terms of the dimensionless velocity vector $\vec{q} = (u,w)$ and the pressure p are,

$$\vec{q}_t + (\vec{q} \cdot \nabla)\vec{q} = -\nabla p + .5\varepsilon\nabla^2\vec{q} - F, \quad \nabla \cdot \vec{q} = 0 \qquad\qquad (2.2,2.3)$$

where $F = gk/\omega^2$, $\varepsilon = 2k^2\nu/\omega$ and $\nabla = (\partial/\partial x, \partial/\partial z)$. These are to be solved subject to the boundary conditions,

$$w = \zeta_t + u\zeta_x \quad \text{at} \quad z = \zeta(x,t) \qquad\qquad (2.4)$$

$$-p + (\varepsilon/L)[w_z - \zeta_x(u_z + w_x) + \zeta_x^2 u_x] = \kappa\sigma \qquad \text{at } z = \zeta(x,t) \quad (2.5)$$

$$\varepsilon(w_z - u_x) + \frac{\varepsilon}{2}(1-\xi_x^2)(u_z + w_x) = L(\sigma_x + \zeta_x\sigma_z) \text{ at } z = \zeta(x,t) \qquad (2.6)$$

where κ is the surface curvature, $L = \sqrt{1+\zeta_x^2}$ and $\sigma(x,\zeta,t)$ is the varible surface tension. The concentrations in the bulk and on the surface are governed by

$$c_t + \vec{q} \cdot \nabla c = D\nabla^2 c, \quad \text{and} \qquad\qquad (2.7)$$

$$\Gamma_t + u\Gamma_x + w\Gamma_z = -(\Gamma/L^2)\{u_x + \zeta_x(u_z + w_x) + \zeta_x^2 w_z\} -$$

$$- (D/2L)(c_z - c_x\zeta_x)|_{z=\zeta} \qquad\qquad (2.8)$$

where D is the diffusion constant.
    Finally at the bottom, $u = 0 = w$, $z = -kd$ \qquad\qquad (2.9)

$$\text{At } t = 0 \quad \zeta(x,0) = \alpha\cos x + 0(\alpha^2) \qquad\qquad (2.10)$$

The imposition of the wave immediately leads to the establishment of potential flow, given by the equations (2.2)-(2.5) with $\varepsilon = 0$. The periodic motion of the fluid then leads to the formation of boundary layers at the bottom and at the free surface with the potential flow outside these layers. Thus the motion starts off with the potential flow and, as is commonly done, the field variables are assumed to admit the following expansions in powers of α:

$$\vec{q} = \sum_{n=1}^{\infty} \alpha^n \vec{q}_n, \quad \zeta = \sum_{n=1}^{\infty} \alpha^n \zeta_n, \quad p = \sum_{n=0}^{\infty} \alpha^n p_n \qquad\qquad (2.11)$$

3. Potential flow and the bottom boundary layer

The results pertaining to the irrotational theory and the bottom boundary layer are well known and are reproducted here from Liu and Davis (1977). Thus we have, for the potential flow,

$$O(\alpha)\zeta_1 = A(\tilde{t})\cos(x - t) \tag{3.1}$$

$$\hat{u}_1 = A(\tilde{t})\,\frac{\cosh(z + kd)}{\sinh\,kd}\,\cos(x - t) \tag{3.2}$$

$$\hat{w}_1 = A(\tilde{t})\,\frac{\sinh(z + kd)}{\sinh\,kd}\,\sin(x - t) \tag{3.3}$$

$$\hat{p}_1 = A(\tilde{t})\,\frac{\cosh(z + kd)}{\sinh\,kd}\,\cos(x - t) \tag{3.4}$$

$$O(\alpha^2)\zeta_1 = \frac{1}{2}A^2(\tilde{t})D\cos 2(x - t) \tag{3.5}$$

$$\hat{u}_2 = 2A^2(\tilde{t})B\,\frac{\cosh\,2(z + kd)}{\sinh\,2\,kd}\,\cos 2(x - t) \tag{3.6}$$

$$\hat{w}_2 = 2A^2(\tilde{t})B\,\frac{\sinh\,2(x + kd)}{\sinh\,2\,kd}\,\sin 2(x - t) \tag{3.7}$$

$$\text{with } D = \frac{2 + \cosh 2\,kd}{\tanh^2 kd\,\sinh 2\,kd},\ B = \frac{3(\coth kd - \tanh kd)}{4\,\tanh^2 kd} \tag{3.8}$$

The amplitude $A(\tilde{t})$, with initial value 1, is the envelope function and $\tilde{t}$, the slow time scale associated with the wave attenuation due to viscosity. Liu & Davis has calculated this by using a two-time asymptotic procedure. Their result, however, is true only in the case of infinite depth. In appendix I, an alternate procedure is presented that yields a uniform asymptotic expansion valid for the entire $O(\alpha)$ flow field. $A(\tilde{t})$ then can be determined therefrom as,

$$A(\tilde{t}) = \exp(-a\tilde{t}),\ a = (2\sinh 2\,kd)^{-1},\ \tilde{t} = \sqrt{\varepsilon}t. \tag{3.9}$$

The corresponding results for the bottom boundary layer, under the assumption that the surface contamination does not affect its structure, are

$$O(\alpha)u_1 = \frac{A(\tilde{t})}{\sinh\,kd}\{\cos(x - t) - e^{-\eta}\cos(x - t + \eta)\} \tag{3.10}$$

$$w_1 = \frac{A(\tilde{t})}{\sinh\,kd}\{\eta\sin(x - t) + \frac{e^{-\eta}}{2}\cos(x - t + \eta) +$$

$$+ .5e^{-\eta}\sin(x - t + \eta) - .5\cos(x - t) - .5\sin(x - t)\} \tag{3.11}$$

$$p_1 = \frac{A(\tilde{t})}{\sinh\,kd}\cos(x - t)\ \text{where } \eta = (z + kd)/\sqrt{\varepsilon}\ \text{for } \varepsilon \to 0. \tag{3.12}$$

The Eulerian steady-drift arises as a result of Reynolds stresses generated by $O(\alpha)$ approximations (3.10)-(3.12). (A comprehensive discussion of this aspect of the theory has been provided by Craik (1982)). The calculations by Liu and Davis show,

$$\bar{u}_2 = \frac{e^{-2a\tilde{t}}}{\sinh\,kd}[\frac{3}{4} - \frac{\eta + 2}{2}e^{-\eta}\cos\eta + \frac{1}{4}e^{-2\eta} - \frac{\eta - 1}{2}e^{-\eta}\sin\eta] \tag{3.13}$$

As $\eta \to \infty$, $\bar{u}_2 \to .75\exp(-2a\tilde{t})/\sinh^2 kd$ \tag{3.14} which is the $O(\alpha^2)$ drift across the bottom boundary layer, obtained by Phillips (1977).

4.   The boundary layer on the contaminated surface

The appropriate equations for the surface boundary layer are adopted from Longuet-Higgins (1953) by using the stretching transformations $(n,\tilde{v}) = \sqrt{\varepsilon}(N,\tilde{V})$.   Here (n,s,t) are curvilinear coordinates where s is measured along the free surface n is normal, drawn positive in the out-ward direction and the free surface is given by $n = 0$. $(\tilde{u},\tilde{v})$ represent the velocity components along the tangent and the normal directions re-spectively.   Furthermore by introducing the approximate relation, $u = -1 + \tilde{U}$, $\tilde{U} \ll 1$, these equations are cast in a frame of reference moving with the wave so that the resulting motion is steady. They are to be solved subject to the following boundary conditions:

the kinematic condition, $\tilde{V} = 0$ $n = 0$ (4.1)

the tangential boundary condition,

$\varepsilon[(1/\sqrt{\varepsilon}) + \sqrt{\varepsilon}\ \tilde{V}s - \kappa\tilde{u}] = \tau$ $n = 0$ (4.2)

where τ is the stress imposed by the film on the surface and $\kappa = \alpha\kappa_1$ is the curvature of the wave profile.   Also the solution must match the potential flow near the lower edge of the layer.
    Aside from these, we must satisfy the equations (2.7) and (2.8). The latter linearized about their equilibrium values have been used in Appendix II to determine

$\tau = - .5\sqrt{\varepsilon}\ (1 - i)K\tilde{U}$ (4.3)

where $\tilde{U} = \alpha\hat{u} + \sqrt{\varepsilon}(\alpha U_2 + \alpha^2 U_3) + \ldots$ (4.4)

$K = \xi[\xi - 1 - i(1 + \mu)]^{-1}$ and $\xi = - 2\tilde{\Gamma}_0\ \sigma_{\tilde{\Gamma}}/\sqrt{\varepsilon}$ (4.5,4.6)

where ξ is a measure of the surface compressional modulus . $\tilde{\Gamma}_0$ is the value in equilibrium of the adsorbed material.   The tangential boundary condition, then, becomes

$(1/\sqrt{\varepsilon})\tilde{u}_s + (\sqrt{\varepsilon}\tilde{V}_s - \kappa\tilde{u}) = - \tilde{\Lambda}\tilde{U}$ (4.7)

$\tilde{\Lambda} = \Lambda/\sqrt{\varepsilon}, \Lambda = .5(1 - i)k = \Lambda_1 + i\Lambda_2$ (4.8)

Substituting the expansions (4.8) in the momentum and the continuity equations, it is easy to check that the $O(\alpha)$ system is identically sat-isfied.
    The terms of the order $O(\alpha\sqrt{\varepsilon})$, yield

$$\tilde{U}_2 = .5A(\tilde{t})e^N \coth kd[\tilde{\Lambda}_1(\sin(s - N) - \cos(s - N)] +$$

$$+ \tilde{\Lambda}_2[\cos(s - N) + \sin(s - N))] + A(\tilde{t})e^N[\sin(s - N) - \cos(s - N] +$$

$$+ A(\tilde{t})N \cos s \qquad\qquad (4.9)$$

$$\tilde{V}_2 = .5A(\tilde{t}) \coth kd[\tilde{\Lambda}_1(\cos s - e^N \cos(s - N)) + \tilde{\Lambda}_2(e^N \sin(s - N) -$$

$$- \sin s)] + A(\tilde{t})[.5N^2 \sin s - e^N \cos(s - N) + \cos s] \qquad (4.10)$$

and the pressure given by the potential theory, is, $p_2 = A(\tilde{t})N \cos s$.

Finally, the steady non-oscillatory component of the $0(\alpha^2\sqrt{\epsilon})$ tangential velocity gradient, induced by $0(\alpha\sqrt{\epsilon})$ Reynolds stress is given by,

$$\tilde{u}_{3N} = .5A^2(\tilde{t}) \cot^2 h\, kd[\tilde{\Lambda}_1\{1 - (N \sin N + (1 - N)\cos N)e^N\} +$$

$$+ \tilde{\Lambda}_2\{(N \cos N + (N - 1)\sin N)e^N] +$$

$$+ A^2(\tilde{t}) \coth kd[1.5 - e^N(N \sin N + (1 - N)\cos N)]$$

$$\rightarrow .5A^2(\tilde{t}) \coth kd\, \tilde{\Lambda}_1 + 1.5A^2(\tilde{t}) \coth^2 kd \text{ as } \eta \rightarrow -\infty \qquad (4.11)$$

The above equation, in conjunction with the tansformation

$$s = x - \alpha z \sin x, \quad n = z - \alpha \cos x$$

yields $\overline{\tilde{u}}_{3n} \rightarrow .5A^2(\tilde{t}) \coth kd\, \tilde{\Lambda}_1 + 2A^2(\tilde{t}) \cot^2 h\, kd$ as $N \rightarrow -\infty$

5. The interior motion

The initial motion in the interior is irrotational and is described by the equations (3.1) to (3.8). The vorticity, diffusing on this domain from the two boundary layers, however, changes its character to that of a rotational flow. Its determination entails solving, the momentum equation, averaged over a wave length,

$$\overline{U}_{et} - \nu\overline{U}_{ezz} = -\frac{1}{\rho}\overline{p}_x \qquad\qquad (5.1)$$

together with the dimensional boundary conditions (3.14) and (4.17), and the initial condition $\overline{U}_e(z,0) = 0$. $\qquad\qquad (5.2)$

$\overline{U}_e$ is the dimensional equivalent of drift velocities found above and the decay rates $\sigma, \tilde{\sigma}$ are given by the equations (I.6) and (6.1) below. Henceforth $\overline{p}_x$ is taken to be zero. This would amount to allowing non-zero mass flow at each instant in an unbounded domain (Craik 1982).

A particular solution of this problem is

$$\overline{U}_p = e^{-2\tilde{\sigma}t} a^2 \omega k^2 (\frac{1}{2}T^2\tilde{\Lambda}_1 + 2T) \frac{\sin \tilde{\delta}(z + d)}{\tilde{\delta} \cos \tilde{\delta}d} + \frac{3}{4} \frac{\alpha^2 k\omega}{\sinh^2 kd} \frac{\cos \delta z}{\cos \delta d} e^{-2\sigma t} \qquad (5.3)$$

where $\tilde{\delta} = \sqrt{2\tilde{\sigma}/\nu}$ and $\delta = \sqrt{2\sigma/\nu}$ (5.4)

This solution is invalid at depths,

$$d = .5\pi(2n - 1)/\tilde{\delta}, \ .5\pi(2n - 1)/\delta \quad n = 1,2,\ldots \quad (5.5)$$

To rectify this situation Craik employed eigenfunctions of the corresponding homogeneous problem. The same procedure is adopted here. The associated homogeneous problem is,

$$\bar{U}_{et} - \nu\bar{U}_{ezz} = 0$$

$$\bar{U}_c(-d,t) = 0 \quad \bar{U}_z(0,t) = 0, \quad \bar{U}_c(z,0) = -\bar{U}_p(z,0)$$

we write $\displaystyle \bar{U}_c(z,t) = \sum_{n=1}^{\infty} A_n \exp(-\lambda_n^2 \nu t)\cos \lambda nz$

where $\displaystyle A_n = -\frac{2}{d} \int_{-d}^{0} \bar{U}_p(z,0)\cos \lambda_n z \ dz$

Integrating, we have

$$A_n = -\frac{\alpha^2 \omega k}{d} \left[\frac{k(T^2\tilde{\Lambda}_1 + 4T)}{\tilde{\delta}^2 - \lambda_n^2} + \frac{3}{2}\frac{(-1)^n \lambda_n}{\sinh^2 kd(\delta^2 - \lambda_n^2)}\right], \ \lambda_n = -\frac{\pi}{d}(\frac{2n-1}{2}) \ n=1,2.$$

The singularities of these coefficients are located at the same depths and have the same order as given in the equation (5.5). As such they can be resolved using L'Hopital's rule. For example, if the coefficient A_M is singular for $\lambda_M = \delta$, then the above procedure i.e., adding the two terms corresponding to $n = M$ together results in the sum to be of the form,

$$\frac{\exp\{-\nu\lambda_M^2 t\}}{\sin \lambda_M d} \ [a_1 zd\lambda_M \sin\lambda_M d + a_2\lambda M^2 t \cos \lambda_M d +$$

$$+ a_3\lambda_M^2 t \cos \lambda_M d \sin \lambda_M d + a_4\lambda_M \cos^2 \lambda_M d]$$

6. The damping rate

The results in Appendix II allow us to calculate the damping rate $\tilde{\sigma}$, where, $\tilde{\sigma}$ = [mean energy dissipation per unit area]/ [total energy density per unit area]. Mean energy dissipation per unit area = $D_1 + D_2$ where D_1 is the rate of dissipation due to viscosity and D_2 is that due to work done against the shearing stress in the film.

The total energy density per unit area = $(2k)^{-1}\rho\omega^2\alpha^2\coth kd$.

$$D_1 = -\rho\nu\int_{-d}^{0} (\overline{\text{vorticity}})^2 dn \approx -\rho\nu\int_{-dk}^{0} (\frac{\partial\tilde{u}}{\partial n})^2 dn, \text{ which implies}$$

$$D_1 \simeq -5/2 \; 2\rho\nu^{1/2}\omega^{5/2}\tilde{\alpha}^2 \coth^2 kd (\Lambda_1^2 + \Lambda_2^2) + 0(\epsilon), \; \tilde{\alpha} = k^{-1}\alpha$$

Also from (II.13) $D_2 = (\text{Re } \tau)\tilde{u}_2 = -.25\tilde{\alpha}^2 \rho\omega\sqrt{2\omega\nu}(\Lambda_1^2 - \Lambda_1\Lambda_2)\coth^2 kd.$

$$\text{Hence } \tilde{\sigma} = \frac{\sqrt{2\omega\nu}}{4} k \coth kd [\Lambda_1 - \Lambda_2)^2 + 2\Lambda_1^2] \tag{6.1}$$

For the insoluble film, we have, $\mu = 0$ in (6.1). Clearly the damping rate in this case is much higher than that in the case of an uncontaminated surface. When $\xi \to \infty$, (6.1) also establishes the result of Phillips (1977) for the case of a thin inextensible slick.

Appendix I

We substitute the expansions (2.1) in (2.2), (2.4)-(2.6) and (2.9), retain $0(\alpha)$ terms and regard, in the resulting equations, time $t = t(t^*, \tilde{t})$ where $t^* = t(1 + \epsilon\omega_1 + \dots)$ is associated with the wave oscillations and $\tilde{t} = t\sqrt{\epsilon}$ is associated with the wave attenuation. The resulting equation to be solved, stated in terms of the stream function ψ, is

$$[\frac{\epsilon}{2}\psi_{zzzz} - \{\epsilon + i(1 + \epsilon\omega_1 + \dots) + \sqrt{\epsilon}\frac{\partial}{\partial\tilde{t}}\}\psi_{zz} + \{i(1 + \epsilon\omega_1 + \dots) +$$

$$+ \sqrt{\epsilon}\frac{\partial}{\partial\tilde{t}} + \frac{\epsilon}{2}\}\psi = 0 \tag{I.1}$$

$$\text{Let } \psi = \psi_0 + \sqrt{\epsilon}\psi_1 \tag{I.2}$$

$$\psi_0 = e^{i(t^*-x)}\{f_0(t)E_1 + g_0(\tilde{t})E_2\} \tag{I.3}$$

$$\psi_1 = e^{i(t^*-x)}\{f_1(\tilde{t})E_1 + g_1(\tilde{t})E_2 + e_1(\tilde{t})e^{-\sqrt{2i\eta}}\}, \text{ etc.} \tag{I.4}$$

$$E_1 = \exp(z + kd), \; E_2 = \exp\{-(z + kd)\} \text{ and } \eta = (z + kd)/\sqrt{\epsilon}$$

(I.1) together with (I.2)-(I.4) and the boundary conditions, yield

$$\psi_0 = A(\tilde{t})\frac{\sinh(z + kd)}{\sinh kd}e^{i(\omega t - x)}$$

where $\quad A(\tilde{t}) = e^{-a\tilde{t}}, \; a = (2\sinh 2kd)^{-1}; \; \omega = 1 - a\sqrt{\epsilon} \simeq 1.$

$$= e^{-\sigma t}, \; \sigma = .5\sqrt{2\nu\omega} \, k \, \text{cosech } 2 kd$$

The solution (3.10) can be obtained from $\psi_0 + \sqrt{\epsilon}\psi$ by first replacing z by η, using (I.4), expanding the resulting expressions in powers of $\sqrt{\epsilon}$.

Appendix II

Writing $c = 1 + \tilde{c}$, $\Gamma = \Gamma_0 + \tilde{\Gamma}$ in the equation (2.7), (2.8), retaining linear terms and changing to the moving coordinates, we obtain,

$$-\frac{\partial \tilde{c}}{\partial s} \approx \frac{D}{2} \frac{\partial^2 \tilde{c}}{\partial n^2}, \text{ and } -\frac{\partial \tilde{\Gamma}}{\partial s} + \Gamma_0 \frac{\partial \tilde{U}}{\partial s} = -\frac{D}{2} \frac{\partial \tilde{c}}{\partial n}\bigg|_{n=0} \qquad (II.1, II.2)$$

We assume that,
    (i) the local equilibrium between the surface and the subsurface
        solution is constantly maintained (Davis and Vos, 1965);
    (ii) the surface tension $\sigma_0 + \tilde{\sigma}$ and the adsorption $\tilde{\Gamma} + \Gamma_0$ are com-
        pletely determined by the surface value c_s of the bulk con-
        centration c;
    (iii) the concentration is very dilute and $D \ll \sqrt{\epsilon}$.
Now Gibb's adsorption isotherm is given by,

$$-d\tilde{\sigma} = \check{R}T\Gamma d \ln c_s \qquad (II.3)$$

where T is the absolute temperature. $\check{R} = Rc_0 k^2/\rho\omega^2$ is the nondimensional
gas constant, R,s signifies the surface values, and $c_0 = c$, in equilibrium position.
    (II.3) together with the Young's isotherm, $\Gamma = \Gamma_\infty c_s/(c_s + a)$ (II.4)

$$\text{yields, } d\tilde{\sigma} = \tilde{R}T\Gamma_\infty d[\ln(1 + c_s/a] \qquad (II.5)$$

where Γ_∞ is the saturation adsorption of surfactant and a is the value
of c_s at which $\tilde{\Gamma}$ attains half its saturation value. Using the assumption (ii) we can write,

$$\tilde{\Gamma} = \lambda \tilde{c}_s, \quad \lambda = \frac{d\Gamma}{d\tilde{c}}\bigg|_s, \quad \nabla \tilde{\sigma} = \frac{d\tilde{\sigma}}{d\Gamma} \nabla \tilde{\Gamma} \qquad (II.6)$$

We define $\chi = -\tilde{\Gamma}_s d\tilde{\sigma}/d\tilde{\Gamma}\big|_s = -\check{R}T\Gamma_\infty c_s/a$, using (II.3) & (II.4).

Also from (II.6) & (II.4), $\lambda = d\tilde{\Gamma}/d\tilde{c} \approx \Gamma_\infty a/(1 + a)^2$.   (II.7)

Writing, $(\tilde{c}, \tilde{\Gamma}) = e^{is}(\tilde{c}, \tilde{\Gamma})$ and using (II.1), (II.2) results in   (II.8)

$$\tilde{\Gamma} = -i\gamma_0 \frac{\partial \tilde{u}}{\partial s}\bigg|_{n=0}, \quad \tilde{c} = -\frac{i\gamma_0}{\lambda} \frac{\partial \tilde{u}}{\partial s} \exp\{\sqrt{-2i/D}\,n\} \qquad (II.9)$$

$$\text{where } \gamma_0 = \Gamma_0[1 + (1-i)\frac{\mu}{2}]^{-1} \qquad (II.10)$$

The constant $\mu = \sqrt{D}/\lambda$ is the measure of solubility (Miles 1967). Now
neglecting dilatational and shear viscosities, the tangential stress is
given by

$$\tau = \frac{\partial \tilde{\sigma}}{\partial s} = \frac{d\tilde{\sigma}}{d\tilde{\Gamma}} \frac{\partial \tilde{\Gamma}}{\partial s} = \frac{d\tilde{\sigma}}{d\tilde{\Gamma}}(-i\gamma_0) \frac{\partial^2 \tilde{u}}{\partial s^2}\bigg|_{n=0} \qquad (II.11)$$

For $0(\alpha)$ linear problem, let the viscous perturbation to the irrotational
velocity $\hat{u}_1$ in the upper boundary layer be $\tilde{u}_1$ and let the actual tangential velocity be $\tilde{u}$ so that $\tilde{u} = \hat{u}_1 - \tilde{u}_1$.
    Then, following Miles, evaluating (II.12) at $z = 0$, and making use
of the fact that the velocity in the film must be parallel to that just

outside the boundary layer, we have $\tilde{u}_s = (1 - K)\hat{u}_1$ and $\tilde{u}_1 = K\hat{u}_1$,   (II.13)
where K is a constant to be determined such that the viscous stress in
the film, given by (II.11) must equal that in the liquid which is,
$\tau = - \sqrt{\varepsilon}/2(1 - i)\tilde{u}_1$.   Thus,

$$\frac{d\sigma}{d\tilde{\Gamma}}(- i\gamma_0)(1 - K) \left.\frac{\partial^2 \hat{u}_1}{\partial s^2}\right|_{n=0} = - \frac{\sqrt{\varepsilon}}{2}(1 - i)K\hat{u}_1\Big|_{n=0}$$

which has the solution, $K = \xi[\xi - 1 - i(1 + \mu)]^{-1}$, where $\exp(is)$ dependence
of u has been used and $\xi = - (2\tilde{\Gamma}_0/\sqrt{\varepsilon})(d\tilde{\sigma}/d\tilde{\Gamma})$ is a measure of surface compressional modulus. Thus to the leading term, $\tau = - .5\sqrt{\varepsilon}(1 - i)K\hat{u}_1$. This
implies, that an appropriate expansion of the tangential velocity is,
$U = \alpha\hat{u} + \sqrt{\varepsilon}(\alpha U_2 + \alpha^2 U_3) + \ldots$.

Acknowledgement

This work is the result of research sponsored by University of Maine
Sea Grant College Program under Grant number NA81-AA-D-00035.

References

Bagnold, R. A. (1947). *Sand movement by waves: small scale experiments
    with sand of very low density.* J. Inst. Civil Engr. London 27,
    447-469.
Craik, A. D. D. (1982). *The drift velocity of water waves.* J. Fluid
    Mech., 116, 187-205.
Davis, J. T. & Vos, R. M. (1965). *On the damping of capillary waves
    by surface films.* Proc. Roy. Soc. A. 288, 218-230.
Dore, B. D. (1975). *Wave induced vorticity in free surface boundary
    layers: applications to mass transport in edge waves.* J. Fluid
    Mech. 70, 257-266.
Dore, B. D. (1978). *A double layer model of mass transport in pro-
    gressive interfacial waves.* J. Engg. Math. 12, 289-301.
Liu, A-K & Davis, S. H. (1977). *Viscous attenuation of mean drift in
    water waves.* J. Fluid Mech. 81, 63-84.
Longuet-Higgins, M. S. (1953). *Mass transport in water waves.* Phil.
    Trans. Roy. Soc. A245, 535-581.
Phillips, O. M. (1977). *The dynamics of the upper ocean.* Camb. Univ.
    Press.
Russel, R. C. H. & Osorio, J. D. C. (1957). *An experimental investi-
    gation of drift profiles in a closed channel.* Proc. 6th Conf.
    Coastal Engg. Miami, 171-193.

MATHEMATICAL MODELING OF BREAKING WAVE STATISTICS

Roman E. Glazman
Graduate School of Oceanography
University of Rhode Island
Narragansett, R.I., 02882
U.S.A.

ABSTRACT. The breaking wave criterion concept is briefly reviewed and a general approach to parametrizing statistics of steep waves is formulated based on the theory of random field's high overshoots. Although no preliminary limitations are imposed on the wave spectrum width, some partial averaging of the original field is found necessary. An averaging procedure is then introduced, which yields an effective solution to the problem of high-order spectral moments.

1. INTRODUCTION

Basic statistics of breaking waves (BW) include the temporal and spatial frequency of BW occurrence, the relative area occupied by whitecaps, etc. Several attempts to model BW characteristics have been undertaken [5,11,13,16]. In doing so, an idea of a BW criterion, put forth in [10], has been used in one form or another. Presently, this concept is reviewed and a general formalism is developed aimed at parametrizing various statistics of events consisting in occurrence of very steep waves. Such events are referred to as BWs, and it is anticipated that at moderate winds they are closely related to actual BWs. We treat the steep waves as rare events occurring in the manner of a Poisson process. The theory pertains to the type of BWs caused by superposition of waves with different wavelengths and phase velocities, which sporadically leads to an extreme steepening of individual waves. The pattern of BWs described in [3] is of reference. It is known that wave slope statistics are determined by high-order spectral moments. Consequently, the problem of evaluating such moments, arising in the present as well as in many other works, e.g. [4,7,8,13,16],is treated in Sec.4.

2. TRUNCATION HYPOTHESIS

The common BW criteria impose a limitation on wave steepness. Due to a dispersion ratio, this limitation can be written either in terms of wave height (amplitude a) and period τ (frequency ω) or in terms of the a and wavelength. For a random sea, such parameters describe a wave envelope,

145

Y. Toba and H. Mitsuyasu (eds.), The Ocean Surface, 145–150.

the notion that is appropriate for a narrow-band spectrum. On the other
hand, the idea of a limiting wave shape, put forth originally for
deterministic plane waves, will be the more adequate, the closer the
wave field resembles such waves. In other words, both the critical
value, Y, and the very existence of the limiting wave slope appear to be
the less certain, the broader the (2-dimensional) wave spectrum.

The method of [10] may be viewed as a truncation approach that
commences by applying a BW criterion to a distribution $p_0(a,\tau)$
characterizing envelope statistics for a sea not yet experiencing BW
activity. Since such distributions are inaccessible to either
experimental or theoretical evaluation, the $p_0(..)$ is replaced with a
$p(..)$ pertaining to the actual sea. A controversy, seemingly caused by
this substitution, is resolved by noting that the BW events are rare, so
that the resultant dissipation of wave energy is negligibly small
compared to the total wave energy (in which case the difference δp
between p_0 and p would be virtually undetectable).

It is instructive to derive a result of [10] by invoking the
solution of Rice's problem of level crossings. Let us also relax the
[10]'s limitation on the spectrum width by employing statistics of the
surface slope record (equivalently, of the vertical acceleration record)
rather than those of the surface displacement one. Recall [9] that the
120° angle at the crest of the progressive Stokes wave corresponds to
$-g/2$ value of the local vertical acceleration. This permits writing the
BW criterion in the form

$$X(t) > u \text{ , where } X(t) = -\partial^2\zeta/\partial t^2 \text{ and } u = Yg \qquad (1)$$

(We do not insist on Y being 1/2, although hope it is a rather constant
quantity). Then the mean, per unit time, number of events consisting in
surpassing level u by random process $X(t)$ is given [15] by

$$n_t = (2\pi)^{-1}(M_{2,x}/M_{0,x})^{1/2}\exp(-u^2/2M_{0,x}) \quad , \qquad (2)$$

where $M_{i,x}$ are spectral moments of the ith order defined with respect to
the record $X(t)$. It is easy to show that $M_{i,x} = M_j$, $j=i+4$, where M_j are
spectral moments defined with respect to the surface displacement, $\zeta(t)$,
record. When the spectrum width is vanishingly small, we have [7]:
$\sqrt{(M_6/M_4)} \approx \sqrt{(M_4/M_2)} \approx \sqrt{(M_2/M_0)}$. In this case, Eq(2) reduces to

$$n_t \approx (2\pi)^{-1}(M_2/M_0)^{1/2}\exp(-\bar{A}^2/2M_0) \quad , \qquad (3)$$

where notation

$$\bar{A} = u/\bar{\omega}^2 \quad , \quad \bar{\omega}^2 = M_2/M_0 \qquad (4)$$

has been used. The exponential factor in Eqs(2),(3) represents the
probability, Q_t, of a BW event in a wave record (eq(6) gives its formal
definition), whereas the factors preceding Q_t form a reciprocal of the
mean wave period. Eqs(3),(4) are equivalent to Longuet-Higgins' [10]
approximation, which means that the latter reduces to imposing a
threshold on the wave height. (Nevertheless, in Sec.4 we find out that
the Longuet-Higgins' estimate of Q_t agrees surprisingly well with the
result of the present, less restrictive, approach.)

Rephrasing the argument of [10], we replace the joint probability density function (p.d.f.) $p(a,\omega)$ by $p(a)\delta(\omega-\bar{\omega})$, which is legitimate to do in a case of a vanishingly narrow spectrum width. Defining

$$\Theta(x) = \begin{cases} 0 & \text{if } x < 0 \\ 1 & \text{if } x \geq 0 \end{cases} \quad , \tag{5}$$

we find the probability Q_t of encountering a wave with $a > A$, where $A = u/\omega^2$ as

$$Q_t = N_b/N_T = \langle\Theta(a-A)\rangle = \int_0^\infty da \int_0^{2\pi\sqrt{(a/u)}} p(a,\tau)d\tau \tag{6}$$

Here, N_b is the mean number of such events in the record, N_T is the total number of all waves in this record, and the record is assumed to be sufficiently long $(N_T\to\infty)$. The $\langle...\rangle$ denotes a probabilistic averaging. Obviously, setting $p(a,\omega) = p(a)\delta(\omega-\bar{\omega})$ leads to the Longuet-Higgins' result:

$$Q_t = \langle\Theta(a-\bar{A})\rangle = \int_A^\infty p(a)da \tag{7}$$

The approach of [5,11] is equivalent to employing the last equality of Eq(6) rather than that of Eq(7).

    With $p(a,\tau)$ specified, one can evaluate statistics of any BW property F, provided the latter is expressible via a and τ: $\langle F^k\rangle = \langle F^k(a,\tau)\Theta(a-A)\rangle$, $k=1,2,..$. In [10] the flux density for energy lost by waves due to breaking was estimated by letting $F=(1/2)\rho g(a-A)$ and $k=1$.

3. FURTHER GENERALIZATION

    Using Eq(2) allows one to avoid the, generally difficult, problem of finding an adequate $p(a,\tau)$. Moreover, once the relationship between steep-wave statistics and that of level crossings is established, further progress can be made by employing appropriate results of the random field theory. Specifically, we are concerned with surface density and surface flux density of BW events. These quantities can be also referred to as spatial frequency, n_s, and temporal-spatial frequency, n_{ts}, of BW events, respectively. The former gives the mean number of BWs per unit area occurring simultaneously at any time instant, whereas the latter gives the mean BW number per unit·area occurring in unit time. Hence, we are interested in the mean rates of excursions by random 2-dimensional and 3-dimensional fields, $X(x_1,x_2)$ and $X(x_1,x_2,x_3)$ with $x_3=t$, above level u.

    If $u \gg \sigma$, with σ being the standard deviation of X, then we can employ asymptotic results for the case when the field's high excursions behave as a Poisson (n-dimensional) process. Such results have been first reported in [12]. They are reviewed in [1,6,17]. We shall write down only the final expressions, and only for a 2-d case, omitting calculations connected with the passage to the acceleration field. The rate of the 2-d field $X(x_1,x_2)$ excursions above level u is given by

$$n_s = (2\pi)^{-3/2}(\sqrt{(\det[\Lambda])}/M_4^{3/2}) \; u \; \exp(-u^2/2M_4) \; , \tag{8}$$

where $[\Lambda] = \{\partial^2 B(x_1,x_2)/\partial x_i \partial x_j\}_{x_1,x_2 \to 0}$,   (9)

with $B(\vec{x})=\langle X(\vec{y})X(\vec{y}+\vec{x})\rangle$, where $\vec{y}$ and $\vec{x}$ are vectors on the (x_1,x_2)-plane. Practical result is obtained by presenting $B(\vec{x})=\int\int\int\omega^4\Psi(\omega,\vec{k})\exp(i\vec{k}\vec{x})d\omega d\vec{k}$, where Ψ is a 3-d wave spectrum. In the simplest, yet quite useful, case of an isotropic sea one finds

$$\det[\Lambda] = \lim_{r\to 0}\{(dB/rdr)(d^2B/dr^2)\} ,   (10)$$

where coordinates have been transformed to polar. Let us express the results in terms of 1-d wave spectrum. Employing a deep-water dispersion ratio, as is done routinely in many works including [4], one finds, after rather lengthy calculations:

$$dB/rdr]_{r=\bar{0}} d^2B/dr^2]_{r=\bar{0}} (1/2)\int_0^\infty S_\zeta(\omega)(\omega^4/g)^2 d\omega   (11)$$

This yields: $\sqrt{(\det[\Lambda])} / M_4 = (2g^2)^{-1}(M_8/M_4)$ (12)

The role of BW probability in this case is played by $Q_s=(u/\sqrt{M_4})\exp(-u^2/2M_4)$. The remaining factors in Eq(8) specify the mean surface area occupied by an individual, 2-d, wave. In a 3-d case, the probability is again multiplied by the factor $u/\sqrt{M_4}$. Therefore, the BW probability notion is determined depending on what is considered to be the wave. If it is the mean oscillation cycle in a 1-d record, Eqs(2),(6) apply; for more complex geometry, statistics become more complicated. Various other parameters of BW events, for example, characteristics of their groupness, can be evaluated within the same theoretical framework. The reviews referenced above give the necessary relationships.

4.   SPECTRAL MOMENTS AND PARTIAL AVERAGING OF SEA SURFACE

    Interested in large-scale features of surface geometry, we must rid our results of the dramatic [2] impact of small-scale roughness on slope statistics. On the other hand, this is necessitated by the fact that any wave data, either experimental or theoretical, are fundamentally inadequate in the high frequency domain. Particularly, computing M_i on the basis of Pierson-Moskowitz (P.-M.), JONSWAP, or any other spectrum with Phillips' equilibrium range, leads to infinitely large M_i for all i $\geq$ 4. In terms of the random field theory this means that the $\vec{X}$-field does not satisfy necessary conditions of mean-square differentiability. Physically though, the discontinuity of the surface slope resulting in $M_4\to\infty$ is unrealizable due to the action of capillary pressure.
    A rather general method of resolving this type of difficulty has been proposed in [17]. It consists in averaging a random process as

$$\overline{\zeta}(t) = (1/T)\int_{t-T/2}^{t+T/2} \zeta(\tau)d\tau   (13)$$

to eliminate an impact of fluctuations with periods T or smaller. It is shown in [17] that the spectrum $S_\zeta(\omega)$ of the averaged wave record is

$$\bar{S}_\zeta(\omega) = S_\zeta(\omega) V(\omega T) \quad , \tag{14}$$

where

$$V(\omega T) = [\sin(\omega T/2)/(\omega T/2)]^2 \quad , \tag{15}$$

and the bar means that a given property pertains to the averaged field. In order to arrive at the spectrum $\bar{S}_x(\omega)$ of $\bar{X}(t)$, we note that the Vanmarcke's procedure, Eq(13), is equivalent to

$$\partial\bar\zeta/\partial t = [\zeta(t+T/2)-\zeta(t-T/2)] / T \tag{16}$$

Furthermore, we find that

$$\partial^2\bar\zeta/\partial t^2 = -[\zeta(t+T)-2\zeta(t)+\zeta(t-T)] / T^2 \tag{17}$$

Substituting Eq(17) into the definition of the autocorrelation function, we ultimately arrive at :

$$\bar{S}_x(\zeta) = \omega^4 S_\zeta(\omega) V^2(\omega T) \tag{18}$$

In a similar manner one relates the spectrum of spatial derivatives with the original wave spectrum. Each spatial differentiation corresponds to $\partial^2/\partial t^2$, and thus leads to adding a power of $V(.)$. As a particular result, the 8th order moment after averaging becomes

$$\bar{M}_8 = \int_0^\infty \omega^8 V^4(\omega T)S_\zeta(\omega)\,d\omega \tag{19}$$

Evidently, a rapid decay of $V(.)$ at $\omega T \gg 1$ ensures convergence of the corresponding integrals, wherein T can be as small as we please. How to specify the averaging period? Let us imply commonly used spectra, such as P.-M. and JONSWAP. Their similarity properties [14] must be preserved in our transformation to avoid conflict with the argument underlying the equilibrium range: we accept that BW probability is governed by the same factors as the Phillips spectrum. Therefore, no new parameters should be introduced into our equations in addition to those already contained in the wave spectra. Then, the smallest T extractable from a wave spectrum is based on the rate of change of the corresponding autocorrelation coefficient in the vicinity of the origin. If $\zeta(t)$ is a stationary random process, this rate is determined by the second-order derivative. Therefore

$$T = \sqrt{[-B(\tau)/(d^2B/d\tau^2)]}_{\tau\to 0} = \sqrt{(M_0/M_2)} \tag{20}$$

One readily recognizes this T to be the Taylor microscale.

　　As a particular result, the 1-D probability of BW events has been found to be $Q_t=\exp(-1/6.5984\beta)$, where the P.-M. spectrum and $Y =1/2$ have been employed to provide comparison with [10], and β is the Phillips constant. The simplification used in [10] leads to $Q_t = \exp(-1/2\pi\beta)$. Therefore, the P.-M. spectrum is sufficiently narrow, as was envisioned in [10], for the approximate Eq(7) to be suitable.

　　Concerning the limitations of the present approach, one expects the requirement of field's normality to be inessential. Indeed, high excursions are expected to "forget" the information about 3rd and higher moments of the original field. Besides, the local averaging, Eq(13), is known to greatly reduce the non-normality of random fields [17]. However, the assumption that Y is a "universal constant" (at low winds?) needs thorough experimental test.

ACKNOWLEDGEMENTS. I thank the U.S. National Science Foundation for covering the cost of my participation in the Symposium, and the University of Rhode Island for partial support of this work.

REFERENCES

1. Adler, R.J., 1981, The Geometry of Random Fields, Wiley, New York, 279
2. Cox, Ch., and W. Munk, 1954, J. Optical Soc. Am., 44, 838-850.
3. Donelan,M., M.S.Longuet-Higgins, and J.S.Turner, 1972. Nature,239,449-451.
4. Glazman, R.E., 1982, Radio Sci., 17, 635-642.
5. Houmb O.G. and T.Overvik, 1976, in Proc. BOSS'76, Vol.1, 144-169.
6. Khusu, A.P., Yu. R. Witenberg, and V.A. Palmov, 1975, Surface Roughness: Theoretical Probabilistic Approach. Nauka Press, Moscow, (in Russian).
7. Longuet-Higgins, M.S., 1957, Phil. Trans. Roy. Soc., A249, 321-387.
8. -- , 1962, The Statistical Geometry of Random Surfaces, in Proc. Symp. Appl. Math. XIII, Providence, R.I., pp. 105-143.
9. -- , 1963, J.Fluid Mech., 16, 138-159.
10. -- , 1969, Proc. Roy. Soc. A310, 151-159.
11. Nath, J.H., and F.L.Ramsey, 1976, Journ. Phys. Oce. 6, 316-323.
12. Nosko, V.P., 1969, Characteristics of Overshoots of a Homogeneous Gaussian Field Beyond High Levels. in Proc. Soviet-Japanese Symp. on Probability Theory, held in Khabarovsk, August, 1969. Nauka Press, Novosibirsk, 209-215 (in Russian).
13. Ochi, M.K., and Cheng-Han Tsai, 1983, Journ. Phys. Oceanogr., 14, 2008-2019.
14. Phillips, O.M., 1977, The Dynamics of the Upper Ocean. 2nd Edition, Cambridge Univ. Press, Cambridge, 336.
15. Rice, S.O.,1945, Mathematical Analysis of Random Noise. Bell System Tech. J.,24,46-156.
16. Snyder, R.L., and R.M. Kennedy, 1983, Journ. Phys. Oce., 13, 1482-1492.
17. Vanmarcke, E., 1983, Random Fields: Synthesis and Analysis, the MIT Press, Cambridge, Mass.,382.

NUMERICAL MODELING OF CURRENT-WAVE INTERACTION

D. T. Chen,[1] S. A. Piacsek,[2] and G. R. Valenzuela[1]
[1]Naval Research Laboratory
Washington, DC 20375, USA

[2]Naval Ocean Research and Development Activity
NSTL Station, MS 39529, USA

Abstract. Numerical models, utilizing field data taken during the Naval Research Laboratory (NRL) Phelps Bank Experiment of July 1982, are used for the evaluation of two-dimensional oceanic current velocity field, swell characteristics, modulation of Bragg resonant wavelet spectral wave energy density, and microwave radar cross-section over Phelps Bank for the simulation of radar imagery (SAR/SLAR) intensity. The objective of this work is to delineate the physical processes responsible for patterns in radar imagery (SAR/SLAR) that correlate with complex bottom topography. The analyses are still in progress and the general methodology being pursued is given here.

1. INTRODUCTION

SEASAT synthetic aperture radar (SAR) images of the ocean surface are known to contain abundant oceanic features ranging in sizes from mesoscale, such as eddies, to macroscale, such as decameter surface waves and internal waves (Beal et al., 1981). Among others, the SEASAT SAR image of Nantucket Shoals contains features related with the bottom topography.

These topography-related features observed in SEASAT SAR images, as well as those observed by de Loor (1981) with an EMI X band side-looking airborne radar (SLAR), cannot be the result of direct probing of the shallow water bottom by the electromagnetic waves because propagating microwave fields decay exponentially in sea water. For SEASAT L band (1.275 GHz) SAR this rate of decay in energy is approximately two-thirds for every centimeter of penetration. Comparing with the water depth at the image features of the order, generally, of meters to tens of meters, the patterns in radar imaging must be the result of surface effects due to hydrodynamic processes coupled to the bottom topography. These surface effects, primarily in the form of current gradients, modulate the amplitudes of the surface ocean waves whose wavelengths are comparable to the radar wavelengths (30 to 40 cm for L band SEASAT illumination angles), and these surface ocean waves are the main contributors (except for contributions from specular scattering) to the microwave backscatter power received.

In order to address the physical processes which are responsible for the topography-related surface features, the Naval Research Laboratory (NRL) conducted an experiment (Chen et al., 1985) over Phelps Bank off the coast of Massachusetts from 6 to 25 July 1982. In-situ measurements of bathymetry, current (≈ 1 m/s at Phelps Bank), density and temperature profiles, tide, directional surface waves, wind velocity, and air-sea temperature profiles and microwave measurements with X band SAR systems, and L and X band Remote Ocean Wave Spectrometer (ROWS) were taken. This work intends to use the collected data as input to a numerical

Y. Toba and H. Mitsuyasu (eds.), The Ocean Surface, 151–159.

circulation model for the computation of the current field. Then this current field is used together with the available wave measurements in a current-wave interaction model for the computation of the perturbed/modulated wave field everywhere. The perturbed wave field will be used to compute the intensity of SAR images by Bragg scattering with which the airborne X band SAR images taken during the experiment can be compared. From these series of comparisons the proper physical mechanism responsible for the topography-related surface features can be identified and, furthermore, the relative importance, to this physical mechanism, of such physical parameters as current, current gradient, air-sea temperature difference, depth, and wind speed can be evaluated.

2. EXPERIMENT DATA SET

The library of data collected during 1982 NRL Phelps Bank Experiment is shown in Table 1. This table indicates the locations, either at deep sea or Phelps Bank, of the experiment and classifies the data set into two categories: microwave measurements made from ship (*USNS Hayes*) or aircraft (Phantom RF-4's and NRL RP-3A) and measurements made in-situ. Cross "x" indicates that data are available and dash "—" indicates that data are not good enough. Detailed information on these data sets can be found in Chen et al. (1985). From in-situ measurements of current by Eulerian current meters array and Lagrangian drogues, tide field over the region has also been obtained.

Table 1 — Library of Data Collected During 1982 Phelps Bank Experiment
July 1982

	7	8	9	10	11	12	13	14	17	18	19	20	21	22	23	24
Remote																
X band rows	X		X			X	X	X			X		X	X	X	X
L band (ship) rows		X	X		X	X		X			X	X	X	X	X	X
L band (ship) ΔK								X			X	X	X	X		
Strip camera	X		X			X	X						X	X	X	
IR thermal scanner	X		X			X	X	X								
Laser profilometer	—		—		—	—	—				—	—	—	—	—	—
X band SAR					X		X									
In-Situ																
Clover-Leaf buoy		X	X		X	X	X	X			X	X	X		X	X
Endeco buoy		X	X		X			X								
Navigation		X	X	X	X	X	X	X	X	X	X	X	X	X	X	X
Bathymetry				X	X	X	X	X	X	X	X	X	X			
Current			E/L	E/L	E	E	CSD/E	E	E/L	E/L	E	FSD				
Temp/Salinity/Densi				X	X	X	X	X	X	X	X	X	X	X	X	
Air-Sea temp. diff.		X	X	X	X	X	X	X	X	X	X	X	X	X	X	X
Wind vector		X	X	X	X	X	X	X	X	X	X	X	X	X	X	X
Location	D.S.	D.S.	D.S.	Bank	Bank	Bank	Front	Bank	Front	Bank	Front	Bank	Front	D.S.	D.S.	D.S.

L = Lagrangian
E = Eulerian
D.S. = Deep Sea
CSD = Controlled Ship Drift
FSD = Free Ship Drift

In order to demonstrate, for example, the look-angle dependence of SAR images, Figure 1 shows three SAR swaths over Phelps Bank taken within 40 minutes of each other on July 11. Each swath ($\approx$18 km wide) is made of four subswaths ($\approx$4.5 km wide each) A. B, C, and D with subswath A the near edge. For additional details see Valenzuela et al. (1985). The arrows at the ends of swaths indicate flight directions. The location of *USNS Hayes* has been marked to the east of Phelps Bank. Tide and wind velocities at both east and west sides of Phelps Bank are also shown. Tide velocities at east and west sides are noticeably different. This difference in tide velocities produces current gradient and, presumably, is related to bathymetry variations due to the presence of Phelps Bank. The highly directional current gradient field modulates Bragg resonant wavelets, masked by air-sea temperature difference as well as by wind stress, whose spatial distribution and energy density yield different SAR images for different look angles of the same scene. Figure 2 shows portions of subswaths A and B of these SAR images

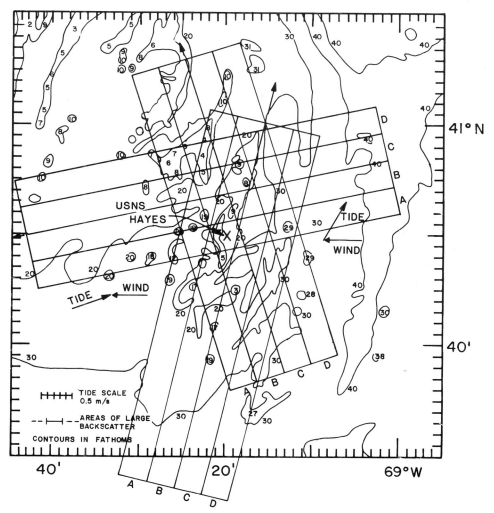

Figure 1. SAR swaths over Phelps Bank taken within 40 minutes on
July 11, 1982 (Valenzuela et al., 1985).

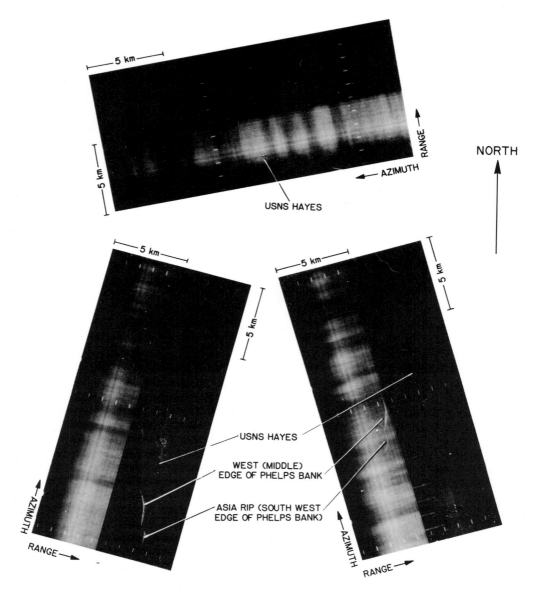

Figure 2. Portions of subswaths A and B of SAR images taken over
Phelps Bank on July 11, 1982 (Valenzuela et al., 1985).

for the swaths shown in Figure 1. These images are not corrected geometrically. The locations
of *USNS Hayes*, west (middle) edge and Asia Rip (southwest edge) of Phelps Bank are all visible. The bottom two images were taken with a difference of 31.5° in look angles. The image at
the top shows quite a different surface feature from those shown by the bottom two images for
the northern portions of Phelps Bank. The look angle of the top image is different by 84.5° and
115° from those of the bottom two images.

3. METHODOLOGY

Referring to Table 1 and our detailed records on environmental conditions, measurements made on July 11 provide the best data set for achieving our scientific objective. Our approach to the problem is divided into three phases:

Phase 1 — Oceanic Circulation Numerical Model,

A two-layer and rigid-lid hydrodynamic ocean model (Holland and Lin, 1975) has been adapted with open boundary conditions to the bottom topography of Phelps Bank. Numerical experiments with both nonstratified (one-layer) and stratified (two-layer) density profiles are being carried out. The tidal forcing is specified from tables and on-site measurements. Initially, the current measurements from the experiment are used to check on the model output. Later, a combined Lagrangian-Eulerian objective analysis scheme is applied to model output, together with temperature measurements, to form a multivariate initial state for the model. A level-type, fixed-vertical-grid-points model (Madala and Piacsek, 1977) is also used to handle horizontally strong inhomogeneous density profiles.

A finite difference numerical scheme is used in the model with spatial resolution of 100-200 meters and time resolution of 400-800 seconds. The result is used for the Phase 2 numerical model as input.

Phase 2 — Current-Wave Interaction,

A nonuniform current field perturbs, kinematically and dynamically, the ocean waves which propagate on the ocean surface and produce changes in wave characteristics. Since SAR images were obtained by X band systems whose electromagnetic wavelength is of the order of 3 cm, the spatial distribution and energy density of oceanic wavelets which satisfy Bragg scattering criteria are obviously of interest (1.5 - 2.1 cm for grazing to 45° incidence at X band). These Bragg resonant wavelets are perturbed (or modulated) by the non-uniform current field as well as, among others, by the energetic long gravity wave components of the random wave field (Phillips, 1981).

The nonuniform current field is provided by the numerical model of Phase 1 and the ocean surface wave field is measured by clover-leaf and Endeco directional wave buoys. It can readily be observed, from the surface wave measurements made on July 11, that there were swells propagating through the experiment area, and that the dominant portion of wind-generated random wave field was very weak energetically. The directional wave buoy cannot provide measurement of small Bragg resonant wavelets. Therefore, it seems logical to investigate the effects of the nonuniform current field on the swell alone first, and then to treat the interaction between swell and Bragg resonant wavelets including the effect due to underlying non-uniform current field. To compute steady state changes in wave characteristics in wavenumber vector and its spectral energy density as the swell propagates across the nonuniform current field we use the kinematic conservation law of wave

$$\nabla n = 0 \tag{1}$$

and the energy balance equation

$$\nabla \cdot E(\mathbf{U} + \mathbf{C}_g) + S_{ij} \frac{\partial U_j}{\partial X_i} = F \quad \text{for } i, j = 1, 2. \tag{2}$$

In Equations (1) and (2), the wave frequency

$$n = \sigma + \mathbf{k} \cdot \mathbf{U} \tag{3}$$

where **k** is the irrotational wavenumber vector, E the spectral wave energy density, $\mathbf{C}_g$ the group velocity, $\mathbf{U}$ the current velocity, S_{ij} the radiation stress, F the forcing function, and σ the intrinsic frequency which satisfies, with depth d and gravitational acceleration g,

$$\sigma^2 = g\,|\mathbf{k}|\,\tanh\,|\mathbf{k}|\,d. \tag{4}$$

The case of $F = 0$ at the right hand side of Equation (2) indicates that net energy gained or lost by the dynamic system is negligible. This assumption is justifiable for the calculation of swell alone but, of course, for the short gravity waves appropriate forcing terms will be included (Hughes, 1978; Valenzuela and Wright, 1979). The method of characteristics (Chen and Bey, 1977), in a finite difference scheme, is used to solve Equations (1) and (2) for the swell in the presence of the non-uniform tidal flow obtained from Phase 1.

Once the characteristics of swell over the whole region are known, the interaction between swell and Bragg resonant wavelets including the effect of underlying non-uniform tidal flow can be determined by using Equations (1)-(4) except now:

(a) The physical parameters are those of the Bragg resonant wavelets of wave number $\mathbf{k}_B$,

(b) $\mathbf{U}$ now contains both the non-uniform tidal flow and the orbital velocity of the swell,

(c) $\sigma^2 = g'|\mathbf{k}_B|$ where g' is the sum of gravitational acceleration and vertical acceleration of the swell, and

(d) F is nonzero, for example, as in the relaxation approximation (Alpers and Hennings, 1984).

We further assume the Bragg resonant wavelets are in equilibrium with a spectral energy density $\psi\,(\mathbf{k}_B)$ indicated by Phillips (1981) as

$$\psi\,(\mathbf{k}_B) = f(\theta)\,|\mathbf{k}_B|^{-4} \tag{5}$$

where $f(\theta)$ is a symmetrical function about the wind direction.

The amplitude of modulation in radar backscatter power (or cross-section) relative to the mean is equal to $\delta\psi/\psi$ in the direction of the radar viewing angle. Valenzuela and Wright (1979) and Phillips (1981) have shown for swell alone that this modulation is proportional to swell slope. Recently Alpers and Hennings (1984), with perturbation theory and the relaxation approximation, have shown for non-uniform current that $\delta\psi/\psi$ is proportional to the product of relaxation time of the Bragg wavelets and the current gradient on the surface in the direction of the radar viewing angle. Furthermore Alpers and Hennings (1984), using simple continuity condition for the current flow over a one-dimensinal bottom feature, found that $\delta\psi/\psi$ is proportional to the ratio of bottom slope to the square of the water depth. In comparison, we are investigating a situation where both two-dimensional current field and swell are present over two-dimensional bottom topography as in the Phelps Bank region. Therefore $\delta\psi/\psi$ in our case will have to be evaluated numerically.

A two-dimensional map of $\delta\psi/\psi$ is produced for Phase 3 as input to the computation of SAR image intensity using Bragg scattering (Valenzuela, 1980).

Phase 3 — Microwave Scattering,

Bragg scattering is used to compute X band SAR image intensity from the two-dimensional $\delta\psi/\psi$ map and the advected phase velocity of the Bragg resonant wavelets. Since θ

and radar viewing angle are known for each swell along its propagation array, the modulation in energy of the Bragg resonant wavelets and their net phase speed may be determined.

4. DISCUSSION

It was suggested by Valenzuela et al., (1983), from what has been observed in NRL Phelps Bank Experiment of 1982, that a residual flow persisted beyond the tidal cycle. This residual flow together with the local tidal flow over the bottom refracted, strained, and blocked the surface ocean waves (Phillips, 1981), in particular the Bragg resonant wavelets. Zimmerman (1978) showed that the irregular bottom topography acts as a catalyst for transferring vorticity from the oscillating tidal flow to the residual flow field. A small-scale vorticity field is produced by the Coriolis and frictional torques over the irregular bottom topograpy and is subsequently transferred to the residual flow field by nonlinear advection. Zimmerman (1980) evaluated further, by assuming a linear frictional force at the bottom, the production of topographic vorticity by vortex stretching in circularly rotating currents in shallow tidal areas as a function of topographic length scale. He showed that the vorticity is transferred from the oscillatory field to the residual flow field and, also, to higher tidal harmonics. For a given slope in the bottom topography the transfer is largest for a topographic wavelength of the order of the horizontal tidal spatial excursion (about 7 km at Phelps Bank). Decreasing the bottom friction parameter increases the absolute value of residual vorticity produced either by topographic stretching of fluid columns having the Coriolis effect or by bottom frictional torques. Residual vorticity produced by the Coriolis effect is positively correlated with the topography. The entropy of vorticity and the energy of tidal current are comparable in magnitudes. Robinson (1981) identified further three important mechanisms which generate vorticity in the presence of bottom topography and oscillating tide: (1) changes in water depth; (2) lateral shearing of the flow (even when the depth is uniform) due to quadratic bottom friction; and (3) bottom friction when there is a depth-distributed force (even if there is no lateral velocity shear).

In-situ current measurements by an Eulerian array of current meters indicate the existence of this residual flow field and higher tidal harmonics. Figure 3 (Greenewalt et al., 1983) shows the progressive current vector diagram or "quasi-trajectory" for the current meter record at 5-m depth approximately 1.2 nautical miles east of the shallowest part of Phelps Bank. Each loop takes one tidal cycle to complete. It can be interpreted that a particular water particle at 5-m depth followed this trajectory from 10 to 20 July 1982. Similar data for the current meter at 13-m depth indicate that the mean drift was in approximately the same direction (220° true north) as that recorded by the near-surface meter, but the speed was slower by about 6 cm/s. The deep meter at 21-m near the bottom recorded the same current speed as the meter at 13-m depth, but the direction of the flow differed by about 10° (210° true north). Current meters at 13-m and 21-m depths did not show the same steady component in their measurements as those by the current meter at 5-m depth shown in Figure 3. Considerable vertical shear or vortex was evident in the steady drift current which implied the existence of residual flow. Some components of tidal harmonics are clearly noticeable in Figure 3 as well.

5. CONCLUSION

In the course of this work, information achieved regarding the characteristics of radar scattering returns can be identified as:

1. More than one physical process is involved in SAR imaging of bottom topography in coastal region;

2. The modulation of surface waves by a current gradient related to bottom topography seems to be the dominant mechanism;

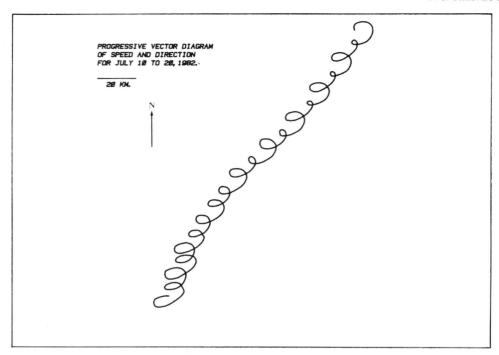

Figure 3. Progressive vector diagram or "quasi-trajectory" for the
current meter record at 5-m depth (Greenewalt et al., 1983).

3. For winds under 7 m/s, modulations of backscattered power are in the order of 20-30
dB at L and X band; and

4. For winds over 7 m/s, modulations decrease to 2-3 dB at L and X band.

Time-space model analyses of circulation, current-wave interaction, and microwave scattering
are still in progress. These model analyses are absolutely necessary to investigate the
parametric dependence of SAR imagery on current-wave-microwave interactions.

References

Alpers, W., and I. Hennings, 1984: A theory of the imaging mechanism of underwater bottom
    topography by real and synthetic aperture radar. *J. Geophys. Res.*, **89**, 10529-10546.
Beal, R. C., P. S. DeLeonibus and I. Katz, 1981: *Spaceborne Synthetic Aperture Radar for
    Oceanography.* Johns Hopkins Univ. Press, 216 pp.
Chen, D. T., and P. Bey, 1977: *Interaction Between Steady Non-Uniform Two-Dimensional
    Currents and Directional Wind-Generated Gravity Waves with Applications for Current Meas-
    urements.* Nav. Res. Lab. Memo. Rep. 3508, Washington, DC, 30 pp.
Chen, D. T., G. R. Valenzuela, W. D. Garrett and J. A. C. Kaiser, 1985: *Naval Research
    Laboratory (NRL) Remote Sensing Experiment of 1982.* Nav. Res. Lab. Memo. Rep., in
    preparation.
de Loor, G. P., 1981: The observation of tidal patterns, currents, and bathymetry with SLAR
    imagery of the sea. *IEEE Trans.*, **OE-6**, 124-129.

Greenewalt, D., C. Gordon and J. McGrath, 1983: *Eulerian Current Measurements at Phelps Bank*. Nav. Res. Lab. Memo. Rep., 5047, Washington, DC, 51 pp.

Holland, W. R., and L. B. Lin, 1975: On the generation of mesoscale eddies and their contribution to the oceanic general circulation, I. A preliminary numerical experiment. *J. Phys. Oceanogr.*, **5**, 642-657.

Hughes, B. A., 1978: The effect of internal waves on surface winds waves: Theoretical analysis. *J. Geophys. Res.*, **83**, 455-465.

Madala, R. V., and S. A. Piacsek, 1977: A semi-implicit model for baroclinic oceans. *J. Comp. Phys.*, **23**, 167-178.

Phillips, 0. M., 1981: The structure of short gravity waves on the ocean surface, *Spaceborne Synthetic Aperture Radar for Oceanography*, R. C. Beal, P. S. DeLeonibus, and I. Katz, Eds., Johns Hopkins Univ. Press, Baltimore, MD, 24-31.

Robinson, I. S., 1981: Tidal vorticity and residual circulation. *Deep Sea Res.*, **28A**, 195-212.

Valenzuela, G. R., 1980: An asymptotic formulation for SAR images of the dynamical ocean surface. *Radio Sci.*, **15**, 105-114.

Valenzuela, G. R., and J. W. Wright, 1979: Modulation of short gravity-capillary waves by longer-scale periodic flows—A higher order theory. *Radio Sci.*, **14**, 1099-1110.

Valenzuela, G. R., D. T. Chen, W. D. Garrett and J. A. C. Kaiser, 1983: Shallow water bottom topography from radar imagery. *Nature*, **303**, 687-689.

Valenzuela, G. R., W. J. Plant, D. L. Schuler, D. T. Chen and W. C. Keller, 1985: Microwave probing of shallow water bottom topography in the Nantucket Shoals. Submitted to *J. Geophys. Res.*

Zimmerman, J. T. F., 1978: Topographic generation of residual circulation by oscillatory (tidal) currents. *Geoph. Astroph. Fluid Dyn.*, **11**, 35-47.

Zimmerman, J. T. F., 1980: Vorticity transfer by tidal currents over an irregular topography. *J. Mar. Res.*, **38**, 601-630.

MEASUREMENT AND ANALYSIS OF SURFACE WAVES IN A STRONG CURRENT

R. J. Lai, R. J. Bachman, A. L. Silver and S. L. Bales
David W. Taylor Naval Ship Research and Development Center
Bethesda, Maryland 20084
U.S.A.

ABSTRACT. Measurements of surface waves in a strong adverse tidal
current were obtained during the SEBEX experiment of 1982. The wave
data were collected from a directional wave tracking buoy which
followed the local currents. The data, which were collected as the
buoy crossed a frontal zone, were analyzed in five consecutive segments
following the local topography and the observed flow pattern. The
spectral shapes changed, and the wave peak frequencies shifted from
segment to segment. The results indicated strong transient wave
phenomena due to wave-current interactions.
      The local current pattern was evaluated using the measured shift
of peak wave frequency in a moving coordinate system. The evaluated
flow pattern agreed well with currents obtained from the ship drift
during the experiment and computation from the conservation of mass.
The current pattern revealed a large velocity gradient in the area.
The effects of current on surface waves were computed for each segment
and the results were compared with measured data. The flow pattern was
also used in a wave-current interaction model to investigate the dynamics of surface waves in a current. The model is linear and takes into
account the bottom bathymetry, radiation stresses and the conservation
of wave number and wave energy. The results from the model showed some
agreement and disagreement with the measured data. Applicability and
limitations of the model are discussed. The operation of a freely
drifting buoy in a strong current is recommended for future studies of
wave-current interaction.

1.   INTRODUCTION

      The variation of surface waves in a strong current is an impor-
tant phenomenon of the ocean surface needing further investigation.
Several analytical and empirical approaches investigating wave-
current interaction have been reviewed elsewhere (Phillips, 1977;
Peregrine and Jonsson, 1983). The critical element missing in these
studies is the lack of field data to substantiate these approaches.
This paper presents some measured surface wave data in a strong current

Y. Toba and H. Mitsuyasu (eds.), The Ocean Surface, 161–169.
© 1985 by D. Reidel Publishing Company.

taken during the SEBEX experiment. The analysis of these data is used
to examine the effects of current on surface waves.

2.   MEASURING SITE AND PROCEDURE

SEBEX refers to the Surface Expression of Bathymetry Experiment of
July 1982 which has been described by Valenzuela, et al., 1983. The
procedure for obtaining surface wave data during SEBEX has been
published previously (Lai and Bachman, 1983). Briefly, wave data were
measured on 14 July 1982 near the Phelps Bank of Nantucket Shoals. The
Bank spans approximately north to south (see Figure 1). On 14 July
1982, a strong tidal current came from the northeast direction and
encountered surface waves which propogated from the south and southwest. A front-like phenomenon appeared near the shallow ridge of the
Bank. An ENDECO directional wave tracking buoy, described by Foley,
Bachman, and Bales (1983), was deployed from the east side of the Bank
and drifted freely following the local current. The buoy was picked up
west of the Bank two hours later. The wave data were continuously
received at the nearby research vessel (USNS HAYES) through a telemetry
system.

After careful examination of the recorded data, they were divided
into five segments, two for the area east of the Bank and three for the
area west of the Bank. Each segment consisted of 17 minutes of data.
Since the buoy was not equipped with a transponder, the exact measurement locations were unknown. However, by analyzing a ship drift curve
and through visual observation, the approximate locations corresponding
to each segment were determined. The wave spectra at the Phelps Bank
together with their corresponding locations are shown in Figure 2. The
variations of the waves between these five segments will be discussed
in the following sections.

3.   RESULTS

3.1 Wave Energy

The measured wave energy clearly varied with bottom topography as
seen in Figure 2. The minimum values of wave energy were located on
the west side of the Bank and the maximum values on the east side. The
spectral shapes show the prevailing bimodal characteristics (e.g., combined swell and wind waves). Variations between swells and wind waves
along the measuring path are apparent. The swell peaks at the ridge of
the Bank while wind waves peak to the east of the Bank. The variations
of spectral shapes between segments indicate the change of local current
patterns and, consequently, the pheonomena of wave-current interaction.

3.2 Doppler Shift

One noticeable phenomenon among these wave spectra is the shift of
peak frequency of wind waves for each different segment. Peak wave

TABLE I - SUMMARIZED WAVE AND CURRENT DATA AROUND THE BANK

Run No.	Local Time	Frequency Peak (Hz)	Wave Direction ($\theta°$, Meas.)	Current Amplitude (U, m/s)	Current Direction ($\theta°_u$, Meas.)	Current* Amplitude ($\overline{U}$, m/s)
7D	1205	0.172	223	1.21	92	1.01
7C	1140	0.164	213	0.62	85	0.75
$7B_2$	1110	0.156	218	0.48	77	0.74
$7B_1$	1050	0.179	225	1.49	70	1.41
7A	1028	0.164	223	0.84	65	0.90

*Currents evaluated from the conservation of mass.

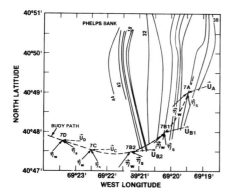

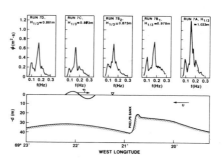

Figure 1. Bathymetry Contours of the Phelps Bank and the Track of Buoy Path

Figure 2. Measured Wave Spectra and the Corresponding Locations Around the Phelps Bank

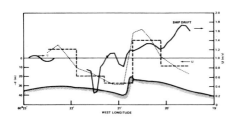

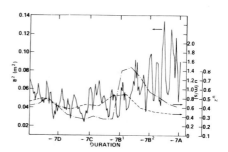

Figure 3. Distribution of Currents Around the Phelps Bank; Solid Line – Ship Drift Velocity; Dashed Line – Computed Currents from Doppler Shift; Dotted Line – Variation of the Computed Currents

Figure 4. Variations of Moving Averaged Wave Energy (Solid Line), Current (Dashed-Dotted Line), and Spectral Bandwidth Parameter (Dashed Line)

frequencies are listed in Table I. Since the buoy was drifting with
the current, the shifts of the peak frequency are the Doppler effects
of the local current. These Doppler shifts have been used as one
method for computing the local current.

3.3 Local Currents

The only current measurements available during the experiment were
a stationary current meter located to the east of the Bank and about
two km north of the site of the buoy's first deployment, and the ship's
drift curve (Greenewalt et al., 1983). Two methods are used to determine the local current. One is the use of the Doppler shift of wave
peak frequency and the other is the conservation of mass. The first
method was presented previously (see Lai and Bachman, 1983). The
results of computed currents, U, based on the first method, are listed
in Table I. During the computation, the current measured by the stationary meter was assigned to Run 7A as the initial value to compute
currents at other segments. The wave direction from the buoy measurement and the current direction from ship drift curve were used in the
computation. All of these values are listed in Table I.

The local current can be computed through the principle of the
conservation of mass (Phillips, 1984). Since the Bank spans north to
south, and the variations of current along the Bank are small compared
to the normal direction to the Bank, variations along the Bank were
assumed to be zero. Consequently, the currents only varied with the
water depth, along the direction normal to the Bank. The currents computed from the conservation of mass are listed as $\overline{U}$ in Table I. The
currents computed from each of the two methods agree well except at
station $7B_2$ which is located near the west side of the Bank. This may
indicate the existence of other current patterns at that location.

These computed currents are plotted in Figure 3 and are compared
with those obtained from the ship drift. The general current pattern
agrees except at the first station (Run 7A). A current pattern which
varied with bottom topography was developed and is indicated by a
dotted line in Figure 3. These local currents are based on the computed results from Doppler shift but varied with the local depth. This
current pattern will be used in the following sections to investigate
wave-current interaction.

4. ANALYSIS

4.1 Correlation of Waves and Currents

The wave data were recorded in both analog and digital form. The
variations of waves throughout the experiment were compared with the
corresponding currents. Wave data, $\overrightarrow{e^2}$, were first condensed by the
mean square average of every 30 seconds and then a moving average was
used to smooth out the large fluctuations. The results are shown in
Figure 4. The variation of wave energy and local current followed
the same trend closely until it reached the far east side of the Bank,

where the wave energy increased and the local current decreased.
    The spectral bandwidth parameter ν^2, is also shown in Figure 4.
The values of ν^2 were determined by,

$$\nu^2 = 1 - \frac{1}{\gamma^2} \tag{1}$$

where γ is the ratio of the number of zero crossing to the number of
peaks.
    The variation of ν showed a consistent trend with wave characteris-
tics. The values of ν reached the peak near the ridge of the Bank
where the strongest current developed and many breaking waves were
observed. At the east side of the Bank, after the waves passed the
ridge, the value of ν declined.

4.2 Transformation of Wave Spectrum

    The waves were measured in a moving coordinate system by the buoy
which drifted with the current. Since the local currents were computed
from the Doppler shift of peak wave frequencies, these currents have
been used to transfer the surface waves to a fixed coordinate system.
The transformation of the wave spectrum from the encountered frequency
domain, $\phi(\omega_e,\theta)$, to a fixed coordinate $\phi(\omega)$, follows,

$$\phi(\omega_e,\theta)d\omega_e = \phi(\omega)d\omega \tag{2}$$

or

$$\phi(\omega_e,\theta)\left\{1/\frac{\partial\omega}{\partial\omega_e}\right\}d\omega = \phi(\omega)d_\omega \tag{3}$$

where ω_e is the encountered frequency in radians/second, θ is the angle
between fixed and moving coordinates, and ω is the frequency of the
fixed coordinates. The Jacobian transformation $\{\partial\omega_e/\partial\omega\}$ in any depth,
including shallow water, is determined from the dispersion relation

$$\omega_e = \omega - kU \cos \theta \tag{4}$$

and

$$\omega^2 = kg \tanh(kh) \tag{5}$$

where k is the wave number, U is the relative velocity between fixed
and moving coordinates, and g is the gravity acceleration. The final
Jacobian transformation is

$$\frac{\partial\omega_e}{\partial\omega} = 1 - \frac{2\omega U\cos\theta}{g}\left(\frac{2 \cosh^2 kh}{\sinh 2kh + 2kh}\right) \tag{6}$$

The wave spectra from the buoy measurements were transformed into fixed
coordinates and the results for Runs $7B_1$, and $7B_2$ are shown in Figure 5.
During these transformations, the mean values of water depths,
currents, and wave directions for each segment were used.

The wave spectra in moving coordinates and at the east and west sides of the Bank are shown in Figure 5b and are compared with spectra transformed to fixed coordinates in Figure 5a. These transformations move all the peak frequencies to the range 0.153 Hz < f < 0.163 Hz. The wave energy in the fixed coordinate system increases as the waves propagate toward the opposing current. Some differences still exist in other frequency ranges, and can be attributed to the dynamic interaction of other physical parameters.

Another wave spectral transformation was performed following the same method. However, in this case, all the waves were analyzed with the moving coordinate system and different current patterns. One of the segments was chosen to be transformed to the fixed coordinate system. The currents of other segments relative to the selected segment were changed accordingly. The results of computation around one segment are shown in Figure 6. In this figure, the measured wave spectrum at the segment of Run 7C was chosen as the fixed coordinate and is shown by a solid line. The other two runs of nearby segments were transformed to the coordinate of the selected segment and are shown by dashed-dotted and dashed lines. The transformations change the peak wave frequencies and move them close together. This assures the accuracy of the computed current values in different segments. The changes of wave energy by these transformations show some discrepancies with measured values. This is further analyzed using wave model computations.

4.3 Wave Model

A wave transformation model in finite water depth has been used here to investigate the dynamic interaction of surface waves at the Bank. A detailed description of the model is not given here, but the model is linear which permits the solution of several wave equations simultaneously (Collins, 1972; Wang and Yang, 1981; Lai et al., 1984). In a steady state and two-dimensional model, these equations are:

$$\nabla (\sigma + \vec{K} \cdot \vec{U}) = 0 \tag{7}$$

$$\nabla \times \vec{K} = 0 \tag{8}$$

$$\frac{\partial}{\partial x_\alpha} \left\{ E \left[U_\alpha + (C_g)_\alpha \right] \right\} + S_{\alpha\beta} \frac{\partial U_\beta}{\partial x_\alpha} = - \varepsilon \tag{9}$$

where σ is the radian frequency, E is the wave energy, x_1 is the coordinate of north-south direction, x_2 is the coordinate of east-west direction, C_g is the group velocity, $S_{\alpha\beta}$ is the radiation stress, and ε is the source term.

The wave model is developed based on the available bottom topography around the Bank (Gordon and Greenewalt, 1982), the prescribed computed currents (see Figure 3), and the bottom friction dissipation term with friction coefficient, $C_f = 0.006$. Since the primary wave trains were coming from the southwest and propagating to the northeast, the

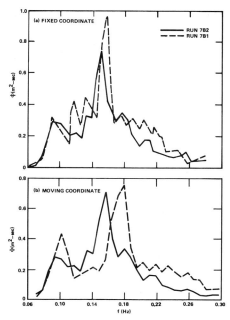

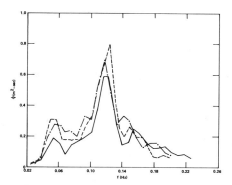

Figure 6. Transformation of
Wave Spectra; Solid Line –
Measured Spectrum of Run 7C;
Dashed Line – Transformed
Spectrum of Run 7D; Dashed-
Dotted Line – Transformed
Spectrum of Run 7B₂

Figure 5. Wave Spectra of Runs
7B₁ (Dashed Line) and 7B₂ (Solid
Line) in (a) Fixed Coordinate,
(b) Moving Coordinate

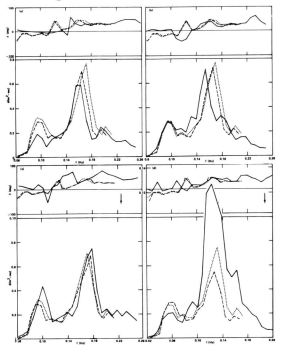

Figure 7. Comparison of
Measured and Computed Wave
Spectra for (a) Run 7C, (b)
Run 7B₂, (c) Run 7B₁,and (d)
Run 7A, where Solid Lines –
Measured Spectra; Dotted
Lines – Computed Spectra
without the Presence of
Currents; Dashed Line –
Computed Spectra with the
Presence of Currents

measured wave spectrum at the far west side of the Bank (Run 7D) was
used as the input to predict the propagation of waves along the buoy
path. All the measured wave spectra from the buoy were in moving
coordinates, so the relative currents between each segment were used in
the computation. The computed results are shown in Figure 7.

Figure 7 consists of four wave spectra of four segments. Both
mean wave direction and wave spectra are shown. The dashed lines and
dotted lines are the computed results from the wave model with and
without currents, respectively, and the solid lines are the corresponding measured values. The computed results are the average values
of several grid points of the model (grid size: $\Delta x_1 = 460$ m,
$\Delta x_2 = 1025$ m). The number of grid points averaged in each segment is
dependent on the amplitude of the local current and ranges from 2 to 4
in the computation. The dotted lines show the shoaling effects around
the Bank, and the dashed lines show the effects of shoaling and
currents from the computation.

The agreement of computed and measured mean wave directions in the
four segments are very good. The effect of currents on wave mean
directions are small. There is some deviation of mean wave directions
for the swells. This suggests that the swells were propagating from
the south and southeast directions and should be treated separately
from the primary wave trains which came from the southwest.

The wave spectra for each segment generally agrees well except for
the last segment (Figure 7d) of Run 7A. For Figures 7a, 7b, and 7c,
the differences between dashed lines and dotted lines clearly indicate
the effect of local currents. The wave model correctly predicted the
variation of surface waves at these segments. However, the magnitude
of the current would have to increase in order to approach a better
agreement. Figure 7d shows somewhat confusing results. On the one
hand, the wave model predicts the shift of peak frequency correctly,
yet the reduction of wave energy from the model is opposite that of the
measurement. Choosing a different current pattern with larger currents
in this segment would cause the wave energy to increase, yet the peak
frequency would shift to the higher range away from the measured
values. Based on these adjustments, the computed results indicate that
large eddy currents around segment 7A may exist. The wave model fails
to predict such strong non-linear interaction. By carefully adjusting
the local current patterns, the agreement between measured and computed
wave spectra will improve.

5. CONCLUSION

Based on these analyses of surface waves and strong currents,
several conclusions are drawn.
    (1) Surface waves are strongly modified by the currents and
         bottom topography
    (2) Using a freely drifting buoy to measure the variation of
         surface waves in a strong current is a practical way to
         investigate wave-current interaction

(3) A linear wave model with the proper grid size and a given
    current pattern will provide a good computation of the
    variation of wave spectra in a current until a phenomenon of
    strong nonlinear interaction takes place.

REFERENCES

Collins, J.I., 'Prediction of Shallow-Water Spectra,' J. of Geophy.
    Res., Vol. 77, 1972.
Foley, E.W., R.J. Bachman and S.L. Bales, 'Open Ocean Wave Buoy
    Comparisons in the North Atlantic,' Proceedings 1983 Symposium on
    Buoy Technology, 1983.
Greenewalt, D., C.M. Gordon and J. McGrath, 'Eulerian Current
    Measurements at Phelps Bank,' NRL Memo. Rep. 5047, 1983.
Gordon, C.M. and D. Greenewalt, 'Bathymetry Measurements at Phelps
    Bank,' NRL Memo. Rep. 4962, 1982.
Lai, R.J. and R.J. Bachman, 'A Field Investigation of Surface Waves
    Across a Frontal Zone,' presented at XVIII General Assembly, IUGG,
    Hamburg, 1983.
Lai, R.J., A.L. Silver and S.L. Bales, 'Wave Statistics in Nearshore
    Zones,' Proceed. of Ocean Structures '84, Corvallis, Oregon, 1984.
Phillips, O.M., The Dynamics of the Upper Ocean, 2nd edition Cambridge
    Univ. Press, 1977.
Phillips, O.M., Private Communication, 1984.
Peregrine, D.H. and I.G. Jonsson, 'Interaction of Waves and Currents,'
    Report No. 83-6, Coastal Eng. Res. Center, Corps of Eng., U.S.
    Army, 1983.
Wang, S. and W.C. Yang, 'Measurements and Computation of Wave Spectral
    Transformation at Island of Sylt, North Sea,' Coastal Eng., Vol.
    5, 1981.
Valenzuela, G.R., D.T. Chen, W.D. Garrett and J.A. Kaiser, 'Shallow
    Water Bottom Topography from Radar Imagery,' Nature, Vol. 303,
    1983.

IN SEARCH OF UNIVERSAL PARAMETRIC CORRELATIONS FOR WIND WAVES[*]

Paul C. Liu
Great Lakes Environmental Research Laboratory/NOAA
2300 Washtenaw Avenue
Ann Arbor, Michigan 48104

Abstract. A large number of wave and wind data recorded from eight NOMAD buoys in the Great Lakes during 1981 were used to examine correlations of wind wave parameters. The results show no precise universal relations among the parameters. The only correlation that shows a universal behavior is that of nondimensional energy versus nondimensional peak-energy frequency.

1. INTRODUCTION

For a given wave field with spectral density $S(f)$, peak energy frequency f_m and total energy $E = \int S(f)df$, along with frictional wind speed $U*$ and fetch distance F, the following parameters, among others, have been used in the literature: $\varepsilon_* = gE/U_*^4$, nondimensional energy; $\nu_* = f_m U*/g$, nondimensional peak-energy frequency; $\xi_* = gF/U_*^2$, nondimensional fetch; $SS = 2\pi f_m^2 E^{1/2}/g$, significant wave slope (Huang et al., 1981); α, equilibrium range constant (Phillips, 1958, 1977); γ, JONSWAP peak enhancement factor (Hasselmann et al., 1973); $\lambda = \varepsilon_* \nu_*^4/\alpha$, JONSWAP shape factor (Hasselmann et al., 1976); and $\varepsilon_* \nu_*/\xi_*$, Wallops fetch factor (Huang et al., 1981).

The basis of using nondimensional parameters is the expectation that by correlating these parameters, universal relations can be found and thereby used in conjunction with theoretical models. Numerous studies have been made and a number of universal correlations among the above parameters have been proposed and widely used. A relevant question seems to have escaped most of the attention in the literature: how universal are these universal correlations? In this paper, we attempt to explore this question by examining wave and wind data recorded from eight NOMAD buoys in the Great Lakes during 1981. These Great Lakes data, averaging over 4,000 data points from each buoy, are ideal for studying the wind wave parametric correlations because the

---

[*]GLERL Contribution No. 426.

Y. Toba and H. Mitsuyasu (eds.), The Ocean Surface, 171–178.
© 1985 by D. Reidel Publishing Company.

lakes are clearly fetch limited and generally free from swell complications.

2. THE DATA

The NOMAD buoys are boat-shaped, 6 m in length, and moored in water
depths ranging from 15 m to 250 m in the Great Lakes. The buoys are
equipped to measure air and surface water temperatures, wind speed and
direction at 5 m above the water surface and wave spectra. The waves
are measured with an accelerometer using an on-board Wave Data Analyzer
System (Steele and Johnson, 1977) that transmits acceleration spectral
data via satellite to a shore collecting station where wave frequency
spectra with 24 degrees of freedom are calculated from 20 min of
measurements hourly. The wind speed and direction, as well as air and
surface water temperatures, are measured with 1 s resolution and
averaged hourly over 8.5 min data. In this study the measured wind
speeds at a 5 m level are converted to $U*$ based on the formula for
overwater roughness length presented by Charnock (1955) and formulation
for stability length presented by Businger et al. (1971), assuming the
neutral drag coefficient to be 1.6×10^{-3}. The data used in this study
were those recorded during 1981 from eastern Lake Superior buoy 45004.
This buoy, moored at a depth of 250 m, has recorded up to 7 m significant wave heights and 18 m s^{-1} wind speeds during 1981 among 3,665 data
points. The resulted correlations from this buoy are substantially
similar to those from other buoys. Hence they are representative of
all the Great Lakes. The data are archived at the NOAA National
Climatic Center in Asheville, N.C.

3. THE CORRELATIONS

3.1. Correlations With Fetch Parameters

One of the main goals of making empirical parametric correlations is to
link internal and external parameters and thereby deduce the influence
and range of external conditions on wave growth processes. The nondimensional fetch ξ is a successful and widely used external parameter.
During JONSWAP, Hasselmann et al. (1973) presented the following power-
law relations based on their own and a variety of other data sources,
both field and laboratory measurements:

$$\nu_* = 1.08\xi_*^{-0.33} \tag{1}$$

and

$$\varepsilon_* = 1.60 \times 10^{-4}\xi_* \tag{2}$$

to show the fetch dependence of the nondimensional peak frequency ν_*
and non-dimensional energy ε_*.

Contending that the balance of dynamical processes are different
between field and laboratory, Phillips (1977) proposed an exponent of

-1/4 in Eq. (1) based on field data only. Huang et al. (1981), however, chose to unify field and laboratory data by proposing the correlation of significant wave slope SS with a combined fetch parameter
$\varepsilon_* \nu_* / \xi_*$ given by

$$SS = (80\pi\varepsilon_* \nu_* \xi_*^{-1}/9)^{4/9}. \tag{3}$$

Correlation of the parameters given in Eqs. (1)-(3) with the 1981
NDBC data is shown in Fig. 1. If the relations are indeed universal,
the 3,665 data points should be clustered around these straight lines.
We found, however, that the points are clustered around galaxy-like
regions instead. The variations within each region can be up to one
order of magnitude. Since the regions generally follow the orientation
of the respective straight lines given by Eqs. (1)-(3) and the lines go
through the main cluster of points, the relation of Eqs. (1)-(3), while
not exactly universal, does represent a crude approximation of the
correlations.

Since the 3,665 data points are applied nondiscriminatingly, we
attempted next to sort the different external conditions to see if we
can reduce the large scatter. We found there are no seasonal effects.
We found, however, that wind speed has a distinctive influence on the
correlations. Waves with wind speeds greater than 10 m s^{-1} have distinctively higher ν_*, lower ε_* and lower ξ_* than those with wind speeds
less than 2 m s^{-1}. Furthermore, we found that over 75% of the data
points are waves with significant wave height under 1 m. In Fig. 2, we
used 880 data points with significant wave height greater than 1 m to
yield fairly agreeable correlations. In subsequent correlations, we
continued the use of these 880 higher wave data points with the
understanding that lower wave data points would only contribute to
increasing the already excessive scatter.

3.2. Correlations with Phillips' α

Phillips (1958, 1977) deduced from similarity considerations that the
deep water wave spectra in the equilibrium range should be of the form

$$S(f) = \alpha g^2 f^{-5} \qquad (f \gg f_m), \tag{4}$$

where α is a universal constant. In JONSWAP, Hasselmann et al. (1973)
computed α by

$$\alpha = (0.65 f_m)^{-1} \int_{1.35 f_m}^{2 f_m} (2\pi)^4 f^5 g^{-2} \exp\left[\frac{5}{4}\left(\frac{f}{f_m}\right)^{-4}\right] S(f) df \tag{5}$$

and again found that α is a function of nondimensional fetch ξ_*

$$\alpha = 0.35 \xi_*^{-0.22}. \tag{6}$$

Hasselmann et al. (1976) deduced that α is also a function of the nondimensional peak frequency ν_*

$$\alpha = 0.033\nu_*^{2/3}. \tag{7}$$

Both α and ν have been used as important parameters in developing numerical wave prediction models (Hasselmann, 1977).

Other forms of the power law α versus ν_* relation have also been proposed. Toba (1978) used

$$\alpha = 1.39\nu_*. \tag{8}$$

Mitsuyasu et al. (1980), on the other hand, developed

$$\alpha = 0.63\nu_*^{6/7}. \tag{9}$$

Fig. 3 illustrates the correlations corresponding to Eqs. (6)-(9). Because the figures contain only data points with significant wave height greater than 1 m, they may indicate that the proposed relations are underestimating the data. Maybe the calculated α's in this study are higher than in previous studies. However, when all the data points are plotted, they are too scattered to confirm the universality of any of the relationships and too scattered to warrant an examination of the differences among the three equations, (7)-(9).

3.3. Correlations with Shape Factors

The shape of the wave spectrum is an important parameter for wave predictions. Hasselmann et al. (1973) introduced the peak-enhancement factor γ, which is the ratio of the peak value of the spectrum to the peak value of the corresponding fully developed spectrum, which can be computed by

$$\gamma = S(f_m)(2\pi)^4 f_m^5 \exp(5/4)(\alpha g^2)^{-1}. \tag{10}$$

The contention is that γ is greater than 1 for growing waves and approaches 1 for fully developed waves. Hasselmann et al. (1976) defined an alternative shape parameter

$$\lambda = \varepsilon_* \nu_*^4/\alpha, \tag{11}$$

which is less affected by individual variations of the peak shape more dependent on average spectral properties. They found $\lambda = 1.6 \times 10^{-4}$ independent of other parameters.

Mitsuyasu et al. (1980) examined these parameters and deduced that

$$\gamma = 4.42\nu_*^{3/7} \tag{12}$$

and

$$\lambda = (2\pi)^{-4} \gamma^{1/3}/5. \tag{13}$$

Fig. 4 represents a comparison of the data with the relationships (11)-(13). We found that the constancy of λ can vary an order of magnitude or more; γ versus ν_* correlates poorly; λ versus γ seems to have some correlation, but differs from Eq. (13). Furthermore the JONSWAP contention that fully-developed waves are those with $\gamma = 1$ and $\nu_* < 0.004$ is unrealized from these data. With many data having $\gamma < 1$ and $\nu_* < 0.004$, the data seem to be indifferent to the significance of these demarcations. These results cast doubt on the existence of fully developed waves, especially in the Great Lakes.

3.4. Correlations Between ε_* and ν_*

Deducing from nonlinear energy transfer calculations with an empirical calibration using JONSWAP data, Hasselmann et al. (1976) give the following relationship between nondimensional energy ε_* and nondimensional peak frequency ν_*

$$\varepsilon_* = 5.30 \times 10^{-5} \nu_*^{-10/3}.$$ (14)

They indicated that Eq. (14) is an equilibrium curve. As the waves grow, the equilibrium state migrates to the left along the curve toward lower frequencies and higher energies.

Toba (1978) proposed a similar relationship with a different coefficient and exponents

$$\varepsilon_* = 2.24 \times 10^{-5} \nu_*^{-3}$$ (15)

Based on their own data, Mitsuyasu et al. (1980) found that

$$\varepsilon_* = 2.16 \times 10^{-5} \nu_*^{-3}.$$ (16)

Based on JONSWAP results, Donelan (1977) deduced still another form with a different coefficient

$$\varepsilon_* = 6.23 \times 10^{-5} \nu_*^{-10/3}.$$ (17)

Hence we have exponents of either $-10/3$ or -3 with different coefficients by different authors. Fig. 5 shows a comparison between the 880 data points and the correlations (14)-(17). We find that the data appear to be well correlated. This is one correlation that may have a universal application. However, the four proposed relations fit the data with varied degrees of closeness. Also the data points still cover a narrow region that does not allow a rational assessment of which exponent it is better to use. By visual observations, Donelan's Eq. (17) seems to fit this data set better.

4. CONCLUDING REMARKS

We have presented a detailed examination of various parametric correlations for their universality using a large number of data points. We found many correlations where data points are scattered widely. Some,

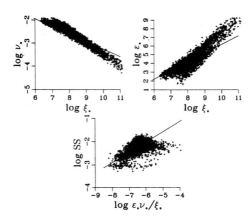

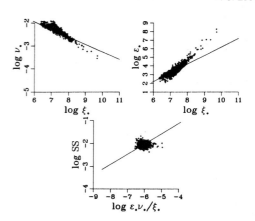

Fig. 1. Corelations with fetch
parameter (all data).

Fig. 2. Same as Fig. 1 for wave
heights greater than 1 m.

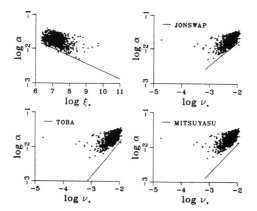

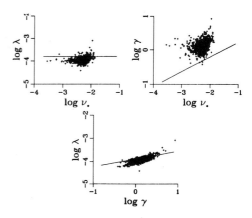

Fig. 3. Corelations
with Phillips' α .

Fig. 4. Corelations with shape
factors.

where points are clustered around a distinctive region rather than a straight line, may be considered quasi-universal correlations. The variations within the region can be up to one order of magnitude or more. Applying one of these relations in practice can be quite accurate at times and erroneous at other times.

We were not able to find clear discernible seasonal effects to sort the correlations. But we found that distinctive correlations are produced under different wind speeds. Clearly, wind speed is the only major factor that controls the correlations.

Among the correlations, the one that consistently exhibits a universal behavior is that of nondimensional energy ε_* versus nondimensional peak frequency ν_*. This confirms the work of Toba (1978) and Mitsuyasu et al. (1980). Application of this correlation in furthering model development has been fruitful. Applying the ε_* versus ν_* power law (15), Toba and his colleagues (Kawai et al., 1979) developed a successful wave prediction model. Using Donelan's correlation (17) for parameterization, Schwab et al. (1984) and Liu et al. (1984) also developed a simple wave prediction model with clearly tractable empiricism that produced quite satisfactory results.

Until truly universal relations are found and if preciseness is not specifically required, the currently available correlations can be used with limited success. However, the task of searching for universal parametric correlations for wind waves continues to confront oceanographers.

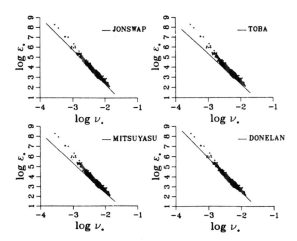

Fig. 5. Correlation between energy and peak frequency.

5. REFERENCES

Businger, J. A., J. C. Wyngaard., Y. Izumi and E. F. Bradley, 1971:
    'Flux-profile measurements in the atmospheric surface layer.' J.
    Atmos. Sci., 28, 181–189.

Charnock, H., 1955: 'Wind stress on a water surface.' Quart. J. Roy.
    Meteorol. Soc., 81, 639.

Donelan, M. A., 1977: A simple numerical model for wave and wind
    stress prediction. Report, National Water Research Institute,
    Burlington, Ont., Canada, 28 pp.

Hasselmann, K., 1977: On the spectral energy balance and numerical
    prediction of ocean waves. Turbulent Fluxes Through the Sea
    Surface, Wave Dynamics and Prediction, A. Favre and K.
    Hasselmann, Eds., Plenum Press, 531–543.

_____, T. P. Barnett, E. Bouws, H. Carlson, D. E. Cartwright,
    K. Enke, J. A. Ewing, H. Gienapp, D. E. Hasselmann, P. Kruseman,
    A. Merrburg, P. Muller, D. J. Olbers, K. Richter, W. Sell and
    H. Walden, 1973: 'Measurements of wind-wave growth and swell
    decay during the Joint North Sea Wave Project (JONSWAP).' Deut.
    Hydrogr. Z., A12, 95 pp.

_____, D. B. Ross, P. Muller and W. Sell, 1976: 'A parametric
    wave prediction model.' J. Phys. Oceanogr., 6, 200–228.

Huang, N. E., S. R. Long and L. F. Bliven, 1981: 'On the importance of
    the significant slope in empirical wind wave studies.' J. Phys.
    Oceanogr., 11, 569–573.

Kawai, S., P. S. Joseph and Y. Toba, 1979: 'Prediction of ocean waves
    based on the single-parameter growth equation of wind waves.' J.
    Oceanogr. Soc. Japan, 35, 151–167.

Liu, P. C., D. J. Schwab and J. R. Bennett, 1984: 'Comparison of a
    two-dimensional wave prediction model with synoptic measurements
    in Lake Michigan.' J. Phys. Oceanogr., 14, 1514–1518.

Mitsuyasu, H., F. Tasai, T. Suhara, S. Mizuno, M. Ohkusu, T. Honda and
    K. Rikiishi, 1980: 'Observation of the power spectrum of ocean
    waves using a cloverleaf buoy.' J. Phys. Oceanogr., 10, 286–296.

Phillips, O. M., 1958: 'The equilibrium range in the spectrum of wind-
    generated waves.' J. Fluid Mech., 4, 426–434.

_____, 1977: The Dynamics of the Upper Ocean, 2nd Ed.,
    Cambridge University Press, Cambridge, England, 336 pp.

Schwab, D. J., J. R. Bennett, P. C. Liu and M. A. Donelan, 1984:
    'Application of a simple numerical wave prediction model to Lake
    Erie.' J. Geophys. Res., 89(C3), 3586–3592.

Steele, K., and A. Johnson, Jr., 1977: Data buoy wave measurements.
    Ocean Wave Climate, M. D. Earle and A. Malahoff, Eds., Plenum
    Press, 301–316.

Toba, Y., 1978: 'Stochastic form of the growth of wind waves in a
    single-parameter representation with physical implications.' J.
    Phys. Oceanogr., 8, 494–507.

SEA-STATE CYCLES

L. SCHMIED
Service Technique des Phares et Balises
Boîte Postale 12
94381 Bonneuil sur Marne Cédex
FRANCE

ABSTRACT. The sea-state cycle is the concept which describes the growth and the decay of a storm. The sea-state cycle is the response of the upper ocean to a low pressure system occuring at a specific point. Approximately 10 000 wave records were analyzed, taken at 3 different points in northwestern Europe over a period of several years. Generally speaking the nondimensional "pseudo-steepness" $\hat{PS} = \dfrac{2\pi(H1/3)}{g(TH1/3)^2}$ remains approximately constant during the growth and the decay of a storm.

For each specific point it was possible to classify the various sea-state cycles according to their pseudo-steepness.

Each type of sea-state cycle was found to correspond to a specific type of atmospheric depression or front.

We deduced a parabolic relation between significant wave height and peak period from Kitaigoroskii's dimensional analysis and a similar equation from PIERSON-MOSKOWITZ' spectrum : $H1/3 = 0.0787 (TM)^2$.

INTRODUCTION. A description of the cyclic behavior of waves is given here. The unity of this cyclic behavior is characterized by the pseudo-steepness of the waves. This sea-state cycle is interpreted from a physical and a climatological points of view. The sea-state cycle is the concept which describes the growth and the decay of a storm, with the every day meaning of this last word. This concept can also be applied to the fully developed sea-state, the non fully developed sea-state and the swell.

The sea-state cycle is the response of the upper ocean to an atmospheric disturbance, occuring at a specific point, which can be far away in the swell case. The duration of a sea-state ranges from 12 hours to 8 days. This period of time corresponds to the duration of the activity of a low pressure system over the upper ocean surface. Roughly 10 000 records were analyzed at specific points in northwestern Europe over a period of several years (Datawell wave rider buoys were used for these experiments) :
- Le Havre (The English Channel - France) 4 year measurements.
- Biarritz (Bay of Biscaye - France) 2 year measurements.

Y. Toba and H. Mitsuyasu (eds.), The Ocean Surface, 179–184.
© 1985 by D. Reidel Publishing Company.

- Stevenson (North East of the Shetland Islands Great Britain) 3 year measurements.

Statistical, meteorological and physical studies were carried out simultaneously.

1.   STATISTICAL ANALYSIS

A sea-state cycle was studied in a 2 dimensional space (TH1/3, H1/3), characteristic parameters of a 20 minute wave record. The parameters were calculated from a zero-up-crossing analysis of the wave signal.

H1/3 is the mean value of the top third of the highest waves, expressed in meters.

TH1/3 is the mean period of the top third of the highest waves, expressed in seconds.

TM is the mean period of all the waves, expressed in seconds.

A sea-state cycle is the time series which starts from the lowest part of the (TH1/3, H1/3) space, reaches the maximum of the storm and decreases to the lowest part, the so-called "calm sea" area as seen in figure 1 (Le Havre), figure 2 (Stevenson) and figure 3 (Biarritz). The storm can be followed from its very beginning, every 3 or 4 hours (two consecutive records (+) are linked by a segment) to its very end.

The nondimensional pseudo-steepness $\tilde{PS} = \dfrac{2\pi(H1/3)}{g\,(TH1/3)^2}$ remains roughly constant during the growth an the decay for both fully developed and non-fully developed sea-state (figures 1 and 2). Generally speaking the nondimensional pseudo-steepness $\tilde{PS}$ is steeper during its growth than during its decay, except for swell cycles.

Swell cycles are characterized by a fairly constant H1/3, the periods growing and decaying as shown in figure 3 (Biarritz).

Interactions between swell and current are specific features of Le Havre - figure 4 : type 3 sea-state cycle is found to be correlated with the periodic tidal current.

At Le Havre 253 sea-state cycles were identified and nearly half of them were found to be swell-current interaction cycles.

At Biarritz 50 sea-state cycles were identified.

At Stevenson 172 cycles were found.

The nondimensional pseudo-steepness $\cdot \tilde{PS} = \dfrac{2\pi(H1/3)}{g\,(TH13)^2}$ is the signature of the storm, $\dfrac{2\pi(H1/3)}{g\,(TM)^2}$ can also be considered.

We were able to classify the various storms according to their pseudo-steepness (CADENAT and SCHMIED, 1980). This classification was made by considering the different modes of the probability distribution of the pseudo-steepness parameter $\tilde{PS}$.

At Le Havre Type 1 sea-state cycles are characterized by their high pseudo-steepness.

Type 2 sea-state cycles are characterized by lower $\tilde{PS}$.

Type 3 sea-state cycles are the result of the swell-current interaction considering the periodic tidal currents.

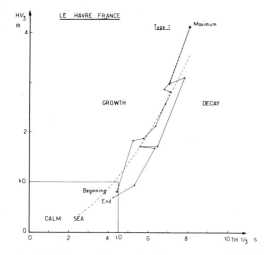

FIGURE 1 : LE HAVRE – TYPE 1 SEA-
STATE CYCLE

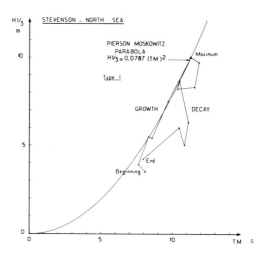

FIGURE 2 : STEVENSON – A FULLY
DEVELOPED SEA-STATE CYCLE

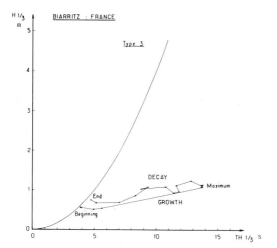

FIGURE 3 : BIARRITZ – A SWELL SEA-
STATE CYCLE

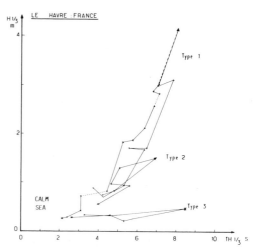

FIGURE 4 : LE HAVRE – 3 DIFFERNT
TYPES OF SEA-STATE CYCLES

Figure 4 illustrates these 3 types.

2.   METEOROLOGICAL ANALYSIS

Coudert (1978) studied the climatology of the sea-state cycles. Generally
it was possible to identify the overall meteorological situation which
generates a sea-state cycle. An overall meteorological situation is
described by the following set of meteorological data : the position and
the intensity of the low and high pressure centers, weather-type, the
position of the atmospheric fronts, geostrophic winds, etc...

Objectives :

First, to identify the overall meteorological situation which generates a
sea-state cycle.
Secondly, to find out if the overall meteorological situations of
all the cycles could be classified into types (1, 2, 3) as defined in the
above section.
Each type of sea-state cycle as described by analytical (pseudo-
steepness) and statistical approaches was found to correspond to a
specific type of overall meteorological situation.
The location of the track of the low pressure centers is the main
feature which characterizes the sea-state type. Generally, the generating
area (fetch zone) could be separated from the propagation area.
In figure 5, each arrow represents the track of the low pressure
center during its wind action towards Le Havre. The fetch area is limited
by the British and French coasts.
In figure 5, type 1 low pressure center tracks are located in a
specific zone delineated by the northern French coastline, the 62nd North
parallel and the 20th West meridian. Fetches are rather long. The maximum
of the cycle is correlated with the cold fronts passing over Le Havre.
In figure 5, the tracks of the low pressure centers for type 2 sea-
state cycles lie outside the type 1 zone except for one track. This type
of weather generates eastwinds in Le Havre area.
The tracks of low-pressure centers for type 3 are slightly different
from type 2. This corresponds to an anticyclonic weather over France. In
this case the time series is correlated with the tide period so we can
say that type 3 is the result of swell-current interaction.

3.   PHYSICAL ANALYSIS

Pseudo-steepness is the main parameter which characterizes the types of
sea-state cycles. We found that this parameter plays a major role in
Kitaigorodskii's dimensional analysis and its application to the Pierson-
Moskowitz model.
Kitaigorodskii (1973) applied dimensional analysis to the fully
developed sea-state. On the basis of his work we deduced a parabolic
relation between H1/3 and Tp (peak period of the wave spectrum).

$$H1/3 = \frac{\sqrt{2}}{2\pi^2} \; g \; C_1C_2{}^2(Tp)^2$$

where C_1 is a constant. $C_1 = \dfrac{g\,\sigma_\eta}{U_\infty^2}$

and C_2 is also constant $C_2 = \dfrac{U_\infty\,\omega_0}{g}$, $\omega_0 = \dfrac{2\pi}{Tp}$, $\sigma_\eta^2 = 2\displaystyle\int_0^{+\infty} S(\omega)\,d\omega$

U_∞ is the average wind speed of the upper edge of the boundary layer. g, the acceleration of gravity.

We also inferred a similar parabolic equation from Pierson-Moskowitz spectrum between H1/3 and T.

$\overline{T}$ is the average period calculated from 0 and 2 order moments of the spectrum. Neumann and Pierson (1966) found the relations : $\overline{T} = 0.81 \dfrac{2\pi U}{g}$

and $H1/3 = 2.12 \times 10^{-2}U^2$. U was elimited in the two above formulae,

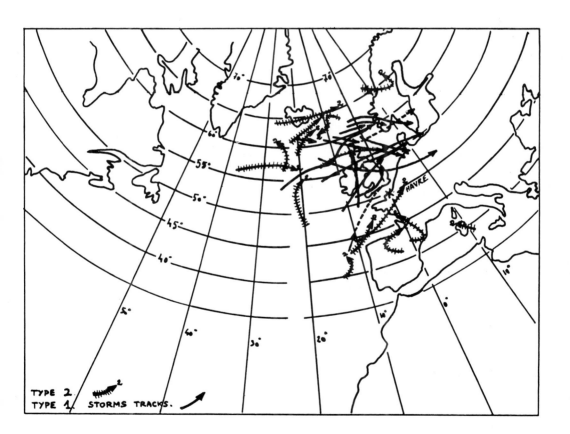

FIGURE 5 : TYPE 1 CYCLONE TRACKS AND TYPE 2 CYCLONE TRACKS

leaving : $\boxed{\text{H1/3} = 0.0787 \, (\overline{T})^2}$ (1) defining the nondimensional pseudo-

steepness $\widetilde{PS} = \dfrac{2 \, \pi \, (H1/3)}{g(\overline{T})^2}$ we have : $\widetilde{PS} = 0.0504$.

According to Kitaigorodskii's analysis and Pierson-Neumann quantification of the spectrum, nondimensional pseudo-steepness $\dfrac{2 \, \pi \, (H1/3)}{g \, (\overline{T})^2}$ remains constant during the fully developed sea-states, regardless of the energy level.

The Pierson-Moskowitz parabola (1) was used as a standard to measure fully developed sea-state cycles : at Stevenson point we found sea-state cycles which follows the Pierson-Mokoswitz parabola upwards (figure 3). These results suggest that, for some cycles, the fully developed sea-state occurs at each step of the cycle. This provides an explanation of the parabolic behavior of the sea-state cycle for the fully developed sea state.

In conclusion, the sea-state cycle concept describes the behavior of fully and non-fully developed sea-states and even a swell considering the time evolution. The pseudo-steepness is the main characteristic parameter for classifying the cycles into different types. This parameter is related to the overall meteorological situation which has been generating the cycle, in particular the position of the low pressure center track.

Kitaigorodskii's dimensional analysis leads to a parabolic relation between Tp and H1/3 which shows the majors role played by the pseudo-steepness for fully developed sea-state.

REFERENCES

CADENAT M. - SCHMIED L., Les cycles d'état de la mer, Revue Technique des Phares et Balises, Février 1980, supplément.

COUDERT E., Etude des relations entre les situations atmosphériques et les états de mer, Mars 1978, Association de Recherche de l'Action des Eléments.

KITAIGORODSKII S.A., The physics of air, sea interactions, Jerusalem : Israel Prog. Sci. Transl. 1973.

NEUMANN G. - PIERSON jr W.J., Principles of Physical Oceanography, Prentice Hall - 1966.

ON THE EFFECT OF BOTTOM FRICTION ON WIND SEA

Peter A.E.M. Janssen[1] and Willem J.P. de Voogt[2]
1 Department of Oceanography, Royal Netherlands
  Meteorological Institute (KNMI), P.O. Box 201,
  3730 AE De Bilt, The Netherlands.
2 Delft Hydraulics Laboratory, De Voorst Laboratory,
  P.O. Box 152, 8300 AD Emmeloord, The Netherlands.

ABSTRACT. We have investigated the effect of bottom friction on the
evolution of wind sea for a constant wind speed. This was done by
integrating the transport equation for the energy density over
frequency and direction, assuming the Kruseman spectrum for wind
sea. The wind input term was derived from empirical data whereas the
bottom friction term was obtained from an average (over frequency
and direction) of the linearized JONSWAP dissipation form. For a
flat bottom and infinite fetch we obtained the following results:
(i)   For infinite duration the results for the nondimensional wave
      height and peak period versus nondimensional depth are in good
      agreement with a number of measurements.
(ii) The saturation level of the wave energy is rather insensitive
      for variations in the dissipation coefficient.
From this we may conclude that our results apply for different types
of bottoms, thereby explaining why at sea approximately the same
saturation level for wind sea is found for different bottoms.

1.   INTRODUCTION

In this note we discuss our attempt to model the behaviour of sea
waves in shallow water. Depth limitations are of great significance
in practical applications concerning offshore constructions and
coastal protection. Typically, extreme storms generate in the
southern part of the North Sea (depth ~ 25 m) wave heights that are
only half as large as in the deeper central part. This sizeable
energy reduction occurs because energy dissipation by bottom effects
can be important. This note only discusses this mild shallow water
effect. We exclude situations of extreme shallow water where bottom-
induced nonlinear effects, such as wave breaking, are significant
for the energy balance. Also the effects of refraction and currents
are disregarded. Thus our study is concerned with processes well
outside the surfzone, where wave heights do not reach the order of
magnitude of the local water depth.

Y. Toba and H. Mitsuyasu (eds.), The Ocean Surface, 185–192.
© 1985 by D. Reidel Publishing Company.

2.   THE PARAMETRIC DESCRIPTION OF WIND SEA

In deep water, wind sea is defined in the ideal case of a uniform
wind field. It is that part of the spectrum that is affected by wind
input (i.e. that feels the effect of the wind). Waves under the
action of wind may be steep, so that nonlinear interactions may be
important. These interactions result in the invariant shape of the
wind-sea energy spectrum that has been observed in field
measurements.

In the wave model GONO the spectral distribution of wind-sea
energy is given by

$$F_{wind-sea} = \frac{2}{\pi} \cos^2 (\theta - \phi) E(f), \quad |\theta - \phi| < \frac{\pi}{2}, \tag{1}$$

where $\phi = \phi(x,y,t)$ is the wind direction and $E(f)$ is the simple
spectral shape (Kruseman, 1976),

$$E(f) = \frac{\hat{\alpha} g^2}{(2\pi)^4} \begin{cases} 0 & , & f < f_{min}, \\ \dfrac{1}{f_p^5} \dfrac{f-f_{min}}{f_p - f_{min}} & , & f_{min} \leqslant f \leqslant f_p, \\ \dfrac{1}{f^5} & , & f > f_p, \end{cases} \tag{2}$$

where f_{min} is the minimum frequency and f_p the peak frequency of the
spectrum. The parameter $\hat{\alpha}$ and the dimensionless peak frequency $\nu =
U_{10} f_p / g$ (U_{10} is the wind speed at 10 m height) depend on the stage
of development parameter

$$\xi = (\varepsilon / \varepsilon_{max})^{1/4}, \quad 0 < \xi < 1, \tag{3}$$

where ε is the total wind-sea energy,

$$\varepsilon = \iint df d\theta \, F_{wind-sea} = \frac{\hat{\alpha} g^2}{(2\pi)^4} \frac{1}{4 f_p^4} (3 - 2\mu), \quad \mu = f_{min}/f_p, \tag{4}$$

and ε_{max} is its maximum value at a given wind speed U_{10} and at
infinite depth. From observations the following relations between
$\hat{\alpha}$, ν and ξ have been obtained (Sanders et al., 1981),

$$\hat{\alpha} = 4.93 \ 10^{-3} \ \xi^{-1.944},$$

(5)

$$\nu = 6.89 \ 10^{-2} \ \beta^{-\frac{1}{2}} \ \xi^{-1.376},$$

where $\beta = 4g\varepsilon_{max}^{\frac{1}{2}}/U_{10}^2$. From (3), (4) and (5) the spectral parameters (and hence the spectral shape) are obtained once the total energy is known.

We now make the important assumption that the spectral shape (1) and (2), and the relations for the spectral parameters (4) and (5) are also valid in shallow water. The only difference with the deep water situation is then that in shallow water the rate of change of the total energy is determined not only by wind input and white capping but also by dissipation through bottom friction. Hence, the behaviour of the energy in time and position (and also the spectral distribution) is different from that in deep water.

3.   THE ENERGY BALANCE EQUATION

If one is interested only in cases of uniform wave fields, neglecting refraction and local currents, the evolution of wind-sea energy follows from the integrated transport equation (cf. Janssen et al., 1984),

$$\frac{\partial}{\partial T} \tilde{\varepsilon} + \frac{\partial}{\partial X} \tilde{v} \ \tilde{\varepsilon} = \langle \tilde{S}_{in} \rangle_{f,\theta} + \langle \tilde{S}_b \rangle_{f,\theta} ,$$

(6)

where $\langle ... \rangle_{f,\theta}$ means integration over frequency and direction, and where we have introduced the nondimensional quantities

$$T = gt/U_{10}, \ X = gx/U_{10}^2, \ \tilde{\varepsilon} = g^2\varepsilon/U_{10}^4, \ \tilde{v} = v/U_{10}.$$

(7)

The transport velocity v depends on μ, and is directly proportional to the group velocity associated with the spectral peak (cf. Janssen et al., 1984). The source term representing energy gain due to wind input and energy dissipation due to wave breaking is obtained from an empirical growth curve, and takes the analytical form

$$\langle \tilde{S}_{in} \rangle = \frac{ab}{8} \ \beta^2 \ \xi^2 \ (1-\xi^4) \ [\frac{1}{2a} \ \ln \ \frac{1+\xi^2}{1-\xi^2} \ ]^{\frac{b-1}{b}} ,$$

(8)

with a = 0.00061, b = 0.75, and β = 0.22. Following JONSWAP (1973) the dissipation of wave energy due to bottom friction is given by the linearized form

$$S_b = -\frac{\hat{\Gamma}}{gD} \ \Phi \ (\omega_D) \ E(f) ,$$

(9)

where $\Phi(\omega_D) = \omega_D^2/\sinh^2 kD$, $\omega_D = \omega(D/g)^{\frac{1}{2}} = (kD \tanh kD)^{\frac{1}{2}}$, D is the water depth, k is the wave number and $\hat{\Gamma}$ is the decay parameter, the average value of which is 0.039 $m^2 s^{-3}$. The dependence of Φ on ω_D is shown in Fig. 1.

4.   AN APPROXIMATION OF THE BOTTOM FRICTION FUNCTION

It is not possible by means of (2) and (9) to obtain an analytical expression for the averaged dissipation source term $\langle \tilde{S}_b \rangle$. However, from Fig. 1 it can be seen that to first order the function $\Phi(\omega_D)$ may be approximated by

$$\Phi = p - q\omega_D \quad , \quad 0 \leqslant \omega_D \leqslant \frac{p}{q}, \ p,q = \text{constant.} \tag{10}$$

Then, from (10) and the Kruseman spectrum we obtain

$$\langle \tilde{S}_b \rangle = - \frac{2}{3} \frac{U_{10} \hat{\Gamma}}{g^2 D} \frac{p}{\delta(3-2\mu)} \quad x$$

$$x \begin{cases} 0 & , \quad \delta < \mu , \\ (\delta-\mu)^3/(1-\mu) & , \quad \mu \leqslant \delta \leqslant 1 , \\ (\frac{9}{2}\delta + \mu - 3\delta\mu + \mu^2 + \frac{1}{2\delta^3} - 4), & \delta > 1 , \end{cases} \tag{11}$$

in which $\delta = f_o/f_p$, and $f_o = (p/2\pi q)(g/D)^{\frac{1}{2}}$.

In order to gain insight in the accuracy of the above approximation we have compared numerical results from (11) with exact numerical solutions from (9), for $\langle S_b \rangle$, for sea states in which bottom effects contribute significantly to the energy balance, i.e. for $0.1 \langle S_{in} \rangle \leqslant |\langle S_b \rangle| \leqslant \langle S_{in} \rangle$. The calculations were performed for wind speeds in the range 10-30 m/s, and for water depths in the range 15-35 m. These ranges cover the most severe realistic situations. By trial and error the values for -p- and -q- in (10) were found to be 1.1 and 0.66, respectively, for satisfying overall results (see also the broken line in Fig. 1). An example of these comparisons is shown in Fig. 2 where graphs of $\langle S_{in} \rangle/\varepsilon$ from (8) (broken line), and $|\langle S_b \rangle|/\varepsilon$ from (9) (exact; full line) are presented for $U_{10} = 20$ m/s and D = 15 m. The approximate values of $|\langle S_b \rangle|/\varepsilon$ from (11) are indicated by crosses in the same figure. It is generally found that within 10% error the JONSWAP form for the attenuation of wave energy by bottom friction is adequately modeled by (11).

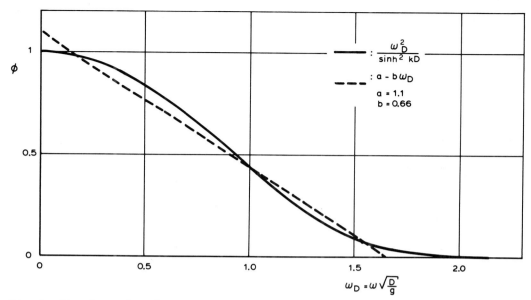

Fig.1   The function Φ versus ω_D.

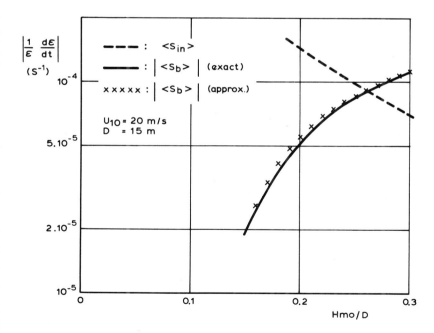

Fig.2   The various source functions versus H_{mo}/D, for U_{10}= 20 m/s and
        D=15 m.

5.   RESULTS

We have simulated depth-limited wave growth by numerically solving
(6), using (8) and (11), for constant uniform wind fields with
speeds 5-30 m/s and constant depths in the range 5-100 m. We started
from an infinite, flat sea so that advection in (6) could be
neglected, and we continued the integration until a stationary state
had been reached. Regarding this stationary state we found that the
saturation level for the zero-moment wave height H_{mo} = $4\sqrt{\varepsilon}$ is rather
insensitive to variations in the values of the dissipation

coefficient $\hat{\Gamma}$, if $\hat{\Gamma} > 0.02$ m^2s^{-3}. This is illustrated in Fig. 3

where we have plotted the maximum nondimensional wave height

gH_{mo}/U_{10}^2 as a function of $\hat{\Gamma}/g^2$, for U_{10} = 10 m/s and D = 5 m. This
result implies that for various bottom conditions (e.g. mud, sand,
pebbles, etc.) giving different dissipation coefficients, approximately the same values for the depth-limited wave heights are
found.
        For infinite fetch and -duration, the computational results for
the dimensionless wave height $\tilde{H}$ = $4\sqrt{\varepsilon}$ and dimensionless period $\tilde{T}$ =

$1/\nu$ as a function of non-dimensional depth $\tilde{D}$ = gD/U_{10}^2 agree well

with similar data originally compiled by Holthuijsen (1980), cf.
Fig. 4. The data are from various sources, i.e. Bretschneider
(1954), Roest (1960), V.d. Molen (1972), and De Reus (1977). It is
not clear whether the measurements were performed under flat or
sloping bottom conditions, and whether the data points really
represent fully grown sea states at the given depth. Irrespective of
this, however, it is clear from Fig. 4 that the data show a strong
dependence of $\tilde{T}$ on $\tilde{D}$, a dependency which is well predicted both
qualitatively and quantitatively by our model, see also SWIM, Part
II (1985), It should be noted that according to our model for depth-
limited wave growth the nondimensional wave height and -period not
only depend on the nondimensional depth, but also explicitly on the
wind speed U_{10}. This follows from the expression for $\langle \tilde{S}_b \rangle$ in (11),

and it is illustrated in Fig. 4 where the results for $\tilde{H}$ versus $\tilde{D}$ are

given for both U_{10} = 10 and 30 m/s. The results for $\tilde{T}$ are for U_{10} =
10 m/s only.

6.   CONCLUSIONS

In this note we have discussed the effect of bottom friction on the
modeling of wind-sea development in shallow water. This we have
studied by supplementing the energy balance equation with an
adequate approximation of the average dissipation function for
bottom friction. Further, the effective transport velocity is taken
depth dependent. Moreover, our model assumes that the spectral
parameters depend in the same manner on the total energy as in deep

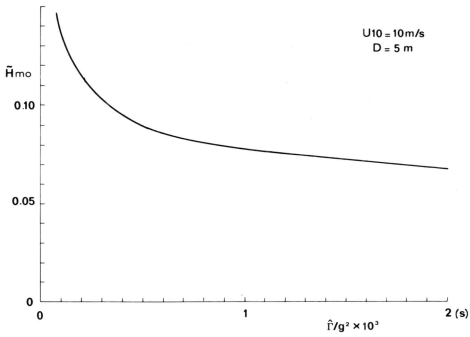

Fig.3 $\tilde{H}_{mo}$ as a function of the decay parameter $\hat{\Gamma}$.

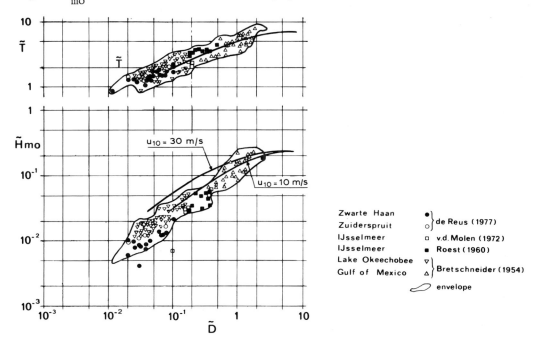

Fig.4   Model and experimental results for $\tilde{H}_{mo}$ and $\tilde{T}$ versus $\tilde{D}$, for infinite fetch and duration.

water. Our results for wave height and peak period appear to be in
agreement with observations. The above technique has been
implemented in the operational version of the wave prediction model
GONO in order to achieve better performance in shallow water.

ACKNOWLEDGEMENT

This work has been performed as part of a joint wave modeling
program of Royal Netherlands Meteorological Institute (KNMI),
Ministry of Transport and Public Works (Rijkswaterstaat) and Delft
Hydraulics Laboratory.

REFERENCES

Bouws, E., J.J. Ephraums, J.A. Ewing, P.E. Francis, H. Günther,
    P.A.E.M. Janssen, G.J. Komen, W. Rosenthal, and W.J.P. de
    Voogt, 1985: Shallow water intercomparison of models (SWIM),
    Part II: Results of three different wave models for idealized
    wind and depth situations. Proc. Symp. on Wave Breaking, Turb.
    Mixing, and Radio Probing of the Ocean Surface, Sendai, Japan.
Bretschneider, C.L., 1954: Field investigation of wave energy loss
    in shallow water ocean waves. Techn. Memo No. 46, Beach Erosion
    Board, Corps of Engineers, Dept. of the Army.
Hasselmann, K., T.P. Barnett, E. Bouws, H. Carlson, D.E. Cartwright,
    K. Enke, J.A. Ewing, H. Gienapp, D.E. Hasselmann, P. Kruseman,
    A. Meerburg, P. Müller, D.J. Olbers, K. Richter, W. Sell, and
    H. Walden, 1973: Measurements of wind-wave growth and swell
    decay during the Joint North Sea Wave Project (JONSWAP), Dtsch.
    Hydrogr. Z., A8(12).
Holthuijsen, L.H., 1980: Methoden voor golfvoorspelling. Report
    Technische Adviescommissie voor de Waterkeringen, Holland.
Janssen, P.A.E.M., G.J. Komen, and W.J.P. de Voogt, 1984: An
    operational coupled hybrid wave prediction model. J. Geophys.
    Res. 89, 3635-3654.
Kruseman, P.J., 1976: Twee practische methoden voor het maken van
    verwachtingen van golfcomponenten met perioden tussen 10 en 25
    seconden nabij Hoek van Holland. Rep. WR 76-1, R. Neth. Meteor.
    Inst., De Bilt, Holland
Molen, H.W.B. van der, 1972: Metingen voor een golfhoogte- en
    lengtediagram. Report B 72-27, Rijkswaterstaat, Dienst der
    Zuiderzeewerken, Afd. Waterloopkunde, Holland.
Reus, J. de, 1977: Toetsing van golfvoorspellingsmethoden aan
    gemeten waarden, afkomstig van golfmetingen in ondiep water.
    Technische Hogeschool Delft, Afdeling der Civiele Techniek,
    Holland.
Roest, P.W., 1960: Wave recording on the IJsselmeer, Proc. 7th
    Coastal Eng. Conf., Den Haag, Council Wave Res., Berkeley, 53-
    58.
Sanders, J.W., W.J.P. de Voogt, and J. Bruinsma, 1981: Fysisch golf-
    onderzoek Noordzee. Sci. Rep. MLTP-2, Raad Overleg Fys.
    Oceanogr. Onderz. Noordzee, De Bilt, Holland

A PARAMETRIC WIND WAVE MODEL FOR ARBITRARY WATER DEPTHS

Hans C. Graber[1] and Ole S. Madsen
Ralph M. Parsons Laboratory
Massachusetts Institute of Technology
Cambridge, MA 02139, USA

ABSTRACT: A hybrid parametric wind wave model for arbitrary water depths is presented. The model is derived from an energy flux transport formulation and includes shoaling, refraction, bottom frictional dissipation, as well as finite depth modifications of the atmospheric input and nonlinear wave-wave interaction source terms. The model is applied to predict the wave characteristics resulting from a complex frontal system which passed over the Atlantic Remote Sensing Land Ocean Experiment (ARSLOE) experimental site on October 25, 1980. The overall agreement between predicted and observed wave characteristics (significant wave height, peak frequency and mean wave direction) in 35 m water depth is considered excellent.

1. INTRODUCTION

Several deep water wind wave models have been developed and applied extensively in the past (Cardone et al., 1976; Günther et al., 1979). Waves produced by severe storm events often have peak frequencies below 0.08 Hz and will "feel" the presence of the bottom in water depths less than 100 m. A logical and appropriate extension of deep water wind wave models would therefore be their explicit inclusion of finite depth effects. In the present paper, the hybrid parametric wind wave model (Hasselmann et al., 1976; Günther et al., 1979) is extended to include shoaling and refraction. The model, briefly discussed here, is based on an energy flux transport formulation rather than the momentum approach taken by SWIM-GROUP (1985). In addition, the present model differs from that of SWIM-GROUP (1985) by its inclusion of bottom frictional dissipation and finite depth modifications of the remaining source terms. The model is applied to predict the wave characteristics resulting from a complex frontal system which passed over the ARSLOE experimental site on October 25, 1980. The overall agreement between predicted and observed wave characteristics (significant wave height, peak frequency and mean direction) in 35 m water depth is considered excellent.

---

1. Presently in the Ocean Engineering Department, Woods Hole Oceanographic Institution, Woods Hole, MA 02543, USA.

Y. Toba and H. Mitsuyasu (eds.), The Ocean Surface, 193–199.
© 1985 by D. Reidel Publishing Company.

2. DISCUSSION OF FINITE DEPTH WIND WAVE MODEL

The inclusion of a mean wave direction as one of the parameters describing the windsea necessitates the adoption of a vector quantity, which we have chosen as the wave energy flux $\mathcal{F}(k,r,t) = c_g(k,r) \, F(k,r,t)$ in which $\underline{k} = (k_x, k_y)$ and $\underline{r} = (x,y)$ are the wave number and position vectors, respectively. $\underline{c_g}$ is the group velocity vector, and F is the energy spectrum. Integration of the energy flux vector over wave number space defines a mean wave direction $\Theta_0 = \arctan(I_x/I_y)$, in which I_x and I_y are the total energy flux components in the x-direction (East) and y-direction (North), respectively, with Θ_0 being positive clockwise from North.

In the derivation of a conservation equation for energy flux we make use of the following identity

$$\frac{D\underline{\mathcal{F}}}{Dt} = \frac{\partial \underline{\mathcal{F}}}{\partial t} + (\underline{\dot{r}} \cdot \nabla_r)\underline{\mathcal{F}} + (\underline{\dot{k}} \cdot \nabla_k)\underline{\mathcal{F}} = \underline{c}_g \frac{DF}{Dt} + F \frac{D\underline{c}_g}{Dt} \qquad (1)$$

Transforming the energy spectrum from wave number space to frequency-direction space, defined by $F(k) = J \cdot F(f,\Theta)$, where J is the Jacobian of the transformation, (1) may be written in terms of the energy flux density $\underline{\mathcal{E}}(f,\Theta) = \underline{c}_g(f,h) \, F(f,\Theta)$

$$\frac{\partial \underline{\mathcal{E}}}{\partial t} + (\underline{c}_g \cdot \nabla_r)\underline{\mathcal{E}} + J^{-1}\underline{\mathcal{E}}(\underline{c}_g \cdot \nabla_r)J + (\underline{\dot{k}} \cdot \nabla_k)\Theta \frac{\partial \underline{\mathcal{E}}}{\partial \Theta}$$

$$= \underline{c}_g T + \underline{\mathcal{E}}(\nabla_r \cdot \underline{c}_g) \qquad (2)$$

The left hand side includes terms accounting for shoaling and depth refraction, while the first term on the right hand side represents the energy flux sources and sinks. By comparison with (1) it is seen that T represents the net addition of energy to the wave spectrum. The processes represented by T are atmospheric input by the wind, T_{in}, dissipation through wave breaking, T_{wb}, nonlinear wave-wave interaction T_{nl}, and bottom frictional dissipation, T_{bf}. In deep water wave models the wave breaking dissipation is generally considered proportional to the atmospheric wind input, which depends strongly on the ratio [wind speed/phase velocity]. In keeping with this formulation we adopt the Snyder et al. (1981) expression for $T_{in} - T_{wb}$, with a depth dependent phase velocity in the finite depth wind wave model. T_{nl} is modified to account for finite depth effects as evaluated by Hasselmann and Hasselmann (1981). The bottom dissipation term is obtained from a linearized bottom friction law, using $\sqrt{2} \, u_{rms}$ as a representative bottom velocity, and a constant wave friction factor (Grant and Madsen, 1979). A detailed derivation of the governing equation (2) and a discussion of the individual terms can be found in Graber (1984).

Adopting the arguments of Bouws et al. (1984), the depth dependent growing wind wave spectrum can be written as $F(f,\Theta;h) = \Phi(kh) \cdot E(f) \cdot \Omega(\Theta - \Theta_0)$ where $h = h(\underline{r})$ is the water depth, and the transformation factor $\Phi(kh)$ is according to Kitaigorodskii et al. (1975).

The one-dimensional windsea spectrum, $E(f)$, is described by the JONSWAP spectral shape (Hasselmann et al., 1976) defined by $E(f)=E_{PM}(f)\exp\{\ln \gamma \exp [-(f/f_m-1)^2/2\sigma^2]\}$ with the set of five free parameters $(f_m,\alpha,\gamma,\sigma_a,\sigma_b)$, and $E_{PM}(f)$, the Pierson-Moskowitz spectrum, hereafter abbreviated as the PM spectrum. Finally, we assume that angular spreading of $F(f,\Theta;h)$ is independent of frequency and water depth and obeys the usual $\cos^2$ law, distributed about the mean wave direction Θ_0.

The concept of a parametric windsea description has originally been proposed by Hasselmann et al. (1976) and was extensively tested in deep water and further developed by Günther et al. (1979, 1981). In order to transform the governing equation (2) from physical space to parametric space we need to construct mapping functionals in such a way that they reveal the JONSWAP parameters. Analogous to the approach of Günther et al. (1981) we define

$$a_i = \phi_i\{\underline{E}\} = \phi_i\left\{\frac{\int[\underline{E}_x^2(f,\Theta;h) + \underline{E}_y^2(f,\Theta;h)]^{1/2} d\Theta}{c_g(f;h) \cdot \Phi(kh)}\right\} = \phi_i\{E(f)\} \quad (3)$$

where $a_i = \{f_m,\alpha,\gamma,\sigma_a,\sigma_b\}$. With this formulation the resulting mapping operators are identical to those previously defined in Günther et al. (1979). For the directional parameter $a_6=\Theta_0$ the operation is carried out directly on the energy flux vector $\underline{E}$ (Graber and Madsen, 1982).

Applying the mapping functionals ϕ_i to the energy flux transport equation one arrives at six equations for the windsea parameters a_i

$$\frac{\partial a_i}{\partial t} + D_{ijx}\frac{\partial a_j}{\partial x} + D_{ijy}\frac{\partial a_j}{\partial y} = S_i + R_i \quad (4)$$

where D_{ij} is the generalized, depth dependent propagation matrix, S_i is the source function and R_i contains all remaining terms such as refraction, shoaling and spatially varying bottom topography. Detailed expressions for the terms in (4) may be found in Graber (1984).

The lowest peak frequency of an active windsea, the PM frequency, expresses a limiting value of the ratio (peak phase velocity/wind speed). In analogy to the previously discussed finite depth modifications of the wind input source term the finite depth PM frequency is modified to read

$$f_{PM} = \frac{0.13 \ g \ \tanh k_{PM}h}{U \cos(\Theta_w - \Theta_0)} \qquad \text{for } |\Theta_w - \Theta_0| \leq \frac{\pi}{2} \quad (5)$$

in which U and Θ_w are the 10m wind speed and direction, respectively, and k_{PM} is the wave number corresponding to f_{PM}.

With the depth dependent PM frequency defined by (5) windsea to swell and swell to windsea transfers are treated following the approach of Günther et al. (1979, 1981). Energy in the swell region is treated in a standard discrete spectral way. Individual energy packets are propagated along straight rays classified by the pair (f,Θ). Currently the hybrid wave model includes shoaling and dissipation by bottom friction in the swell model, whereas refraction effects are neglected.

3. MODEL APPLICATION TO ARSLOE STORM

During the ARSLOE experiment a sharp frontal system passed directly over the measurement site. Ahead of the frontal passage the prevailing north-easterly winds generated a rapidly growing windsea which propagated towards the shore. In the warm sector of the frontal system winds briefly turned to the south. Within a few hours after the front had passed, the winds essentially completed a 180° shift in direction and were now blowing offshore. The geostrophic model winds were obtained from 3-hourly synoptic weather maps and corrected to surface making use of surface wind measurements at NOAA buoys in the model area which extended from 29° to 41° North and from 61° to 80° West.

A summary of model results, obtained using a grid spacing of 80 km and a time step of 1 hour, is presented in Fig. 1 for the grid point at which the XERB buoy was located. The wind history shown in Fig. 1a demonstrates the relatively rapid turning of the winds around noon on October 25, 1980 associated with the frontal passage. As illustrated by the results presented in Figs. 1b and 1c, the wave conditions were exclusively windsea prior to frontal passage and predominantly swell arriving from the south at later times. The significant lag of windsea direction relative to that of the local wind during frontal passage is illustrated in Fig. 1d.

A comparison of measured and hindcasted wave parameters is shown in Fig. 2. The hindcasted wave heights (Fig. 2a) just hug the upper side of the data, following quite accurately the trend leading up to the maximum wave height. On the backside of the storm, the calculated H_S values are generally too high. This discrepancy is attributed to the neglect of refraction in the swell model. In fact, a close inspection of the model results revealed that the swell present at the XERB buoy was generated elsewhere. This observation allows us to approximately correct the predicted swell energies for refraction by a simple application of Snells Law. The corrected wave heights, shown by the dashed line in Fig. 2a, are in reasonable agreement with observations. Model runs with and without bottom friction showed little difference in the computed wave heights. The reason for this is the relatively large water depth, 35 m, at the XERB buoy. Were the model extended to cover shallower areas, bottom friction would become important. Observed and predicted peak frequencies are depicted in Fig. 2b. Here the solid line corresponds to the JONSWAP parameter fm and the dashed line indicates the frequency of the absolute spectral peak as determined from calculated model spectra. The wave directions are compared in Fig. 2c.

A detailed comparison of predicted and observed wave spectra is performed in Fig. 3. Fig. 3a represents a comparison of pure wind-sea spectra prior to peak conditions, while Fig. 3b corresponds to composite windsea-swell spectra one hour after peak wave conditions.

4. CONCLUSIONS

The finite depth wind wave model discussed and applied in the present paper was derived by extensions of methodologies and concepts developed

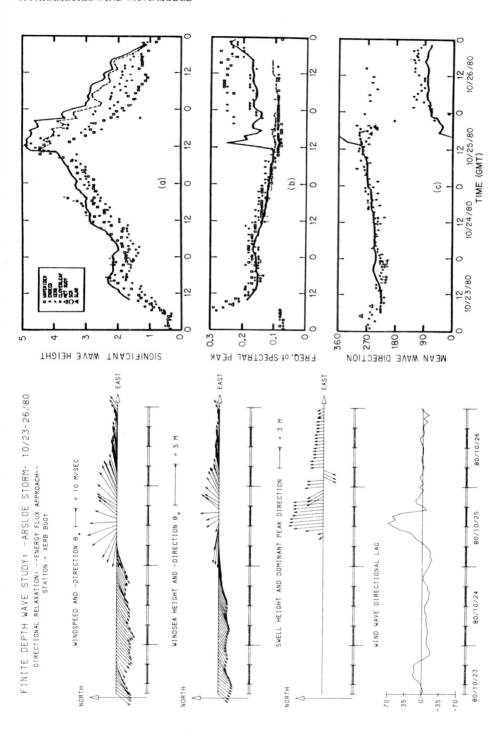

Fig. 2: Comparison of measured and predicted wave parameters. (From Szabados and Esteva, 1983).

Fig. 1: Time series of predicted spectral wave parameters at XERB BUOY.

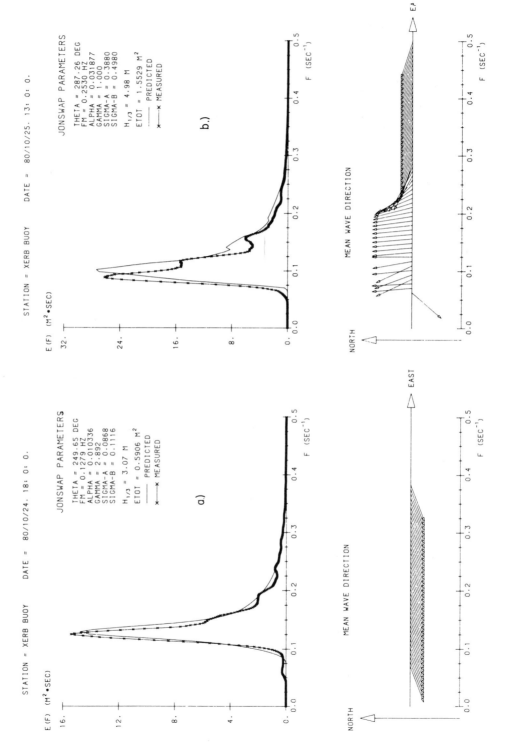

Fig. 3: Comparison of predicted and measured wave spectra for the clover leaf buoy.

in the context of deep water wind wave models to include finite depth effects. In particular, it should be emphasized that these extensions were accomplished without the introduction of "new" parameters in need of tuning prior to model application. In view of this fact, the over-all agreement between model predictions and observations exhibited by Figs. 2 and 3 is considered excellent.

Since swell is treated in a discrete spectral fashion, this portion of hybrid parametric wave models requires most of the computational time. Inclusion of swell refraction in the present finite depth wind wave model is recognized to be important, but would considerably increase the computational requirements. Efforts are currently under way to improve the swell aspects of the finite depth wind wave model.

ACKNOWLEDGEMENTS

The study reported here was supported by the Office of Sea Grant and the Coastal Waves Program through NOAA Grant NA81AA-D-00069.

5. REFERENCES

Bouws, E., et al., 1984: Similarity of the wind wave spectrum in finite depth water. Part I - Spectral Form. To appear in J. Geophys. Res.

Cardone, V.J., et al., 1976: Hindcasting the directional spectra of hurricane generated waves. J. Petroleum Techn., **28**, 385-394.

Graber, H.C. and O.S. Madsen, 1982: The directional relaxation of a windsea spectrum using an energy flux approach. EOS, **65**, No. 45, 970.

Graber, H.C., 1984: A parametric wind-wave model for arbitrary water depths. Sc.D. Thesis, Mass. Inst. of Techn., Cambridge, USA.

Grant, W.D. and O.S. Madsen, 1979: Combined wave and current interaction with a rough bottom. J. Geophys. Res., **84**, 1797-1808.

Günther, H., et al., 1979: A hybrid parametrical wave prediction model. J. Geophys. Res., **84**, 5727-5738.

Günther, H., et al., 1981: Response of surface gravity waves to changing wind direction. J. Phys. Oceanogr., **11**, 718-728.

Hasselmann, K., D.B. Ross, P. Müller and W. Sell, 1976: A parametric wave prediction model. J. Phys. Oceanogr., **6**, 200-228.

Hasselmann, S. and K. Hasselmann, 1981: A symmetrical method of computing the nonlinear transfer in a gravity wave spectrum. Hamburger Geophysikalische Einzelschriften, Heft **52**.

Kitaigorodskii, S.A., et al., 1975: On Phillip's theory of equilibrium range in the spectra of wind-generated gravity waves. J. Phys. Oceanogr., **5**, 410-420.

Snyder, R.L., et al., 1981: Array measurements of atmospheric pressure fluctuations above surface gravity waves. J. Fluid. Mech., **102**, 2-59.

SWIM-GROUP, 1985: Shallow water intercomparison of wave models. Proc. Sym. Wave Breaking, Turbulent Mixing and Radio Probing of the Ocean Surface, Sendai, Japan.

Szabados, M.W. and D.C. Esteva, 1983: Comparison of offshore wave measurements. IEEE, J. Oceanic Eng., **OE-8**, No. 4, 206-211.

SHALLOW WATER INTERCOMPARISON OF WAVE MODELS - PART I
THREE DIFFERENT CONCEPTS TO MODEL SURFACE WAVES IN FINITE WATER DEPTH

The SWIM Group:

E. Bouws[1], J.J. Ephraums[2], J.A. Ewing[3], P.E. Francis[2],
H. Günther[4], P.A.E.M. Janssen[1], G.J. Komen[1], W. Rosenthal[4],
W.J.P. de Voogt[5].
1 Royal Netherlands Meteorological Institute, KNMI, De Bilt,
  The Netherlands.
2 Meteorological Office, Bracknell, United Kingdom.
3 Institute of Oceanographic Sciences, Wormley, United
  Kingdom.
4 Institut für Meereskunde and Max-Planck-Institut für
  Meteorologie, Hamburg, F.R.G.
5 Delft Hydraulic Laboratory, Delft, The Netherlands.

ABSTRACT. The SWIM-project is an extension into water of finite depth
of the SWAMP-study (1982, 1984), which investigated the deep water
performance of different wave prediction models.
    In this paper we review the deep water parts of three of the
SWAMP models and describe their extension into water of finite depth.
Main emphasis is given to a comparison of the different concepts to
reduce wave energy.
    The results of the intercomparison for artificial cases and a
real hindcast are presented in the following two papers.

INTRODUCTION

The shallow water intercomparison of wave models (SWIM, 1984) project
has to be seen as a follow up project of the successful intercomparison of deep water wave models, the Sea Wave Modeling Project
(SWAMP, 1982, 1984).
    In the SWIM project three numerical wave model participated. The
model of the Meteorological Office, UK, (BMO) (Golding, 1983), the
GONO-model of the Royal Netherlands Meteorological Institute (Janssen,
Komen, de Voogt, 1984) and the HYPAS model of the University of
Hamburg, FRG (Günther, Rosenthal, Richter, 1984). All these models
are in operational use for wave prediction in the North Sea. An
extensive intercomparison of quasi-operational predictions of HYPAS
and GONO has been carried out for a couple of months (Günther,
Komen, Rosenthal, 1984), and the statistical results do not show any
significant differences between the model performances. A similar
comparison was made between GONO and the BMO model (Bouws et al., 1980).

Y. Toba and H. Mitsuyasu (eds.), The Ocean Surface, 201–205.
© 1985 by D. Reidel Publishing Company.

In the artificial SWAMP-cases quite different behaviour of the models
was discovered. From the results many open questions on the underlying physics of wave development, dissipation and propagation were
formulated and suggestions for further research could be deduced.
This motivated the three groups to investigate the state of the art
of shallow water wave modelling in a similar way.

      The full intercomparison project is described elsewhere (SWIM,
1984). Therefore this paper summarizes the basic model concepts. The
basic results are presented in the following papers.

2. DEEP WATER MODEL CHARACTERISTICS

The basic equation of all models in deep water is the transport
equation of the two-dimensional energy density spectrum $F(\underset{\sim}{x},t,f,\theta)$,
in which $\underset{\sim}{x},t,f$ and θ are the location vector, time, frequency and
direction, respectivly,

$$\frac{\delta F}{\delta t} + \underset{\sim}{v}_g \cdot \underset{\sim}{\nabla} F = S$$

where $v_g = \frac{g}{4\pi f}$ is the group velocity and S is the sourcefunction,

composed of the input S_{in}, the dissipation S_{dis}, and the weak non

linear wave-wave interaction sourceterm S_{nl}.

      Following the SWAMP 1984 nomenclature, the deep water wave
model realiation can be summarized as in Table 1.

3. INCORPORATION OF SHALLOW WATER EFFECTS

The three wave models are developed for seastate prediction in the
North Sea, where depths in the southern part are about 30 m. Extreme
shallow water is excluded, because in that case bottom induced non
linear effects (such as wave breaking) are significant for the
energy balance. But even well outside the breakerzone, a sizeable
energy reduction occurs, because in mild shallow water the dispersion
relation is already sensitive to the depth and energy dissipation
by bottom effects can be important.

      Basic equation for all models is the energy balance equation

$$\frac{\delta F}{\delta F} + \underset{\sim}{v}_g \cdot \underset{\sim}{\nabla}_x F + \underset{\sim}{\dot{k}} \cdot \underset{\sim}{\nabla}_k F = S$$

where

$$\underset{\sim}{v}_g = \nabla_{\underset{\sim}{k}}\omega \qquad\qquad \underset{\sim}{\dot{k}} = -\nabla_{\underset{\sim}{x}}\omega$$

$$\omega^2 = gk \tanh kd$$

The depth dependence of the group velocity $\underset{\sim}{v}_g$ is included in all three

models. The refraction term $\underset{\sim}{\dot{k}} \cdot \nabla_{\underset{\sim}{k}}F$ is applied throughout in the BMO

model and only the wind wave part of HYPAS.
      To achieve energy reduction in finite water depth all models
assume bottom friction as additional source term in the decoupled
discrete spectral part of the wave spectrum. The attenuation
coefficients are quite similar to the empirical JONSWAP value. In
the wind-wave part of the spectrum (coupled part) the BMO and GONO
model apply the same bottom friction term and do not change the
"deep water" source functions. The reduced energy of wind-waves is
reshaped by the same diagnostic formulae as in deep water to the
JONSWAP or Kruseman spectrum in the BMO or GONO-model, respectively.
The spectral form itself is not depth dependent.
      In the wind-wave part of the HYPAS-model the depth dependent
TMA (Texel-Marsen-Arsloe)-spectrum is used. This spectrum is an
extension of the deep water JONSWAP form to water of arbitrary

depth by considering a k^{-3} dependence instead of f^{-5} (Bouws et al.,
1984). The basic assumption is that the parameters of the TMA-spec-
trum develop as in deep water, and a fast process exists, which
reshapes the spectral energy to the TMA form. The physical process
that increases dissipation and leads from the deepwater JONSWAP
shape to the less energy containing depth dependent TMA-shape is
unknown. From the experimental evidence that the shallow water
seastate establishes to the TMA-shape it can be said that this
dissipation process is a monotonously increasing function of
wavenumber.
      These quite different approaches to model shallow water waves
show up in model behaviour. Therefore in the next paper of this
issue the model results for two artificial cases, a flat and a
sloping bottom, are discussed. The third paper about SWIM will
present the results of a realistic hindcast in the North Sea.

SWAMP - TYPE	BMO	GONO	HYPA
	coupled discrete (CD)	coupled hybrid (CH)	coupled hybrid (CH)
prognostic variables	energy in frequency and direction bands $E(f_i, \theta_j)$	parametric part: total energy mean, direction decoupled part: $E(f_i, \theta_j)$	parametric part: JONSWAP parameters, mean direction decoupled part: $E(f_i, \theta_j)$
advection	grid finite differences	parametric part: grid finite differences decoupled part: rays	parametric part : grid finite differences decoupled part: rays
source functions	input: Snyder-Cox, dissipation: white capping, non-linear interaction; reshaping to JONSWAP spectrum by diagnostic formulae	input/dissipation: empirical growth curve for total energy non-linear interaction: diagnostic formulae for spectral parameters	source functions for JONSWAP parameters out of dimensional analysis of non-linear interaction, Snyder-Cox input and empirical fetch laws. Directional relaxation tuned to measurements.
wind wave spectral form	JONSWAP spectrum	Kruseman spectrum	JONSWAP spectrum

REFERENCES

Bouws, E., B.W. Golding, G.J. Komen, H.H. Peeck, and M.J.M. Saraber, 1980: Preliminary results on a comparison of shallow water wave predictions. Koninklijk Nederlands Meteorologisch Instituut (KNMI), Scientific Report W.R. 80 - 5.

Bouws, E., H. Günther, W. Rosenthal, C.L. Vincent, 1984: Similarity of the wind wave spectrum for finite depth water, Part I: Spectral form. Accepted for publication in J. Geophys. Res.

Golding, B.W., 1983: A wave prediction system for real time sea state forecasting. Quart. J.R. Met. Soc. 109 393 - 416.

Günther, H., W. Rosenthal and K. Richter, 1984: A hybrid parametrical wave prediction model for water of finite depth. In preparation.

Günther, H., G.J. Komen and W. Rosenthal, 1984: A semi- operational comparison of two parametrical wave prediction models. Accepted for publication in Deutsche Hydrogr. Zeitschr.

Janssen, P.A.E.M., G.J. Komen and W.J.P. de Voogt, 1984: An operational coupled hybrid wave prediction model. J. Geophys. Res. 98, 3635 - 3645.

The SWAMP Group: J.H. Allender, T.P. Barnett, L. Bertototti, J. Bruinsma, V.J. Cardone, L. Cavaleri, J. Ephraums. B. Golding, A. Greenwood, J. Guddal, H. Günther, K. Hasselmann, S. Hasselmann, P. Joseph, S. Kawai, G.J. Komen, L. Lawson, H. Linné, R.B. Long, M. Lybanon, E. Maeland, W. Rosenthal, Y. Toba, T. Uji, and W.J.P. de Voogt, 1984: Sea Wave Modeling Project (SWAMP), An intercomparison study of wind wave prediction models, Part I: Principal results and conclusions. Ocean Wave Modeling. Plenum Press, New York.

The SWAMP Group: J.H. Allender, T.P. Barnett, L. Bertotti, J. Bruinsma, V.J. Cardone, L. Cavaleri, J. Ephraums, B. Golding, A. Greenwood, J. Guddal, H. Günther, K. Hasselmann, S. Hasselmann, P. Joseph, S. Kawai, G.J. Komen, L. Lawson, H. Linné, R.B. Long, M. Lybanon, E. Maeland, W. Rosenthal, Y. Toba, T. Uji, and W.J.P. de Voogt, 1982: Sea Wave Modeling Project (SWAMP), An intercomparison study of wind wave prediction models, Part II: A compilation of results, KNMI publication 161.

The SWIM Group: E. Bouws, J.A. Ewing, J. Ephraums, P. Francis, H. Günther, P.A.E.M. Janssen, G.J. Komen, W. Rosenthal, and W.J.P. de Voogt, 1984: Shallow Water Intercomparison of Wave Prediction Models (SWIM). Submitted for publication in Quart. J.R. Met. Soc.

SHALLOW WATER INTERCOMPARISON OF WAVE MODELS – PART II
RESULTS OF THREE DIFFERENT WAVE MODELS FOR IDEALIZED WIND AND DEPTH
SITUATIONS

The SWIM Group: E. Bouws[1], J.J. Ephraums[2], J.A. Ewing[3],
P.E. Francis[2], H. Günther[4], P.A.E.M. Janssen[1], G.J. Komen[1],
W. Rosenthal[4], W.J.P. de Voogt[5].

1. Royal Netherlands Meteorological Institute, KNMI,
   P.O. Box 201, 3730 AE De Bilt, The Netherlands.
2. Meteorological Office, Bracknell, Berkshire RG12 2SZ,
   United Kingdom.
3. Institute of Oceanographic Sciences, Wormley,
   Surrey GU8 5UB, United Kingdom.
4. Institut für Meereskunde and Max-Planck-Institut für
   Meteorologie, Bundesstrasse 55, 2000 Hamburg 13, FRG.
5. Delft Hydraulics Laboratory, P.O. Box 152,
   8300 AD Emmeloord, The Netherlands.

ABSTRACT

In SWAMP (1982, 1984) the model performance in deep water was
investigated for ideal fetch limited cases. We use similar simple
cases with finite water depth and horizontal or slightly sloping
bottom to exemplify the spectral properties built into the models to
simulate shallow water features. The large differences of the output
from different models signalize the poor knowledge of wave growth and
decay processes in shallow water. Since all models use to a large
extent field data for tuning their internal parameters, the outcome of
this exercise also indicated simple experimental conditions where data
about surface waves are badly needed both from wave tank and field
experiments. Two of the models (BMO and GONO) behave much like the
Sverdrup–Munk–Brettschneider type model (SMB-model). The other model
(HYPA) shows a behaviour which is qualitatively different from the
SMB-model. Since some parts of the North Sea show depth configurations
that resemble the considered ideal cases and all models show
satisfactory results for routine purposes, a closer inspection of the
underlying physical mechanisms is needed.

1. INTRODUCTION

From the SWAMP (1982, 1984) study it is known that useful insight in
the behaviour of wave models can be obtained by running them for
idealized configurations. In the preceeding contribution to this

Y. Toba and H. Mitsuyasu (eds.), The Ocean Surface, 207–214.
© 1985 by D. Reidel Publishing Company.

symposium it is explained how we decided to extend SWAMP to the study of shallow water aspects. There, also the participating models were introduced. Here we give a brief description of the two idealized cases as well as a discussion of the results. For a fuller account we refer to the SWIM report (SWIM, 1984).

2. CASE I

In this case we study ideal generation by a constant off-shore wind (see fig. 1). The wind speed is taken to be 20 m/s. For the bottom

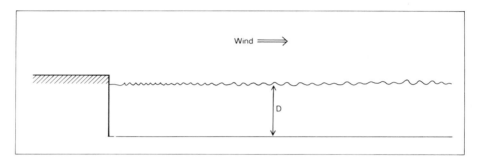

Fig. 1.   Configuration of case I: a constant offshore wind blowing
           over a basin of constant depth D.

depth we have chosen D = 15, 30, 60 and 120 m. This case was selected because it allows us to study the effect of depth limitation in a simple context. In all three models the bottom influence leads to a frequency dependent reduction of spectral level. As discussed in the previous section the resulting shallow water spectra differ from the corresponding deep water ones. To characterize the model differences it is useful fo study the behaviour of the total energy and the peak frequency both as a function of fetch and duration. As usual we start from a flat sea at t = 0, and the models until they became stationary to get fetch curves. For a depth of· 120 m the growth curves are effectively deep water curves. These deep water curves have already been discussed in the SWAMP report: BMO and HYPAS behave very similarly for deep water; GONO reaches approximately the same asymptotic level, but more slowly, whereas for short fetches it has higher energy values than the other models. A striking effect is the reduction of wave energy E_∞ in the fully grown state due to bottom effects. The reduction is relatively weak in HYPAS ($E_\infty^{HYPAS}(120) = 5.5$ m², $E_\infty^{HYPAS}(15) = 2$ m²) and rather strong in GONO ($E_\infty^{GONO}(15) = 0.6$ m²) and BMO (0.7 m²). We also studied the dependence of the dimensionless peak frequency on fetch. Here, also, there are differences between BMO and GONO on the one hand, and HYPAS on the other hand. In BMO and GONO the peak frequency in the asymptotic steady state is much higher in shallow water than in deep water. In

HYPAS the peak frequency does not vary much with depth. The relevant curves can be found in the SWIM (1984) report. Here we illustrate the differences between the models with fig. 2, in which we show the asymptotic spectra for each of the models (infinite fetch and infinite duration) as a function of depth.

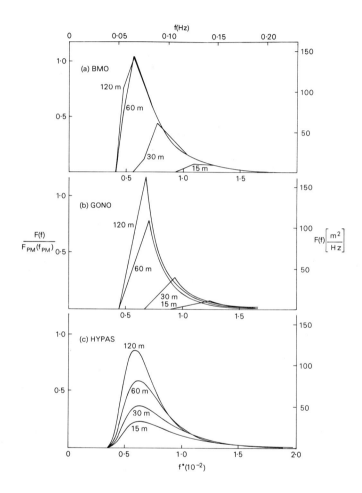

Figure 2. The fully developed wave energy spectra as a function of depth (D = 120, 60, 30, 15 m) from the 3 models, (a) BMO, (b) GONO, (c) HYPAS, for case I. The wave energy density F(f) (m^2/Hz) is plotted against frequency f(Hz) and non-dimensional frequency f* = fu*/g. Also shown is the energy density scaled by the peak energy density of the Pierson-Moskowitz spectrum F_{PM} (f_{PM}) = Af^{-5} exp($-B$ ($f_{PM}/f)^4$), A = 5 x m^2 s^{-4}, B = 1.25, f_{PM} = 0.13g/u_{10}.

3. CASE II

Here we investigate wave evolution by a constant wind field over a sloping bottom. The geometry is depicted in fig. 3. Wind is blowing

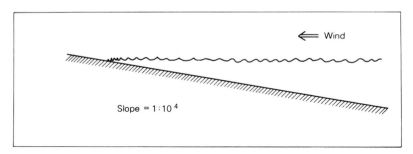

Figure 3   Configuration of case II: a constant onshore wind blowing over an idealized sloping coastal shelf.

towards shore with a speed of 20m/s. The bottom was chosen to rise linearly with a slope of 1: 10^4. This slope is typical for the southern North Sea. Case II was selected as an idealization of situations encountered frequently in coastal areas. As in case I the models start from a flat sea, and are run until a stationary state has been reached, with the fully developed deep water spectrum as a boundary condition in deep water. We analysed this stationary solution only.

The very different outcome for the three models is shown in fig. 4, 5 and 6. In fig. 4 the dimensionless energy is given as a function of $X_* = Xg/u_*^2$, with X now representing the distance to shore. All models show a reduction in wave energy, from maximum values of wave height in deep water of about 9.5 m to values around 5.5 m in a depth of 20 m. BMO has the strongest attenuation rate for depths of less than 20 m while HYPAS has the weakest rate. For the HYPAS model the wave heights at a given depth are similar to the saturation values obtained for the same depth in Case I. However for BMO and GONO wave heights are greater than the corresponding saturation values in Case I, the margin increasing with decreasing depth for GONO and having amaximum at about 30 m depth in BMO.

The peak frequencies are given in fig. 5. The behaviour of GONO and HYPAS is very similar, both models exhibit a peak frequency that does not deviate much from the deep water value. In contrast the BMO result shows a strong increase of peak frequency with decreasing depth. All models have a local peak frequency wich is lower than the saturation peak frequency at the same depth exhibited in Case I.

Fig. 6 represents the wave spectra for different depths, given sufficient fetch and duration to reach equilibrium conditions. The earlier discussion concerning the behaviour of peak frequency and total energy can be extended to these figures in a straightforward manner.

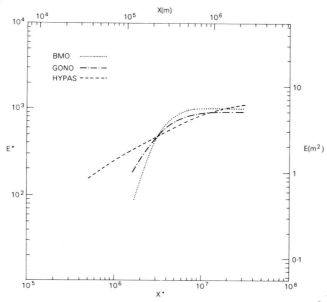

Figure 4   The stationary distribution of wave energy $E(m^2)$ and non-dimensional wave energie E* as a function of non-dimensional distance to shore from the 3 models for case II. Depth in meters is $D = X/10^4$.

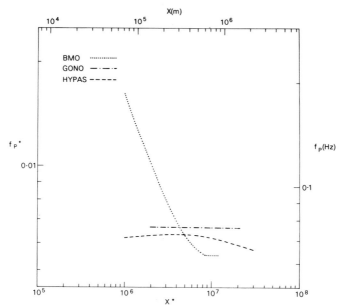

Figure 5   The stationary distribution of peak frequency f_p (Hz) and non-dimensional peak frequency f_p* from the 3 models. As Fig. 4.

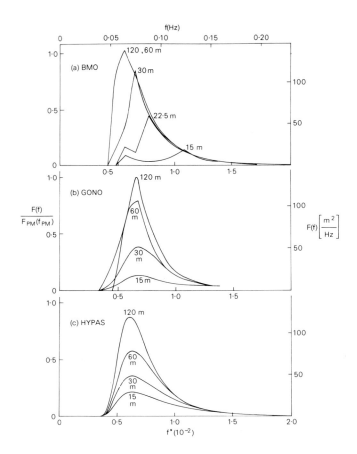

Figure 6   The stationary wave energy spectra as a function of shelf
depth from the 3 models, (a) BMO, (b) GONO, (c) HYPAS for
case II. As Fig. 2.

4. DISCUSSION OF RESULTS

The results of the two cases can be understood in a rather
straightforward way on the basis of the model descriptions given in
the preceeding paper. The bottom friction term in the wind sea
algorithm of GONO and BMO selectively damp long waves, and this leads
to the increasing peak frequency with decreasing depth in Case I.
HYPAS has approximately the same prognostic equation for the peak
frequency in deep and shallow water. Therefore, peak frequency does
not significantly depend on depth.

In Case II HYPAS results for a given depth are not very different
from Case I. GONO and BMO have wind sea spectra as in Case I, but in
addition low frequency energy is advected from deeper water. This
leads to a shift in the mean frequency. In GONO the low frequency
energy is larger than in BMO, in fact so large that the peak of the

spectrum is dominated by it. For that reason GONO behaves like HYPAS in Case II.

The analysis of the idealized cases indicated the existence of a fair amount of uncertainty with respect to depth limited wave modelling. This is partly due to a lack of reliable and uncontaminated observations, and partly to the unavailability of exact solutions of the full energy balance equation.

There is only limited experimental material available on which to base theories for wave growth in shallow water. Some of this evidence has been summarised by Holthuijsen, (1980), and broadly speaking the behaviour of the GONO and BMO models in Case I is supported by these data. However, details of the experiments imply some degree of doubt as to such important aspects as the consistency of the associated wind field, the definitions of peak frequency and the flatness of the sea bed.

Another major data set has been constructed from the results of the MARSEN experiment in 1979 and the ARSLOE experiment in 1980, together with a number of spectra from the so called Texel storm in 1976. A total number of about 3000 spectra have been collected to derive data for growing waves in water of finite depth, (Bouws et al, 1984). The HYPAS model has been developed from the evidence of this TMA (Texel-Marsen-Arsloe) data set. Again, there is some doubt as to homogeneity of the collected data, in this instance advected energy from deeper water may be present in some of the spectra thus removing similarities with Case I.

5. CONCLUSION

We found significant differences in the output from the different models. This signalizes the poor knowledge of wave growth and decay processes in shallow water. This exercise indicates simple experimental conditions where data about surface waves are badly needed. Detailed numerical studies resolving all the individual source terms would also be most welcome.

REFERENCES

Bouws, E., H. Günther, W. Rosenthal and C.L. Vincent, 1984. Similarity of the wind wave spectrum in finite depth water, Part I -Spectral from. J. Geophys. Res. (in press).
Holthuijsen, 1980. "Methoden voor Golfvoorspelling", Technische Adviescommissie voor de Waterkeringen. The Hague, 2 vols.

SWAMP Group: J.H. Allender, T.P. Barnett, L. Bertotti, J. Bruinsma, V.J. Cardone, L. Cavaleri, J.J. Ephraums, B. Golding, A. Greenwood, J. Guddal, H. Günther, K. Hasselmann, S. Hasselmann, P. Joseph, S. Kawai, G.J. Komen, L. Lawson, H. Linné, R.B. Long, M. Lybanon, E. Maeland, W. Rosenthal, Y. Toba, T. Uji and W.J.P. de Voogt, 1982. Sea Wave Modelling Project (SWAMP). An intercomparison study of wind-wave predictionmodels, Part 2 - A compilation of results, KNMI Pub. 161.

SWAMP Group: J.H. Allender, T.P. Barnett, L. Bertotti, J. Bruinsma, V.J. Cardone, L. Cavaleri, J.J. Ephraums, B. Golding, A. Greenwood, J. Guddal, H. Günther, K. Hasselmann, S. Hasselmann, P. Joseph, S. Kawai, G.J.Komen, L. Lawson, H. Linné, R.B. Long, M. Lybanon, E. Maeland, W. Rosenthal, Y. Toba, T. Uji and W.J.P. de Voogt, 1984. An intercomparison study of wind-wave prediction models, Part 1 - Principle results and conclusions, Proc. IUCRM Symp. on Wave Dynamics and Radio Probing of the Ocean Surface, Miami, Plenum Press.

SWIM Group: E. Bouws, J.A. Ewing, J. Ephraums, P. Francis, H. Günther, P.A.E.M. Janssen, G.J. Komen, W. Rosenthal and W.J.P. de Voogt (1984). Shallow water intercomparison of wave prediction models (SWIM). Submitted for publication in Quart. J. Roy. Met. Soc.

SHALLOW WATER INTERCOMPARISON OF WAVE MODELS - PART III
A hindcast storm in the North Sea

The SWIM Group:

E. Bouws[1], J.J. Ephraums[2], J.A. Ewing[3],
P.E. Francis[2], H. Gunther[4], P.A.E.M. Janssen[1],
G .J. Komen[1], W. Rosenthal[4], W.J.P. de Voogt[5].
1   Royal Netherlands Meteorological Institute, KNMI,
    De Bilt, The Netherlands
2   Meteorological Office, Bracknell, United Kingdom
3   Institute of Oceanographic Sciences, Wormley,
    United Kingdom
4   Institut fur Meereskunde and Max-Planck-
    Institut fur Meteorologie, Hamburg, F.R.G.
5   Delft Hydraulics Laboratory, Delft, The Netherlands

ABSTRACT.   This paper describes the intercomparison and verification
of results from three operational shallow water wave prediction models
in their hindcasts of a severe storm period.   The three models, BMO
(Meteorological Office), GONO (KNMI) and HYPAS (Max-Planck-Institut)
have already been compared theoretically and via idealised experiments
with constant winds in Parts I and II of the SWIM project (SWIM Group,
1985).   The work reported here attempts to unravel the complicated
processes involved in the prediction of shallow water waves in a
complex synoptic situation by running all three models in parallel with
common wind fields and similar computation grids and bottom topography.
The case chosen, two North Sea storms in November 1981, is fully
described in SWIM (1984).   Results presented here show that all models
produce broadly similar results and acceptable shallow water energy
levels at three verification sites, despite the differences in
formulation and results shown in Parts I and II of this project.

1.   THE HINDCAST PERIOD

Two storms in the North Sea during 20-26 November 1981 were chosen for
the hindcast, since for this period there were sufficient spectral wave
data for verification, and synoptic data were readily available for
numerical wind analyses.
    The first storm crossed the North Sea eastwards during 20 and 21
November giving generally strong westerly winds over the southern
areas.   After a spell of calm south-westerlies the second storm between
23 and 25 November brought very strong winds which veered from west to
north-west in the same general area.   Further north there were

215

Y. Toba and H. Mitsuyasu (eds.), The Ocean Surface, 215–220.

persistent strong northerly winds later on in this period which combined with a long fetch to generate unusually large waves in the central North Sea and considerable swell further south.

2.   THE HINDCAST WIND FIELDS

The production of the wind fields was done in two stages. For the first stage, three-hourly wind fields from Meteorological Office NWP model twelve hour forecasts from this period were reformatted as background fields. These surface winds had already been calculated from 900 mb winds, employing well proven temperature dependent empirical relationships (Findlater et al, 1966). At this second stage all archived surface wind observations and measurements for the period were numerically analysed in conjunction with the background fields from stage 1. The analysis was performed on a 50 km polar-stereographic grid over the area of the North Sea, the Norwegian sea up to 70°N and the N.E. Atlantic out to 20°W.

We verified the windfields at several locations in the North Sea and found a very good agreement with measurements in these data dense areas. In data sparse areas it was not possible to check the accuracy of the forecast model background fields

3.   MEASURED WAVE DATA

Although data were collected from seven locations, only results from three sites will be discussed here. Measurements of one-dimensional wave energy spectra were obtained from waverider buoys adjacent to the three North Sea platforms K-13 (53.2°N, 3.2°E), EURO (52.0°N, 3.5°E) and ELD (53.2°N, 4.6°E). These sites are in shallow waters ranging in depth from 28 m to 23 m and thus are ideally suited for the experiment. It is worth noting that at these sites the tidal depth range of about 2 m together with a probable surge of 2-3 m at the second storm peak introduced some uncertainty in the exact value of water depth at each position.

The significant wave height, H_s and mean period T_{01} referred to in this paper are defined by $H_s = 4 M_0$ and $T_{01} = M_0/M_1$, where the moments of the one-dimensional wave energy spectrum F(f) are given by $M_n = \int f^n F(f) df$, where f is the frequency in Hz.

4.   THE MODELS

The HYPAS model grid area is very similar to that of BMO, both covering the windfield area, whereas GONO terminates its left hand boundary at about 5°W. In HYPAS and BMO there was zero energy flux at the open sea boundaries, and although GONO had a symmetrical radiation condition for the wind sea none of the models were able to model swell which might have propagated in from areas outside of their grids.

Although computation sea depths are model specific, all models accurately represent the basic seabed topography in the North Sea. Some other model details are listed in Table I; despite small variations in resolution the only important model differences in this study resolve

themselves at the level of the internal specifications of wave
modelling mechanisms.

TABLE I.   Model Details

	BMO	GONO	HYPAS
Gridlength (km)	25	75	50
Timestep (minutes)	30	90	30
Directional resolution (degr)	$22^1/_2$	30	15
Lowest frequency (Hz)	0.04	0.05	0.0425

5.   STATISTICAL RESULTS

Table II shows the verification of each model at the three shallow
water sites. At both ELD and K-13 the average measured wave height
($\bar{H}_S$) during the period was about 3.4 m and the maximum about 5.9 m.
Waves at EURO were rather lower (mean 2.4 m, maximum 4.0 m) due to a
shorter fetch under the prevailing westerly type winds.

TABLE II.   Verification and Intercomparison of Models during
20-26 November 1981. D= Depth (metres), N = Number of measurements.
H_S = Significant wave height (metres), T_{o1} = mean period (seconds)
$\bar{H}_S$ and $\bar{T}_{o1}$ are the average of the measured data during this period.

Measured data				Model	D	H_S Errors		T_{o1}	
D	N	$\bar{H}_S$	$\bar{T}_{o1}$			Mean	S.D.	Mean	S.D.
K-13 28	50	3.3	6.6	BMO	27	-0.20	0.44	-0.16	0.60
				GONO	25	0.39	0.46	-0.12	0.69
				HYPAS	31	-0.08	0.53	-0.32	0.67
EURO 25	51	2.4	5.8	BMO	27	-0.05	0.48	0.18	0.76
				GONO	25	0.00	0.30	-0.24	0.43
				HYPAS	28	-0.26	0.54	-0.47	0.71
ELD 23	49	3.4	7.0	BMO	22	0.16	0.45	0.03	0.53
				GONO	30	0.55	0.38	-0.04	0.57
				HYPAS	19	-0.39	0.69	-0.57	0.68

In general GONO has higher energy levels than those measured, or
simulated by the other models, and HYPAS has lower values. This trend
at ELD is however exaggerated by the unavoidable choice of a gridpoint
which is slightly too deep in GONO. More specifically GONO was found
to overpredict only at lower energy levels and not at storm peaks, and
this model has the lowest standard deviation of errors about the mean
(S.D.), with HYPAS possessing the greatest values here.

Verification of wave periods shows the BMO model having slightly
higher values compared against measurements and the other models; HYPAS
has the lowest set of figures overall here. In general we find that
the models show fewer differences in wave period compared to
significant wave height and that altogether these verification figures
show all models performing well during the complete hindcast.

6.   AN EXAMPLE OF STATIONARY WIND FIELDS

We can examine results from the models in the case of a stationary
moderate ($\approx$ 15 m/s) south-westerly wind at location K-13 during 23rd.
The situation is characterised by fetch limited waves in fairly shallow
water and results should therefore be broadly in line with those
discovered in the experiments described in Part II of this project.

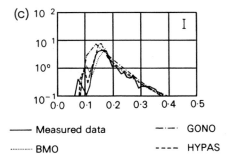

——— Measured data   —·— GONO

············ BMO   - - - - HYPAS

Figure 1.   Measured and modelled logarithmic wave energy spectra from
location K-13 at 09Z 23rd November 1981.   The horizontal axis is
frequency (HZ).

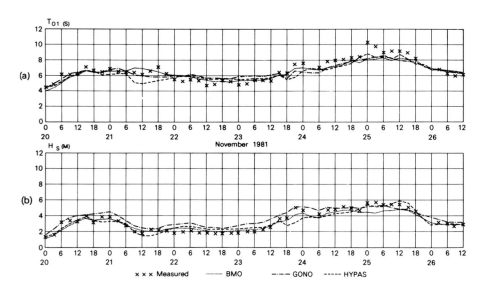

Figure 2.   Timeseries of a) mean period and b) significant wave
height, measured and modelled at location K-13 during 20-26 November
1981.

The measured spectra were fairly stationary between 03Z and 09Z on this date, with $H_s \approx 2$ m and the trace presence of swell at about 0.1 Hz. As can be seen from Fig. 1 both BMO and, especially, HYPAS models produce good descriptions of the spectral shape and total energy, whilst GONO overpredicts the total energy and has only one spectral peak. These results confirm the findings from Part II, namely that GONO has higher wave growth at short fetches ($\approx$ 100 km) and shallow depths ($\approx$ 30 m) compared with the other models.

7.   AN EXAMPLE OF WAVE HEIGHT AND PERIOD TIMESERIES

Fig. 2 shows a timeseries plot of model and measured wave heights and periods for the whole hindcast event at K-13, the position also discussed above. Here we can see the different energy levels in the three models quite clearly, and whilst GONO appears consistently higher there are still exceptions to the general trend. Between 18Z on 23rd and 00Z on 24th BMO has higher wave heights than HYPAS, the situation being reversed between 18Z on 24th and 18Z on 25th, and indeed between 06Z and 18Z on 25th HYPAS has the highest values of all three models.

The differences between the wave periods from each model are not so striking, with the exception being beween 06Z on 21st and 00Z on 22nd when the wind speed fell rapidly after the first storm and all models were simulating the separation of the wave energy spectrum into wind sea and swell and attempting to model the correct dissipation rates.

8.   DISCUSSION

It was difficult to pinpoint more precisely the exact strengths and weaknesses of each model in this study because of some uncertainties which remain even after the lengthy preparation of the experiment. There were problems in interpreting the large number of measured and modelled spectra, especially in the absence of measured directional information at most stations. The separation of wind sea from swell in the measured spectra was not always obvious, and it was not possible to confidently verify the spatial evolution of the wave field.

Further north in the hindcast area it was apparent that either the reconstructed winds were too light in this data sparse region during the second storm (Caudwell, 1984) or that the models had insufficient northerly fetch to represent incoming swell.

It may also be hasty to judge or modify each model on the basis of only two storm events, and in any case the results may not be directly transferrable to the models in their operational use where they use wind fields derived in a different manner.

Despite these comments the experiment was highly successful as a verification exercise, even though we were disappointed that we could not infer more detailed diagnostic information about our respective models in shallow water. What did become clear, however, was that the models are weakest in their representation of the lesser understood processes (e.g. directional relaxation, wind sea and swell separation) which are usually modelled on the basis of intuition and computational

efficiency or simplicity in the absence of sound theoretical or
experimental data.

9.   CONCLUSIONS

a)   The accuracy of reconstructed wind fields is still the most
important parameter in hindcasting extreme events.
b)   It was possible to attribute differing energy levels between
models to different depth dependent growth rates described in Part II.
c)   For this particular hindcast, GONO slightly overpredicted lower
energy levels and BMO and HYPAS slightly underpredicted peak values in
the second storm.   These results are, however, not indicative of the
performance of each model in operational use.
d)   There were, as well, inconsistent differences in model behaviour
which are probably due to the different properties of the three models
in treating swell in turning or decreasing winds.
e)   Either a higher density or a longer timeseries of directional
measured wave data is required for unambiguous diagnostic information
on modelling complicated wave fields. It may be necessary to develop
objective techniques for verifying large quantities of model wave
spectra.

10.   ACKNOWLEDGEMENTS

We thank the Shell Development Company, Houston, and the
Rijkswaterstaat for providing measured wave data.

11.   REFERENCES

Caudwell, W.D. and L. Draper, 1984: Case study of the North Sea Storm
        of 23/24 November 1981. Unpublished Report.
Findlater, J., T.N.S. Harrower, G.A. Howkins and H.L. Wright, 1966:
        Surface and 900 mb wind relationships, Met. O. Sci. Pap. 23,
        HMSO, London.
SWIM Group: E. Bouws, J.J. Ephraums, J.A. Ewing, P.E. Francis,
        H. Gunther, P.A.E.M. Janssen, G.J. Komen, W. Rosenthal and
        W.J.P. de Voogt, 1985: A shallow water intercomparison of wave
        models – Part I: Three different concepts to model surface waves
        in finite water depth, Symposium on wave breaking, turbulent
        mixing and radio probing of the ocean surface, Sendai, Reigel.
SWIM Group: E. Bouws, J.J. Ephraums, J.A. Ewing, P.E. Francis,
        H. Gunther, P.A.E.M. Janssen, G.J. Komen, W. Rosenthal and
        W.J.P. de Voogt, 1985: A shallow water intercomparison of wave
        models – Part II: Results from three different wave models for
        idealised wind and depth situations, ibid.
SWIM Group: E. Bouws, J.J. Ephraums, J.A. Ewing, P.E. Francis,
        H. Gunther, P.A.E.M. Janssen, G.J. Komen, W. Rosenthal and
        W.J.P. de Voogt, 1984: A shallow water intercomparison of wave
        models. Submitted for publication in Quart. J.R. Met. Soc.

A COUPLED DISCRETE WAVE MODEL MRI-II

T. Uji
Meteorological Research Institute
1-1 Nagamine Yatabe-machi Tsukuba-gun
Ibaraki-ken 305
Japan

ABSTRACT. An ocean wind-wave prediction model MRI-II is developed on the basis of the energy balance equation which contains five energy transfer processes, namely, the input by the wind, the non-linear transfer among the components of windsea by resonant wave-wave interactions, wave breaking, frictional dissipation and the effect of opposing winds. The non-linear energy transfer is expressed implicitly together with the wind effect by Toba's one-parameter representation of windsea, but neither swell-swell nor swell-windsea resonant interactions are considered. Hypothetical assumptions are introduced to describe wave breaking effects. The numerical constant required in the assumptions of wave breaking is determined through trial test runs for a hindcast performed on the Northwestern Pacific Ocean. The results of the hindcast and numerical experiments of the SWAMP test cases show that MRI-II outforms our old model MRI both in describing growth stages of windsea and in application to actual wind fields.

1. INTRODUCTION

The wave model MRI-II was developed to overcome the weaknesses of our old model(Uji and Isozaki, 1972, Isozaki and Uji, 1973, Uji, 1975) called MRI(Meteorological Research Institute), which has been used for routine wave forecasting at the Japan Meteorological Agency(JMA) since 1977, by incorporating recent physical pictures of windsea growth and a hypothetical representation of the .wave breaking effect(Uji, 1984). Here, follows a condensed version of it. Following MRI, the spectrum of waves is also represented in MRI-II as a two-dimensional discrete array of frequency-direction energy "packets" to inherit the advantage of MRI. Thus, MRI-II is classified as a Coupled Discrete wave model.

2. FUNDAMENTAL EQUATIONS

The evolution of a surface wave field in space and time is governed by the energy balance equation

Y. Toba and H. Mitsuyasu (eds.), The Ocean Surface, 221–226.

$$dF/dt = \partial F/\partial t + Cg \cdot \nabla F = S_{net} = S_{in} + S_{nl} + S_{ds}$$

where $F(\sigma,\theta';x,t)$ is the two-dimensional wave spectrum, dependent on angular frequency σ and propagation direction θ', $Cg = Cg(\sigma,\theta')$ is the group velocity, ∇ is the gradient operator in the horizontal plane and the net source function S_{net} is represented as the sum of the input S_{in} by the wind, the non-linear transfer S_{nl} and the dissipation S_{ds} (Hasselmann et al.,1973).

Variables which are suffixed * at the upper right are non-dimensionalized in terms of the friction velocity u_* and the acceleration of gravity g. Let us express Pierson-Moskowitz spectrum in the form $\phi_{pm}(\sigma;\sigma_{pm})$, where σ_{pm} is the peak frequency of the spectrum and assume that the one-dimensional spectrum of windsea is similar in form to Pierson-Moskowitz spectrum. Then, the two-dimensional spectrum of windsea is expressed as

$$F*(\sigma*,\theta';\sigma_p*) = (\sigma_p/\sigma_{pm})\phi_{pm}(\sigma*;\sigma_p*)\Gamma(\theta), \tag{1}$$

where σ_p is the peak frequency of windsea and $\Gamma(\theta)$ is the directional distribution function for the windsea which is expressed by the angle θ between wind and θ' as $(2/\pi)\cos^2\theta$. This spectrum follows Toba's(1978) minus third power law between the total energy E* and σ_p^*. The evolution of the parameter σ_p^* is assumed to follow Toba's(1978) stochastic growth equation arranged for our spectral form, i.e.

$$d\sigma_p*^{-2}/dt* = 1.783 \cdot 10^{-3}\{1 - erf(4.59 \cdot 10^{-2}\sigma_p*^{-1})\}. \tag{2}$$

Since we cannot apply Eqs.(1) and (2) to over-saturated components, hypothetical assumptions are introduced to obtain the expression of S_{ds} for these components. The basic idea on the assumptions is that wave breaking is a process in which a water mass at a wave crest with a mass proportional to the square of the wave height loses its wave motion energy. The expression of S_{ds} thus obtained is

$$S_{ds} = -Br \cdot F = -\{C_b \cdot Pi \cdot \sigma_p \cdot E^2(1 + (\sigma/2\sigma_p)^4)/E_n\} \cdot F, \tag{3}$$

where Br is the damping ratio, Pi is the probability of breaking determined from data collected by Toba(1979), $E = \iint(1 + (\sigma/2\sigma_p)^4)F\,d\sigma d\theta$ and C_b is a constant whose value was determined through the case study of waves due to typhoon ORCHID as $C_b = 1/600$ m^2.

For the waves in opposing wind, three energy dissipation effects are considered: (1) wave breaking, (2) the suppressing effects of wind for waves against the wind and (3) the energy dissipation due to viscosity. The non-linear energy transfer between swell and windsea due to resonant wave-wave interactions is neglected.

After all the net source function S_{net} is expressed as

$$S_{net} = \{F(\sigma_p + \Delta\sigma_p) - F(\sigma_p)\}/\Delta t, \qquad \theta < 90° \text{ and } F_\infty > F$$

$$S_{net} = \{1 - (F/F_\infty)^2\}Br \cdot F, \qquad \theta < 90° \text{ and } F_\infty < F < 1.414 \cdot F_\infty,$$

$$-Br \cdot F, \qquad \theta < 90° \text{ and } \qquad F > 1.414 \cdot F_\infty$$

$$S_{net} = -(B \cdot \Gamma(\theta)+Df^4+Br)F, \qquad \theta > 90°$$

where Δt^* is the time interval of numerical integration, $\Delta \sigma_p^*$ is the amount of change of σ_p^* given by Eq.(2) for Δt^*, B is the growth rate of waves by wind, whose numerical value is taken over from MRI, $F_\infty = \phi_{pm}(\sigma_{pm})\Gamma(\theta)$ and $D=1/3600$ s^3.

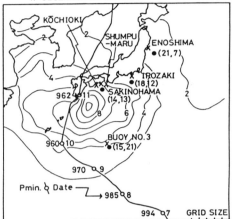

o LOCATION OF THE TYPHOON AT 00Z
• GRID POINT
X OBSERVATION STATION

3. NUMERICAL SCHEME

We use the same expression of the two-dimensional spectrum and the same scheme for wave energy propagation as in MRI(Uji and Isozaki , 1972). Our calculation consists of the following seven steps: 1) each two-dimensional energy spectral component propagates for a given time interval Δt; 2) the damping coefficient Br due to wave breaking is determined by Eq. (3); 3) the value of a frequency spectrum $\phi(\sigma)$ is determined to minimize the value of the integral

$$\int_{-\pi/2}^{\pi/2} (F(\sigma,\theta') - \phi(\sigma)(2/\pi)\cos^2\theta)^2 d\theta ;$$

4) the peak frequency σ_p is determined by the best fit of

Figure 1. The track of typhoon ORCHID and hindcasted wave height(m) distribution at 00z 11 Sept. with grid size 127 km.

$(\sigma_p/\sigma_{pm})\phi_{pm}(\sigma;\sigma_p)$ to $\phi(\sigma)$, where the swell components are disregarded; 5) the peak frequency σ_p is transformed to the corresponding duration t using the duration-limited growth curve of σ_p obtained from Eq.(2); 6) the new windsea is given for the new time step $t+\Delta t$; 7) the evolution of non-growing components are calculated by simple time integration using the damping coefficient Br.

4. A CASE STUDY AND DETERMINATION OF THE MAGNITUDE OF THE WAVE BREAKING EFFECT

The wind fields of typhoon ORCHID were analyzed every 3 hours and every 127 km from 15:00 GMT(15z) 06 Sept. 1980 to 12z 13 Sept. 1980 at JMA. The two-dimensional wave spectra at the boundary are determined by use of the results of the routine operation of wave prediction at JMA. The observation stations and the path of the typhoon in our computation area are shown in Fig.1.

For various values of C_b, hindcast calculations of sea waves caused by the typhoon were repeated. Obtained wave heights were compared with observed ones with ship(buoy)-borne wave recorders on Buoy No.3 and on the R.V. Shumpu-maru. The value, $C_b = 1/600$ m^{-2}, is obtained as the best value of C_b which minimized the sum of the two root mean squares;

one concerns differences in wave height between observed at Buoy No.3
and hindcasted waves and the other root mean squares is between data
collected by the R.V. Shumpu-maru and hindcasted waves.

Figure 2 compares wave
heights observed with an ultrasonic
wave gauge at Sakinohama and with
wave recorders on a buoy moored
off Kochi and on the R.V. Shumpu-
maru during the period from 21z 08
Sep. to 12z 13 Sep. with those
hindcasted by three wave models:
(1) MRI (open circles), (2) MRI-II
with no wave breaking effects (
solid circles) and (3) MRI-II with
the wave breaking effects (crosses)
MRI-II with no wave breaking effects
consistently gives much higher wave
heights than the other two wave
models and actual observations
through the whole period except
in the early growth stages. This
shows that the energy dissipation
for over-saturated components
plays an important roll in complicated wind fields, such as
typhoons.

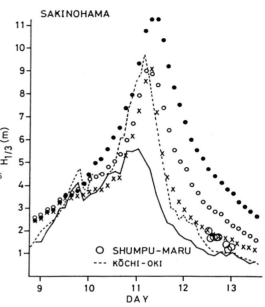

Figure 2. Observed and computed
significant wave heights at
Sakinohama, Koch-oki, the grid
point(14,13) and those observed on
the Shumpu-maru.

One-dimensional spectra
computed by MRI and MRI-II(from
now on MRI-II refers to MRI-II
with wave breaking effects
included) at the grid point(14,
13) at 18z 12 September are shown
in Fig.3.   For the spectra at 18z
MRI gives a large energy peak in
the frequency range from 0.11
to 0.14 Hz, but MRI-II yields
no significant energy peak.
The spectra observed by the
R.V. Shumpu-maru and at
Sakinohama at the same time are
shown in Fig.3.   It is remarkable that both one-dimensional
spectra observed R.V. Shumpu-
maru and obtained with MRI-II
have almost the same form.

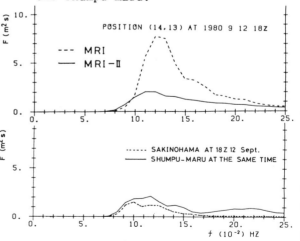

Figure 3. One-dimensional spectra
for MRI and MRI-II and those observed
by the Shumpu-maru and at Sakinohama
during the decay stage of waves.

5.   INTERCOMPARISON

Here, typical two results of
SWAMP experiments are shown.

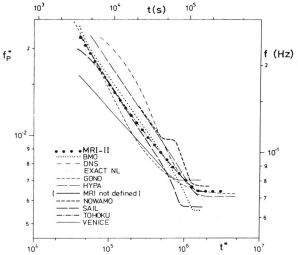

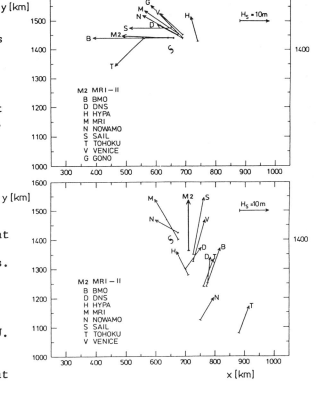

Figure 4. Non-dimensional duration limited growth curves for the peak frequency f_p^* (reproduction of Fig. 7.7 in SWAMP Part 1).

Figure 5. Positions of $(H_{1/3})_{max}$ for different models for the stationary hurricane (upper) and moving hurricane (lower). Arrows point in the mean wave direction and are proportional to $(H_{1/3})_{max}$ in length (reproduction of Fig. 12.4 in SWAMP Part 1).

Fig.4 is the duration limited growth curve of the peak frequency of windsea for Case II. MRI is inferior in predicting the spectral form for early growth stage. This figure shows that the weak point of the MRI is overcome by MRI-II. Since the growth equation of the peak frequency of MRI-II is same that of TOHOKU (Toba et al., 1981), the growth curve of MRI-II and that of TOHOKU are identical.

Fig.5 shows positions, mean wave directions and wave heights of the highest waves in the hurricane wave fields obtained with different wave models both for stationary and moving hurricanes. The shift of the position of MRI-II from that of MRI is small. However, the position of MRI-II largely different from that of TOHOKU. The highest wave of MRI-II points the direction between the direction for MRI and that for TOHOKU. The wave height of MRI-II is slightly larger that of MRI. The wave height of MRI-II is much larger than that of TOHOKU for the moving hurricane.

6. CONCLUSIONS

MRI-II is designed to overcome a weak point of MRI . For this
purpose, a one-parameter representation of windsea is incorporated into
MRI and the weak point of MRI is overcome. Hypothetical assumptions are
also introduced into MRI-II to describe wave breaking effects for
over-saturated components. As a result of incorporation of these wave
breaking effects, MRI-II gives more reasonable estimates of waves than
MRI. Taking into account the accuracy and the time and spatial
sparseness of available wind data for wave prediction, the performance
of MRI-II is remarkably good.

ACKNOWLEDGEMENTS

The author would like to acknowledge the continuing encouragement of
Prof. Y.Toba of Tohoku University. He is grateful to the officials of
the Maritime Meteorological Division of JMA for providing the wind data
on typhoon ORCHID and their hindcast results obtained using MRI.

REFERENCES

Hasselmann,K.,T.P.Barnett,E.Bouws,H.Carlson,D.E.Cartwright,K.Enke,
    J.A.Ewing,H.Gienapp,D.E.Hasselmann,P.Kuseman,A.Meerburg,P.Muller,
    D.J.Olbers,K.Richter,W.Sell and H.Walden(1973): Measurements of
    wind-wave growth and swell decay during the Joint North Sea Wave
    Project(JONSWAP). Dt. Hydrogr. Z., Suppl. A. 8, No.12, 95pp.
Isozaki,I. and T.Uji(1973): Numerical prediction of ocean waves. Papers
    in Met. and Geophys., 24(2), 207-232.
The SWAMP Group(1984): The Sea Wave Modelling Project, Part 1; Principal
    results and conclusions. Proc. Symp. on Wave Dynamics and Radio
    Probing of Ocean Surface, Miami, 1981, Plenum Press.
The SWAMP Group(1982): The Sea Wave Modelling Project, Part 2;
    Compilation of results. KNMI Publication 161.
Toba,T.(1978): Stochastic form of the growth of wind waves in a
    single-parameter representation with physical implications. J.
    Phys. Oceanogr., 8(3), 494-507.
Toba,Y.(1979): Study on wind waves as a strong non-linear phenomenon.
    12th Symp. on Naval Hydrodynam., National Acad. of Sci.,
    Washington, D.C., 521-540.
Toba,Y., S.Kawai and P.S.Joseph(1984): The Tohoku Wave Model. Proc.
    Symp. on Wave Dynamics and Radio Probing of Ocean Surface, Miami,
    1981, Plenum Press.
Uji.T. and I.Isozaki(1972): The calculation of wave propagation in the
    numerical prediction of ocean waves. Papers in Met. and Geophys.,
    23(4), 347-359.
Uji.T.(1975): Numerical estimation of the sea waves in a typhoon area.
    Papers in Met. and Geophys., 26(4), 199-217.
Uji.T.(1984): A coupled disctete wave model MRI-II. J. Oceanogr. Soc.
    Japan, 40(4), 303-313.

THE TOHOKU-II WAVE MODEL

Y. Toba[1], S. Kawai[2], K. Okada[3] and N. Iida[4]
1 Department of Geophysics, Tohoku University, Sendai 980 Japan
2 Deceased
3 Japan Weather Association, Tokyo 100 Japan
4 Data Processing Center, Kyoto University, Kyoto 606 Japan

ABSTRACT. The TOHOKU-II Wave Model is a revised version of the TOHOKU
Wave Model which participated in the SWAMP. Main points of the revision
are described, and a hindcasting case study for a typhoon ORCHOID (8013)
is presented, with the observed wave data for comparison.

1. INTRODUCTION

The TOHOKU Wave Model (Joseph, Kawai and Toba, 1981; Toba, Kawai and
Joseph, 1984) belonged to the coupled-hybrid wave prediction models. In
its participation of the SWAMP (The SWAMP-GROUP, 1984), it was proved
that in the case of steady simple winds, this model gave results that
lay near the average of many models. However, some shortcomings were
also recognized. These deficits have been eliminated from the original
model, to construct the TOHOKU-II Wave Model.

2. MAIN POINTS OF THE REVISION

2.1. Energy allocation and adjustment of source function

The old model used a vector interpolation of the wind-wave energy with
direction. This caused some energy loss e.g., in crossing a sharp
front. This deficit has been overcome by adopting energy allocation, at
every step, to three or four grid points from the new position of the
wave packet which started from each grid point.
    Figure 1 shows the explanation of the energy allocation procedure.
The wave energy that has advanced from A to W in Δt, e.g., is first
allocated to B and A in a proportion of AW/AB and WB/AB, respectively.
The energy allocated to B is then allocated to two or three points as
shown in the right part of Figure 1.
    When the growth equation of wind waves was derived for the
nondimensional momentum (Toba, 1978), a kind of source function G,
defined by $G \equiv (1/\tau)(dM/dt)$, was obtained from fetch-graph formulas by
Wilson, where τ was the wind stress, M the wave momentum. This growth

227

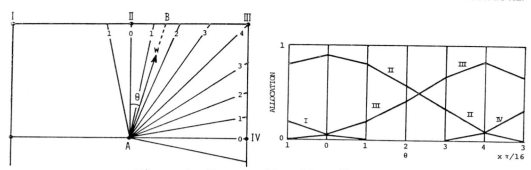

Figure 1. Energy allocation diagram.

equation corresponded to the prognostic equation of the form

$$\frac{dE}{dt} = \frac{\partial E}{\partial t} + c_g \frac{\partial E}{\partial x} = S \tag{1}$$

for the wave energy E and the wave group velocity c_g. If we apply (1) to E of representative waves, then c_g varies with E and generally (1) should have a form

$$\frac{dE}{dt} = \frac{\partial E}{\partial t} + \frac{\partial (c_g E)}{\partial x} = S' \tag{2}$$

In the case of energy interpolation, a term of $\partial c_g/\partial x$ is normally cancelled out and (1) is correct. However, if we adopt energy allocation, we should use c_g in the form corresponding to (2), and we should adjust the source function G as corresponding to S'.

 We now revert to the fetch-graph procedure to obtain the source function, which has no $\partial E/\partial t$ term, and so

$$\frac{\partial (c_g E)}{\partial x} = E \frac{\partial c_g}{\partial x} + c_g \frac{\partial E}{\partial x} = S' \tag{3}$$
$$\text{(I)} \qquad \text{(II)}$$

We know already that, from the 3/2 law, (II)/(I) = $c\delta E/E\delta c$ = 3, where c is the phase speed (Tokuda and Toba, 1982). Consequently S' in (2) should be (I) + (II) = (4/3)S = S'. However, since we are using the growth equation for the nondimensional energy E_* in the form

$$\frac{d(E_*^{2/3})}{dt_*} = G = GoR[1 - \text{erf}(bE_*^{1/3})] \tag{4}$$

the correction factor will not necessarily be 4/3. We have adopted a new value of the factor GoR empirically which is 1.2 times larger than the original value, i.e., GoR = 2.9×10^{-4}, by a test of numerical integration for a simple case, to give the same result as the original case.

2.2. Conversion between wind wave and swell, swell divergence

 The new model uses a threshold value for the swell emission E_m which is two times that of the original model. This reflects some recent data showing that swells continue to have the characteristics of

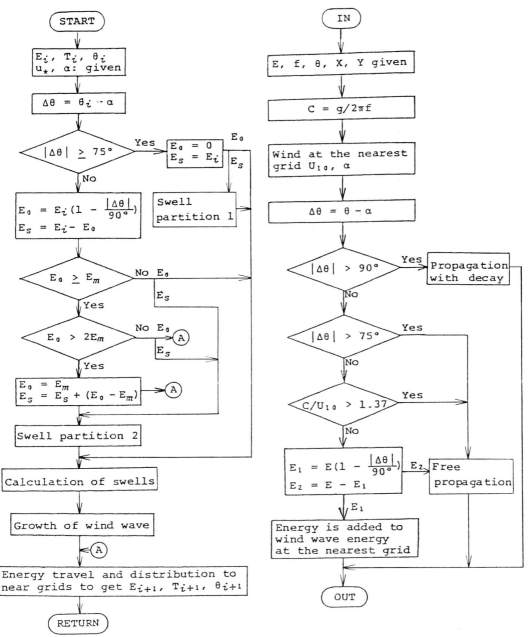

Figure 2. Main flow diagram of
    TOHOKU-II Wave Model. For swell
    partitions 1 and 2, see Fig. 19.4
    of Toba et al. (1984).

Figure 3. Calculation of swells of
    TOHOKU-II Wave Model.

wind waves (e.g., 3/2-power law) up to a large value of E^* for
decreasing winds, e.g., as Mitsuyasu et al. (1980) data show.

For the wind direction change $\Delta\theta$, the new model uses a linear
function of $\Delta\theta$ as shown in the main flow diagram, Figure 2. This new
form gives the same results for different values of Δt, for a change in
wind direction which is linear with time.

The threshold of conversion from swells to wind waves is still
$c/U_{10} = 1.37$ in the new model, based on the physical consideration,
using the same linear function of $\Delta\theta$, as seen in Figure 3.

When swells propagate 500 km, the swell energy splits into two rays
100 km distance apart, in the new model. The direction of propagation
itself is conserved.

3.   COMPARISON OF CALCULATION BY TOHOKU AND TOHOKU-II MODELS

This is shown, in Figure 4, for Case 4 and Case 5 of SWAMP, since in
these cases the TOHOKU Model showed conspicuous deficits. There is

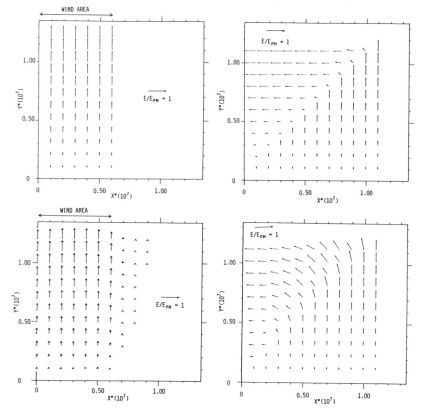

Figure 4. Comparison of calculations by the TOHOKU and TOHOKUI-II
    models. Left: Case 4 of SWAMP, right: Case 5 of SWAMP; upper:
    TOHOKU, lower: TOHOKU-II. Δx = 25 km, Δt = 30 min.

still a $\Delta x/\Delta t$-dependence for the adjustment of the representative
direction of the wave energy to the change in wind direction; we
consider $\Delta x/\Delta t$ of the order of $2c_g$ as a typical condition for our model.

4. HINDCASTING CASE STUDY FOR TYPHOON ORCHOID (8013)

A hindcasting study for a typhoon ORCHOID (8013), which passed through
the western North Pacific Ocean hitting Japan (Figure 5), has been made
for the new model. Successive wind field data from Sept. 6 to Sept. 13,
1980 were used, which were the same as the data of Uji (1984).
     Figure 6 shows the weather chart of Sept. 10. Figure 7 shows the
hindcasted wave height ($H_{1/3}$) by TOHOKU-II Model (Δx = 127 km, Δt = 1.5
hrs). Figure 8 is the wave chart ($H_{1/3}$) compiled from ships reports
(JMA). Figure 9 shows comparison of the time series of the observed and
computed $H_{1/3}$ at two stations shown in Figure 5.
     The general trend of the hindcasted $H_{1/3}$ distribution is similar to

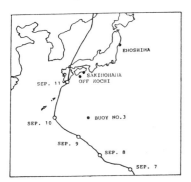

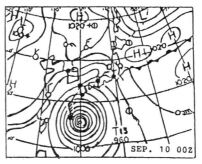

Figure 6. Weather chart of Sept. 10,
1980.

Figure 5. Track of the typhoon ORCHOID
     (8013) wave observation stations
     and the area of the hindcast.

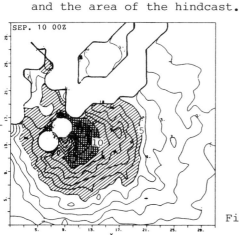

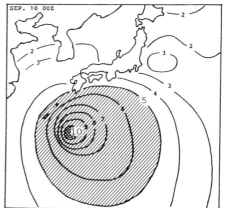

Figure 8. Wave chart ($H_{1/3}$) complied
     from ships reports (data from JMA).

Figure 7. Hindcasted wave height
     ($H_{1/3}$ in m) by TOHOKU-II Model.

that of the observation, though the maximum wave height, which occurred
to the northeast of the centre of the typhoon, was overpredicted by
about 30 %, and the position of the maximum wave height predicted was
northeast of that in the wave chart compiled from ships records.

The authors thank Dr. T. Uji who kindly provided data and help for
the hindcasting case study. The intercomparison of the TOHOKU-II Model
and the MRI-II Model by Uji (1984) will be published elsewhere.

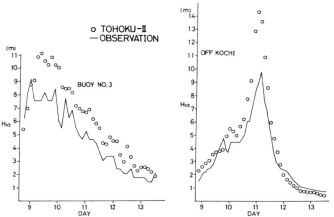

Figure 9. Comparison of time series of the observed and computed $H_{1/3}$ by
TOHOKU-II Model.   Stations are shown in Fig. 5.

REFERENCES

Hasselmann, K., D.B. Ross, P. Müller and W. Sell, 1976: A parametrical
    wave prediction model. *J. Phys. Oceanogr.*, **6**, 200-228.
Joseph, P.S., S. Kawai and Y. Toba, 1981: Ocean wave prediction by a
    hybrid model--Combination of single-parameterized wind waves with
    spectrally treated swells. *Tohoku Geophys. J.*, **28**, 27-45.
Mitsuyasu, H., F. Tasai, T. Suhara, S. Mizuno, M. Ohkusa, T. Honda and
    K. Rikiishi, 1980: Observation of the power spectrum of ocean waves
    using a cloverleaf buoy. *J. Phys. Oceanogr.*, **10**, 286-296.
The SWAMP-GROUP, 1984: Sea Wave Modelling Project (SWAMP), Part I.   In
    Ocean Wave Modeling, (Eds. K. Hasselmann and O.M. Phillips),
    Plenum, New York.
Toba, Y., 1978: Stochastic form of the growth of wind waves in a single-
    parameter representation with physical implications. *J. Phys.
    Oceaongr.*, **8**, 494-507.
Toba, Y., S. Kawai and P.S. Joseph, 1984: The TOHOKU Wave Model.   In
    Ocean Wave Modeling (Eds. K. Hasselmann and O.M. Phillips), Plenum,
    New York.
Tokuda, M. and Y. Toba, 1982: Statistical characteristics of individual
    waves in laboratory wind waves. II. Self-consistent similarity
    regime. *J. Oceanogr. Soc. Japan*, **38**, 8-14.
Uji, T., 1984: A coupled discrete wave model MRI-II.   *J. Oceanogr.
    Soc. Japan*, **40**, 303-313.

MICROWAVE SENSING OF THE OCEAN SURFACE

Gaspar R. Valenzuela
Naval Research Laboratory
Washington DC 20375 USA

Abstract. A review of electromagnetic (e.m.) and hydrodynamic processes pertinent to the backscatter of microwaves from water/ocean waves is presented. The understanding of e.m. scattering from the ocean has increased rapidly in the past two decades. The enhanced knowledge has allowed the probing of ocean waves with radio/microwaves, paving the way for the success of the first oceanographic satellite in 1978, SEASAT. In the future, as the calibration of microwave sensors improve, satellite measurements of the world oceans should become routine. In particular, the forthcoming 1989-1994 World Ocean Circulation Experiment (WOCE) will rely on microwave measurements from space for wind stress and geostrophic currents information.

1. INTRODUCTION

Crombie's (1955) observations on the Doppler spectrum of radar sea echo at 13.56 MHz paved the way to the present understanding of the scattering physics for electromagnetic (e.m.) waves for the ocean surface. For example, Crombie deduced correctly that radio waves are reflected back (backscattered) by ocean waves one-half the radio wavelength (for near grazing incidence) travelling in the radial direction. The mean Doppler frequency of the main peak of the spectrum was approximately 0.38 Hz (Figure 1), the frequency of ocean waves 22.1 m long. In effect the sea surface behaved like a "diffraction grating", and the return was principally contributed by the (presently called Bragg) resonant water waves. From the shape of the Doppler spectrum Crombie deduced that the scattering ocean waves travelled in "trains" superimposed on the crests of longer waves. Therefore, Crombie recognized the presence of "parasitic" waves usually present in water waves. Crombie even realized the possibility of "higher order" contributions to the scatter from waves an integer number of half-radio wavelengths ($\sqrt{2} \times 0.38$Hz $\approx$ 0.54 Hz for a gravity wave twice the resonant length), and predicted a continuous spectrum for higher radar frequencies where the resonant water waves are much shorter than the dominant waves of the ocean (i.e. the Doppler spectrum for a two-scale or composite rough surface).

Later most of these pioneering observations by Crombie for HF radio frequencies were to be verified and quantified by Wright (1966, 1968, 1978) and Bass et al. (1968) for microwaves, and the complete hierarchy of Doppler spectra from Bragg scattering were obtained under controlled wavetank conditions (Figure 2). Initially the line-spectrum of first-order Bragg scattering is evident, its intensity increasing with wind speed as the Bragg wave grows and the water spectrum develops toward smaller wavenumbers. Subsequently, when the air friction velocity u_* (about 5 percent of the wind speed) becomes greater than 30 cm/s, the second and higher

Y. Toba and H. Mitsuyasu (eds.), The Ocean Surface, 233–244.

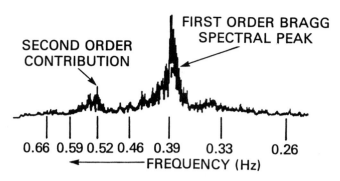

SECOND ORDER
CONTRIBUTION

FIRST ORDER BRAGG
SPECTRAL PEAK

0.66 0.59 0.52 0.46 0.39   0.33   0.26

◄——————— FREQUENCY (Hz)

Figure 1. Doppler spectrum of radar sea-echo at 13.56 MHz.
(Crombie, 1955)

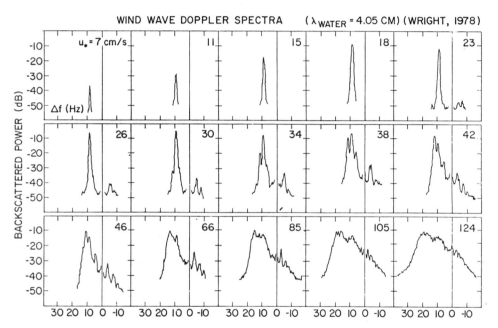

Figure 2. Hierarchy of Doppler spectra of radar backscatter (C band) obtained in a
wavetank for wind-waves. $u*$ is the air friction velocity about 5 percent of the wind speed.

order contributions appear on either side of the first-order Bragg line (these latter spectra are
similar to those observed at HF frequencies from the ocean), until for $u_* > 85$ cm/s the continuum spectrum for a two-scale rough surface (typical of microwave backscatter from the
ocean) predicted by Crombie develops.

2. MICROWAVE SCATTERING

Comprehensive reviews on the state of scattering physics for e.m. waves from the ocean surface
prior to 1978 were given by Wright (1978), Valenzuela (1978a, 1978b), and Barrick (1978).
Here only a brief review and update will be given. Presently it is well accepted that for angles

of incidence between 15° and 80° (the exact range of angles is a function of the surface rough-ness), Bragg scattering is the main mechanism contributing to the microwave backscatter from the ocean surface. More specifically, the scattering is from ocean waves of wavenumber $k = 2k_o \sin \theta$ travelling along the line-of-sight, where θ is the angle of incidence of the propa-gation vector of the e.m. radiation in relation to the normal vector to the mean surface, and k_o is the wavenumber of the e.m. radiation in free space.

In first-order Bragg scattering the backscatter power is proportional to the energy, and the doppler frequency of the return is equal to the frequency, of the Bragg resonant wave (if both receding and approaching waves are present, two spectral lines appear on either side of zero Doppler frequency). The return for vertical polarization is larger than for horizontal polariza-tion, and the ratio is in agreement with perturbation scattering theory.

However, on the ocean surface the large dominant waves "tilt" the "patches" of Bragg resonant waves away from the horizontal plane causing a change in the local plane of incidence of the e.m. radiation for each "patch". Since the microwave return from "patch" to "patch" is decorrelated (this neglects the long term residual correlation introduced by the dominant waves), a reasonable approximation for the resulting microwave backscatter is an incoherent addition of the local returns from all "patches", or equivalently the net microwave backscatter power (or cross-section) from the ocean may be obtained by averaging the first-order Bragg microwave backscatter power for one "patch" over the distribution of slopes of the dominant waves of the ocean. This two-scale or composite surface scattering model for the ocean was proposed and developed by Wright (1968) and independently by Bass et al. (1968).

Therefore it is clear the two-scale model applies for relatively large illuminated areas on the ocean that include at least several wavelengths of the dominant wave. As the illuminated area is reduced to the point where less than one wave period of the dominant wave is included, the nature of the scatter changes. For example it has been known for some time that the statis-tics of sea-clutter (radar backscatter) change from Rayleigh to Ricean (or Log-normal) when the resolution on the ocean is less than 20m (Trunk, 1972; Lewis and Olin, 1980); also see Fig-ure 3, which applies for small grazing angle but does demonstrate how the return signal changes drastically as the radar resolution is decreased.

For radar resolutions less than the dominant wavelength on the water, Bragg scattering still applies locally, but if the resolution is too small (the antenna is focused on the water) the amount of specular scattering (contributed by locally smooth facets normal to incoming e.m. radiation) is enhanced. In the extreme case the illuminated area is small enough to include only one specular point; then the Bragg scattering contribution is nil. In this regard Kwoh and Lake (1981), using an antenna focused on mechanically generated (deterministic) water waves found the specular contribution (identified by the unity ratio between vertically and horizontally polarized returns) was more important than previously thought. Wright and co-workers had already identified three basic scattering mechanisms in controlled wavetank studies on wind-waves (Duncan et al., 1974; Keller et al., 1974). These mechanisms are specular scattering near normal incidence, Bragg scattering from free waves, and Bragg scattering from trapped (parasitic) short gravity-capillary waves travelling at the phase speed of longer gravity waves; there was no indication of unusually large specular scattering. Kwoh and Lake's (1981) results could thus be an indication that deterministic water waves contribute more specular scattering than predicted from the probability density function of slopes of the rough surface. The assess-ment of how present scattering theories should be modified to account for the additional scattering is an area that deserves further research.

Generally speaking, specular scattering dominates near the specular direction (near nadir for backscatter), while toward grazing incidence other mechanisms also become important: sha-dowing, diffraction by wave crests, and refraction and trapping by the marine boundary layer (Wetzel, 1978).

HANSEN AND CAVALERI (1982)

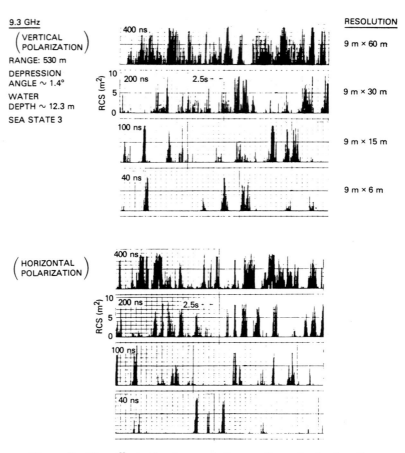

Figure 3. The effect of radar resolution on the radar backscatter.

Wave breaking has also been shown to be a major contributor to microwave backscatter (Keller et al., 1981), where it is shown that Bragg scatterers on the ocean are advected by the orbital motion of the dominant waves, and during periods of wave breaking the scatterer speed increases toward the phase speed of longer waves (Figure 4). Hence specular scattering may also be identified from the Doppler spectrum of the return signal in a coherent microwave radar. Figure 4 contains evidence of wave breaking on one crest of the dominant wave.

Advances in scattering theory have been able to combine specular scattering, Bragg scattering and shadowing in a single formulation (Brown, 1978), and recently Bahar and Barrick (1983) have applied the "full-wave" approach for rough surface scattering to a composite sur-face. The "full-wave" solution to the e.m. boundary value problem is posed in terms of the equivalent Generalized Telegraphists Equations; it is useful for one-dimensional rough surfaces and includes lateral waves, surface waves, and radiation fields. Explicit results are obtained by iteration (only terms up to second order are retained) and steepest descent integration is used.

The boundary conditions in the "full-wave" approach are satisfied rigorously and no res-triction is imposed on the amplitudes and slopes of the surface roughness. The only validation

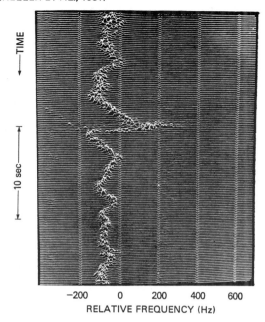

Figure 4. Doppler spectra at 9.375 GHz (Vertically polarized) microwave backscatter from the Nordsee Tower, German Bight (0.2 s averages) obtained September 18, 1979. One case of strong wave breaking at a wave crest the doppler frequencies exceed 300 Hz (100 Hz corresponds to a radial velocity of 1.6 m/s, zero doppler = −116 Hz).Radar depression angle = 350°, the wind speed is 16.5 m/s, the wave period = 9 s and the RMS wave height = 1.12 m. Note the advection of the Bragg resonant waves by the orbital velocity of the dominant waves.

for the "full-wave" scattering results is that they are in agreement with previous results in the limiting cases of gently undulating surfaces (physical optics) and slightly rough surfaces (first-order Bragg scattering). However, no detailed analysis exists on the accuracy of the scattering results, which obviously contain a number of approximations. Future efforts to improve scattering theory should probably start with the more exact Integral and Green's Function methods.

3. HYDRODYNAMICS OF GRAVITY-CAPILLARY WAVES.

Clearly an exact scattering theory for the ocean is useless, unless the proper kinematics and dynamics of ocean wind waves are used. In particular the properties of the short gravity-capillary waves which are the Bragg scatterers of the microwave radiation, and their interaction with other ocean waves, is required.

From present experimental evidence it seems that Hasselmann's (1968) picture of weakly interacting sinusoidal water waves offers an adequate description of wind waves for most purposes (Plant and Wright, 1977, 1979, 1980; Plant, 1980). Although in cases of stronger non-linearity involving finite amplitude waves the "soliton" theory of Lake and Yuen (1978) should have merit. However, the "soliton" model of a nonlinear wave train of waves travelling at the speed of a carrier wave may have limited application for the ocean surface (Segur, 1982).

To describe the development of wind waves in the weak interaction approximation (Hasselmann et al., 1973) with the Boltzmann transport (or energy balance) equation one needs the source functions for the input from the wind, the energy transfer by nonlinear interactions and the dissipation by viscosity and wave breaking. Wright and Keller (1971) found in a wavetank that the development of the wind waves is intimately related to the highly sheared wind-drift established in the water (about 4 percent of the wind speed). Using microwave backscatter Larson and Wright (1975) obtained the temporal growth rates of wind waves in a wavetank for much higher winds (up to 15 m/s) and much shorter waves (down to 0.7 cm) than before. These growth rates are due to the linear instability of the viscous boundary-layers of the coupled shear flow in the air and water. Kawai (1979) investigated the initial generation and development of wind waves (wavelets) in a wavetank and obtained close agreement between measurements and numerical calculations with a coupled shear flow model originally suggested by J. W. Wright and developed by Valenzuela (1976). The initial wavelets after generation, in about 10 s, develop into a state of stronger nonlinearity including intrinsic turbulence. A comparison of growth rates from wavetanks and ocean field measurements has been made by Plant (1982).

The energy transfer by resonant nonlinear interactions for water waves is well known (Hasselmann et al., 1973; Valenzuela and Laing, 1972). Plant (1980) included the wind-drift in the dispersion relation of gravity-capillary waves and found the energy fluxes for gravity waves (third-order) and gravity-capillary waves (second-order) are of the same magnitude for about 3.5 Hz waves, with the third-order transfer mechanism for gravity waves dominating for the longer waves and the second-order transfer for gravity-capillary waves dominating for the shorter waves. Of course the transfer mechanisms described apply to homogeneous wave fields; for inhomogenous fields (waves of different wavenumber are correlated) Crawford et al. (1980) has shown that the energy transfer for gravity waves will occur at the lower second-order.

Hasselmann (1974) has suggested a source function for waveıbreaking which is a quasi-linear function of the spectrum of the wave field. Wave breaking is a very important dissipative and mixing process, not only from the hydrodynamic point of view, but as a contributor of strong microwave backscatter. A number of investigations have been performed on wave breaking; here we mention the most recent, on parametric modeling by Longuet-Higgins (1982, 1983), and the photographic measurements of Koga (1981, 1982). Koga (1981) has observed that for wave breaking to occur not only must a stagnation point be present, but in addition Kelvin-Helmholtz instability must develop to overcome the restoring forces of gravity and surface tension. Wave breaking is one class of strong nonlinear wave phenomena, other investigations on nonlinear wave processes have been performed by Toba et al. (1975), Okuda et al. (1976, 1977), and Toba (1978, 1979) which are characterized by a statistical similarity structure.

Extensive work on the form of the high frequency part of the spectrum of wind waves has been performed for example, by Mitsuyasu and Honda (1975). However, the two-dimensional wavenumber spectrum, which is needed to predict microwave backscatter, cannot be obtained from the frequency spectrum because the short gravity waves are advected by the orbital velocity of the dominant waves and under this condition there is no one-to-one relation between frequency and wavenumber spectra. Therefore, still there is important work remaining in the relation between frequency and wave number spectra of high frequency ocean waves, and that microwave sensing combined with in situ observations should play an important role in this problem.

The short gravity-capillary waves are also "strained" by the orbital velocity of the long waves, causing a modulation of their amplitude, direction, and wavenumber. The modulation of short gravity waves by long waves was investigated in a wavetank by Keller and Wright (1975), who developed a linear relaxation model to predict these modulations. Since then, modulation transfer functions of gravity-capillary waves have also been obtained in the field (Plant et al., 1978, 1983; Wright et al., 1980) as a function of a number of parameters (i.e. wind speed, wave direction, wave height, air-sea temperature difference, etc.). Generally the modulation transfer functions in the field can be larger than those in wavetanks, the reason why, is an active area of research.

The modulation of short gravity waves by long waves plays an important role in the energy and momentum transfer from the atmosphere to the ocean waves, and also has a direct part in the formation of a radar image of the ocean. A higher order perturbation theory for the prediction of these short wave modulations was developed by Valenzuela and Wright (1979).

In addition to "straining" there are other sources of hydrodynamic modulations of gravity-capillary waves. Among these we have the long wave induced flow in the air (Wright et al., 1980) and random fluctuations in the air that do not correlate with the long waves (Plant et al., 1983).

4. APPLICATIONS TO REMOTE SENSING OF THE OCEAN SURFACE

From the exposition of the previous sections, it should now be clear that coherent radars can make ideal probes of ocean waves. At near normal incidence the backscatter power is of a specular nature, with the amplitude of the radar return being related to the probability density function of slopes of the sea surface and the doppler of the return related to the speed of the ocean waves. Of course this assumes the radar system is stationary. For a moving radar such as that in a satellite the frequency of the return also will be a function of platform velocity and antenna beamwidth.

However, for intermediate angles of incidence, away from normal and not too close to grazing, the backscatter is due to Bragg scattering and the power is proportional to the energy of the Bragg resonant waves. The Doppler of the return is related to the intrinsic phase speed of the Bragg resonant waves plus advection by the orbital velocity of dominant waves and currents (Wright, 1978). Then for radar footprints small compared to the dominant wave length, the phase of the return is proportional to the orbital velocity of the long ocean waves, so the wave height spectrum may be obtained by proper normalization (Keller et al., 1982). Figure 5 is a comparison of spectra derived with an L-band (1.5 GHz) radar versus the nondirectional wave spectra obtained with a *Baylor* gauge at CERC pier, N.C., US. Hence, for radar calibration purposes it should be worthwhile to obtain better in situ measurements of water velocities, such as: mean, orbital, and turbulent, under ocean waves.

Dual-frequency coherent radars have also been developed which resonate with the difference Bragg resonant wave (Plant and Schuler, 1980); the spectrum of the ocean waves may then be obtained by processing a number of frequency separations. However, the signal-to-noise ratio in dual-frequency radars is low because the background clutter also contains fluctuations at the frequency of interest. Recently Schuler et al. (1984a) have developed a three-frequency radar (TRIFAR) with constant frequency separation which is able by judicious processing to remove the background clutter almost completely. An improvement in the signal-to-noise ratio of up to 13 dB in relation to the conventional dual-frequency radar has been achieved. Figure 6 is the ocean spectrum derived from a de-multiplexed (using 25 frequency triplets) three-frequency radar (Schuler et al., 1984b).

Increased understanding of the physics of e.m. scattering from the ocean has resulted in the development and operation of a number of microwave instruments from tower, aircraft, and satellite platforms. For example SEASAT in 1978 carried four microwave instruments: altimeter, scatterometer, scanning multi-channel microwave radiometer, and synthetic aperture radar. With these instruments a number of ocean parameters may be recorded: wave height, surface currents, wind vector/wind stress, sea-surface temperature, atmospheric water, and surface waves.

(KELLER ET AL., 1982)

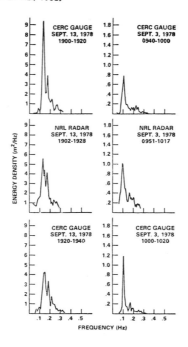

Figure 5. Ocean wave spectra obtained with an X band, vertically polarized, CW radar wave probe. For comparison the wave spectra obtained with a standard Baylor gauge are given. Measurements performed at CERC pier, Duck, North Carolina, USA. (All spectra have 40 degrees of freedom). Wind speeds were 6 m/s on September 3, 1978 and about 12.5 m/s on September 13, 1978.

Figure 6. Ocean wave spectrum derived from de-multiplexed (25 frequency triplets) L band three-frequency scatterometer (TRIFAR). Data taken on June 5, 1984 at CERC pier, Duck, North Carolina, USA. The grazing angle = 2.5°, the radar footprint on the ocean = 105 m × 300 m at a range of 300 m from the radar sight. The radar was mounted 12.5 m above sea-level.

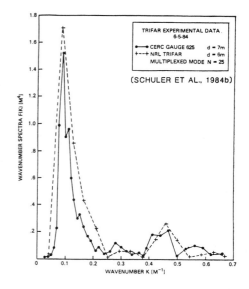

An excellent review of space oceanography has been given by Huang (1982), and a number of the satellite results may be found in Gower (1981), Beal et al. (1981) and Allan (1983).

5. CONCLUSIONS

Although the overall understanding of scattering physics for e.m. waves from the ocean increased rapidly after 1968, a decade passed until SEASAT demonstrated that microwave sensors on board satellite can give near synoptic coverage of the world oceans. The standard achieved with the SEASAT results is given in Table 1 (Born et al., 1982). The quality of these results has created a great deal of interest by meteorologists, oceanographers, and geophysicists in applying satellite microwave measurements to climatology studies.

(Born et al., 1982)

Table 1 — SEASAT Results

Sensor	Observable	Demonstrated* Accuracy (1)	Demonstrated Range of Observable
Altimeter	Altitude	8 cm (precision)	$H_{1/3} \leq 5$ m
	Significant Wave Height $(H_{1/3})$	10% or 0.5 m	0 to 8 m
	Wind Speed	2 m/s	0 to 10 m/s
Scatterometer	Wind Speed	1.3 m/s	4 to 26 m/s
	Wind Direction	$16°$	0 to $360°$
Scanning Multichannel Microwave Radiometer	Sea–Surface Temperature	$1.0°C$	10 to $30°C$
	Wind Speed	2 m/s	0 to 25 m/s
	Atmospheric Water	10% or 0.2 g/cm^2	0 to 6 g/cm^2
Synthetic Aperture Radar	Wavelength	12%	Wavelength $\geq$ 100 m
	Wave Direction	$15°$	0 to $360°$

*"Demonstrated Accuracy" is a measure of the agreement of coincident satellite and in situ measurements.

A great deal of improved understanding and better calibration of satellite microwave measurements is still needed, however, the IOC/SCOR Committee on Climatic Changes and the Ocean (CCCO) and the ICSU/WMO Joint Scientific Committee for the World Climate Research Programme (JSC) are already placing a strong emphasis on satellite microwave measurements. For example, in the World Ocean Circulation Experiment (WOCE) for 1989-1994, wind-stress over the world oceans will be needed as input to the climate models and this may be obtained from scatterometer measurements. Table 2 (Gregg, 1983) gives a tentative listing of ocean related spacecraft for the next decade.

Acknowledgment. The helpful comments and suggestions by Fred Dobson that improved the paper are appreciated.

(Gregg, 1983).

Table 2 — Ocean-Related Spacecraft Planned for the Next Decade

Satellite	Sponsor	Ocean-Related Sensors/Comments	Launch	Status
DMSP	USAF	MR	1984	Approved
	NASA	Data Processing Facility		Proposed
GEOSAT	USN	ALT	1984	Approved
NOSS	USN/NOAA/NASA	ALT, CS., MR, SCAT	1986	Cancelled
MOS-1	Japan	CS, IR, MR	1986	Approved
NOAA-H,1	NOAA	IR	1987	Approved
	NASA	Contribute Piggyback CS		Proposed
SPOT-2,3	CNES (France)	Piggyback ALT	1987 ?	Proposed
ERS-1	ESA	ALT, SAR, SCAT, IR	1989	Approved
	NASA	SAR Data Receiving/Processing Facility		Proposed
*TOPEX	NASA	ALT + Option for SCAT	1988 ?	Proposed
*N-ROSS	USN	ALT, MR, SCAT	1988 ?	Proposed
	NOAA	Contribute Bus (NOAA-D Spacecraft)		Proposed
	NASA	Contribute SCAT		Proposed
*ERS-1	Japan	SAR	1991 ?	Proposed
	NASA	Utilize SAR Data Facility		Proposed
GRM	NASA	Satellite-to-Satellite Tracking	1989 ?	Proposed
*ERS-2	ESA	ALT, SAR, SCAT ?	1990+ ?	Tentative
RADARSAT	Canada	SAR	1990+ ?	Proposed
	NASA	Contribute Launch & Piggyback SCAT		Proposed
*MOS-2	Japan	ALT, CS, MR, SCAT	1990+ ?	Tentative

*World Ocean Circulation Experiment (WOCE) satellites

References

Allan, T. D., 1983: *Satellite Microwave Remote Sensing.* Halsted Press, 526 pp.

Bahar, E., and D. E. Barrick, 1983: Scattering cross sections for composite surfaces that cannot be treated as perturbed-physical optics problems, *Radio Sci.,* **18**, 129-137.

Barrick, D. E., 1978: HF radio oceanography—A review. *Bound-Layer Meteorol.,* **13**, 23-43.

Bass, F. G., I. M. Fuks, A. E. Kalmykov, I. E. Ostrovsky and A. D. Rosenberg, 1968: Very high frequency radiowave scattering by a disturbed sea surface. *IEEE Trans.,* **AP-16**, 554-568.

Beal, R. C., P. S. DeLeonibus and I. Katz, 1981: *Spaceborne Synthetic Aperture Radar for Oceanography.* Johns Hopkins Press, 215 pp.

Born, G. H., et VI al., 1982: Seasat data utilization project report. NASA/JPL Rep. 622-325, Nat. Space Aeronaut. Ad./Jet Propul. Lab., Pasadena, Cal., 72 pp.

Brown, G. S., 1978: Backscattering from a gaussian-distributed perfectly conducting rough surface. *IEEE Trans.,* **AP-26**, 472-482.

Crawford, D. R., P. G. Saffman and H. C. Yuen, 1980: Evolution of a random inhomogeneous field of nonlinear deep-water gravity waves. *Wave Motion,* **2**, 1-16.

Crombie, D. D., 1955: Doppler spectrum of sea echo at 13.56 Mc/s. *Nature,* **175**, 681-682.

Duncan, J. R., W. C. Keller and J. W. Wright, 1974: Fetch and wind speed dependence of doppler spectra. *Radio Sci.,* **9**, 809-819.

Gower, J. F. R., 1981: *Oceanography from Space.* Plenum, 978 pp.

Gregg, W. W., 1983: *NASA Oceanic Processes Program.* NASA Tech. Mem. 85632, Nat. Aeronaut. Sp. Ad., Washington, DC, 154 pp.

Hansen, J. P., and V. F. Cavaleri, 1982: *High-resolution radar sea scatter, experimental observations, and discriminants.* NRL Rep. 8557, Nav. Res. Lab., Washington, DC, 49 pp.

Hasselmann, K., 1968: Weak-interaction theory of ocean waves. *Basic developments in fluid dynamics*, M. Holt, Ed., Academic, 117-182.

Hasselmann, K., 1974: On the spectral dissipation of ocean waves due to white capping. *Bound-Layer Meteorol.*, **6**, 107-127.

Hasselmann, K., and XV al., 1973: Measurements of wind-growth and swell decay during the joint north sea wave project (JONSWAP). *Dtsch. Hydrogr. Z.*, **A12**, 95 pp.

Huang, N. E., 1982: *Survey of remote sensing techniques for wave measurements. Measuring Ocean Waves*, Marine Board, Nat. Res. Council, Acad. Sciences, Washington, DC, 38-79.

Kawai, S., 1979: Generation of initial wavelets by instability of a coupled shear flow and their evolution to wind waves. *J. Fluid Mech.*, **93**, 661-703.

Keller, W. C., and J. W. Wright, 1975: Microwave scattering and the straining of wind-generated waves. *Radio Sci.*, **10**, 139-147.

Keller, W. C., T. R. Larson and J. W. Wright, 1974: Mean speeds of wind waves at short fetch. *Radio Sci.*, **9**, 1091-1100.

Keller, W. C., J. W. Plant and J. W. Johnson, 1982: Microwave measurement of sea surface velocities from pier and aircraft. *Proc. 1982 Oceans Conf.*, 909-913.

Keller, W. C., W. J. Plant and G. R. Valenzuela, 1981: Observation of breaking ocean waves with coherent microwave radar, in *IUCRM Symp. on Wave Dynamics and Radio Probing of the Ocean Surface*, Miami Beach, Fl.. (Also in *Proc. IUCRM Symp.*, O. M. Phillips and K. Hasselmann, Eds., Plenum, 1985).

Koga, M., 1981: Direct production of droplets from breaking wind-waves—Its observation by a multi-colored overlapping exposure photographing technique. *Tellus*, **33**, 552-563.

Koga, M., 1982: Bubble entrainment in breaking wind waves. *Tellus*, **34**, 481-489.

Kwoh, D. S. W., and B. M. Lake, 1981: Microwave backscatter from short gravity waves: deterministic, coherent, dual-polarized study of the relationship between backscatter and water wave properties, in *IUCRM Symp. on Wave Dynamics and Radio Probing of the Ocean Surface*, Miami Beach, Fl.. (Also in *Proc. IUCRM Symp.* O. M. Phillips and K. Hasselmann, Eds., Plenum, 1985).

Lake, B. M., and H. C. Yuen, 1978: A new model for nonlinear wind waves; Part I Physical model and experimental evidence. *J. Fluid Mech.*, **88**, 33-62.

Larson, T. R., and J. W. Wright, 1975: Wind-generated gravity-capillary waves: laboratory measurements of temporal growth rates using microwave backscatter. *J. Fluid Mech.*, **70**, 417-436.

Lewis, B. L., and I. D. Olin, 1980: Experimental study and theoretical model of high-resolution radar backscatter from the sea. *Radio Sci.*, **15**, 815-828.

Longuet-Higgins, M. S., 1982: Parametric solutions for breaking waves. *J. Fluid Mech.*, **121**, 403-424.

Longuet-Higgins, M. S., 1983: Bubbles, breaking waves and hyperbolic jets at a free surface. *J. Fluid Mech.*, **127**, 103-121.

Mitsuyasu, H., and T. Honda, 1975: The frequency spectrum of wind-generated waves. *Rep. Res. Inst. Appl. Mech.*, Kyushu Univ., **17**, 327-355.

Okuda, K., S. Kawai, M. Tokuda and Y. Toba, 1976: Detailed observation of the wind-exerted surface flow by use of flow visualization methods. *J. Oceanogr. Soc. Japan*, **32**, 53-64.

Okuda, K., S. Kawai and Y. Toba, 1977: Measurement of skin friction distribution along the surface of wind waves. *J. Oceanogr. Soc. Japan*, **33**, 190-198.

Plant, W. J., 1980: On the steady-state energy balance of short gravity wave systems. *J. Phys. Oceanogr.*, **10**, 1340-1352.

Plant, W. J., 1982: A relationship between wind stress and wave slope. *J. Geophys. Res.*, **87**, 1961-1967.

Plant, W. J., and J. W. Wright, 1977: Growth and equilibrium of short gravity waves in a wind-wave tank. *J. Fluid Mech.*, **82**, 767-793.

Plant, W. J., and J. W. Wright, 1979: Spectral decomposition of short gravity wave systems. *J. Phys. Oceanogr.*, **9**, 621-624.

Plant, W. J., and D. L. Schuler, 1980: Remote sensing of the sea surface using one- and two-frequency microwave techniques. *Radio Sci.*, **15**, 605-615.

Plant, W. J., and J. W. Wright, 1980: Phase speeds of upwind and downwind traveling short gravity waves. *J. Geophys. Res.*, **85**, 3304-3310.

Plant, W. J., W. C. Keller and J. W. Wright, 1978: Modulation of coherent microwave backscatter by shoaling waves. *J. Geophys. Res.*, **83**, 1347-1352.

Plant, W. J., W. C. Keller and A. Cross, 1983: Parametric dependence of ocean wave-radar modulation transfer functions. *J. Geophys. Res.*, **88**, 9747-9756.

Schuler, D. L., W. J. Plant, A. B. Reeves and W. P. Eng, 1984a: Removal of clutter background limitations in dual-frequency scattering from the ocean: the three frequency scatterometer. *Int. J. Rem. Sens.*, to be published.

Schuler, D. L., W. J. Plant and H. C. Miller, 1984b: Experimental verification of the three-frequency scatterometer concept. *Proc. 1984 Oceans Conf.*, 145-149.

Segur, H., 1982: Solitons and the inverse scattering transform. *Topics in Ocean Physics* (LXXX Corso), Soc. Ital., di Fisica, Bologna, 235-277.

Toba, Y., 1978: Stochastic form of the growth of wind waves in a single-parameter representation with physical implications. *J. Phys. Oceanogr.*, **8**, 494-507.

Toba, Y., 1979: Study on wind waves as a strongly nonlinear phenomenon. *Proc. Twelfth Symp. Naval Hydrod.*, Nat. Acad. Sci., 529-540.

Toba, Y., M. Tokuda, K. Okuda and S. Kawai, 1975: Forced convection accompanying wind waves. *J. Oceanogr. Soc. Japan*, **31**, 227-234.

Trunk, G. V., 1972: Radar properties of non-Rayleigh sea clutter. *IEEE Trans.*, **AES-8**, 196-204.

Valenzuela, G. R., 1976: The growth of gravity-capillary waves in a coupled shear flow. *J. Fluid Mech.*, **76**, 229-250.

Valenzuela, G. R., 1978a: Theories for the interaction of electromagnetic and oceanic waves— A review. *Bound-Layer Meteorol.*, **13**, 61-83.

Valenzuela, G. R., 1978b: Scattering of electromagnetic waves from the ocean. *Surveillance of Environmental Pollution and Resources by Electromagnetic Waves*, T. Lund, Ed., Reidel, 199-226.

Valenzuela, G. R., and M. B. Laing, 1972: Nonlinear energy transfer in gravity-capillary wave spectra, with applications. *J. Fluid Mech.*, **54**, 504-520.

Valenzuela, G. R., and J. W. Wright, 1979: Modulation of short gravity-capillary waves by longer-scale flow—A higher-order theory. *Radio Sci.*, **14**, 1099-1110.

Wetzel, L., 1978: On the origin of long-period features in low-angle sea backscatter. *Radio Sci.*, **13**, 313-320.

Wright, J. W., 1966: Backscattering from capillary waves with application to sea clutter. *IEEE Trans.*, **AP-14**, 749-754.

Wright, J. W., 1968: A new model for sea clutter. *IEEE Trans.*, **AP-16**, 217-223.

Wright, J. W., 1978: Detection of ocean waves by microwave radar; the modulation of short gravity-capillary waves. *Bound-Layer Meteorol.*, **13**, 87-105.

Wright, J. W., W. J. Plant, W. C. Keller and W. L. Jones, 1980: Ocean wave-radar modulation transfer functions from the West Coast Experiment. *J. Geophys. Res.*, **85**, 4957-4966.

ON THE HYDRODYNAMICS OF SMALL-SCALE BREAKING WAVES AND THEIR MICROWAVE
REFLECTIVITY PROPERTIES

M.L. Banner[1] and E.H. Fooks[2]
[1] School of Mathematics, Univ. of N.S.W., Kensington, 2033
Australia
[2] School of Elec. Eng. and Computer Science, Univ. of N.S.W.,
Kensington, 2033, Australia

ABSTRACT.   This contribution is aimed at advancing our understanding
of local microwave reflectivity properties of small-scale breaking
water waves, a widespread feature of the wind-driven sea surface.
In this contribution a detailed laboratory study of the disturbances
generated in the breaking zones of short breaking gravity waves is summarized.*   The response of microwaves to these hydrodynamic disturbances was investigated at X-Band, for which large local reflected
power coherent with the disturbances was found.   The wavenumber
structure of the breaking zone disturbances was found to be consistent
with Bragg scattering.   It is concluded that further investigation is
warranted of the contribution to the backscattered cross-section of the
small-scale breaking wave components on the sea surface in order to
improve understanding of the relationship between radar backscattter
and the desired sea surface properties of interest such as the wind-
stress vector and ocean wave spectrum.

1.   INTRODUCTION

With the advent of satellite-borne active microwave remote sensing devices, such as the scatterometer for windstress determination and synthetic aperture radar for ocean wave spectra, intensive surface truth
studies have revealed unforeseen complexities in relating the average
level and modulation levels of the backscattered microwave power to
the underlying oceanic property of interest.   Wright et al (1980)
found inexplicably high radar modulation levels coherent with the long
wave spectrum while in a more extensive survey, Plant et al (1983)
found modulation levels incoherent with the long waves of the same
order of magnitude as the modulation coherent with the long waves.
More recently, Plant et al (1985) reported substantial dependence of
the average radar cross-section on the total long wave slope, at a
given wind speed.   These recent findings suggest an incomplete under-

---

* A full version of this study has been accepted for publication in the
  Proceedings of the Royal Society of London, Series A.

245

Y. Toba and H. Mitsuyasu (eds.), The Ocean Surface, 245–248.
© 1985 by D. Reidel Publishing Company.

standing of the sources contributing to the microwave backscatter and
its local variability in the oceanic situation.   This is the context
of the present study, which is a detailed laboratory investigation of
the local hydrodynamics of breaking zones of small-scale breaking gravity waves and the associated microwave reflectivity properties.
This work is seen as complementary to the previous studies of Wright
et al (1972) and of Kwoh and Lake (1983).

2.   EXPERIMENTAL ARRANGEMENT AND RESULTS

A small laboratory wind-wave tank was used to conduct initial studies
of microwave reflectivity characteristics of small-scale breaking waves.
3Hz mechanically-triggered waves were strongly wind-driven to maintain
a state of near-continuous breaking.   A small X-band microwave apparatus was deployed looking upwind in the V-V mode at a depression angle
of 55°.   The field of view was limited by a microwave absorbing aperture to about 100 mm × 100 mm.   A sidewall-mounted optical wave height
probe was located at the same axial position along the tank as the
microwave boresight intersection with the mean water level.   The results obtained clearly associated strong local reflectivity with the
breaking crest region.   Non-breaking waves of similar period, obtained
by reducing the wind stress, produced negligible backscatter.
   To understand the phenomenon responsible for the strong backscatter, an alternative configuration was used in which the wave tank was
operated as a flume, supplied from a constant head tank.   The flume
allowed the generation of a *stationary*, steadily breaking wave at the
test site by suitable placement of a subsurface airfoil obstacle.   The
wavelength and degree of breaking were adjustable via the water speed
and airfoil orientation respectively.   A wavelength of 20 cm was chosen to match the 3Hz propagating waves and the degree of breaking was
adjusted to provide the minimum level necessary for sustained, quasi-
steady breaking.
   The microwave apparatus and aperture were traversed axially to determine the mean spatial microwave reflectivity characteristics of the
stationary, airfoil-generated breaking wave and for comparison, similar
profiling of a very steep non-breaking wave of the same wavelength.
The latter was generated by reducing the attack angle of the airfoil.
The results show that a strong local mean power return arises only from
the *breaking* zone region near the crest, very similar to the propagating case described earlier.   This strong correspondence provided justification for a more detailed probing of the stationary breaking wave
in preference to the more difficult transient propagating wave.
   It was observed that the reflected microwave power time series produced by the stationary breaking wave was characterized by a strong
temporal periodicity.   Frequency spectra at various depression angles,
looking 'upwind' confirmed a locally enhanced peak (at a frequency of
around 14Hz) with a close correspondence with the frequency spectrum of
the hydrodynamic disturbances in the breaking zone, as detected by a
capacitance wave gauge sited in the breaking zone.

Having confirmed the strong correspondence between the hydrodynamic breaking zone disturbances and microwave power fluctuations, it was
of interest to determine further salient properties of the breaking
zone disturbances with the aim of identifying the scattering mechanism.

It was noted firstly that the frequency characteristics of the disturbances were preserved axially along the wave profile, with the peak
amplitude located just ahead of the crest.   The axial and transverse
coherence characteristics were determined for the peak and half power
disturbance frequencies using two identical wave height gauges.   The
coherence data revealed the laterally compact nature of the disturbances which persist coherently in the longitudinal direction over about
one half of the breaking wave length.   The lateral coherence length
was an order of magnitude less than the breaking wave length.

Again, using two identical wave height gauges, with one fixed just
ahead of the breaking crest, two-sided wavenumber-spectra were determined for various disturbance frequencies.   The phase information
associates positive wavenumbers with rearward travelling waves.   The
most noteworthy feature of the wavenumber-frequency spectra is the
existence of significant levels of hydrodynamic disturbance energy in
the wavenumber band appropriate for Bragg backscatter at X-Band (micro-
wavelength $\simeq$ 32 mm) at the various depression angles used in this study.
Similar determinations were carried out for a 33 cm wavelength breaking
wave.   Here the characteristic hydrodynamic disturbances were found to
peak at a somewhat lower frequency with correspondingly lower wavenumbers constituting the energetic peak of the wavenumber spectrum.   However, the wavenumber spectra were sufficiently broad as to contribute
significant backscatter at X-Band for the 33 cm breaking wave.
Finally, it should be noted that most of the measurements were carried
out for H-H polarization as well, with H-H reflected power levels always found to be less than, but within 3db of the V-V power levels.

3.   DISCUSSION AND CONCLUSIONS

The investigation summarized here leads to the following view of the
disturbances generated in breaking zones of short breaking waves : at
any instant during the breaking cycle, the breaking zone produces a
fairly sharp frequency spectrum of disturbances with corresponding
wavenumbers which are in the Bragg resonance range for X-Band and C-
Band radars for moderate depression angles.   These disturbances can
lead to significant local backscatter from the breaking zones.   It remains to assess their absolute importance in contributing to the backscattered cross-section from the sea surface in an effort to improve
the existing calibrations used for correlating microwave backscatter
from the sea surface with oceanic properties of interest.   In any
event, this study has provided new physical insight into the organized
structure of breaking zones of spilling breaking waves.   These properties are believed to characterize spilling breaking zones on all
scales.

4.   REFERENCES

Plant, W.J., Keller, W.C. and Cross, A., 1983 : Parametric dependence
    of ocean wave-radar modulation transfer functions, *J. Geophys.*
    Res. 88, 9747-56.
Plant, W.J., Keller, W.C. and Weissman, D.E., 1985 : The dependence of
    the microwave radar cross section on the air-sea interaction and
    the wave slope.   *Presented at the Symposium on Wave Breaking, Tur-*
    bulent Mixing and Radio Probing of the Sea Surface, 19-25 July,
    1984, Sendai, Japan.
Kwoh, D. and Lake, B.M., 1983 : Microwave backscattering from short
    gravity waves: a deterministic, coherent and dual-polarized labo-
    ratory study.   *TRW Report No. 37564-6001-UT-00,* Calif.90278.
Wright, J.W., Plant W.J., Keller, W.C. and Jones, W.L., 1980 : Ocean
    wave modulation transfer functions from the West Coast Experiment.
    J. Geophys. Res. 85, 4957-66.
Wright, J.W., Duncan, J.R. and Keller, W.C., 1972 : Wind wave studies :
    Part 1 - Doppler Spectra.   *N.R.L. Report 7473,* Nov.8, 1972,
    Washington D.C.

THE NATURE OF MICROWAVE BACKSCATTERING FROM WATER WAVES

D. S. Kwoh and B. M. Lake
Fluid Mechanics Department
TRW Space and Technology Group
One Space Park
Redondo Beach, California 90278

ABSTRACT. Our recent laboratory study shows that Bragg scattering by itself is not an adequate description for microwave backscattering from water waves. It may account for part of the scattering, but reflection from specular facets and wedge-like diffractive scattering from small radius crests of waves can predominate. Our first experiment was performed on wave pedal-generated short gravity waves. Using a scanning laser slope gauge to measure the surface and the moments method to compute the scattering, we found that the small radius crests of waves can be the more dominant source of scattering and better described by wedge diffraction than Bragg scattering. Bragg scattering does describe the scattering from the parasitic capillaries. We also found that specular reflection is more important than generally expected. Our second experiment was performed on wind waves. We found that at low wind, the Doppler spectrum is narrow peaked but it gradually evolves to become a double peaked spectrum at high wind. Analysis shows that low wind scattering is indeed Bragg scattering. At high wind, the lower frequency peak again is due to Bragg scattering from rough patches, whereas the higher frequency peak is due to scattering from waves not unlike our pedal-generated short gravity waves.

1. INTRODUCTION

In the last decade the "composite model" has become increasingly accepted as the correct theory for describing microwave scattering from the ocean surface. When we started our study of microwave scattering from water waves, it seemed unsatisfactory to us that there was no clear concept of what the "Bragg waves" really are. Are they corrugation-like wavelets on a windblown surface, or are they a Fourier component of a fairly sharp crest? Or could they be a Fourier component in a random rough patch? To answer these questions, the obvious thing to do would be to examine a scattering surface in great detail. More specifically, we attempted to measure the exact surface profile while microwave scattering is taking place. Paddle-generated and wind-gener-

Y. Toba and H. Mitsuyasu (eds.), The Ocean Surface, 249–256.
© *1985 by D. Reidel Publishing Company.*

ated short gravity waves were studied in the laboratory in this
deterministic fashion. The results are briefly presented here.

2. EXPERIMENTAL SETUP

The experiment was performed in a wave tank 12 m long, 92 cm wide, and
90 cm deep. On top is a wind tunnel with a cross section of 122 cm x
92 cm. A microwave absorber covers the wind tunnel's inside surface.
Water waves are either wave paddle or wind generated.
    The X-band radar is a 9.23 GHz cw superhet coherent system with
100 mw transmitted power and dual transmitting and receiving channels,
each providing individual phase and amplitude outputs. Each channel
can be nulled to cancel stray reflections. The antenna is a corrugated,
conical horn with a 22.9 cm aperture fitted with a matched dielectric
lens with a focal length of 45.7 cm. The 3-dB beamwidth is 8.4 cm for
both vv and hh polarizations at the focal plane.
    To measure an almost instantaneous profile of the water surface,
we have developed a scanning laser slope gauge (SLSG). The laser beam
measures the slope of the water surface over 13.3 cm at 39.063 Hz. The
SLSG has an angular range of 60° and an accuracy of ± 0.3°.
    Besides the SLSG, another optical sensor detects specular facets
normal to the microwave incidence. A projector lamp is placed on one
side of the corrugated horn and a camera on the other side. A photo-
diode detector at the camera's film plane produces spikes whenever
specular facets appear in the camera's field of view. This separates
the "nonspecular" from the "specular" scattering events.
    For comparison purposes, a capacitance level gauge was also in-
stalled close to the side of the wave tank.

3. EXPERIMENTAL APPROACH

We first studied mechanically-generated short gravity waves with frequencies of 2.5 Hz (~ 25 cm) and amplitude ~ 1 cm (ka ~ 0.17).
Microwave backscattering amplitude and phase for both vv and hh
polarizations were measured for 40°, 55° and 70° incidence angles.
    Specular scattering events were first separated from the non-
specular events by the specular reflection sensor. Completely nonspecular events were then analyzed deterministically. The scanned
slope profiles were first recorded. Together with the measured antenna
pattern, the moments method was applied to compute the induced surface
electromagnetic current and the scattering amplitudes in all directions,
in particular, the backscattering direction. Figure 1(a) shows an
example of a scanned slope profile and its integrated displacement profile. Moments method is applied to compute the scattered power for
both the TE mode (vv) and TM mode (hh) [Figure 1(b)]. The computed
backscattered power, polarization ratio and phase were then compared
with the measured values to validate both our experimental and numerical procedures. Numerical modeling was then performed to determine
which water surface features contribute significantly to the scattering
and which approximate theories best describe these scatterings.

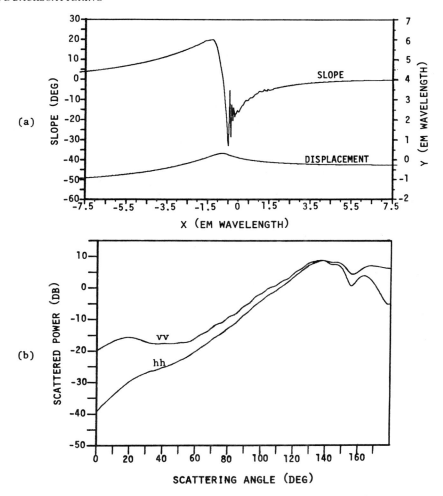

Figure 1. (a) Slope and displacement profile of event in a scan.
          (b) vv and hh scattered power ($\equiv 20 \log_{10}|E_z^S|$) as a function
              of angle as computed by moments method.

For wind-wave scattering, a similar approach was taken. In addition, the Doppler spectrum was also measured. The task was to understand the shape and evolution of the Doppler spectrum with increasing
wind speed as a function of the evolution of the water surface with
increasing wind speed.

4.   BACKSCATTERING FROM MECHANICALLY-GENERATED WAVES

Figure 2 shows a comparison between computed and measured results in
backscattered power, polarization ratio and Doppler shift. The fact

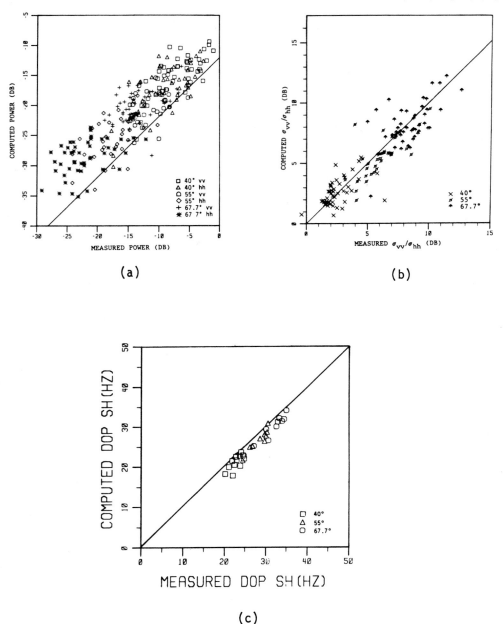

Figure 2.   Computed results vs. measured results for backscattering
            events at 40°, 55°, 67.7° from 2.5 Hz wavetrains.
            (a) Backscattered power. Solid line is calibration line.
            (b) Polarization ratio.   Solid line is 45° line.
            (c) Doppler shift.   Solid line is 45° line.

that the measured backscattered power is ~ 4 dB less than the computed
power may be because the scattering region is less than the antenna
bandwidth. Otherwise, the agreement is good. We are now justified in
doing numerical modeling. For example, scattering from the previously
shown slope profile [Figure 1(a)] can be thought of as a direct sum of
scattering from the capillary waves and scattering from the wave with
the capillaries removed. Analysis shows that whereas scattering from
the capillary wave is indeed Bragg scattering, scattering from the wave
with the capillaries removed is closer to wedge diffraction [as in the
Geometric Theory of Diffraction (GTD)].

So far, we have concentrated on nonspecular scattering. There is
occasionally specular reflection from small facets which are either
steep capillaries or are in the turbulent wake behind a breaking wave.
One good indication of specular reflection (other than the specular
reflection sensor) is that vv and hh amplitude are almost equal.

The following table summarizes a description of scattering from
mechanically-generated short gravity waves.

BACKSCATTERING FROM MECHANICALLY GENERATED SHORT GRAVITY WAVES	
SPECULAR (15-30%)	NONSPECULAR (85-70%)
o From steep parasitic capillaries o From turbulent wake	o From capillary SPT but wrong frequency o From crest "rounded wedge" ~ GTD o From turbulent wake

5. BACKSCATTERING FROM WIND WAVES

Figure 3 shows the Doppler spectrum in both polarizations from 2.5 m/s
to 8.8 m/s wind speed at θ_i = 55°. At low wind, the spectra in both vv
and hh are narrow peaked, centered at 16 Hz and have polarization ratios
of ~ 11 dB. At high wind, the spectra in both vv and hh are double
peaked. The high frequency peak occurs at ~ 37 Hz and has a polarization ratio of ~ 3 dB. The lower frequency peak is broad and occurs at
~ 20 Hz, with a polarization ratio of ~ 9 dB.

To understand the spectra, we have to look at the slope recording.
Figure 4 shows wind-wave scattering at 3.0 m/s and 7.9 m/s. At low
wind speed [Figure 4(a)], the scattering has the following characteristics: (i) The microwave amplitude has longer time duration compared
with the water wave period, implying that scattering may be associated
with a large patch rather than with individual waves. (ii) The large
polarization ratio is evident. (iii) The phase channel shows uniform
Doppler frequency. These strongly suggest the water surface is
"slightly rough" so that small perturbation theory (SPT) is applicable.
At high wind [Figure 4(b)] the scattering shows the following characteristics: (i) The slope gauge shows two distinct features (a) rough
patches, as in A and C in Figure 4(b), (b) waves with parasitic capillaries, as in B in Figure 4(b). (ii) The rough patches are associated

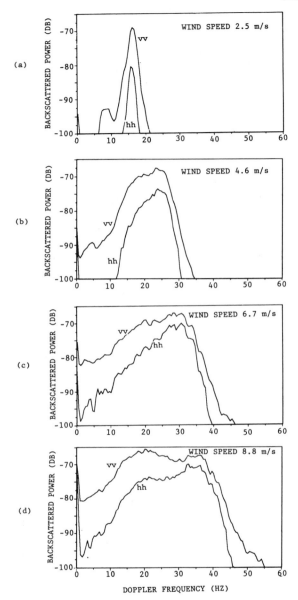

Figure 3.   Wind-wave backscattering Doppler spectrum at
θ_i = 55°.

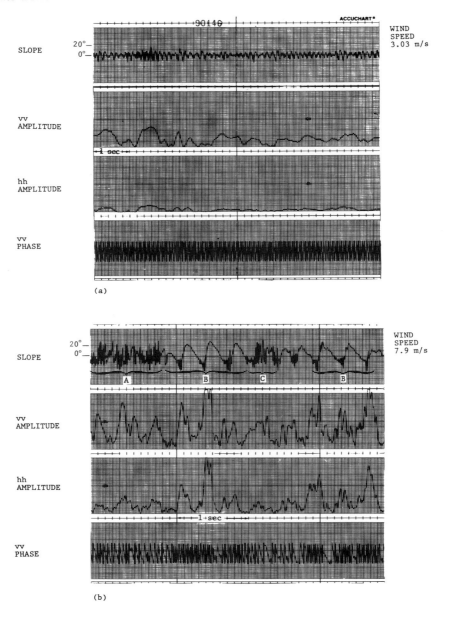

Figure 4.   Wave slope gauge data.

with large polarization ratio and small Doppler shift. (iii) The waves
with capillaries are associated with a small polarization ratio and
large Doppler shift. In this case the microwave amplitude also shows
individual bursts corresponding to individual waves. The large polarization ratio associated with the rough patches suggests the scattering
may again be described by SPT. The waves with capillaries look similar
to the paddle-generated waves that we have studied, in which case there
would be specular reflection from facets, wedge diffraction from crests
and Bragg scattering from the capillaries. The large Doppler shift
simply corresponds to the phase velocity of the dominant wave. Referring back to the Doppler spectrum in Figure 3, it is now evident that
for low wind speed, the narrow peak and the large polarization ratio
of ~ 11 dB are because the water surface is "slightly rough" and Bragg
scattering is the appropriate description. For high wind speed, the
double peak in the spectrum suggests there are two distinct scattering
features. The lower frequency peak with a large polarization ratio
corresponds to scattering from the "rough patches", which again may be
described as "Bragg scattering". The high frequency peak with a small
polarization ratio corresponds to the "waves with parasitic capillaries", which scatters the microwave by a combination of specular reflection, wedge diffraction and Bragg scattering.

The above offers a qualitative scenario for microwave scattering
from wind waves. Subsequent quantitative analysis confirms this picture. The following table summarizes the description.

WIND-WAVE SCATTERING

LOW WIND		HIGH WIND	
EM: SPT HYDRO: FREE WAVES ON WIND DRIFT LAYER	ROUGH PATCHES EM: SPT HYDRO: TURBULENCE	WAVES WITH CAPILLARIES EM: SPECULAR FACETS CREST $\sim$ "ROUNDED WEDGES" CAPILLARIES $\sim$ SPT HYDRO: BOUND WAVE	

6. CONCLUSIONS

Our conclusions based on these laboratory measurements can be summarized as follows:

1. Electromagnetically, we have identified three mechanisms for
microwave backscattering from water waves: (i) specular reflection
from turbulent wakes or steep capillaries, (ii) rounded wedge
diffraction from sharp crests of waves, and (iii) Bragg scattering from
parasitic capillaries or turbulent patches.

2. Hydrodynamically, we have identified three sources for wind-
wave scattering: low wind free waves, high wind bound waves and turbulent patches.

3. The implications for microwave backscattering from ocean waves
are: (i) specular reflection at low θ_i (e.g., 20°) may be more important than generally expected, and (ii) the relative frequency of specular facets, sharp crests and turbulent patches will determine the character of ocean scattering.

THE WAVE FIELD DYNAMICS INFERRED FROM HF RADAR SEA-ECHO

P. FORGET
L.S.E.E.T., Equipe de Recherche Associée du C.N.R.S.,
Université de Toulon et du Var
639, Boulevard des Armaris
83100 Toulon
France

ABSTRACT. We present and discuss some characteristics of wind waves under fetch-limited conditions, as remotely sensed by two HF radars. The methods of data processing are based on the theories of radio scatter and radio propagation. The behaviour of the peak frequency of the wave spectrum as a function of wind speed and fetch, as well as the shape of the spectrum are in agreement with previous results like JONSWAP. Some disagreement is found concerning the directional spreading function in that it seems to be more or less invariant with respect to wave frequency and wind velocity. Also, our wave steepnesses are greater than in other experiments and may be related to the specific properties of the wind field.

1. INTRODUCTION

The observation of a wave field with two HF ground-wave radars is concerned here. The study is based on two major properties of the radio technique, i.e. first, the synoptic coverage over a fairly large surface area; second, the selective nature of the interactions between the radio waves and the sea which lets one observe ocean waves at a fixed wave number. This property is revealed by the spiked first and second-order spectra of the Doppler echoes (Fig. 2). The main goals of the experiment described in Sec. 2 were to obtain the up- and cross-wind radar signatures of the sea surface under fetch-limited conditions and to infer from the Doppler spectra the characteristics of the wave directional spectrum in terms of wind speed U and fetch X. The results could then be compared with those obtained by in-situ devices for similar sea states. Results are given in Sec. 4 after the description of the methods used (Sec. 3).

2. EXPERIMENTAL CONDITIONS

A sea zone ranging from 10 to 34 Km off the coast is viewed simultaneously by two radar systems S and C (Fig. 1) at frequencies f(~6.7 MHz)

Y. Toba and H. Mitsuyasu (eds.), The Ocean Surface, 257–262.

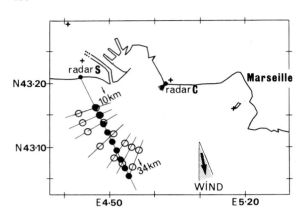

Figure 1 : General conditions of the experiment (o, ● : radar cells of stations C and S respectively). Crosses indicate the location of the weather stations used for determining the synoptic wind conditions.

and f'($\sim$ 13.5 MHz). Radar S is fixed in azimuth whereas radar C can scan over seven azimuths. The procedure of data collection as well as the evaluation of off-shore wind conditions are described in detail in [1]. The weather conditions during the ten days of the experiment consist mainly of three periods of northerly winds with velocities generally greater than 10 m/s and ranging up to 25 m/s. For these periods, the wind direction is nearly parallel to the beam axis of radar S so that, as is desirable, the sea cells have progressively increasing fetch off the shore and are looked at radially by station S and transversally by station C.

3. METHODOLOGY

Analysis is confined to some of the more energetic features of the Doppler spectra (Fig. 2). These are, first, the Bragg lines A_1^{+-} for station C and the line A_1^- for station S; and second, the second-order line A_2^- at frequency $-\sqrt{2}f_B$, where f_B is the Bragg frequency (Doppler shifts due to currents are disregarded), of the spectra of station S. In order to refer our radar data to fetch-limited rather than duration-limited conditions, each radar run is associated with a wind parameter which is the wind speed averaged over the three preceding hours. For the velocities usually encountered, this interval represents a sufficient delay for the wind to establish a stationary sea state in the area of observation. Radar data are classified to give a set of mean values dependent on the two parameters U and X; U is the mean wind speed grouped in intervals of about 2 m/s (U = 11,13,15,17,20 m/s) and X is the distance to the shore of the sea cells along the wind direction.
   The processing of radar data is based on the following equations :

$$R(\Theta) \equiv A_1^+/A_1^- = \phi(\pi-\Theta)/\phi(\Theta) \qquad (1)$$

$$A_1^- = K\ G\ k_i^6\ S(2\ \vec{k}_i) \qquad (2)$$

$$A_2^- = K\ G\ \alpha k_i^{10}\ S^2(\vec{k}_i) \qquad (3)$$

Θ is the angle of wind/wave direction to radar bearing; ϕ is the directional spreading function for Bragg waves; S is the directional wave spectrum; $\vec{k}_i$ is the radar wave number; K is an unknown gain factor coming from the radar equation and G is the dimensionless attenuation factor [2] modified by an increase in scattering area proportional to range. Equations 1-2 are derived from the first-order theory of HF scatter[3]; equation 3 is the expression for the energy of the second-order line integrated over one interval of spectral resolution and results from the second-order theory[1].

Measurements of R by station C gives directional information on the wave field. The study of the frequency spectrum uses the data of station S and is carried out in the following steps :

(i)   The attenuation factor G is determined theoretically[2,1] so as to dispose of relative values of $S(\vec{k})$ for each radar frequency.

(ii) We obtain the relation between the gain factors K and K' at radar frequencies f and f'(≈ 2f) by virtue of the fact that, k_i(f') being twice k_i(f), A_2^-(f') is proportional to A_1^-(f) squared. Observationally we find : K(f') = 2K(f)-243.3 dB (s.d.= 1.3 dB). At this stage of the calculations we have obtained relative but comparable directional spectral amplitudes $S(\vec{k})$ for waves propagating radially towards radar S at frequencies F(Hz) : 0.186, 0.263 and 0.375 corresponding respectively to A_2^-(f), A_1^-(f) and A_2^-(f'), and A_1^-(f')).

(iii) The relation between the peak frequency F_m, U and X is deduced from the variations with fetch of these spectral amplitudes. Indeed F and F_m coincide in principle at the transition between either growth and decay or growth and saturation (which are effectively observed), characterised by parameters U, X and F_m.

(iv) Finally, the frequency spectrum is parametrized as follows :
$$S(F) = S_m \, s(x) \qquad (S_m = S(F_m) \; ; \; x = F/F_m)$$
$$s(x) = x^n \; (x \geqslant 1) \text{ or } s(x) = x^{n'} (x \leqslant 1) \tag{4}$$
So the system of equations 2-3 used in conjunction with all radar data (except A_2^-(f') which has been already used with A_1^-(f) for getting the relation between gain factors) is linear on the logarithmic scale. The unknowns of the system are the peak amplitudes S_m, parameters n and n' and the still unknown gain factor K(f). The system is solved by a least-squares method.

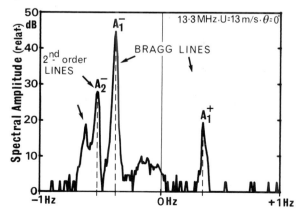

Figure 2 : Example of experimental Doppler spectrum. The second-order line on the left is attributed to an e.m. effect ("corner reflector") and is not taken into account in the present analysis.

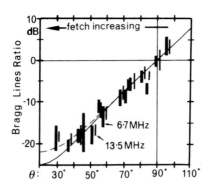

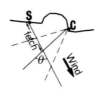

Wave frequency F:

6·7MHz → F=0·26Hz
13·5MHz → F=0·38 Hz

Figure 3 : Bragg lines ratio measurements at the two radar frequencies from station C. Curves
correspond to the function
$\phi(\Theta) \sim a+(1-a)\cos^5\Theta/2$ with a=.003
(.006) at 6.7 MHz (13.5 MHz).

4.   RESULTS

The directional properties of the wave field

It has been often observed (e.g. in[4]) that the angular distribution of
wave energy is quite sensitive to changes in the parameter $x=F/F_m$ and in
the wind speed referred to the velocity of dominant waves. Since waves
of two frequencies are involved in the radar observations from station
C and since U varies over a large interval, we might expect the radar
parameter R to vary significantly between the two data sets as well as
show a scatter of values at both frequencies. Analysis of radar measurements does not reveal such tendencies (Fig. 3). The data are consis-
tent with a functional ϕ of the form : $\phi(\Theta) \approx a+(1-a)\cos^\rho\Theta/2$ with $\rho=5$,
both for the ratios and, not shown here, for each wave spectral compo-
nent associated with the first-order lines. This solution is not unique
and the parameter ρ may also vary over some interval that includes the
value of 5. However, the results contrast in all cases with the results
obtained in previous experiments in that the angular width of the spreading function of wave energy varies little with wave frequency and wind
speed.

The peak frequency

Estimates of F_m are normalized by $\tilde{F}_m = F_m U/g$ and are plotted on Fig. 4
versus the dimensionless fetch $\xi = gX/U^2$. The data fit the relation :
$$\tilde{F}_m = 3.1 \; \xi^{-0.33} \tag{5}$$
The value of the exponent of ξ is in remarkable agreement with previous
experimental results[5-7]; also, the magnitude of the multiplicative
coefficient is comparable to other estimates that, in general, differ
slightly from one author to another (e.g. 3.5 for JONSWAP, referred here
to as J, 3.18 in[7], 2.92 in[6]). The discrepancies are often explained as
due to differences in the drag coefficient C_D. Indeed, a universal
relation of the form of Eq. 5 should be valid if $\tilde{F}_m$ and ξ are scaled
using the friction velocity rather than the wind speed at some height
above the surface. Compared to J for which C_D was equal to 10^{-3} Eq. 5
gives : $C_D = 2.10^{-3}$. This value is quite realistic with regard to the

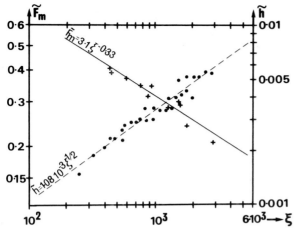

Figure 4 : Observations of
the normalized peak frequency (crosses) and rms wave
height (circles) versus the
dimensionless fetch.

high wind speeds encountered.

The function s(x)

The least-squares procedure is applied first within each class of wind
speed in order to test the validity of the implicit assumption that the
quantities n and n' (Eq.4) and K (gain factor)are constants. It is
found indeed that the variabilities of n, n' and K are rather small
(with s.d. of 1.35, 1.03 and 2.02 dB respectively) and uncorrelated
with U. Then, for providing a homogeneous set of data of the unknowns
S_m, the procedure is applied to all the data. We thus obtain 29 estimates of S_m associated to pairs of U and X, and estimates of the constants which are : K = 247.6 dB (implying : K' = 251.9 dB), n' = 2.93
and n = -5.19. Figure 5 compares our experimental function s(x) with
the form deduced from J. The agreement is good at frequencies greater
than F_m. The fact that the forward face of s(x) is less steep than for
J could be brought about by the presence in the low frequency part of
the spectrum, of waves of low amplitude advected into the area of observation from elsewhere.

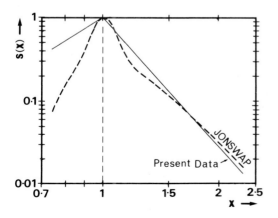

Figure 5 : Normalized wave
frequency spectrum s(x) of
Eq.4.

The rms wave height

The values of S_m are used to obtain the rms wave height h through :

$$h = \left[\int_F S(F)dF \right]^{1/2} = \left[S_m \; F_m \int_X s(x)dx \right]^{1/2} \tag{6}$$

Estimates of h are given in Fig. 4 in a diagram with coordinates ξ and $\tilde{h} = hg/U^2$, and are found to fit the relation :

$$\tilde{h} = 10.8 \; 10^{-4} \; \xi^{0.5} \tag{7}$$

As in the case of F_m, the exponential term of ξ in Eq. 7 agrees well with other experiments[5-7]. On the other hand, the multiplicative factor is typically twice the previous experimental results. The reasons for the large values of h found here are believed to be related to atmospheric layer parameters that are not taken into account in the scaling laws used, especially those relating to stability. Indeed there is a reason to expect that wave generation under unstable stratification is more active than under stable or neutral conditions[8]. The wind regimes encountered in the present case consist of northerly cold air flows channelled by the Rhône valley (known as the "Mistral"). In that situation, the sea surface temperature T is greater than the air temperature T' and thus instability occurs. As a check, a measurement of T and T' in the area of observation is available at the end of the third period of Mistral and shows a difference of 7°C between T and T'.

REFERENCES

1. Forget, P., 1983 : Télédétection des vagues par radar HF:développements de la méthode et application à l'observation des vagues levées par le Mistral. Thèse d'Etat, Univ. de Toulon et du Var, France.
2. Forget, P., Broche, P. and J.C. De Maistre, 1982 : Attenuation with distance and wind speed of HF surface waves over the ocean. *Radio Sci.*, 17, 599-610.
3. Barrick, D.E., 1972 : Remote sensing of sea state by radar. In *Remote Sensing of the Troposphere*, Edited by V.E. Derr, Chap. 12, U.S. Gvt Printing Office, Washington D.C.
4. Hasselmann, D.E., Dunckel, M. and J.A. Ewing, 1980 : Directional wave spectra observed during JONSWAP 1973. *J. Phys. Oceanogr.*, 10, 1264-1280.
5. Hasselmann, K. and al., 1973 : Measurements of wind-wave growth and swell decay during the Joint North Sea Wave Project (JONSWAP). *Deut. Hydrogr. Z.*, Suppl. A,8,N°12.
6. Mitsuyasu, H., Tasai, F., Suhara, T., Mizuno, S., Ohkusu, M., Honda, T. and K. Rikiishi, 1980 : Observation of the power spectrum of ocean waves using a cloverleaf buoy. *J. Phys. Oceanogr.*, 10, 286-296.
7. Kahma, K.K., 1981 : A study of the growth of the wave spectrum with fetch. *J. Phys. Oceanogr.*, 11, 1503-1515.
8. Liu, P.C. and D.B. Ross, 1980 : Airbone measurements of wave growth for stable and unstable atmospheres in Lake Michigan, *J. Phys. Oceanog.*, 10, 1842-1853.

DOPPLER SPECTRA OF MICROWAVE RADAR ECHO RETURNED FROM CALM AND ROUGH SEA SURFACES

A.Shibata[1],T.Uji[1],I.Isozaki[1],K.Nakamura[2] and J.Awaka[2]

1 Meteorological Research Institute, Tsukuba, 305 Japan
2 Radio Research Laboratories, Kashima, 314 Japan

ABSTRACT. The returns of C band (λ= 5.6cm) Doppler radar of horizontal polarization whose beam is directed to Kashima Nada near grazing angles are investigated. On the sea surface there is a variety of water waves, such as capillary waves, wind waves, and swell, which have their own velocities of several tens of cm/s to several m/s respectively. The scatterer's velocity measured with Doppler radar is helpful to classify scatterers. In most cases the velocities of the scatterers on the sea surface measured with Doppler radar are enough large as compared with phase velocities of capillary waves. Those velocities appear to be about phase velocities of wind waves and swell. Then the sea echoes of the radar are discussed, relating to the breaking waves.

1. INTRODUCTION

In the microwave region the returns from open sea surfaces are less understood near grazing angle. It is said that different scattering mechanisms are likely to exist in the angular region between 0° and 5° from grazing (Fung and Ulaby,1983). For instance, Bragg scattering ($\sigma_{vv} > \sigma_{hh}$) by capillary waves is proposed to be valid in the microwave region as well as in the HF region. But, intuitively, at these extreme angles most of the sea surface is in shadow, and isolated scatterers that pop into the radar beam may be visible (Wetzel,1977). Kalmykov and Pustovoytenko (1976) found the returns, called 'bursts,' for the horizontal polarization. They assumed that the burst source came from the sharp crests of waves (wedge scattering, $\sigma_{hh} >> \sigma_{vv}$), and that the burst velocity coincides with the group velocity of waves. Even if not rigorous, Wetzel (1981) provided a self-consistent explanation of scattering ($\sigma_{hh} \cong \sigma_{vv}$) by breaking waves behaving like 'plumes'. But at the present stage of investigation the question remains unsolved on what the radar is seeing near grazing angle.

On the sea surface there is a variety of water waves, such as capillary waves, wind waves, and swell, which have their own velocities of several tens of cm/s to several m/s respectively. The scatterer's velocity measured with Doppler radar is helpful to classify scatterers.

Y. Toba and H. Mitsuyasu (eds.), The Ocean Surface, 263–268.

For instance, if Bragg scattering by capillary waves is occurring, then their Doppler velocity will be no more than about 1 m/s, except for the capillary waves trapped on the crests of breaking waves (or steepest waves) which travel at about the phase velocity of the long waves. This paper reports experimental results of backscatter of C band Doppler radar, whose beam is directed to Kashima Nada, that is, the Pacific ocean. To understand the scattering mechanism more clearly, a dual polarized radar will be necessary. Unfortunately the polarization of the radar used here is only horizontal.

2. METHODS

Observations were made on 8 days in 1982 with radar facilities of the Radio Research Laboratories (RRL), Kashima, Japan. The nominal radar characteristics are as follows: radar wavelength 5.6 cm, peak power 250 KW, minimum receiving power −102.7 dBm, pulse repetition frequency 900 pps, pulse width 0.5 μsec, antenna gain 40.8 dB, and polarization HH. The radar is installed at 47 m above sea level. The radar beam can be directed to Kashima Nada in range at azimuthal angles between 50° and 80°. The distance is about 3 km from the radar site to the shoreline, and grazing angles are less than 0.5°. The water depth of illuminated area increases monotonically offshore from 22 m to 27 m. Wind velocities and directions were measured near the radar site of RRL. Sea truth data were lacking except on 26 Jan.

One sample of radar observation consists of 4096 pulses, and nine consecutive samples are averaged to obtain a Doppler spectrum shown in the following section. Nyquist's frequency is 450 Hz, and frequency resolution is 0.21 Hz. A Doppler shift 100 Hz is equivalent to a Doppler velocity 2.8 m/s.

3. RESULTS

We will examine the Doppler spectra under three wind conditions, that is, upwind, downwind, and light air. Fig.1 shows the upwind case, namely, the wind of 4 m/s from NNE on 26 Jan. The right hand part of this figure represents the backscatter from approaching targets, and also the left hand part represents the backscatter from receding targets. The reason why the spectrum is symmetrical about zero Hz is either because the separation of I signal and Q signal is incomplete (system error), or because the backscatter really occurs from both approaching and receding targets. The former is more probable because of the fact that ship echo is also symmetrical.

The significant wave height of 2.0 m and period of 8.0 sec were observed on 26 Jan. in Kashima harbor (sensor depth 21 m), so the following velocities are obtained: the orbital velocity of the swell 0.8 m/s, the group velocity of the swell 6.5 m/s, and the phase velocity of the swell 11.4 m/s. If Bragg scattering is occurring by free capillary waves of wavelength 2.8 cm, then the Doppler shift has a value of about

8 Hz. But the spectrum shown in Fig.1 appears to have two echoes at
frequencies of about 100 Hz and 300 Hz, respectively.
    First we consider the lower frequency echo. Fig.2 is a map for
this measurement. The length of the arrows shown in Fig.2 is
proportional to the peak frequency and the direction of the arrows is of
the higher power in Fig.1 (toward the radar site). In this figure the
direction of the arrow on the land shows the wind direction. The
Doppler velocity 2.8 m/s is much larger than the orbital velocity of the
swell. This Doppler velocity applies to the wind waves. Now comes up
the question on which mechanism works on the backscattering. Fig.3
shows a schematic figure of breaking waves (right) and non-breaking
waves (left). The echoes mentioned above cannot be explained by Bragg

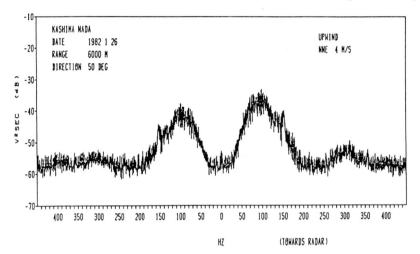

Fig.1   Doppler spectrum in the upwind case (the wind of 4 m/s
         from NNE) on 26 Jan. 1982, at the range 6 km in the radar
         beam direction 50°.

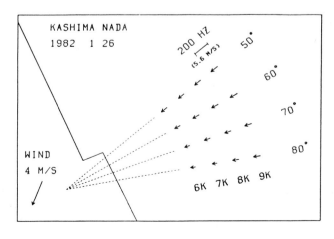

Fig.2
Map of the frequency
(velocity) of echoes
returned from the wind
waves on 26 Jan. 1982.

scattering, even if taking account of the orbital velocity of
non-breaking waves. On the other hand the capillary waves trapped on
the crests of breaking waves (or steepest waves) travel at about the
phase velocity of the long waves. Therefore Bragg scattering by such
capillary waves may explain this lower frequency echo. As another
speculation, the breaking part on the crests of breaking waves may have
many facets which are perpendicular to the radar beam and reflect the
radio wave effectively. It should be solved in future which mechanism
is the case.

Fig.3
Schematic figure of
breaking waves (right)
and non-breaking waves
(left).

        Next we comment on the higher frequency echo shown in Fig.1. The
peak frequency 300 Hz decreases to 250 Hz when the radar beam direction
is changed from 50° to 80°. This largest Doppler velocity applies to
the swell. Its velocity 8.4 m/s is of magnitude nearer to the group
velocity (6.5 m/s) rather than the phase velocity (11.4 m/s) of the
swell.
        Fig.4 shows the Doppler spectrum under the downwind case, that is,
the wind of 8 m/s from WNW on 25 Nov.. The spectrum has an echo at a
frequency of about 200 Hz, but the receding part (away from the radar
site) has a higher power. Fig.5 shows the map of the arrows in the
downwind case, and the wind direction is also shown in this figure. The
Doppler velocity of about 5.6 m/s relates to the wind waves.

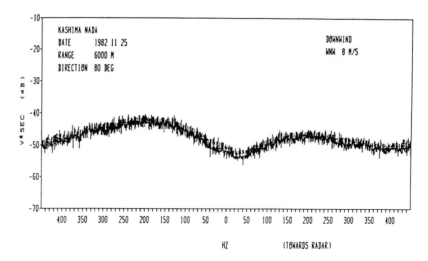

Fig.4   Doppler spectrum in the downwind case (the wind of 8 m/s
         from WNW) on 25 Nov. 1982, at range 6 km in the radar beam
         direction 80°.

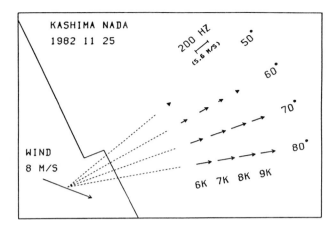

Fig.5
Map of the frequency
(velocity) of echoes
returned from the wind
waves on 11 Nov. 1982.

It is known that the receiving power depends on the angle between
the wind direction and the radar beam direction. But in our case there
is little correlation between them because obstacles such as trees
obstruct the radar beam in some azimuthal angles. The receiving power
ranges from -67 dBm to -62 dBm in the upwind case on 26 Jan., and from
-84 dBm to -75 dBm in the downwind case on 25 Nov. at the range 6 km.

On 19 Nov. a wind of 3 m/s from the west blew until noon, and after
noon a wind of 3 m/s from the east blew. Fig.6 shows the Doppler
spectrum obtained at noon. The receiving power decreases to -91 dBm at
the range 6 km. The spectrum shown in Fig.6 has a peak at a frequency
of about 5 Hz. But a ground clutter also has a peak at zero Hz. As a
result the interpretation of the echo shown in Fig.6 becomes difficult.
The echo in Fig.6 has a power of about 30 dB higher than the noise

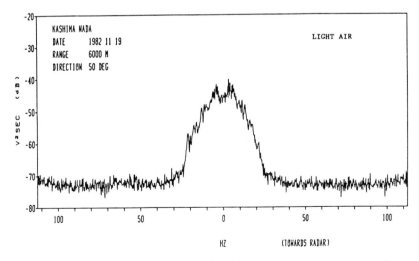

Fig.6   Doppler spectrum in light air at noon on 19 Nov. 1982, at
        the range 6 km in the radar beam direction 50°.

level. On the other hand the ground clutter has usually a power of
about 15 dB higher than that. Then non-zero echo in Fig.6 is probably a
sea echo. If the Bragg resonant scattering is occurring by the
capillary waves of wavelength 2.8 cm, the Doppler shift has a value of
about 8 Hz. The difference of about 3 Hz may be due to tidal currents
or orbital velocities of swell.

4. DISCUSSIONS

It is found that that in most cases the velocities of scatterers on the
sea surface measured with Doppler radar are enough large as compared
with phase velocities of free capillary waves. Those velocities appear
to be about phase velocities of wind waves and swell. Such sea echoes
of the radar may relate to the breaking waves. The capillary waves
trapped on the crests of the breaking waves (or steepest waves) are
travelling at about phase velocities of the long waves. If such
capillary waves would scatter the radio wave, Bragg scattering would be
valid near grazing angle even for rough sea surfaces. On the other hand
the breaking part on the crests of breaking waves may have many facets
which are perpendicular to the radar beam and reflect the radio
wave effectively. At The present stage, we can't say which mechanism is
the case.

Acknowledgments   The authors wish to thank T.Furuhata for his
considerable efforts in gathering data, M.S.Longuet-Higgins for
comments, and G.R.Valenzuela for his critical review of the manuscript.

References

Fung,A.K., and F.T.Ulaby,Matter-energy interaction in the microwave
        region, Manual of Remote Sensing vol.1,R.N.Colwell,Ed., American
        Society of Photogrammetry,115-164,1983.
Kalmykov,A.I., and V.V.Pustovoytenko,On polarization features of
        radio signals scattered from the sea surface at small grazing
        angle, J.G.R.,81,1960-1964,1976.
Wetzel,L.B.,A model for sea backscatter intermittency at extreme
        grazing angles, Radio Science,12,749-756,1977.
Wetzel,L.B.,On microwave scattering by breaking waves, in IUCRM
        Symposium on wave dynamics and radio probing of the ocean
        surface,Miami,1981.

MEASUREMENTS OF DIRECTIONAL SEA WAVE SPECTRA USING A TWO-FREQUENCY MICROWAVE SCATTEROMETER

A. Takeda, M. Tokuda and I. Watabe
National Research Center for Disaster Prevention
Hiratsuka Branch
9-2 Nijigahama, Hiratsuka, Kanagawa 254
Japan

ABSTRACT. An experiment of two-frequency scatterometry measuring directional spectra of sea waves was made from a marine observation tower in the sea. It is shown that an X-band, CW, two-frequency scatterometer can measure a spectrum which corresponds qualitatively to the wind wave spectrum observed by a wave gauge, as it is expected from the theory of two-frequency scatterometry. Relative magnitude of the spectrum is larger when the azimuth of the scatterometer antennae is equal to the wind direction than when it is not. The results indicate that at higher wind speeds the scatterometry has the capability to detect directional differences of sea wave spectra.

1. INTRODUCTION

The theory of two-frequency scatterometry measuring sea waves was first proposed by Ruck(1972) and Hasselmann(1972), and was described in detail by Plant(1977) and Alpers and Hasselmann(1978). The technique features an active microwave remote sensing which can measure a spatial spectrum of sea wave vectors parallel to the horizontal line of sight with a single small antenna system. A directional spectrum of sea waves are estimated from the measurements with changing the azimuth angle of the antenna system.

In our laboratory, a study developing the active microwave technique to measure sea waves over a large sea surface has started in 1981 as a part of the Ocean Remote Sensing Program of Science and Technology Agency, Japan. This paper reports an experiment of the first step of the study to examine the capability of two-frequency scatterometry to detect wave directionalities.

2. INSTRUMENTATION

An X-band, continuous wave(CW), two-frequency microwave scatterometer was used in the experiment. It transmits a mixture of two single microwaves of slightly different frequencies f_1 and f_2 continuously; f_1 is

Y. Toba and H. Mitsuyasu (eds.), The Ocean Surface, 269–274.
© *1985 by D. Reidel Publishing Company.*

fixed at 10.502GHz, while f_2 is variable between 10.502GHz and 10.548GHz. Consequently the frequency difference between f_1 and f_2, Δf, is 46MHz at the maximum and is equivalent to 0MHz at the minimum. These micro-wave signals are generated by Gunn diodes and are very stable. Because of the radio regulation the transmitted power is limited to 0.01W. Two identical horn type antennas of a rectangular cross-section were used respectively for transmitting and for receiving. Each antenna has 28 degree beam width. The two antennas are fixed in a frame side by side. A 30cm separation between them does not cause any serious trouble of interference.

3. EXPERIMENTAL CONDITIONS

The experiment was carried out from a marine observation tower in the sea 1km off the Hiratsuka Beach where the water depth is 20m(Figure 1). The beach draws an arc along the east-west direction and faces on the Pacific Ocean in the south direction. The tower has preferable condi-tions for measurements of waves propagating from the south-easterly to south-westerly directions. The antennas of two-frequency scatterometer were mounted on the tower at a height of 20m above the mean sea level so that their azimuth angle may be variable in the southern hemicircle, and that their depression angle may be variable between 10degree and 60degree.

Figure 1. The experimental site: Hiratsuka Marine Observation Tower of National Research Center for Disaster Prevention.

Figure 2.   The antenna system of two-frequency microwave scatterometer
installed on the marine observation tower.

    Many measurements were made under various conditions of winds and
waves. At wind speeds lower than 7m/s, the measurements were very much
difficult because of weak microwave returns and consequently of large
S/N values. The measurements reported here done on January 5th, 1983,
at high wind speeds between 13m/s and 17m/s from the southwest or
west-southwest, with parameters of 45degree depression angle, of HH
polarization and of several azimuth angles. Wave heights were measured
at the three apex points of a regular triangle around the tower by
three conventional capacitance type wave gauges, but because of some
circuit troubles the records of two gauges were not available.

4.   RESULTS AND DISCUSSION

The data obtained in the experiment were analized by the method
detailed by Plant(1977). The power spectrum of the product of two
measured return signals of transmitted microwaves of f_1 and f_2 was
calculated by an FFT analyzer(ONO SOKKI CF-500).
    It is found that the power spectrum has a sharp peak at the
frequency close to that of a sea wave component whose wave length is,
by the Bragg's relation, in resonance with that of the beat wave of Δf
generated by the two transmitted microwaves of f_1 and f_2. In Figure 3
a typical example is illustrated.
    In a series of measurements where Δf was changed successively, the
azimuth angle was constant and the wind and wave conditions were
supposed to be almost steady, it is indicated that the peaking component shifts its position toward higher frequencies with increasing Δf.
Figure 4 shows the above situation by a three-dimensional display of
the product spectrum vs Δf.
    According to the theory the product spectrum consists of two parts.
One is the background spectrum and another is the resonance component
which relates to the sea wave spectrum. If it is assumed here that the
background spectrum changes little within a series of measurements, the
resonance component of a relative value may be estimated from the peak

in a product spectrum. A curve connecting the peaking values of product spectra for various Δf plotted agaist Δf implies roughly a directional component of a sea wave spectrum parallel to the horizontal line of sight. In Figure 5, such curves obtained from the experiment at higher wind speeds are presented together with azimuth and wind directions and frequencies of peaking component in the sea wave spectrum measured by a wave gauge. It is indicated that a curve derived from measurements

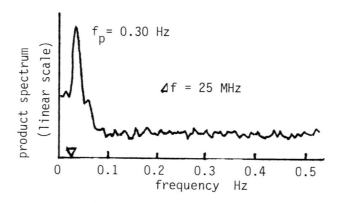

Figure 3.   A typical resonance peak in the product spectrum of two-frequency returns.

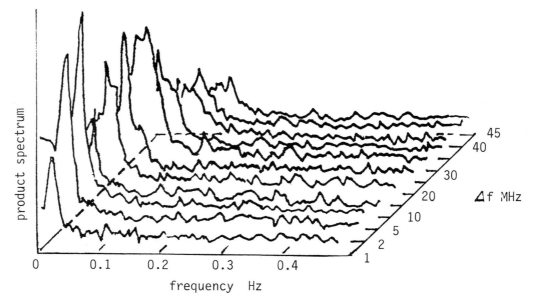

Figure 4.   A three dimensional display of the product spectrum vs Δf and the corresponding frequency of sea waves.

on the azimuth angle equal to the wind direction has significantly a
larger value in contrast to that on the azimuth angle out of the wind
direction. The curves also have similar trends to those of sea wave
spectra in the higher frequency range. But in the lower frequency
range, significant results were not obtained because of the poor resolution probably due to a small footprint of a radar beam. It was very
much trouble that any directional information about sea wave spectra
was not collected by wave gauges.

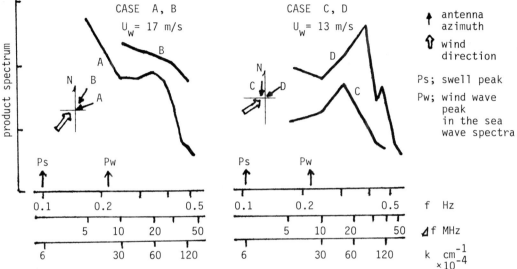

Figure 5.   Curves connecting the peaks of product spectra. They
correspond to sea wave spectra.

5.   CONCLUSIONS AND REMARKS

The results suggest that the two frequency microwave scatterometry has
the ablity to measure directional differences of wave spectra at least
at higher wind speeds. There remain, however, several problems to be
solved. The first is inconsistencies between frequencies of observed
peaking components and those expected from Δf. The descrepancy may not
be explained by the effects of the Doppler shift due to drift currents.
The second is the problems to establish the modulation transfer function relating the results of two-frequecy scatterometry to sea wave
spectrum. The third is quantitative evaluations of the effects of
higher-order returns on the product spectrum. It was pointed out that
in the case of tower experiment, the effects of tower should be
accounted.

ACKNOWLEDGEMENTS.   The authors appreciate suggestive comments of Dr. Gasper Valezuela, Dr. Fred Dobson, and Professor David Weissman. This study was supported by Research Coordination Bureau, Science and Technology Agency of Japanese Government.

REFERENCES

Alpers, W., and K. Hasselmann (1978);   The two-frequency microwave technique for measuring ocean-wave spectra from an airplane or satellite. *Boundary Layer Met., 13,* 215-130.
Hasselmann, K. (1972);   The energy balance of wind waves and the remote sensing problem. *NOAA Tech. Rep. ERL 228 AOML-7,* 25.1-25.55.
Plant, W.J.,(1977);   Studies of backscattered sea return with a CW, dual-frequency, X-band radar. *IEEE Trans. Ant. Prop. AP25,*28-36.
Ruck, G.T., D.E. Barrick, and T. Kaliszewski,(1972);   Bistatic rader sea state monitoring *Battelle, Columbus Laboratories Tech. Rep.*

MEASUREMENTS OF OCEAN WAVE SPECTRA AND MODULATION TRANSFER FUNCTION
WITH THE AIRBORNE TWO FREQUENCY SCATTEROMETER

D.E. Weissman[1] and J.W. Johnson[2]
1 Hofstra University, Hempstead, NY 11550
2 NASA Langley Research Center, Hampton, VA 23665

ABSTRACT. The results of this research show that the directional
spectrum and the microwave modulation transfer function of ocean waves
can be measured with the airborne two frequency scatterometer-microwave
resonance technique. The results here are favorable to the future
application of this or similar techniques from airborne or spaceborne
platforms. Similar to tower based observations, the aircraft measurements of the modulation transfer function show that it is strongly
affected by both wind speed and sea state. Also detected were small
differences in the magnitudes of the MTF between downwind and upwind
radar look directions, and variations with ocean wavenumber. Unexpected results were obtained that indicate the MTF measured from an
aircraft is larger than that measured using single frequency, wave
orbital velocity techniques such as tower based radars or "ROWS"
measurements from low altitude aircraft. Possible reasons for this are
discussed. The ability to measure the ocean directional spectrum with
the two frequency scatterometer, with supporting MTF data, is demonstrated.

1. INTRODUCTION

This study is advancing the ability of active microwave radar to
measure ocean wave spectra from high altitude aircraft. The experimental data analyzed here was acquired during the Atlantic Remote
Sensing Land Ocean Experiment (ARSLOE) during November 1980. The NASA
Langley two frequency scatterometer participated onboard the Wallops
P-3 aircraft, and was able to receive supporting ocean data from the
surface contour radar (operated by the NASA Wallops Flight Center),
almost simultaneously. Directional spectra and non-directional spectra
were available from the XERB buoy. The spectrum of the sea surface
reflectivity can be measured directly from the backscattered signals at
the two closely spaced microwave frequencies. The cross product of
these signals displays a resonance whose intensity is analyzed using
theoretical relationships and models that have been developed independently by three groups of researchers (Alpers and Hasselmann, 1978;

275

Y. Toba and H. Mitsuyasu (eds.), The Ocean Surface, 275–282.

Plant and Schuler, 1980; and Johnson and Weissman, 1984). The validity of this equation was established in an earlier phase of this experimental data analysis.

The microwave radar frequency is K_u-band (14.6 GHz), and operates with two "simultaneous" (interrupted CW using time multiplexed long pulses) frequencies, that can be separated by a variable amount from 1 to 20 MHz. The polarization is horizontal.

Data was collected during the many high altitude flight lines between altitudes of 2800 to 7000 feet. Complete details of the aircraft radar, its flight patterns, data collection and processing, and the supporting surface contour radar results can be found in the recent paper by Johnson and Weissman (1984).

Another important feature of this experiment was the directional discrimination capability of the radar, as a result of its large illuminated surface area, relative to the dimensions of the long gravity waves. In effect, all MTF results presented in this study are directional because of this spatial filtering effect. The data taken on Nov. 13 observed the wave spectrum from 4 different directions. On this day, the sea was a combination of swell and a wind driven spectrum, differing in direction by 45°. Flight directions A and B in Fig. 1 show the directions in which the two frequency scatterometer made measurements. The directional spectra measured by the buoy in these directions is shown as the solid curve in Fig. 2. The directional resolution of the XERB buoy is 135° (with smoothing coefficients) while that of the surface contour radar is 40° (Walsh et al., 1981).

2.   EXPERIMENT DESCRIPTION

The data analyzed in this study was acquired from about 20 to 70 km offshore near Duck, N.C. The flight operations, environmental conditions and details of the illumination geometry are presented in the paper by Johnson and Weissman (1984). Of strong interest in this experiment was the behavior of the modulation transfer function as a function of illuminated area and the incidence angle of the radar wave. The illuminated area and the dimensions relative to the ocean wavelength was varied over a substantial range as the aircraft altitude varied from 2800 to 7000 ft, and as the incidence angle ranged from 16° to 50°. These conditions were met with the Nov. 12 data, which also included upwave and downwave flight directions. The four different flight directions employed on Nov. 13 were at a fixed altitude and incidence angle; 4500 feet and 25° from nadir. Support of the Nov. 12 radar data was given by the surface contour radar and the single frequency wave spectrometer referred to as the "ROWS" technique.

3.   SYSTEM EQUATION AND MODULATION TRANSFER FUNCTION

The equation that relates the surface elevation spectrum to the two-frequency resonance response is:

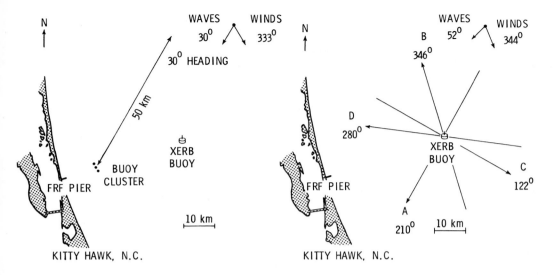

Fig. 1 ARSLOE Experiment Site, left: Nov. 12 flight paths; right:
Nov. 13 flight paths.

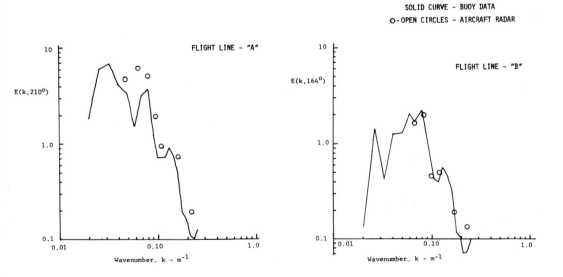

Fig. 2 Comparison of directional wave spectra from buoy data and the two
frequency radar – Nov. 13 flights "A" & "B"

$$\chi(k) = \frac{2\pi^2 |m(k)|^2 \cdot k^2 E(k,\theta)}{A} \tag{1}$$

where   k = wavenumber of ocean wave that is in resonance with
              two-frequency EM wave
          X(k) = ratio of resonance intensity to background spectrum energy
              (modulation strength)
          E(k,θ)= directional wave spectrum in "θ" direction
          m(k)= modulation transfer for ocean wave of wavenumber, k.

A critical assumption in this model is that the reflectivity variation
sensed by the radar at each wavenumber is moderately coherent with that
part of the height and slope spectrum. A more general relation, on
which the above equation is based, is:

$$\chi(k) = \frac{2\pi^2 \Phi_R(k,\theta)}{A} \tag{2}$$

where $\Phi_R(k,\theta)$ is the instantaneous two-dimensional reflectivity
spectrum. The concept of modulation transfer function is then based on
an input-output relation point of view, in which the surface slope
spectrum, $k^2 E(k,\theta)$ is the input and $\Phi_R(k,\theta)$ is the output. Then:

$$|m(k)|^2 = \frac{\Phi_R(k,\theta)}{k^2 E(k,\theta)} \tag{3}$$

Detailed studies of this relation and the modulation function can be
found in the recent paper by Plant, Keller and Cross (1983).
        A key contribution of these ARSLOE results is to use Equation (1)
either:

A. determine the MTF across a range of conditions of radar parameters
    (incidence angle, flight direction, altitude and wavenumber) and
    with different types of surface conditions, using X(k) obtained
    from the two-frequency resonances.

B. determine the directional surface spectrum using estimates of the
    MTF from non-directional spectrum measurements and the values of
    X(k) mentioned above.

        As discussed in the paper by Johnson and Weissman (1984), support-
ing measurements of the MTF, its spectral variation and its coherence
properties, were made with a single frequency radar that receives and
correlates the backscattered power and Doppler variations (related to
the surface orbital velocity) to achieve an independent measurement of
this quantity.

4. MEASURED MTF RESULTS

The MTF results are those derived from Eq. (1) for the various flight
parameters of Nov. 12. A value of the MTF can be calculated at each k
(or difference electromagnetic wavenumber) so that each flight line

yields the functional dependence of the MTF vs. the matching ocean
wavenumber, at a fixed altitude and angle, and direction relative to
the wind. Twelve of these functions have been computed for the Nov. 12
data (and supported by the SCR derived wave spectra) and two from the
"A" and "B" lines of Nov. 13 (based on buoy derived spectra). The
functions obtained from the Nov. 12 data have been plotted individually
for each upwave and downwave path; a typical plot is Fig. 3. The 90%
confidence interval about computed value of m is plus or minus 12%,
based on statistics of E(k). Therefore, error bars can be drawn at
each plotted point with $\pm$12% limits.

Analysis of these results was done from several points of view.
Almost all share the following characteristics: the magnitudes start
high at the lowest wavenumbers, then decrease to a definite minimum
about k=.06 to .08 m^{-1}, then usually rise by at least 25% or up to 100%
above this minimum. This is often followed by a gradual change at the
higher k values, either an increase or decrease. Another definitive
and general result is that the magnitudes differed strongly on these
two days, in accordance with the environmental conditions. On Nov. 12,
the winds were high, about 12 m/s, with accompanying high waves. The
MTF values averaged in the range from 8 to 16. In contrast, on Nov. 13
the values were usually between 20 and 40. The factor of two increase
shows a good correlation with the inverse of the wind which, on this
day was below 6 m/s when the flights were made. This is in good
agreement with measurements of the MTF conducted from towers (Plant et
al., 1983; Weissman, 1983) where the magnitude was observed to depend
inversely on the local wind, and other environmental parameters.

Special attention was also given to other characteristics of the
MTF, such as its dependence on incidence angle, upwind vs. downwind
look directions, and any variations caused by changes of the flight
path relative to the wind direction. Across the spectrum of wavenumbers from k=.05 to .14, ratios of the MTF looking downwave vs. upwave
have been computed to test for significant differences. Another
important dependence of the MTF studied is that due to incidence
angle. Flight operations on Nov. 12 encompassed a range from 16° to
50°, substantial enough to test conditions of interest for the remote
sensing of ocean waves. Considering the large assortment of wavenumber
values in the data set, a simplification was performed to work with the
average of a subset of values of the MTF measured along each flight
line. The data was averaged, but only values of MTF whose wavenumbers
are in the range .05 < k < .09 were included in the average. These
twelve average values are plotted in Fig. 4. The upwind/downwind
condition creates a small "random" fluctuation, but they still show a
definite trend, downward with increasing incidence angles. For the 16°
case, the MTF ranges from 9.7 to 13.2, while at 40° it's from 8.3 to
10.4, and the 50° value is 9.2. Numerically, this is not a large
effect, but detecting its presence will be helpful in sorting out other
dependencies in future applications.

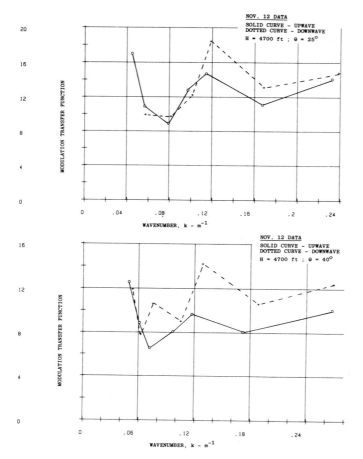

Fig. 3 Two frequency
Scatterometer
inferred modulation
transfer function
versus ocean
wavenumber

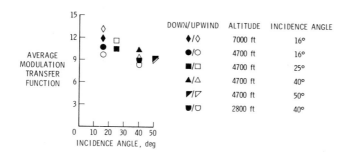

Fig. 4 Modulation
Transfer Function versus
incidence angle

5. CONCLUSIONS

Our major findings are the functional dependencies of the MTF on ocean
surface wave number, flight direction and incidence angle. The abundant data set allows the quantitative description of these dependencies. Small incidence angle dependence means that future remote
sensing systems need not be limited in the choice of incidence angle of
the radar beam. A difference in magnitude between the aircraft and
tower results was also detected and found to be plausible, on physical
grounds. In addition, the effects of environmental conditions have
also been seen and analyzed: increases in wind speed and sea state
cause strong decreases in the MTF. With this new knowledge, the two
frequency scatterometer can now be considered a useful instrument for
the measurement of ocean surface spectrum by aircraft. For the MTF, no
theoretical explanation of these scattering effects based on electromagnetic scattering theory and air-flow over ocean waves (including all
short capillary generation mechanisms) has yet been achieved.
    Much progress has been made on accumulating tower based radar
results and wave follower measurements under a variety of air-sea
conditions (Plant, et. al., 1983; Hsiao and Shemdin, 1983), but more is
necessary in order to explain the aircraft results. Such questions as:
 what are the physical sources generating the short capillary waves,
modulating the long ocean waves?, and why should the MTF observed from
an aircraft be different from what is measured on an ocean platform?,
need to be addressed.
    It is recommended that two topics be focussed on to advance this
technique:

(1) Learning more about the mechanism for the modulation of ocean waves
    (their origin, linearity, and coherence with the orbital velocity)
    that is detected by remote sensing radars.
(2) Combined (dual) sensor capabilities to improve the measurement
    accuracies of each separate instrument. Simultaneous two frequency
    scatterometry and conventional scatterometry would measure wave
    spectra plus wind speed. Since both quantities affect each sensor,
    the accuracy of each sensor could be improved.

    At very high aircraft and spacecraft altitudes there would be
considerable benefit to the measurement approach of using the three
frequency technique (that sharpens the selectivity of the two frequency
resonance and greatly improves the signal-to-background ratio) being
developed by the Naval Research Laboratory (Schuler, Plant and Miller,
1984).

6. ACKNOWLEDGMENTS

The excellent results obtained in this study were made possible by the
supporting ocean spectrum data from the Surface Contour Radar provided
by Dr. Edward J. Walsh of the NASA Wallops Flight Center. His interest
and cooperation in this study is greatly appreciated. We also appreci-

ate the advice of Mr. Kenneth Steele of the NOAA Data Buoy Office, who provided us with the best available data set for Nov. 13. We thank Dr. William J. Plant for enlightening and stimulating discussions in the course of this study.

Support for this work, which was provided by the NASA Oceanic Processes Program, Dr. W. Stanley Wilson, through Grant NAGW-468, is gratefully acknowledged.

7. REFERENCES

Alpers, W. and K. Hasselmann, (1978), The two frequency microwave technique for measuring ocean surface wave spectra from an airplane or satellite. Boundary Layer Met., 13, 215-230.

Hsiao, S.V. and O.H. Shemdin, (1983), Measurements of wind velocity and pressure with a wave follower during MARSEN, J. Geophys. Res., 88, C14, 9841-9849.

Johnson, J.W. and D.E. Weissman, (1984), The two frequency microwave resonance technique from an aircraft: a quantititave estimate of the directional ocean surface spectrum. Radio Sci., 19, 841-854, May-June.

Plant, W.J. and D. Schuler, (1980), Remote sensing of the sea surface using one or two frequency microwave techniques. Radio Sci., 15, 605-615.

Plant, W.J., W.C. Keller and A. Cross, (1983), Parametric dependence of ocean wave-radar modulation transfer function. J. Geophys. Res., 88, C14, 9747-9756.

Schuler, D.L., W.J. Plant, W.P. Eng, W. Alpers and F. Schlude, (1982), Dual-frequency microwave backscatter from the ocean at low grading angles: Comparison wth theory. Int. J. Remote Sensing, 3(4), 363-371.

Schuler, D.L., W.J. Plant and H.C. Miller, (1984), Experimental verification of the three-frequency scatterometer concept. OCEANS 84 Conference Record, 10-12 Sept., Washington, D.C. IEEE Oceanic Engg Society and Marine Technology Society.

Walsh, E.J., D.W. Hancock III, D.E. Hines and J.E. Kenney (1981), Surface contour radar remote sensing of waves. Proceedings of the Conference on Directional Wave Spectra Applications, 14-16 Sept. 1981, R.L. Weigel, Editor; American Society of Civil Engineers, New York.

Weissman, D.E., (1983), The dependence of the radar modulation transfer function on environmental conditions and wave parameters, Final Report, ONR Contract N00014-83-M-0081, August 1983.

AN ALGORITHM OF MICROWAVE BACKSCATTERING FROM A PERTURBED SEA SURFACE

N. Iwata
Tokyo University of Mercantile Marine
Etchujima,Koto-ku,Tokyo 135 Japan

ABSTRACT. A model to calculate backscatter cross sections of microwaves from a perturbed sea-surface for the range of incident angles between 30^0 and 70^0 is proposed. This model depends on (1) the power spectrum of short gravity-capillary waves,(2) the angular spreading of energy of these short waves,(3) the probability density function of slope of the larger waves on which the gravity-capillary waves are superposed,and (4) nonlinear hydrodynamical interactions between gravity-capillary waves and the larger waves. Some numerical results by this proposed model are compared with SASS-model functions(Schroeder et al.,1982) and it is found that the difference between them remains at most within 2.5 dB for 10.2 m/sec < U < 22.4 m/sec and $36^0 < \theta < 56^0$.

1. INTRODUCTION

Recently an empirical model function has been proposed by Schroeder et al.(1982) to relate the normalized radar cross section σ^0 of the ocean and the wind vector at a height of 19.5 m above the surface, assuming neutral stability. The relationship is specified in the form of a table which gives two coefficients G and H in the equation

$$\sigma^0 (dB) = 10(G(\theta,\phi) + H(\theta,\phi)\log_{10}U) , \qquad (1)$$

where θ is the radar incident angle, ϕ is the angle between wind direction and radar azimuth, and U is the wind velocity in m/sec. The G and H coefficients are tabulated separately for V and H polarizations every 2^0 in incidence and every 10^0 in azimuth angles.
    On the other side several semi-empirical models for backscatter cross sections have been proposed (Chang & Fung,1977;Wentz,1978;Fung & Lee,1982). All of these models have in common (1) the directional spectrum of gravity-capillary waves,(2) the probability density function of the surface slope of the longer gravity waves. However,the modulation of gravity-capillary waves due to the longer waves has not been taken into account except in Wentz's paper.
    In the following sections we propose another model based upon

283

Y. Toba and H. Mitsuyasu (eds.), The Ocean Surface, 283–288.

recently exploited experimental results of gravity-capillary waves (
Mitsuyasu & Honda,1974;Fujinawa,1980),as well as upon the modulation of
short waves by nonlinear hydrodynamical interactions with longer waves.

2.   ALGORITHM OF BACKSCATTER CROSS SECTIONS

For incident angles $\theta > 30^0$, the backscatter cross section per unit area
for a moving rough surface is given by

$$\sigma^0(k^i) = T(\theta)\{F(2\underline{k}^i) + F(-2\underline{k}^i)\} \, , \tag{2}$$

where $k^i = (\underline{k}^i, k_3^i)$ is the wave vector of the radar, $F(\underline{k})$ is the surface
wave spectrum and $T(\theta)$ is a function of polarization and incident angle

$$T(\theta) = 4\pi k^4 \cos^4\theta |\alpha_{pp}|^2 \, , \tag{3}$$

where $k = |k^i|$, and for horizontal and vertical polarization p = H and
p = V respectively, so that

$$\alpha_{VV} = \frac{(\varepsilon-1)\{\varepsilon+(\varepsilon-1)\sin^2\theta\}}{\{\varepsilon\cos\theta+\sqrt{\varepsilon - \sin^2\theta}\}^2} \quad \text{and} \quad \alpha_{HH} = \frac{\varepsilon - 1}{\{\cos\theta + \sqrt{\varepsilon - \sin^2\theta}\}^2} \, , \tag{4}$$

where ε is the relative dielectric constant of sea water.
     We assume that the frequency spectrum of gravity-capillary waves
can be given by (Toba,1973)

$$\Psi(f) = \frac{\alpha_g}{(2\pi)^3} \hat{g} \, \hat{u} \, f^{-4} \quad \text{with} \quad \hat{g} = g + \gamma k_B^2, \, k_B = 2k\sin\theta \, , \tag{5}$$

where $\hat{u}$ denotes the friction velocity of air and γ represents the ratio
of surface tension to water density. Toba(1973) assumed α_g as constant
but Mitsuyasu & Honda(1974) found that α_g varies with friction velocity
as well as fetch. We now assume(see Fig.1)

$$\alpha_g = 0.04(\frac{\hat{u}}{80})^{3/4} \, , \tag{6}$$

fitting the observed values for maximum fetch. From the frequency
spectrum we can obtain wave number spectrum $S(k_B)$ by means of the
dispersion relation

$$S(k_B)k_B dk_B = \Psi(f)df \tag{7}$$

so that

$$S(k_B) = \frac{\alpha_g}{2k_B^4}(1 + 2\gamma \frac{k_B^2}{\hat{g}})(\frac{k_B \hat{u}^2}{\hat{g}})^{0.5} \, . \tag{8}$$

     The angular spreading factor of the wave energy may be expressed in
the form(Longuet-Higgins,1962)

$$F(k,\phi) \sim \cos^{2s}(\frac{1}{2}\phi) \, , \tag{9}$$

where ϕ shows the direction of propagation of each harmonic component,
so the spectrum reduces to(Iwata,1983)

$$S(\underline{k}_B) = F(2\underline{k}^i) + F(-2\underline{k}^i) = S(k_B)H(\phi) , \tag{10}$$

where

$$H(\phi)= \frac{1}{4\sqrt{\pi}} \frac{\Gamma(s+1)}{\Gamma(s+1/2)} \{\cos^{2s}(\frac{\phi}{2})+\sin^{2s}(\frac{\phi}{2})\}$$

$$\stackrel{\sim}{=} \frac{1}{2\pi}\{1 + \frac{s(s-1)}{4+s(s-1)}\cos2\phi\} . \tag{11}$$

Introducing (8) and (11) into (10) and (2) we can calculate σ^0 of the agitated water surface. However, in order to apply (2) to the real ocean the surface tilt effect due to the large-scale waves must be taken into account,

$$\sigma^0 = \frac{1}{2\pi}\int \stackrel{\sim}{T}(\mu,\nu,\theta,\phi)\stackrel{\sim}{S}(\stackrel{\sim}{\underline{k}}_B)\exp\{-\frac{1}{2}(\mu^2+\nu^2)\}d\mu d\nu , \tag{12}$$

where the tilde denotes quantities defined with respect to the local large-scale wave. The function $\stackrel{\sim}{T}(\mu,\nu,\theta,\phi)$ includes modification of $T(\theta)$ through $\theta \rightarrow \stackrel{\sim}{\theta}$ and $k^i \rightarrow \stackrel{\sim}{k}^i$ in addition to a rotation of the polarization axes (see Fig.2). The probability density function of a large-scale surface wave slope is assumed as Gaussian and we let

$$\mu = \frac{n_1}{\sqrt{<n_1^2>}} , \qquad \nu = \frac{n_2}{\sqrt{<n_2^2>}} .$$

Furthermore assume that the hydrodynamical modulation of the spectrum is small and can be represented as

$$\stackrel{\sim}{S}(\stackrel{\sim}{\underline{k}}_B) = S(\stackrel{\sim}{\underline{k}}_B) + \Delta S(\stackrel{\sim}{\underline{k}}_B) , \tag{13}$$

where $\stackrel{\sim}{k}_B=2\stackrel{\sim}{k}^i$ and the first term on the right-hand side represents modulated spectrum due to the purely geometric effect that the facets are tilted by the carrier waves and the second term describes the nonlinear hydrodynamical modulation.

When we expand $\stackrel{\sim}{T}(\mu,\nu,\theta,\phi)S(\stackrel{\sim}{\underline{k}}_B)$ in a power series of n_α, $(\alpha=1,2)$ and substitute them in (12), we get the leading terms

$$\sigma^0=\stackrel{\sim}{\sigma}_0+ \frac{1}{2}(\frac{\partial^2\stackrel{\sim}{\sigma}}{\partial n_\alpha \partial n_\beta})_0<n_\alpha n_\beta> + \{\frac{\partial\stackrel{\sim}{T}(\mu,\nu,\theta,\phi)}{\partial n_\alpha}\}_0<\Delta S(\stackrel{\sim}{\underline{k}}_B)n_\alpha> +\cdots, \tag{14}$$

where the subscript zero refers to values at $n_\alpha=0$ and

$$\stackrel{\sim}{\sigma}(\mu,\nu) = \stackrel{\sim}{T}(\mu,\nu,\theta,\phi)S(\stackrel{\sim}{\underline{k}}_B) \tag{15}$$

$$<n_\alpha n_\beta> = \frac{1}{2\pi}\int_{-\infty}^{\infty} n_\alpha n_\beta\exp\{-\frac{1}{2}(\mu^2+\nu^2)\}d\mu d\nu . \tag{16}$$

Next we replace the differential dn_α by the finite difference $\Delta n=\sqrt{<n_\alpha^2>}$. We then obtain $\mu=1$ for $\alpha=1$, $\nu=1$ for $\alpha=2$, and the second term of the right-hand side of (14) turns out to be

$$\frac{1}{2}(\frac{\partial^2\stackrel{\sim}{\sigma}}{\partial n_1^2})_0+\frac{1}{2}(\frac{\partial^2\stackrel{\sim}{\sigma}}{\partial n_2^2})_0=\frac{1}{2}\{\stackrel{\sim}{\sigma}(1,0)+\stackrel{\sim}{\sigma}(-1,0)+\stackrel{\sim}{\sigma}(0,1)+\stackrel{\sim}{\sigma}(0,-1)\}-2\stackrel{\sim}{\sigma}(0,0), \tag{17}$$

where $\overset{\sim}{\sigma}(1,0)$ means $\overset{\sim}{\sigma}(\mu,\nu)|_{\mu=1,\nu=0}$.
For the third term representing the hydrodynamical modulation we let

$$\Delta S(\overset{\sim}{\underline{k}}_B) = B\mu S(\underline{k}_B) \quad , \tag{18}$$

and similarly we have

$$(\frac{\partial \tilde{T}}{\partial n_\alpha})_0 <\Delta S(\overset{\sim}{\underline{k}}_B)n_\alpha> = \frac{B}{2} S(\underline{k}_B)\{\tilde{T}(1,0,\theta,\phi) - \tilde{T}(-1,0,\theta,\phi)\} \quad . \tag{19}$$

We are now ready to calculate σ^0 by introducing (17) and (19) into (14). However, in order to simplify the algorithm we expand $S(\overset{\sim}{\underline{k}}_B)$

$$S(\overset{\sim}{\underline{k}}_B) = S(\underline{k}_B) + \{ \frac{\partial S(\overset{\sim}{\underline{k}}_B)}{\partial n} \}\cdot n = (1+A_1\mu+A_2\nu)S(\underline{k}_B) \quad . \tag{20}$$

The final form of the backscatter cross section is given by introducing (20) into (17) as follows,

$$\sigma^0 = \frac{S(\underline{k}_B)}{2}\{-2\tilde{T}(0,0,\theta,\phi)+\tilde{T}(1,0,\theta,\phi)(1+A_1+B)+\tilde{T}(0,1,\theta,\phi)(1+A_2)+$$

$$+\tilde{T}(-1,0,\theta,\phi)(1-A_1-B)+\tilde{T}(0,-1,\theta,\phi)(1-A_2)\} \quad , \tag{21}$$

where $A_1=\sqrt{<n_1^2>}\Omega\cos\phi$,

$$\Omega=(3.5- \frac{4\gamma\frac{k_B^2}{\hat{g}}}{1+2\gamma\frac{k_B^2}{\hat{g}}})\cot\theta \quad .$$

$A_2=\sqrt{<n_2^2>}\Omega\sin\phi$,

The modulation coefficient B in (18) may be a function of a characteristic relaxation time of gravity-capillary waves and this time may be regarded equall with inverse of the temporal growth rate (Valenzuela & Wright,1979). Incidentally we assume

$$B = 4.62\sqrt{<n_1^2>}(\frac{\hat{g}}{k_B\hat{u}^2})^{0.4} \quad . \tag{22}$$

The form of $\tilde{T}(\mu,\nu,\theta,\phi)$ is now(Iwata,1983)

$$\tilde{T}(\mu,\nu,\theta,\phi)=4\pi k^4\frac{\{(n_1\cos\phi+n_2\sin\phi)\sin\theta+\cos\theta\}^4}{(1+n_1^2+n_2^2)^{3/2}} \times$$

$$\times\frac{1}{D^4}|\{\sin\theta-\cos\theta(n_2\sin\phi+n_1\cos\phi)\}^2\{\overset{\sim}{\alpha_{VV}}\}+(n_2\cos\phi-n_1\sin\phi)^2\{\overset{\sim}{\alpha_{HH}}\}|^2,$$

where
$$D^2=\sin^2\theta+\cos^2\theta(n_1^2+n_2^2)-\sin2\theta(n_1\cos\phi+n_2\sin\phi)+\sin^2\theta(n_1\sin\phi-n_2\cos\phi)^2.$$

This form of $\tilde{T}(\mu,\nu,\theta,\phi)$ can also be derived from Valenzuela et al.(1971, eqs.1 and 2) by using the relation, $-\tan\Psi= n_1\cos\phi+n_2\sin\phi$. The upper line corresponds to the vertical and the lower line to the horizontal polarization and $\overset{\sim}{\alpha}_{VV}$ as well as $\overset{\sim}{\alpha}_{HH}$ can be obtained from (4) by using $\overset{\sim}{\theta}$ instead of θ.

$$\cos\overset{\sim}{\theta}= \frac{1}{\sqrt{1+n_1^2+n_2^2}}\{(n_1\cos\phi+n_2\sin\phi)\sin\theta+\cos\theta\};\sin\overset{\sim}{\theta}= \frac{D}{\sqrt{1+n_1^2+n_2^2}} \quad .$$

3.   CALCULATED RESULTS AND CONCLUDING REMARKS

In order to compare the computed results with SASS function of (1) we assume the frequency of the incident microwaves to be 14.6 GHz and the dielectric constant ε = 40.1 - i39.1. The angular spreading factor of wave energy in (11) is taken as s = 2.2. The mean-squared slope of the sea surface are taken from Cox & Munk(1954) for a clean surface

$$<n_1^2> = 3.16 \times 10^{-3} U, \quad <n_2^2> = 0.003 + 1.92 \times 10^{-3} U, \tag{23}$$

where U is the wind velocity in m/sec at a height of 12.5 m above the sea surface. For each polarization, calculations are carried out for $\hat{u}$ = 17,27,37$\cdots$107 cm/sec. They correspond (Cardone,1969) to wind velocities at a height of 19.5 m of U = 5.3, 8.1, 10.2$\cdots$22.4 m/sec, respectively,using a logarithmic wind profile in neutral state with roughness length

$$z_0 = \frac{0.684}{\hat{u}} + 4.28 \times 10^{-5} \hat{u}^2 - 4.43 \times 10^{-2}. \tag{24}$$

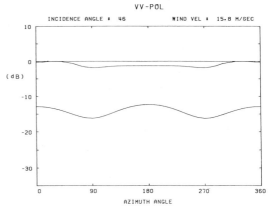

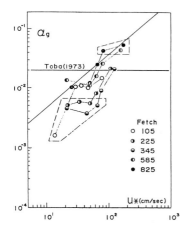

Fig.1: Coefficient   in (6) versus friction velocity(after Mitsuyasu and Honda 1974)

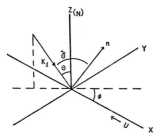

Fig.2: Diagram of the coordinate system

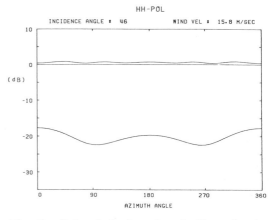

Fig.3: Calculated   for both polarizations corresponding to U=15.8 m/s

Azimuth angles ϕ = 0,90,180 correspond to up-, cross- and down-wind direction respectively. Incident angles are resricted to θ > 30^0 in order to avoid specular scattering.

Fig.3 represents an example of calculations for û=47 cm/sec(U=15.8 m/sec) and the differences,

SASS function by (1) $-$ σ^0 by (21).

The general characteristic features from the calculations are:
(1) Differnces remain at most within 2.5 dB for 10.2 m/sec < U < 22.4 m/sec as well as for 36^0 < θ < 56^0. (2) In the range 38^0 < θ <60^0 SASS function for VV-polarization are generally smaller than the computed values and vice versa for HH-polarization.(3) In the range 66^0 < θ <70^0 the computed values of both polarizations are generally smaller than the SASS function, especially for HH-polarization at incident angles larger than 60^0.

Acknowledgments. The author acknowledges G.R.Valenzuela who brought up a number of crucial points and made suggestions to clarify them.

REFERENCES

Cardone,V.C.,1969:Specification of the wind distribution in the marine boundary layer for wave forcasting.*TR-59-1,Dept.Meterol.and Oceanogr.*,New York Univ. 131 pp.

Chang,H.L. and A.K.Fung,1977:A theory of sea scatter at large incident angles.*J.Geophy.Res.*,82,3439-44.

Cox,C. and W.Munk,1954:Measurement of the roughness of the sea surface from photographes of the sun glitter.*J.Opt.Soc.Amer.*,44,838-50.

Fujinawa,Y.,1980:Directional spreading of short gravity waves.*Rep.Nat'l. Res.Ctr.Disaster Prevention*(in Japanese).23,185-92.

Fung,A.K. and K.M.Lee,1982:A semi-empirical sea-spectrum model for scattering coefficient estimation.*IEEE OE-7*,166-76.

Iwata,N.,1983:Backscatter of microwaves from the sea surface and the modulation of spectra of short gravity waves.*J.Oceanogr.Soc.Japan*, 39,43-52.

Longuet-Higgins,M.S.,1962:The directional spectrum of ocean waves and processes of wave generation.*Proc.Roy.Soc.London*,Ser.A,256,286-315.

Mitsuyasu,H. and T.Honda,1974:The high frequency spectrum of wind generated waves,*J.Oceanogr.Soc.Japan*,30,29-42.

Schroeder,L.C.,D.H.Boggs,G.Dome,I.M.Halberstam,W.C.Jones,W.J.Pierson and F.J.Wents,1982:The relationship between wind vector and normalized radar cross section used to derive SEASAT-A satellite scatterometer wind. *J.Geophys. Res.*,87,C-5,3318-36.

Toba,Y.,1973:Local balance in the air-sea boundary process III. *J.Oceanogr.Soc.Japan*,29,209-20.

Valenzuela,G.R.,M.B.Laing andJ.C.Daley,1971:Ocean spectra for the high-frequency waves as determined from airborne radar measurements. *J: Marine Res.*29,69-84.

Valenzuela,G.R. and J.W.Wright,1979:Modulation of short gravity-capillary waves by longer-scale periodic flows- Ahigher-order theory.*Radio Science*,14,1099-110.

Wentz,F.J.,1978:Estimation of the sea surface's two-scale backscatter parameters.*NASA Contractor Rep.* 145255,122 pp..

THE DEPENDENCE OF THE MICROWAVE RADAR CROSS SECTION ON THE AIR-SEA
INTERACTION AND THE WAVE SLOPE

W.J. Plant,[1] W.C. Keller,[1] and D.E. Weissman[2]
1 U.S. Naval Research Laboratory, Washington, DC   20375
2 Hofstra University, Hempstead, NY   11550

ABSTRACT. Backscatter measurements from a tower in the Gulf of Mexico
by the Naval Research Laboratory show that an X-band radar cross
section of the surface depends on wind speed and air-sea temperature
difference, and also on the total RMS slope of the large gravity waves
that modulate the short centimetric capillary waves. These were
extensive measurements with an X-band, vertically polarized, CW radar
and were conducted over a one month period. The environmental conditions spanned a wide range of wind speeds, air-sea temperature and wave
conditions. A selective study of the dependence of the cross section
values (obtained from a twenty minute average) on the winds, the
Monin-Obukhov length and the slope has been made. This will enable the
development of empirical relationships and models that will advance the
remote sensing capabilities of microwave radars for the measurement of
the ocean surface winds and other surface quantities, from airborne and
space platforms. Under some conditions, the dependence of the average
radar cross section on the total slope (at a given wind speed) is
substantial. The implications of this data set are: 1) that two or
more quantities should be measured simultaneously in order to acquire
the most accurate air-sea parameter information and 2) these results
contribute to the study of the mechanisms of energy transfer from long
to short waves.

1.   EXPERIMENTAL CONDITIONS AND DATA ANALYSIS

This paper reports an observation of a dependence of both X-Band cross
sections and modulation transfer functions on atmospheric stability and
long wave slope. The measurements reported here were obtained in the
Gulf of Mexico during November and December 1978. Determination of the
stability and slope dependence was possible because of a favorable
range of wave and weather conditions during the measurement period that
permitted comparison of cross sections and MTFs within a large dynamic
range of the environmental parameters. The measurements were made from
an oceanographic tower, known as Stage I, operated by the Naval Coastal
Systems Center of Panama City, Florida. This tower functions as a

Y. Toba and H. Mitsuyasu (eds.), The Ocean Surface, 289–296.

residence and research facility with living accommodations and support
facilities. The platform, located 12 miles offshore in 32 meters of
water, is 32 meters square and about 18 meters above the surface. The
ocean bottom slope was small: 0.001. Throughout the period of observation, wave height, sea surface and air temperature, wind speed and
direction, and other meteorological variables were recorded by Calspan
Corporation (Niziol and Mack, 1979). Measurements of true wind speed
and direction were made at a height of 24.7 meters with a Beckman-
Whitley wind system. Data on wind speed and direction were continuously recorded. Sea surface temperature and air temperature at heights
of 4.4, 9.3 and 24.7 meters were continuously monitored with calibrated
Foxboro thermistors. Three cold fronts passed the tower during the
experiment and caused environmental conditions over the period as a
whole to be extremely variable. Wind speeds ranged from 1 to 15 m/sec.
More importantly, the air temperature varied from a minimum of 2 °C up
to 24 °C. Since the sea surface temperature stayed between 20 °C to
23 °C, both stable and unstable conditions were encountered. Wind
speed, air temperature at 4.4 meters, and sea surface temperature
recorded during passage of the strongest cold front are presented in
Figure 1. Approximately 130 hours of data were recorded during the
experiment. The measure of long wave slope used in this paper is somewhat unique since its properties were set operationally by parameters
of the microwave system. Plant, et. al., 1983 showed that ocean wave
variance spectra may be determined from the FM output of a CW radar
illuminating a small spot by tracking the centroid of the Doppler
spectrum.

The data were blocked into 21 1/3 minute segments and analyzed on
a Digital Equipment Corporation MINC-11CA computer using the same
program as the MARSEN data reported by Plant, et. al., 1983. Output
functions for these segments are: FM spectra, AM spectra, orbital
velocity spectra, wave height spectra, wave slope spectra, magnitude-
squared coherence function, and magnitude and phase of the MTF.
Additional output quantities include wind speed and direction, long
wave height and slope from the FM channel, and relative cross section.
All output data are stored on hard and floppy discs prior to subsequent
averaging. In order to search for the dependence of MTFs or received
power on a single parameter, all files with a chosen, limited range of
the other parameters were averaged. Then the single parameter was
stepped through a sequence of values. For example, in the case of wind
speed dependence, the MTF versus frequency was computed from all
records having wind speeds of 4-6, 6-8, 8-10, and 10-12 m/sec while
other parameters were held within specific ranges.

2. VARIATION OF SURFACE STRESS WITH ATMOSPHERIC STABILITY

In the following section, we present results of our measurements of
cross sections and MTFs as a function of atmospheric stability. Since
it is known that surface wind stress is a function of atmospheric
stability and since this stress variation is expected to affect the
measured backscattering parameters, we shall briefly summarize here the

equations used to calculate the variation of surface stress over the ocean due to changes in atmospheric stability. We used the equations presented here to calculate values of friction velocity, u_*, for use in Fig. 2. Note that the wind stress is simply ρu_*^2, where ρ is the density of air.

We use primarily the results presented in Haugen, 1973 and Large and Pond, 1981. The friction velocity u_* may be related to the drag coefficient C_D and the wind speed U measured at a height z by

$$u_*^2 = C_D U^2 \tag{1}$$

Furthermore, according to the similarity theory of Monin and Obukhov (1954) wind speed profiles have the form

$$U = \frac{u_*}{K}\left[ln\left(1 + \frac{z}{z_0}\right) - \psi \right] \tag{2}$$

where K (=.41) is the von Karman constant, z is height, z_0 is the roughness length and ψ is a universal function of z/L where L is the Obukhov length, originally defined as the height at which the production of mechanical energy by turbulence equals the production of buoyant energy. Standard assumptions of present boundary layer models are that u_* and L are constant throughout the surface layer. The detailed derivation for u_* can be found in Keller, et. al., 1984.

3. RADAR CROSS SECTION EFFECTS RELATED TO ATMOSPHERIC STABILITY

The average of the backscattered power measured with a tower-based microwave system over a time interval much longer than the wave period yields a measure of the average surface roughness and is proportional to the radar cross section measured from an aircraft or satellite with a scatterometer (Schroeder, et. al., 1982). For X-Band, the wind speed dependence has been shown to be strong but considerable variation and data spread is always observed when wind speed is used as the independent variable. Atmospheric conditions during this experiment provided an excellent opportunity to observe the role of air-sea temperature difference on the radar cross section. If the possible role of long wave slope is ignored (results are not categorized by slope in this section) then the wind speed dependence of the relative cross sections measured during this experiment appears as shown in Figure 2(a). Here again, antennas pointed into wind and waves. Each point in Figure 2(a) represents a 21 1/3 minute average of the relative backscattered power. An increase in receiving power at a constant wind speed with decreasing air minus sea temperature is obvious from Figure 2(a). We replotted the data of Figure 2(a) versus u_* as determined in Section II in order to see if a more unique relation existed between u_* and cross section than between U and cross section. The result is shown in Figure 2(b). Obviously, the data still exhibit a stability dependence at a constant value of u_*.

Thus it appears that our cross section data are not unique functions of either U or u_* but rather depend on atmospheric stability in a

manner not explained by stability variations of u   calculated with
contemporary surface layer models.

4.   BACKSCATTERING EFFECTS RELATED TO SEA STATE

Earlier aircraft-based measurements of the dependence of the radar
cross section on the air-sea interaction (Krishen, 1971; Guinard, et.
al., 1971; Ross and Jones, 1978) presumed, like classical steady-state
spectrum models, that there is a one-to-one relationship between wind
speed and wave height (or slope) so that only wind speed parameterization is necessary. However, actual ocean conditions usually do not
satisfy this assumption. Depending on fetch, wind direction and duration, and remote sources of waves, a wide range of possible wave
heights and slopes is possible at a given wind speed. The data
analysis in this study indicates that both MTFs and cross sections at
X-Band have a dependence on long wave slope when the other parameters
can be kept within a narrow range. Also important is the fact that the
dependence on long wave slope, wind speed, and atmospheric stability
are not "separable." Unstable atmospheric conditions produce very
different dependences on long wave slope and wind speed than do stable
conditions. Circumstances during this experiment were very fortuitous
in that a wide range and mixture of conditions influenced the cross
section and MTF data.
     Examination of the received power as a function of long wave slope
revealed that the cross section increases somewhat with long wave slope
under near-neutral conditions. This behavior is the opposite of that
of the MTF for the same wind speed ranges. In order to ascertain
whether the observed increase of cross section with slope might be due
to a coincidental change of wind speeds, 7.5-7.9 m/s, for which a substantial amount of data existed and plotted cross section versus slope
for this range. The result shown in Figure 3 provides assurance that
we are not accidentally observing a wind speed effect. In Fig. 4, the
stable data set was split into 3 separate groups, based on three adjacent slope regimes. Here again we see the effect of slope magnitude on
the radar cross section vs. friction velocity relationship.

5.   DISCUSSION AND CONCLUSIONS

The results of this study indicate that microwave backscattering from
the sea surface at X-Band can be quite complicated when operating in
the Bragg regime of incidence angles. Both backscattering cross sections and modulation transfer functions can depend on wind speed, long
wave slope, and atmospheric stability. This study has not addressed
questions of absolute cross section magnitude or its dependence on
incidence angle or azimuth angle relative to wind and wave directions.
However, for the case of a 45° incidence angle and antennas pointed
into wind and waves, we find that MTFs are lower in unstable than in
near-neutral conditions while cross sections are higher. The increase
in cross section is more rapid than the increase in surface stress

calculated with present surface layer models. Furthermore, in near-neutral atmospheric conditions, MTFs decrease with long wave slope while cross sections increase. No dependence of either quantity on wave slope could be detected in unstable conditions.

The increase in relative cross section with long wave slope observed here is consistent with wind stress measurements in falling winds reported by Large and Pond (1981). They found that in such situations wind stress was higher than under more steady or rising wind conditions and postulated that effects of the previously higher wind were "remembered" by the sea surface through higher wave conditions. Since higher wind stresses imply higher radar cross sections, this effect agrees with the increase of cross section with wave slope which was observed here. Attempts to check this result in NRL's wind wave tank are underway. Preliminary indications are that X-Band cross

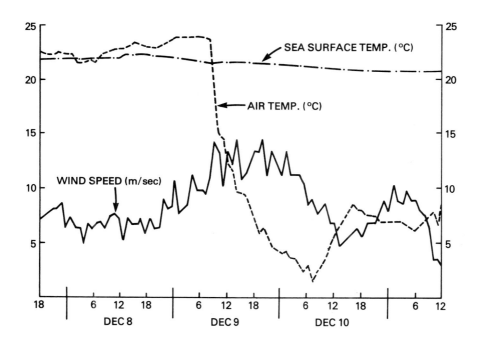

Fig. 1 Temporal record of wind speed, air temp-
        erature, and sea surface temperature during
        the passage of the strongest cold front to
        occur during the experiment

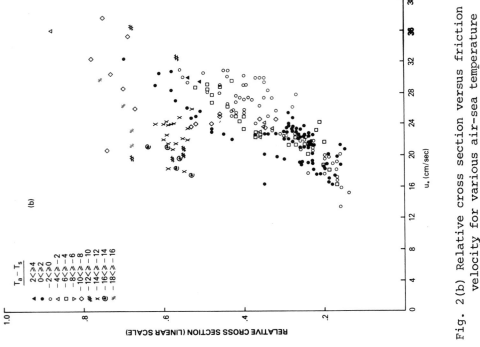

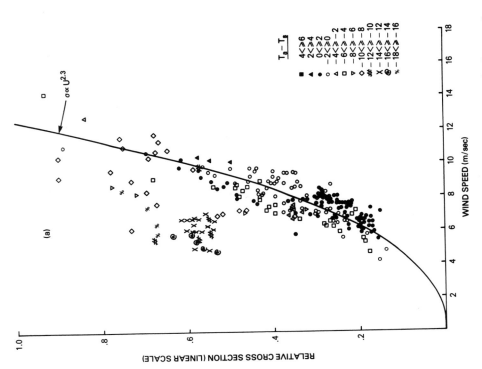

Fig. 2(a) Relative cross section versus wind speed
for various air-sea temperature differences

Fig. 2(b) Relative cross section versus friction
velocity for various air-sea temperature
differences

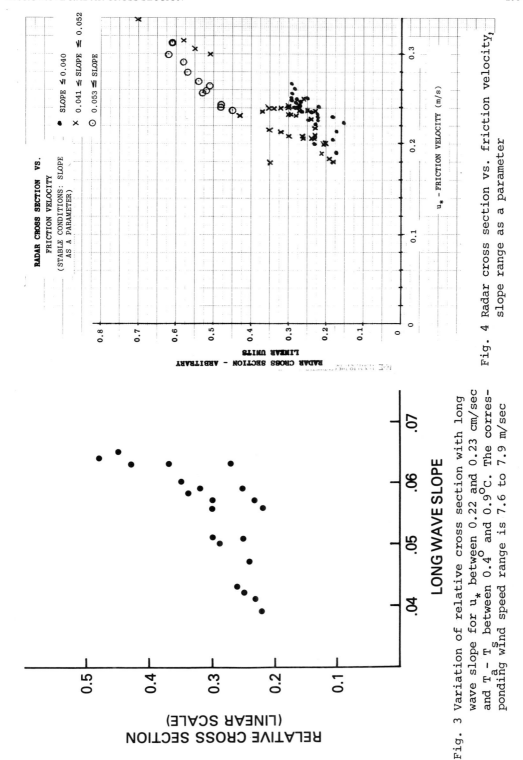

Fig. 4 Radar cross section vs. friction velocity, slope range as a parameter

Fig. 3 Variation of relative cross section with long wave slope for u_* between 0.22 and 0.23 cm/sec and $T_a - T_s$ between 0.4° and 0.9°C. The corresponding wind speed range is 7.6 to 7.9 m/sec

sections are also dependent on the height of long, paddle-generated waves under the steady wind conditions in the tank.

The implications of this study for scatterometry and microwave imagery of ocean waves (either SAR or SLAR) at X-Band are clear. In order to determine accurate wind speeds with an X-Band scatterometer supplementary measurements of atmospheric stability and wave slope are necessary. Similarly, the inversion of SAR or SLAR imagery at X-Band to obtain ocean wave spectra will be hampered by the a priori need to know the long wave slope, a quantity closely related to the desired output. Of course, SAR has similar nonlinearity problems at any frequency. (See, for instance, Hasselmann, et.al., 1984.) The dependence of cross section on stability, though, will restrict the operation of both SAR and SLAR at X-Band.

6. REFERENCES

Guinard, N.W., J.T. Ransone, Jr. and J.C. Daley, (1971), Variation of the NRCS of the Sea with Increasing Roughness, J. Geophys. Res., 78, 1525-1538.

Hasselmann, K., R.K. Raney, W.J. Plant, W. Alpers, R.A. Schuchman, D.R. Lyzenga, C.L. Rufenach, M.J. Tucker, (1984), Theory of SAR ocean wave imaging: a MARSEN view, submitted to J. Geophys. Res.

Haugen, D.A., ed, (1973), Workshop on Micrometerology, American Meteorological Society.

Keller, W.C., W.J. Plant, and D.E. Weissman, (1984), The dependence of X-band microwave sea return on atmospheric stability and sea state, accepted for publication, J. Geophys. Res.

Krishen, K., (1971), Correlation of radar backscatter with ocean wave height and wind velocity, J. Geophys. Res., 76, 6528-6539.

Large, W.G. and S. Pond, (1981), Open ocean momentum flux measurements in moderate to strong winds, J. Phys. Ocean., 11, 324-336.

Monin, A.S. and A.M. Obukhov, (1954), Basic laws of turbulent mixing in the atmosphere near the ground, Tr. Akad. Nauk SSSR Geofix. Inst., No. 24(151), 163-187.

Niziol, T.A. and E.J. Mack, (1979), Reduced data from CALSPAN's participation in the Panama City II Field Experiment in the northern Gulf of Mexico during Nov. - Dec. 1978, CALSPAN Corp., Report #6467-M-1.

Plant, W.J., W.C. Keller and A. Cross, (1983), Parametric dependence of ocean wave - radar modulction transfer functions, J. Geophys. Res., 88, C14, 9747-9756.

Ross, D. and W.L. Jones, (1978), On the relationship of radar backscatter to wind speed and fetch, Boundary Layer Meteor., 13, 45-54.

Schroeder, L.C., W.L. Grantham, J.L. Mitchell and J.L. Sweet, (1982), SASS measurements of the Ku-Band radar signature of the ocean, IEEE, J. Ocean Engr., OE-7, 3-14.

PRELIMINARY RESULTS OF DUAL POLARIZED RADAR SEA SCATTER

Dennis B. Trizna
Propagation Staff, Radar Division
Naval Research Laboratory
Washington, D.C. 20375

Abstract. Preliminary results are presented for dual polarized X-band radar sea scatter measurements made at the U.S. Army Field Research Facility instrumented pier at Duck, N.C. Ranges measured extended from the surf zone to deep water. The 30-ns pulse radar data indicated that the radar return was localized about the wave crests, which were tracked in range from relatively deep water through breaking at the shoreline. This ability to track waves allowed a measure of their phase velocity under shoaling conditions. Radar cross sections are presented as a function of depression angle for simultaneous dual-polarized data. Radar cross sections are found to be very similar for plunging breakers, presumably due to specular and volume scatter. Vertically polarized scatter from wave crests over a shoal 300 m off shore was found to be 20 dB or more than that for horizontal polarization, not in disagreement with predictions of the two-scale model.

I. INTRODUCTION

Simultaneous dual-polarized radar scatter from ocean waves offers a useful diagnostic tool for determining radar scattering mechanisms. The cumulative distribution function of sea scatter radar cross sections has been shown to be a useful method to describe differences in such data [1]. Those results suggested that two different sources of scatter are present for horizontal polarization, while just a single scattering mechanism may be sufficient to describe vertically polarized sea scatter.

We have operated a short-pulse dual-polarized radar at the U.S. Army Coastal Engineering Research Center Field Research Facility field site at Duck, North Carolina. The site features a research pier with the capability for measurement of ocean currents and wave spectra at points along the 600 m length perpendicular to the coastline. Measurements are also made by Waverider buoys offshore. The radar was situated on a dune, with the antenna at 9.85 m above mean sea level, and could simultaneously illuminate ranges from the water's edge to a mile offshore. The radar was operated using a dual-polarized antenna feed on alternating pulses into a single antenna reflector, resulting in a virtually identical antenna gain and pattern for both polarizations. A thirty nanosecond pulse length was used. A general review of preliminary results is now presented.

RADAR MEASUREMENTS OF SHALLOW WATER PHASE VELOCITY

A very important measurement available at Duck is that of the subsurface shoal contour, made over the area bi-monthly in the study of beach erosion by the Coastal Engineering Research

297

Y. Toba and H. Mitsuyasu (eds.), The Ocean Surface, 297–302.

Center (CERC). These underwater profiles are useful for radar purposes, since they can be used to model the change of the ocean wave profile as it progresses into shallow water. The shore lies perpendicular to an aspect of about 70 deg east of north at Duck. The radar was generally pointed into the incoming waves, typically 20 deg south of the pier. The bottom profile along the 90 deg incoming wave propagation bearing was calculated from the measured bathymetry contour made on October 1 and is shown in Figure 1. It is similar to a typical offshore profile, with a shoal at about 300 m from the shoreline, peaking at 4 m below mean sea level. Grazing angles are indicated in the figure for several ranges of interest.

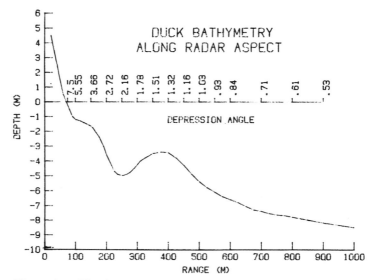

Figure 1 — The bottom contour along the radar aspect is shown relative to zero mean sea level. Radar depression angles are labeled along the zero depth line drawn. A typical offshore shoal is seen centered near 400 meters range which caused enhanced radar scatter for vertical polarization, but not for horizontal.

When illuminating waves parallel to their propagation direction along 1.5 km of range, the general character of the radar return was one of periodicity in range. Echo spacings were less than the wavelength determined by measuring the wave period of the breakers on the beach. The radar returns were observed to be associated with the peaks of the waves, which were not necessarily crests, considering the relatively low amplitude of the waves. By pulse-to-pulse tracking of the radar echo peaks in time, we found that the speed of the perpendicularly-illuminated waves agreed reasonably well with the predicted shallow-water gravity-wave phase velocity. In addition, as the waves passed the offshore shoal, the radar return typically disappeared, then reappeared with a slight phase shift before breaking at the shoreline.

An example of these effects is shown in Figure 2, for September 30, 1983, a day after the offshore passage of storm Dean. Each cross indicates the position of a peak in the radar echo above a chosen threshold just above the noise level. A lack of such peaks is seen near the range of the subsurface offshore shoal of Figure 1, showing the effect of the disappearance of the radar echo, and indicating perhaps some hydrodynamic interaction of the wave with the shoal. A horizontal line is drawn at 300-m range, and intersects five consecutive waves at a period of ten seconds. Since the wave period does not vary with depth, these waves are ten

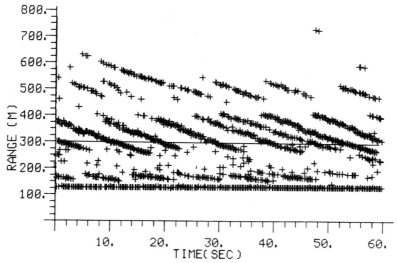

Figure 2 — A plot of the temporal development of threshold radar
returns is shown for September 30, 1983, for a one minute period.
A horizontal line drawn at 300 meters range is seen to intersect
several time traces each ten seconds, indicating a ten-second wave
period as responsible for the radar returns. A change in slope
represents a gravity wave phase velocity change, due to shallow
water bottom effects.

second waves. The corresponding deep water wavelength is 156 m, with a phase velocity of
15.6 m/s. If one draws straight lines at the longer ranges through tracks of these radar echo
peaks, and measures the distance they advance over a ten second period, one gets values
between 87.5 and 92.5 m, or phase velocities of 8.75 to 9.25 m/s, much less than the deep
water values.

As a deep water wave begins to shoal the wave frequency, Ω, remains fixed, but the wave
number changes according to the equation:

$$\Omega = [(gK) \tanh(KD)]^{1/2} \tag{1}$$

where $g = 9.81$ m/s; K is the shallow-water wave number; and D is the water depth. Since
the phase velocity is the ratio of the depth-independent wave frequency to the depth-dependent
wave number, it is also depth-depdendent. The hyperbolic-tangent becomes unity for deep
water, and one can solve for the depth-dependent wave length by setting the deep-water frequency equal to the shallow-water frequency using Equation 1. One then gets the following
relationship between the shallow-water wavelength, L_s, and the deep water-wavelength, L_d:

$$L_d = L_s/[\tanh(2\pi D/L_s)] \tag{2}$$

For depths of 7, 8, and 9 m, one gets from Eq. 2 shallow water wavelengths of 79.2, 84.2, and
92.1 m, in quite good agreement with the observed radar return spacings and rates of advancement toward the beach from Figure 2.

Finally, the rapid advancement of the wave as it breaks is seen at the very shortest ranges,
just beyond the continuous echo from the water's edge. Such a radar measurement appears to
offer a useful diagnostic tool toward identifying sub-surface offshore shoals and studying wave-
bottom interactions.

HIGH RESOLUTION SEA SURFACE RCS VERSUS DEPRESSION ANGLE

The behavior of the wind and waves during the radar measurements made on September 30, 1983, was as follows. Swell from the passage on the previous day of storm Dean far offshore was of the order of 1.5 to 2 m in height, and of ten second period. Winds were blowing out from shore against the incoming swell during this data collection period, at speeds about 10 m/s.

The range-time data of Figure 2 were determined by thresholding a one minute data record for a fixed radar gain setting, which provided about 25 dB of usable dynamic range. The threshold was set a few decibels above the noise level and peaks above it were identified and stored. Several similar one minute records with different gain settings were required to cover the full dynamic range of the discrete sea scatter echos observed for radar cross section studies, all being collected within a ten minute period under similar wave conditions. The radar equation was solved for each peak in range, and these were plotted versus depression angle for both polarizations, with some points eliminated so as not to cause overlap in amplitude from file to file. The results are shown in Figures 3 and 4, for vertical and horizontal polarization.

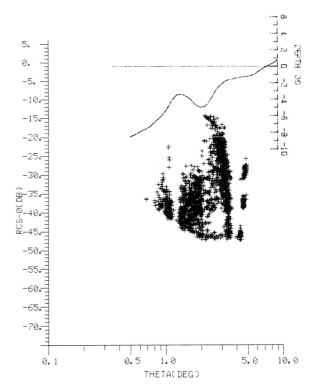

Figure 3 — Three one-minute records of normalized radar cross section derived from data similar to the previous figure are plotted versus radar grazing angle for vertical polarization. The bands along the 5-deg mark are artifacts to be ignored. The band of returns between 2.5 and 3.5 are due to specular and volume scatter from collapsing breaking waves, while the returns near 2 deg are due to Bragg scatter from small scale wave structure generated by interactions of the long wave with the offshore shoal.

The dark vertical band of data points lying between 2.5 to 3.5-deg depression angle represent scatter from breaking waves due to the last rapid change in the slope near the shoreline between 150 and 250 m off shore. Results for horizontal and vertical polarization are nearly identical in this region. Radar returns at these ranges are suggested as primarily due to specular scatter from plunging breakers, which would be expected to be essentially polarization

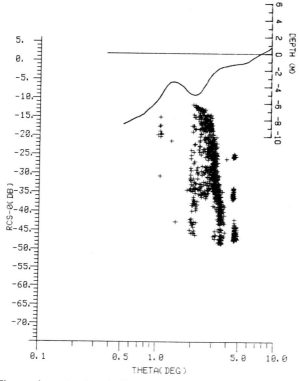

Figure 4 — A plot similar to the previous one, but for horizontal polarization. The band of returns about 3-deg is nearly identical to that for vertical polarization, but the returns due to the offshore shoal are below the limit of the radar sensitivity, in agreement with the cross section predicted by the two-scale model.

independent. In fact, the amplitude behavior with depression angle (or range from the shore-line) could be explained by a model which is dependent upon the area of the specularly scatter-ing surface. That is, larger amplitude waves would be expected to break farther from shore than smaller waves, and present a larger concave surface to the radar beam. The smaller the wave, the closer to shore the breaking occurs, and the smaller the scattering cross section. Space does not allow a more detailed comparison of radar scatter models with breaking wave models at this time.

The region between 1.5 and 2 deg shows many more returns for vertical polarization than for horizontal, in agreement with the two-scale model. That is, small scale structure is expected to be formed over the underwater shoal near 400 m by wave-bottom interactions, which would provide Bragg scatterers on the long waves responsible for the observed returns. The nature of the scattering elements might be expected to be different than wind-driven capil-lary waves, however. Nonetheless, such small scale randomly distributed elements are expected to provide stronger scattering for vertical polarization than for horizontal by more than 50 dB (3), much greater than our measurement capability for this occasion.

In summary, it would appear that an extended program of radar measurements presented here under a variety of wind, wave, and profile conditions would allow a quantitative comparison with scattering theories. With additional measurements and theoretical foundation, the understanding of the interaction of gravity waves with subsurface features could be greatly improved, perhaps providing a diagnostic tool for remotely monitoring offshore and near shore bathymetry.

ACKNOWLEDGEMENT

I wish to acknowledge the contributions of Melvin Lehman, Roger Pilon, and Ed Shepler to the experimental aspects of the problem, and Donna Donovan and Rod MacLean on the software development and data processing. I also wish to thank the U.S. Army Coastal Engineering Research Center, under the direction of Curt Mason, for their co-operation and the use of the pier facilities at Duck.

References

Hansen, J.P. and V.F. Cavaleri, High-resolution radar sea scatter, experimental observations and discriminants, NRL Report 8557, March 1982.

Kalmykov, A.I. and V.V.Pustovoytenko, On polarization features of radio signals scattered from the sea surface at small grazing angles, Journal of Geophysical Research, Vol 81, pp. 1960-1964, April 1976.

Wright, J.W., A new model for sea clutter, IEEE Trans. Ant & Prop., AP-16, 217-223, 1968.

SLAR AND IN-SITU OBSERVATIONS OF WAVE-CURRENT INTERACTION ON THE
COLUMBIA RIVER BAR

F. I. Gonzalez[1]
E. D. Cokelet[1]
J. F. R. Gower[2]
M. R. Mulhern[1]

[1] NOAA/Pacific Marine Environmental Laboratory, Seattle,
   WA 98115
[2] Institute of Ocean Sciences, Sidney, B.C. V8L 4B2, Canada

ABSTRACT.   Observations at the Columbia River entrance have been
com-pared to wave height amplification factors predicted by linear,
one-dimensional wave-current interaction theory. A previous study
found good agreement between this theory and observations, with the
ex-ception of one so-called "severe event" which was seriously underpredicted. The present analysis utilizes SLAR and in-situ data to
demonstrate that the probable cause of this failure is two-dimensional
current refraction induced by lateral current shear. The conclusion
is reached that such two-dimensional effects must be better understood
if these "severe events" are to be accurately predicted on the
Columbia River Bar.

1.   INTRODUCTION

Wave-current interactions at the Columbia River entrance make this one
of the most hazardous navigational regions in the world. Significant
wave height can easily double on the Bar in just the few hours from
slack to ebb, as ocean swell and local wind waves approach the river
entrance and meet opposing currents occasionally in excess of 6 knots.
The sudden increase in steep and breaking waves frequently catches the
mariner unaware, sometimes with tragic consequences. The U.S. Coast
Guard station at Cape Disappointment conducts hundreds of search and
rescue missions annually and reports an average of 10 fatalities a
year [1].
       The National Oceanic and Atmospheric Administration (NOAA) is
responsible for forecasting navigational conditions at hazardous sites
along the U.S. coast and has sponsored research in this area for
several years. Enfield [2] made the first thorough-going study of
hazardous wave conditions on the Bar, but he was hampered by lack of
direct wave measurements in the river entrance. In a more recent
study, observations of wave height amplification on the Columbia River
Bar were found to agree surprisingly well with simple one-dimensional

Y. Toba and H. Mitsuyasu (eds.), The Ocean Surface, 303–310.
© 1985 by D. Reidel Publishing Company.

wave-current interaction theory [3]. However, that study also documented one "severe event" in which this simple theory failed and the
observed Bar wave height was seriously underpredicted. The observed
discrepancy was attributed to two-dimensional effects not accounted
for by the theory. Specifically, it was hypothesized that wave energy
was being focused on the Bar through wave refraction induced by
lateral current shear. The present report supports this hypothesis,
and describes the results of an experiment conducted at the Columbia
River entrance during the period 10-13 September 1981 [4].

2. DATA COLLECTION

The bathymetry and configuration of the Columbia River entrance are
presented in Figure 1. Coast Guard navigational buoys mark the entrance channel, which is maintained at a nominal depth of 15 m by the
U.S. Corps of Engineers. Local mariners refer to the region from
between the North and South jetty tips to buoys 1 and 2 as "the Bar"
(technically a misnomer, since shoals are not exposed at low tide);
the "inner Bar" is the area near the jetties, and the "outer Bar" is
farther seaward [2].

The NOAA Data Buoy Center (NDBC) environmental buoy 46010 collected hourly wind speed, wind direction, and wave spectral estimates
at a point 5 nm southwest of the entrance. Waverider data and surface
current drifter data were also collected on the Bar during two observation periods on each of the four days; the intent was to complement
observations of peak ebb conditions with similar data collected at
slack or peak flood on the same day. The Waverider was deployed from
a surface vessel between buoys 8 and 10, tethered to a 300-foot line
by a 3-point bridle arrangement.

Coincident with such wave rider deployments, SLAR imagery of the
region was acquired by the Oregon Army National Guard (OANG) on approximately eight separate flight lines in a box-shaped pattern centered
on the entrance. Examples of such imagery are presented in Figure 2.
Portions of the 11 and 12 September SLAR imagery were digitized; they
corresponded to the location of buoy 46010, 10 km southwest of the
entrance, and another location a similar distance northwest of the
entrance. Two-dimensional Fourier transforms of these subscenes gave
estimates of the wavelength and direction of ocean swell incident on
the tidal current at the entrance.

3. THE OBSERVATIONS

On 10 September 1981, a strong low pressure system in the Gulf of
Alaska reached full development. A long, clearly defined fetch region
in the southeast sector was characterized by 30 to 40 knot winds and
oriented toward the northwest U.S. Coast. The arrival at the Columbia
River entrance of swell from this storm was marked by a ridge in the
wave energy spectrum which displayed an increase in frequency from
an initial low of 0.07 Hz at 1300 PDT 11 September to a value of
0.1 Hz at 1500 PDT on 13 September. As a consequence, significant

wave height at buoy 46010 rose from 1.5 m at 0000 PDT on 10 September
to a peak of 3.0 m at 0000 PDT on 12 September, then decayed gradually
to 2.2 m at 1500 PDT on 13 September.

A summary is presented in Table 1 of *in-situ* observations collected during the eight observation periods. In three instances,
conditions at buoy 8 were too hazarous for Waverider deployment, and
we estimated significant wave height H_s there by a least-square-fit
extrapolation from Waverider data collected east of buoy 8. These
data displayed a linear decay in H_s east of buoy 8 with very high
correlation coefficients of 0.9 or greater. Examples of this H_s
distribution on the slack and ebb of 11 September are presented in
Figure 2. The SLAR images in this figure also display strong refraction of the waves east of buoy 8 toward the shoals north and south of
the navigation channel, and this effect must contribute to the
observed decrease in wave height upriver.

SLAR-derived estimates of the offshore wavelength and direction
are summarized in Table 2. The measurements are consistent with the
interpretation of the changes in buoy 46010 spectral characteristics
presented above. Relatively short waves of 240 m were incident on the
entrance at slack tide on 11 September. Three hours later on the ebb,
longer waves, evidently from the 10 September storm, have arrived
from a somewhat different direction. The next day, on the ebb of 12
September, the wavelength displays the expected decrease as the wave
energy spectrum shifts to higher frequencies.

4. COMPARISON WITH THEORY

In this section, we compare the observed wave height amplification on
the Bar at buoy 8 with the simple linear one-dimensional wave-current
interaction theory of Longuet-Higgins and Stewart [5]. If no lateral
variations are permitted in any of the variables, then the theoretical
wave height amplification is given by

$$T(\omega,U;\theta_0,d,d_0,U_0) \equiv \left(\frac{E}{E_0}\right)^{\frac{1}{2}} = \frac{a}{a_0} = \left(\frac{U_0 + c_{go} \cos\theta_0}{U + c_g \cos\theta}\right)^{\frac{1}{2}} \left(\frac{\sigma}{\sigma_0}\right)^{\frac{1}{2}} \quad (1)$$

where unscripted parameters refer to the river entrance and the subscript "o" indicates initial conditions, or offshore values. Here,
E is the spectral energy density value, a is the wave height amplitude, U is the surface current velocity, C_g is the wave group
velocity, θ is the relative wave-current direction, σ is the intrinsic
wave frequency, ω is the observed wave frequency, and d is the water
depth. The parameters in (1) are also related by

$$\omega = \sigma + kU\cos\theta \quad (2a)$$

$$\sigma = (gk \tanh kd)^{\frac{1}{2}} \quad (2b)$$

Table 1. *Summary of the in-situ observations*

Date (Sep)	Time (PDT)	Buoy 8 H$_s$ (m)	Buoy 46010[1] H$_s$ (m)	Buoy 46010[1] f$_p$ (H$_z$)	Current[2] U (m/s)
10	1528	2.2	1.7	0.09	-2.1
11	1445	3.1	2.7	0.07	-1.3
11	1615	6.5[3]	2.9	0.07	-2.0
12	1700	4.8[3]	2.8	0.08	-2.2
13	1825	2.9[3]	1.9	0.09	-2.3
10	1030	1.5	1.8	0.08	1.3
12	1127[4]	1.0	1.4	0.08	1.5
12	1127[4]	1.2	2.4	0.11	1.5
13	1117	1.5	2.1	0.09	1.4

[1] Values were interpolated from observations bracketing the time of Waverider observations.

[2] From drifter speed estimates. Ebbs are negative, floods are positive

[3] Extrapolated from data east of buoy 8 (e.g., see Fig. 2).

[4] Two distinct energy peaks were apparent in the wave spectra.

Table 2. *SLAR-derived offshore wavelength and direction observations.*

Date (Sep)	Tidal Current	Wavelength (m)	Direction (deg true)
11	Slack	240	122
11	Ebb	283	107
12	Ebb	248	101

Table 3. *Comparison of observed wave height amplification T', with computed values, T$_c$. In Eq. (1), we used $d = 60$ m, $d_o = 15$ m, and the Table 1 values for u and f.*

Date (Sep)	Time (PDT)	T' (H$_s$ ratio)	θ (SLAR)	T'/T$_c$ Eq.(1)	T$_o$/T$_c$
Ebbs					
10	1528	1.3	45°	1.2	1.1
11	1445	1.1	30°	1.2	0.9
11	1615	1.8[1]-2.2	30°	1.4	1.3[1]-1.6
12	1700	1.7	25°	1.4	1.2
13	1825	1.5	25°	1.4	1.1
Floods					
10	1030	0.8	30°	0.8	1.0
12	1127	0.7	10°	0.9	0.8
12	1127	0.5	10°	0.8	0.6
13	1117	0.7	10°	0.8	0.9

[1] Lower value corresponds to the assumption that wave height was limited to H$_b$, given by Eq. (4).

$$C_g = \frac{1}{2}(1 + \frac{2kd}{\sinh 2kd}) \frac{\sigma}{k} \quad . \qquad (2c)$$

Here k is the wave number, and an analogous set of equations can be written for the subscripted variables. A final relationship

$$k \sin\theta = k_0 \sin\theta_0 \qquad (3)$$

closes this coupled system of transcendental equations, which can be solved iteratively for values of T as given by equation (1).

Table 3 summarizes the comparison of these theoretical values with observed values, obtained from Table 1 by forming the ratio of H_s at buoy 8 to H_s at buoy 46010. The mean direction of surface drifters on the ebb was approximately 255°, while that on the flood was 90°. The values of θ_0 in Table 3 are thus the positive acute angles formed by the SLAR-derived wave direction and the current direction, rounded to 5°. SLAR data for 10 September and 13 September are not yet analyzed, so wave direction was assumed to be the same on the 10th as on the slack tide of the 11th, and the same on the 13th as on the ebb of the 12th. We also found that at 1615 on 11 September, the es-timated wave height of 6.5 m at buoy 8 exceeded the breaking wave height value $H_b = 5.2$ m computed by the semi-empirical criteria of Battjes [6]

$$H_s \leq H_b = \frac{0.4}{k} \tanh kd \quad . \qquad (4)$$

Computations using both wave height estimates are entered in Table 3 for this case.

There are a number of potential errors in these computations. For example, the depth on the Bar is a subjective mean which worked well in the previous study [3]. Also, because the frequency resolution of NDBC 46010 is only 0.01 Hz, the observed peak frequency f_p must be considered a rough estimate. Nonetheless, the linear one-dimensional theory does reasonably well on 10 and 13 September (agreement to within 10%). But the 11 September ebb, and to some extent the 12 September ebb and flood, are characterized by discrepancies of the same order as the error for the extreme event reported previously [3].

Why is Eq. (1) inappropriate in the case of the 11 September ebb? An examination of the SLAR imagery in Figure 2b shows significant curvature in the wave crests at the entrance and energy is thereby focused on the Bar through a two-dimensional mechanism not accounted for by Eq. (1). This observed refraction must be due to lateral shear in the current, since the slack scene for that same day (Fig. 2a) does not display the same curvature. In contrast, the SLAR images for 10 and 13 September (not shown), are characterized by little if any wave crest curvature at the entrance, and observed values of H_s agree very well with one-dimensional theory. Finally, the SLAR imagery for 12 September (not shown) displays less curvature, than the ebb of 11 September. Not surprisingly then, it appears that the greater the degree of two-dimensional focusing the greater the discrepancy between observations and one-dimensional theory.

In Figure 2b, there also appears a pattern on the Bar which is
strongly suggestive of either crossing wave crests, or two-dimensional
structure induced by nonlinear instabilities and documented experimentally by Su [7] in a series of wave tank experiment which did not
involve currents. Further analysis will be required for a more detailed understanding of these complicated wave patterns in the SLAR
imagery.

5. SUMMARY AND CONCLUSIONS

A preliminary analysis has been made of SLAR and *in-situ* observations
of wave-current interaction at the Columbia River entrance. On comparison with simple linear, one-dimensional theory, we have found that
observed and predicted wave height amplification agree to within 10%
for those cases in which the SLAR imagery displays little or no wave
crest curvature that would focus energy on the Bar. In contrast, an
underprediction of 30 to 60% occurred for the case in which such

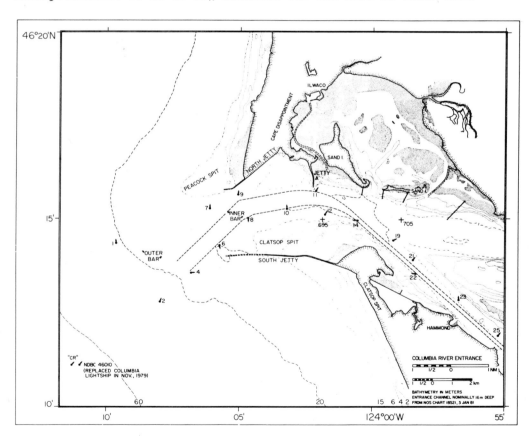

Figure 1. Reference map of Columbia River entrance.

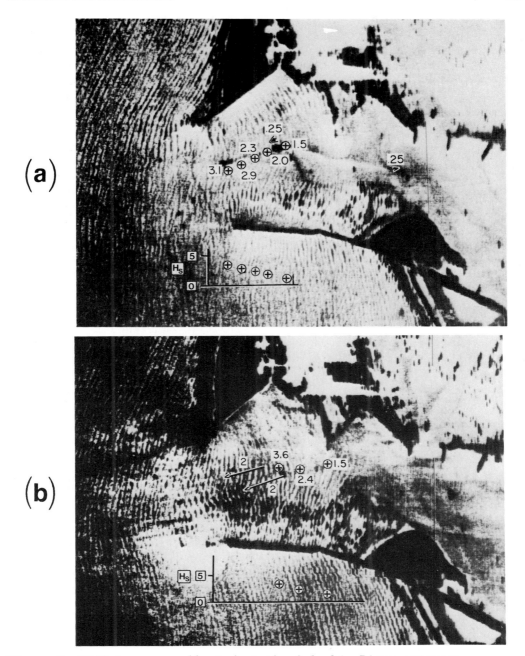

Figure 2. SLAR imagery collected at the Columbia River entrance, showing Waverider H_s estimates in meters (at circled crosses) and surface drifter current estimates in m/sec (arrows). (a) Slack current on 11 September 1981. (b) Ebb current on 11 September 1981.

focusing is most apparent. We believe this observed refraction is due
to the presence of significant lateral current shear.

We conclude that the "severe event" documented in a previous
study [3], was probably also due to such two-dimensional effects, and
that a more detailed understanding of these processes is needed if we
expect to accurately forecast the occurrence of such severe events.

ACKNOWLEDGEMENTS

The Ocean Services Division of the National Ocean Service, NOAA,
provided partial support for the data reduction and analysis reported
here. We also gratefully acknowledge the support of the Coastal
Science Program of the Office of Naval Research, which partially
funded the four-day field experiment.

The cooperation and assistance of the U.S. Coast Guard is also
gratefully acknowledged. The National Marine Fisheries Service provided surface vessel support; our thanks to the crews of the R/V EGRET
and NERKA for their cheerful assistance. The meteorological data were
provided by the NWS Seattle Ocean Services Unit and the NWS Astoria
field office. The SLAR imagery was provided by the 1042nd MICAS unit
of the Oregon Army National Guard; the excellent quality of these data
testifies to the professionalism and technical competence of this
organization. This work is a contribution of the Marine Services
Research Division at the Pacific Marine Environmental Laboratory.

REFERENCES

[1] Department of Transportation, 1982: United States Coast Guard
         1981 Search and Rescue Statistics. U.S. Govt. Printing
         Off., Washington, D.C., 142 pages.
[2] Enfield, D.B., 1973: Prediction of hazardous Columbia River Bar
         conditions. Ph.D. thesis, Dept. Oceanography, Oregon State
         University, 204 pages.
[3] Gonzalez, F.I., 1984: A case study of wave-current-bathymetry
         interactions at the Columbia River entrance. J. Phys. Ocean.,
         14, 1065-1078.
[4] Gonzalez, F.I., M.R. Mulhern, E.D. Cokelet, T.C. Kaiser, J.F.R.
         Gower, J. Wallace, 1984: Wave and Current Observations at
         the Columbia River Entrance, 10-13 September 1981, NOAA
         Tech. Memo ERL PMEL-58, 217 pages.
[5] Longuet-Higgins, M.S. and R.W. Stewart, 1961: The changes in
         amplitude of short gravity waves on steady non-uniform
         currents. J. Fluid Mech., 10, 529-549.
[6] Battjes, J.A., 1982: A case study of wave height variations due
         to currents in a tidal entrance. Coastal Eng., 6, 47-57.
[7] Su, M.-Y., 1982: Three-dimensional deep-water waves. Part 1.
         Experimental measurement of skew and symmetric wave pat-
         terns. J. Fluid Mech., 124, 73-108.

ON THE EFFECTS OF OCEAN SURFACE ROUGHNESS ON EMISSIVITY AND REFLECTIVITY OF MICROWAVE RADIATION

Y. Sasaki, I. Asanuma, K. Muneyama
Japan Marine Science and Technology Ceneter
2-15 Natsushima Yokosuka 237
Japan

G. Naito
National Research Center for Disaster Prevension, Hiratsuka
Branch
9-2 Nijigahama Hiratsuka 254
Japan

and

Y. Tozawa
Tokyo Scientific Center, IBM Japan
5-19 Sanban-cho Chiyoda-ku Tokyo, 102
Japan

ABSTRACT. We discuss the effects of ocean surface roughness and foams on the reflectivity of the sky brightness temperature and the inherent surface emissivity. The interesting conclusion to be drawn is that both polarizations of the reflected sky brightness temperature showed the dependences on wind velocity, that is, wind-induced roughness and on observation angle, but the horizontal polarization is of higher reflectivity and angular dependence than the vertical polarization. It was also found that the horizontal polarization showed the slight change by 0.03 in emissivity when the surface is roughened by the wind of velocity of 10.5 m/sec, but almost no change was to be seen for the wind of velocity of 8.5 m/sec for the vertical polarization.

INTRODUCTION

It has been known that the emissivity and the reflectivity of the ocean surface vary with roughness and foam. Nowadays much efforts are actively made for deeper and more exact knowledge for the final goal of their application to ocean observations.

For this purpose we have to identify the relations between the variations of intensity of microwave radiations emitted from the ocean surface and the roughness. But it's too difficult to express exactly

Y. Toba and H. Mitsuyasu (eds.), The Ocean Surface, 311–318.
© *1985 by D. Reidel Publishing Company.*

what the surface roughness and the foam are. So is to estimate the intensity of inherently emitted microwave radiation which varies with roughness and foam coverage. Partly because we can't refuse the sky radiations reflected at the surface, which also depend on the surface roughness and the foam coverage.

In this work we studied the dependence of the reflectivity of the sky brightness temperature, then inferred the inherent surface brightness temperature from the apparent one observed to identify the dependence of the emissivity on the surface roughness and the foam coverage.

We introduced the statistical theory developed by C. Cox and W. Munk[1],[2] and Stogryn's method[3] for the estimation of the surface roughness and the foam coverage. Discussions are made in terms of the probability density in which the surface slopes appear.

2. ANALYSES

2.1. Correction of Antenna Pattern and Derivation of Emissivity Change

In most cases the brightness temperature of the sea averaged over the main beam of the antenna is adopted for the comparison among the data from many instruments. But it might not neccessarily be a reasonable way, since each instrument has its own characteristic pattern of the antenna. Then we tried to derive the sea surface emissive properties along the direction of the main axis of the antenna for more reasonable comparison.

The apparent surface brightness temperature $T_a(\theta,\phi)$ is given as (see Fig.1 for notations)

$$T_a(\theta,\phi) = T_{SST} \cdot \varepsilon(\theta,\phi) + R(\theta,\phi) \tag{1}$$

where

T_{SST} ; actual sea surface temperature (K)

$\varepsilon(\theta,\phi)$; emissivity in the direction (θ,ϕ)

$R(\theta,\phi)$; reflected sky brightness temperature (K)

Emisivity $\varepsilon(\theta,\phi)$ may be expressed as

$$\varepsilon(\theta,\phi) = \overset{\circ}{\varepsilon}(\theta,\phi) + \overset{r}{\varepsilon}(\theta,\phi) \tag{2}$$

where

$\overset{\circ}{\varepsilon}(\theta,\phi)$; emissivity for the specular surface

$\overset{r}{\varepsilon}(\theta,\phi)$; emissivity change due to the surface roughness and the foam

If we expand $\overset{\circ}{\varepsilon}(\theta,\phi)$, $\overset{r}{\varepsilon}(\theta,\phi)$ and $R(\theta,\phi)$ in Taylor series at the point (θ_0, ϕ_0) on the main axis and assume their linearity, we find that

$$\overset{\circ}{\varepsilon}(\theta,\phi) = \overset{\circ}{\varepsilon}(\theta_o,\phi_o) + (\theta-\theta_o)\cdot\frac{\partial\overset{\circ}{\varepsilon}(\theta_o,\phi_o)}{\partial\theta} + (\phi-\phi_o)\cdot\frac{\partial\overset{\circ}{\varepsilon}(\theta_o,\phi_o)}{\partial\phi} \tag{3}$$

$$\varepsilon'(\theta,\phi) = \varepsilon'(\theta_o,\phi_o) + (\theta-\theta_o)\cdot\frac{\partial\varepsilon'(\theta_o,\phi_o)}{\partial\theta} + (\phi-\phi_o)\cdot\frac{\partial\varepsilon'(\theta_o,\phi_o)}{\partial\phi} \tag{4}$$

$$R(\theta,\phi) = R(\theta_o,\phi_o) + (\theta-\theta_o)\cdot\frac{\partial R(\theta_o,\phi_o)}{\partial\theta} + (\phi-\phi_o)\cdot\frac{\partial R(\theta_o,\phi_o)}{\partial\phi} \tag{5}$$

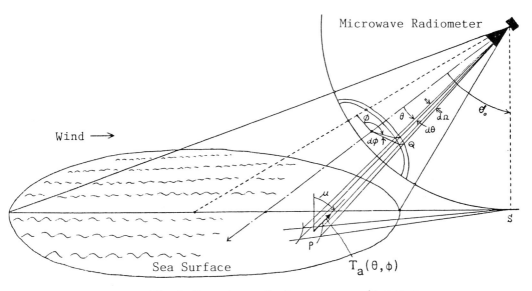

Fig.1 Geometry of microwave radiometry

On the other hand the apparent surface brightness temperature observed with a radiometer which has the antenna pattern $G(\theta,\phi)$ is given in the form

$$T = \frac{\int T_a(\theta,\phi)\cdot G(\theta,\phi)\,d\Omega}{\int_{\Omega} G(\theta,\phi)\,d\Omega} \tag{6}$$

Then we have a following expression for the apparent surface brightness temperature from eq. (1) through (6)

$$T = T_{SST}\{\overset{\circ}{\varepsilon}(\theta_o,\phi_o) + \varepsilon'(\theta_o,\phi_o)\} + R(\theta_o,\phi_o) +$$

$$+ T_{SST}\{\frac{\partial\overset{\circ}{\varepsilon}(\theta_o,\phi_o)}{\partial\theta}\cdot\overline{G}_\theta + \frac{\partial\overset{\circ}{\varepsilon}(\theta_o,\phi_o)}{\partial\phi}\cdot\overline{G}_\phi + \frac{\partial\varepsilon'(\theta_o,\phi_o)}{\partial\theta}\cdot\overline{G}_\theta\}$$

$$+ \frac{\partial R(\theta_o,\phi_o)}{\partial\theta}\cdot\overline{G}_\theta \tag{7}$$

where

$$\overline{G}_\theta = \frac{\int (\theta - \theta_c) \cdot G(\theta, \phi) \, d\Omega}{\int G(\theta, \phi) \, d\Omega} \quad , \quad \overline{G}_\phi = \frac{\int (\phi - \phi_c) \, G(\theta, \phi) \, d\Omega}{\int G(\theta, \phi) \, d\Omega}$$

$\overline{G}_\theta$ and $\overline{G}_\phi$ can be obtained from the antenna pattern. If we assume that $\partial \varepsilon'(\theta_c, \phi_c)/\partial\theta$ and $\partial R(\theta_c, \phi_c)/\partial\theta$ are of little contribution we can derive the value of $\varepsilon'(\theta_c, \phi_c)$ from eq. (7).

2.2. Estimation of the Sky Brightness Temperature Reflected at the Surface

2.2.1. For Low Winds

In the case of low winds in which no foams are formed the ocean surface may be assumed to consists of a large number of facets which have almost flat plane. We assume that the ocean surface can be described by the statistical theory of the surface slope by C. Cox and W. Munk. Let us consider the situation in which a microwave radiometer views the ocean surface toward the up-wind direction. If the further assumption is made that we can apply the optical reflection rules to the radiations incident on the facet planes the reflected sky brightness temperature T_{w_p} may be given in the form"

$$\frac{\int_\Omega \int_{-\cot\theta_c'}^{\infty} \int_{-\infty}^{\infty} T_{atm}(\theta', \phi') \cdot \{R_V \cdot \sin\beta_p + R_H \cdot \cos\beta_p\} \cdot P(z_x \cdot z_y) \cdot G_p(\theta, \phi) \cdot dz_x \cdot dz_y \cdot d\Omega}{\int_\Omega \int_{-\cot\theta_c'}^{\infty} \int_{-\infty}^{\infty} P(z_x \cdot z_y) \cdot G_p(\theta, \phi) \cdot dz_x \cdot dz_y \cdot d\Omega} \tag{8}$$

where

$T_{atm}(\theta', \phi')$; sky brightness temperature in the direction (θ', ϕ'). (θ' and ϕ' are different from θ and ϕ. They are the zenith angle and the azimuth angle respectivly).

θ_c'; observation angle

p; type of polarization

β_p; angle which the normal line of the facet makes with that of the plane in which p type of polarized radiation oscillates.

$P(z_x, z_y)$; probability density of facet with the slopes z_x and z_y

$$\frac{1}{2\pi\sigma_u\sigma_c} \cdot \exp\left[-\frac{1}{2}(\xi^2 + \eta^2)\right] \cdot \left[1 - \frac{1}{2} C_{21}(\xi^2 - 1)\eta - \frac{1}{6} C_{03}(\eta^3 - 3\eta)\right.$$

$$+\frac{1}{24} C_{40}(\xi^4 - 6\xi^2 + 3) + \frac{1}{4} C_{22}(\xi^2 - 1)(\eta^2 - 1)$$

$$\left. +\frac{1}{24} C_{04}(\eta^4 - 6\eta^2 + 3)\right]$$

$$\xi = \frac{z_x}{\sigma_c}$$

$$\eta = \frac{z_y}{\sigma_u}$$

$$\sigma_c^2 = 0.003 + 1.92\times10^{-3}\cdot W \pm 0.002$$

$$\sigma_u^2 = 0.000 + 3.16\times10^{-3}\cdot W \pm 0.004$$

$$\sigma_c^2 + \sigma_u^2 = 0.003 + 5.12\times10^{-3}\cdot W \pm 0.004$$

$$C_{21} = 0.01-0.0086\cdot W \pm 0.03$$

$$C_{03} = 0.04-0.0033\cdot W \pm 0.12$$

$$C_{40} = 0.40 \pm 0.23$$

$$C_{22} = 0.12 \pm 0.06$$

$$C_{04} = 0.23 \pm 0.41$$

W ; wind velocity

2.2.2. For High Winds

For the winds of velocities higher than about 8 m/sec foams are observed. They make the emissivity and the reflectivity change. In this case the following expression is given for the reflected sky brightness temperature[3].

$$T_{s_p} = (1-F)\cdot T_{w_p} + F\cdot R_{f_p}\cdot T'_{atm} \tag{9}$$

where

T'_{atm} ; sky brightness temperature in the direction (θ , 0).

R_{f_p} : reflectivity of the foam-covered area

F ; foam coverage $7.751\times10^{-6}\cdot W^{3.231}$

3. OBSERVATIONS

The observations of the brightness temperatures of the sea surface and the sky at 18.6 GHz have been performed on the stationary tower situated in Sagami Bay 1.3 Km off Hiratsuka since September 1983. Wave direction, wave height, wind velocity, wind direction, water temperature, air temperature, salinity, cloud amount, etc. have been also measured. The data has been automatically sampled by a microcomputer and stored in its floppy discs. The microwave radiometer is at the height of 20 m above the sea level and directed in the south. Much care has been taken not to be affected by surroundings through the main lobe and the side lobes.

4. RESULTS AND DISCUSSIONS

4.1. Effects of the Reflected sky Brightness Temperature on the Apparent Surface Brightness Temperature

Fig.2 shows the relationship between the sky brightness temperature in the direction of 25 degrees from the horizon and the apparent surface brightness temperature at the observation angle of 50 degrees (measured from the nadir). The wind velocities and the actual surface water temperatures corresponding to the data in Fig.2 lie in the range between 2.5 and 11.5 m/sec and 17.5 and 22.5 C. We can easily notice from Fig.2 that there are obvious relationships between these two brightness temperatures and that the reflected sky brightness temperature is of great contribution to the apparent surface brightness temperature.
    Fig.3 shows the observed dependence of the sky brightness temperature on zenith angle. The sky brightness temperature measurements were made under various cloud conditions. From our observations it was noticed that the sky brightness temperature is closely related to the cloud amount, that is, the liquid water content but we could not derive any quantitative relations bwtween them. Furthremore the sky brightness temperature showed only slight variations in the azimuth directions. Finally we had the following empirical expression for the sky brightness temperature $T_{atm}(\theta')$ at an arbitrary angle θ'.

$$T_{atm}(\theta') = T_{atm}(\overset{\circ}{0}) \cdot 10^{a \cdot 10^{-7} \cdot \theta'^b} \tag{10}$$

where

$T_{atm}(\overset{\circ}{0})$; sky brightness temperature at zenith angle $\theta' = 0°$

a , b ; constants (a $\simeq$ 1.5~1.7, b $\simeq$ 3.4~3.6)

4.2. Effects of Observation Angle and Wind-induced Roughness on the Reflectivity of the Sky Brightness Temperature

Fig.4(a) shows the dependence of the reflected sky brightness temperature on observation angle. Curves are drawn after eq. (8) or (9). In both calculations 25 K is employed for $T_{atm}(\overset{\circ}{0})$ for comparison. For the horizontal polarization some improvements are made on the algorithms for the reflected sky brightness temperature previously reported[?]. The interesting conclusions are derived that both polarizations show the similar patterns of angular dependence, that is, they gradually increase till the observation angle goes up to the neighborhood of 60 degrees and then they decrease with this angle. The horizontal polarization is more sensitive than the vertical polarization.
    Fig.4(b) shows the wind velocity dependence of the reflected sky brightness temperature. Both polarizations show little dependence on the wind velocity, in other words, the wind-induced roughness but only a slight dependence of the vertically polarized temperature for the lower winds.

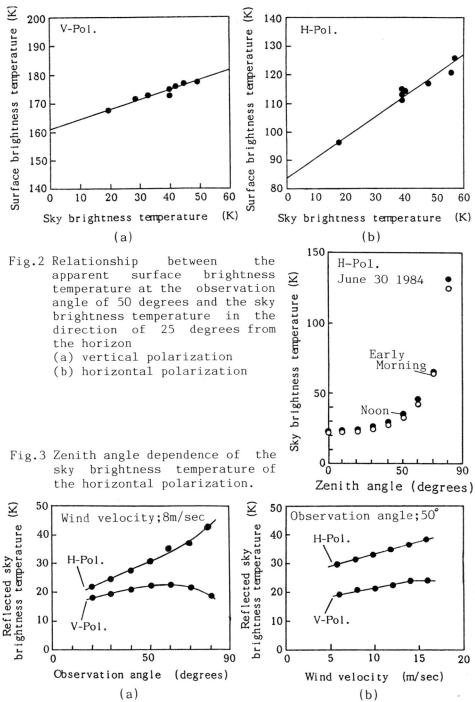

Fig.2 Relationship between the apparent surface brightness temperature at the observation angle of 50 degrees and the sky brightness temperature in the direction of 25 degrees from the horizon
(a) vertical polarization
(b) horizontal polarization

Fig.3 Zenith angle dependence of the sky brightness temperature of the horizontal polarization.

Fig.4 Dependences of the reflected sky brightness temperature on the observation angle and the wind velocity

4.3. Effects of Wind-induced Roughness on Sea Surface Emissivity

Fig.2 may suggest that the deviation of the points from the line is due to the wind-induced roughness. If this is true we may be able to come to an interesting conclusion that the horizontally polarized temperature is slightly more sensitive than the vertically polarized one. Then we tried to deduce the emissivity values for both polarizations at the observation angle 50 degrees from eq. (7). The data adopted here are from the cases of the mean wind velocities of 10.5 and 8.5 m/sec for the horizontal and the vertical polarizations respectively. As the result, the emissivity of the horizontally polarized radiation appeared 0.03 higher than a specular surface. On the other hand almost zero value of emissivity change was found for the vertical polarization. Further studies need to be made especially for more exact estimation of the reflected sky brightness temperature, since it might be a casting vote to the precise prediction of emissivity change.

REFERENCES

1) C. Cox and W. Munk, 'Statistics of the Sea Surface Derived from Sun Glitter', J. Marine Res., 13(2), 198, 1958
2) C. Cox and W. Munk, 'Measurement of the Roughness of the Sea Surface from Photographs of the Sun's Glitter', J. Opt. Soc. Am., 44(11), 838, 1958
3) A. Stogryn, 'The Apparent Temperaturer of the Sea at Microwave Frequencies', IEEE Trans. Antennas and Propag., AP-15(2), 278, 1967
4) A. Stogryn, 'The Emissivity of Sea Foam at Microwave Frequencies', J. Geophys. Res., 77(2), 1658, 1972
5) W. Nordberg, J. Conway and P. Thaddeus, 'Microwave Observations of Sea State from Aircraft', Quart. J. Res. Met. Soc., 95, 408, 1969
6) J. Hollinger, 'Passive Microwave Measurements of Sea Surface Roughness', IEEE Trans. Geos. Elec. GE-9(3), 165, 1971
7) J. P. Claassen and A. K. Fung, 'The Recovery of Polarized Apparent Temperature Distributions of Flat Scenes from Antenna Temperature Measurements', IEEE Trans. Antennas and Propag., AP-22(3), 433, 1974
8) F. J. Wentz, 'A Model Function for Ocean Microwave Brightness Temperature', J. Geophys. Res., 88(C3), 1892, 1983
9) Y. Sasaki, I. Asanuma, K. Muneyama, G. Naito and Y. Tozawa, 'Effects of Sea Surface Roughness on Surface Brightness Temperature', Proc. 1984 Int'l Symp. on Noise and Clutter Rejection in Radars and Imaging Sensors, 135

EXPERIMENTAL RESULTS OF SEA-SURFACE SCATTERING BY AIRBORNE MICROWAVE SCATTEROMETER/RADIOMETER

H. Masuko[1], K. Okamoto[1], T. Takasugi[1], M. Shimada[2], H. Yamada[2], and S. Niwa[2]

1 Remote Sensing Division, Radio Research Laboratories, Ministry of Posts and Telecommunications, Tokyo 184 Japan
2 Tsukuba Space centre, National Space Development Agency of Japan, Ibaraki 305 Japan

ABSTRACT. The Normalized Radar Cross Section (σ^0) of the ocean was measured as combined functions of microwave wavelength, polarization, wind speed, azimuth angle, and incident angle, using an airborne dual-frequency microwave scatterometer/radiometer system. The azimuth anisotropic signatures and the wind speed dependence of the σ^0 are observed for Ka-band similar to those for X-band. It is confirmed that the surface wave spectrum is anisotropically developed in the capillary spectrum region in the same manner as those in the gravity-capillary spectrum region. The wave number spectra for these waves are derived using the incident angle dependences of the σ^0 for VV-polarization. The growth of the surface wave spectrum with wind speed is obtained under the open sea condition. The decreasing rate of the spectral density for wave number derived from the X-band data is larger than those derived from the Ka-band data, and the rate becomes small with increasing the wind speed.

1. INTRODUCTION

Microwaves have been applied in the wide area of the remote sensing of the ocean. The interesting one is the measurement of the wind vector on the sea surface, which becomes possible on account of the fact that the signatures of the microwave scattering are connected with the wave conditions caused by the surface wind. However, the behaviors of the short surface waves interacting with microwaves have not been understood so well. On the other hand, there is a possibility to know the wave spectrum of the short surface waves by analyzing the dependence of the scattering coefficient on the incident angle (Wright, 1968, and Valenzuela et al., 1971).

The Radio Research Laboratories and the National Space Development Agency of Japan made joint experiments to measure the σ^0 of the ocean surfaces in 1980 and 1981, using an airborne dual-frequency microwave scatterometer/radiometer system. A set of data of the σ^0 was obtained and analyzed to get experimental verification of the complex dependences of the scattering on the wavelength of microwave (λ: 3.00 cm and 0.87 cm), polarization (p: HH and VV), azimuth angle (ϕ: 0-360°), incident

319

Y. Toba and H. Mitsuyasu (eds.), The Ocean Surface, 319–327.

angle (θ: 0-70°), and wind speed (U: 3.2-17.2 m/s). The purpose of this paper is to discuss the signatures of the high frequency wave spectrum of the sea surface in relation to the microwave scattering based on the data obtained in the flight experiment.

2. EXPERIMENTS

The system used in the experiments consists of the X-band (10.00 GHz) and Ka-band (34.43 GHz) scatterometers, which are operated in pulse mode using pencil beam antennae with the same beam pattern. The dual-wavelength system can simultaneously observe nearly the same sea surface region, so that the system is preferable to study the wavelength dependence of the microwave scattering signatures of the ocean. In the present experiments, for the purpose of obtaining the data for large incident angles all over the azimuth directions, all measurements without the case of vertical incidence (θ = 0°) were made using the circle flight technique with fixed incident angle. The general concept of the technique is shown in Figure 1. The detailed descriptions about the experiments are shown by Masuko et al.(1984).

The most part of the flight measurements were performed on the open ocean where the influences of the coast lines and bottom of the sea on the surface conditions may be neglected. Two main experimental sites are shown in Figure 2. The one was around the large meteorological buoy in the middle of the Sea of Japan settled by the Japan Meteorological Agency. The position of the buoy is about 180 km away from the nearest coast and the depth of the neighboring sea is more than 2000 meters. The vicinity of the position is the region where the warm and cold currents (Tsushima and Liman currents, respectively) are mixing with each other and surface current speed is generaly small and its

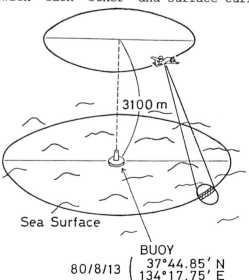

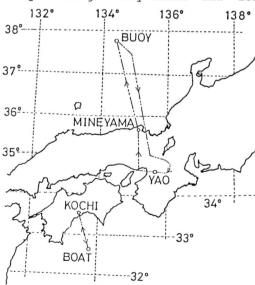

Figure 1. The general concept of the flight experiment.

Figure 2. The flight courses and experimental sites.

direction is unstable.   The data obtained by the buoy were used as sea
truth data.   The other experimental site was in the Pacific Ocean off
the city of Kochi.   The position is about 70 km distant from the
nearest coast and the depth of the surrounding sea is more than 1000
meters.   The sea truth data in the area were collected using a boat.
In the experiment, the current speed and direction around the area were
measured to be 1.2 m/s and 41°, respectively.   It is considered from
the strong current speed and its direction that the area was on the main
part of the Kuroshio, the large warm current.   The surface wind speeds
measured by the buoy and boat are converted to the values at 19.5 meters
above the mean sea level, using the formula proposed by Cardone (1969).

3.   RESULTS

3.1. Dependence on azimuth angle

Figure 3 shows the examples of the dependence of the σ^0 on azimuth angle
for VV-polarizations of X-band and Ka-band.   The abscissa is the radar
azimuth relative to the wind direction.   A bar accompanied on each
point indicates the standard deviation, whose averaged values $(\overline{\sigma})$ all
over the azimuth angle exists almost between 0.5 and 1.0 dB as shown
below each figure.   For both bands, the azimuth anisotropic signatures
of the σ^0 can be observed.   The maximum values appear when the radar
beams are directed to the up-wind and down-wind directions, and the σ^0
takes the minimum values when the radar beams are crossing the wind
direction.   These periodic variations are approximated by the cosine
expansion regression:

$$\sigma^0 (\lambda,p,\theta,\phi,U) = \sum_{n=0} A_n (\lambda,p,\theta,U) \cos(n\phi) \qquad (1)$$

The solid curves in Figure 3 show the results of the 2nd-order

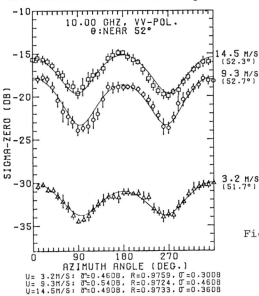

U= 3.2M/S: $\overline{\sigma}$=0.46DB, R=0.9759, σ=0.30DB
U= 9.3M/S: $\overline{\sigma}$=0.54DB, R=0.9724, σ=0.46DB
U=14.5M/S: $\overline{\sigma}$=0.49DB, R=0.9733, σ=0.36DB

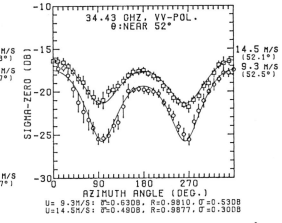

U= 9.3M/S: $\overline{\sigma}$=0.63DB, R=0.9810, σ=0.53DB
U=14.5M/S: $\overline{\sigma}$=0.49DB, R=0.9877, σ=0.30DB

Figure 3.   The dependences of the σ^0 on
azimuth directions and the results
of the 2nd-order cosine expansion
regressions in the incident angle
near 52°.

regression.   The   values   R and σ below each figure show the   multiple
correlation coefficient and the averaged deviation from the experimental
values for each approximation formula, respectively.   The Values of R
are more than 0.95 and those of σ are in the range from 0.3 to 0.7 dB in
the   present   measurements.   In   this   experiment,   such   cosine-like
signatures are observed for the incident angles from 20° to   70°,   where
the resonant scattering (the Bragg scattering) mechanism is predominant.
        Jones and Schroeder (1978) have indicated using the data up to 13.9
GHz that the ratio of the up-wind to down-wind peaks and that of the up-
wind peak to cross-wind minimum, which are the measures of the azimuth
anisotropy, generally increase with increasing the reciprocal of the
Bragg wavelength (Λ = λ/2sinθ).   However, in the present results, these
ratios don't increase as expected by them in the range from X-band to
Ka-band.   That is, the azimuth anisotropy in the Ka-band is almost the
same as or a little larger than those in the X-band.   It seems that the
azimuth anisotropy of the σ⁰ tends to be gradually   saturated   in the
frequency region above X-band.

3.2. Dependence on wind speed

Figure   4 shows the examples of the wind speed dependence of the σ⁰ for
X-band and Ka-band as a parameter of incident   angle.   Generally,   the
dependence is expressed with a power-law formula:

$$\sigma^0(\lambda,p,\theta,\phi,U) = g(\lambda,p,\theta,\phi)U^{H(\lambda,p,\theta,\phi)} \qquad (2)$$

The   quantity $H(\lambda,p,\theta,\phi)$ is known in the name of wind speed   exponent.
The   solid lines in each figure show the results of the regression using
the   formula   (2)   except the case of the 70°   incident   angle,   in   the
vicinity   of   which   the   wind speed dependence   is   more   complicated.
Probably, the simple power-law formula cannot be applied in the incident
angles   larger than about 70°,   because of the influences of   shadowing,
diffraction,   trapping   by   atmospheric dacts   and   others   (Valenzuela,
1978).   The   dotted lines in the figure for 10.00 GHz show the results
obtained   by   Moore and Fung (1979) using the 13.9   GHz   data,   and   the
broken   lines   show   the results of the SEASAT (14.6   GHz)   reported   by
Schroeder et al.(1982).   In spite of the difference of the frequencies,
the   results obtained in the present data agree well with those obtained
by Moore and Fung (1979).   The wind speed exponent is negative in the
incident   angles below 10°,   and generally increases as incident   angles
increase as large as about 70°.
        Table   1 shows the comparison between the wind speed   exponents   by
various experiments and the theory as a parameter of the incident angle.
The   wind   speed exponents are almost constant in the   frequency   region
above 10 GHz.   In the region below 10 GHz, where the reliable data are
a   few,   the   wind   speed   exponents become small   with   decreasing   the
frequency.   The   facts   agree well with the semi-empirical   estimation
based on the composite rough surface model made by Fung and Lee (1982).

3.3. Dependence on incident angle

Figure 5 shows an example of the incident angle dependence of the σ⁰ as a

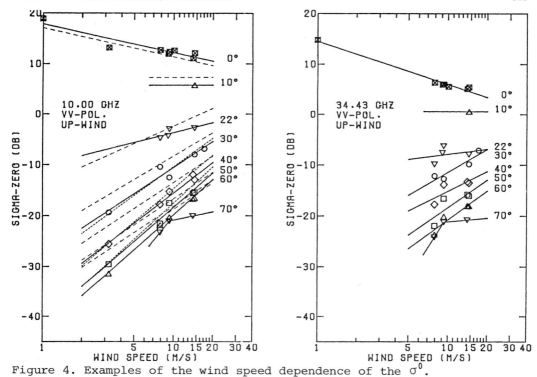

Figure 4. Examples of the wind speed dependence of the σ^0.

Table 1. Frequency dependence of the wind speed exponent for the case of HH-polarization and up-wind. The theoretical values used here are obtained by Fung and Lee (1982).

Frequency (GHZ) Wave Length (cm)		0.428 70.1	1.228 24.4	4.455 6.73	8.910 3.37	10.00 3.00	13.9 2.16			14.599 2.05	34.43 0.87
Incident Angle	Experimenter & Analyst	NRL				RRL/NASDA	SKYLAB	LaRC	Moore & Fung	SEASAT	RRL/NASDA
20°	Experiment	–	–	–	–	0.88 (22°)	–	1.03	–	0.96	0.69 (22°)
	Theory							1.22			
30°	Experiment	0.97 0.47	1.12 1.54	1.30 1.40 1.35	1.20 0.95 0.70	1.72	1.32	1.65 (32°)	1.63	1.48	1.18
	Theory			1.21	1.37			1.50			
40°	Experiment	–	–	–	–	2.08	1.89 1.31 (43°)	1.98	2.05	1.91	1.95
	Theory		2.12 (45°)	2.21 (45°)	2.29 (45°)			2.03			
50°	Experiment	–	–	–	–	2.36	1.81 1.15	1.93	2.40	2.24	2.54
	Theory							2.65			
60°	Experiment	–	–	1.50	1.45	2.49	–	–	–	2.30	2.21
	Theory		1.97	2.36	2.75						

parameter of frequency. The σ^0 values decrease with increasing the incident angle. In the incident angles smaller than 20°, where the quasi-specular scattering is predominant, the σ^0 for Ka-band are smaller than those for X-band. According to the specular point model

(Valenzuela, 1978), this is because the Fresnel reflection coefficient
is small and the variance of the surface slope is probably large for Ka-
band compared with those for X-band.

In the incident angles larger than 20° (the resonance scattering
region), the values of the σ^0 for Ka-band are a little smaller than or
almost equal to those for X-band, which are also recognized in Figure 3,
and these two kinds of the σ^0 values also agree with those obtained for
13.9 GHz and 14.6 GHz as shown in Figure 4. Therefore, it is concluded
that the σ^0 are almost saturated in the frequency region above X-band,
which is good agreement with the results of the semi-empirical theory
(Fung and Lee, 1982).

Considering the dependences of the azimuth anisotropy, wind speed
exponent, and the values of the σ^0 itself on microwave frequencies,
together with the microwave propagation characteristics, the frequencies
from X-band to Ku-band are more preferable to apply in the measurements
of the sea surface wind vector.

4. DISCUSSIONS ON SURFACE WAVE SPECTRUM

The azimuth variations of the microwave back-scattering from the sea
surface as shown in Figures 3 can be explained by the anisotropic
developments of the short surface waves, which are the origin of the

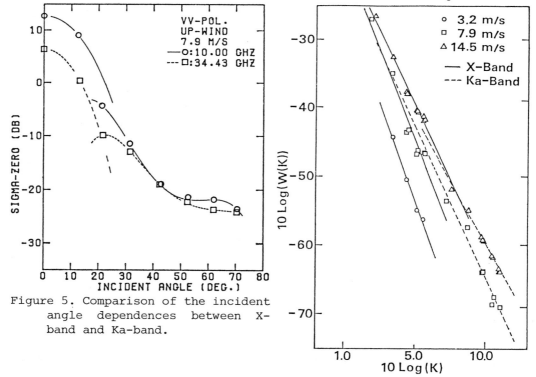

Figure 5. Comparison of the incident
    angle dependences between X-
    band and Ka-band.

Figure 6 (the right). The wave number spectra as a function of wind
speed derived from the incident angle dependence of the σ^0.

Bragg scattering, owing to the nonlinear modulation by the larger
gravity waves (Moore and Fung, 1979). The Bragg wavelengths for 10.00
GHz and 34.43 GHz are from 4.39 to 1.60 cm and from 1.27 to 0.46 cm for
the incident angles from 20° to 70°, respectively. From the dispersion
relation for sea surface waves, the Bragg wavelengths for 34.43 GHz
extends to the pure capillary spectrum region. As a result, it is
concluded that the surface wave spectrum is anisotropically developed in
the capillary spectrum region similar to that in the gravity-capillary
spectrum region. However, considering the azimuth anisotropy of the σ^0
for Ka-band are not so increased as expected from the tendency in the
lower frequencies as mentioned before, the growth of the azimuth
anisotropy of the surface wave spectrum tends to be saturated in the
wavelength region below a few centimeters.

In the resonant scattering region, the two scale theory based on
perturbation techniques for a slightly rough surface can provide good
estimates to the experimental data. In the theory, the surface is
modeled as ripples (short surface waves), which are the dominant
scatterers and are developed by the surface winds, superimposed on
large-scale waves (gravity waves), which are taken to have a tilting
effect: the composite rough surface model (Moore and Fung, 1979).
Therefore, according to the model, the net σ^0 is given by an average of
each σ^0 of a single slightly rough facet with local incident angle over
the distribution of slopes of the large-scale waves. However, Wright
(1968) has shown that, for VV-polarization, the tilting effect has
little influence on the averaging process in the incident angles from
15° to 60°. In the case, the σ^0 is approximated by

$$\sigma^0 (\lambda,p,\theta,\phi,U) = 4\pi k^4 |\alpha_{VV}| W(K,\phi,U) \tag{3}$$

where α_{VV} is the quantity related to the Fresnel reflection coefficients.
The $W(K,\phi,U)$ is the anisotropic wave number spectrum of the sea surface.
Therefore, the wave number spectrum in the short surface waves can be
obtained from the incident angle dependence of the σ^0 using the formula
(3). Figure 6 shows the wave number spectrum in the wavelength region
from 4.5 to 0.5 cm for up-wind direction as a function of surface wind
speed obtained from the present data for X-band and Ka-band; the
wavelengths coresspond to the frequency region from 6 to 63 Hz
neglecting the effect of surface currents. Generally, the wave
spectrum is expressed with a power-law formula of the wave number K:

$$W(K,\phi,U) = B(U,\phi) K^{\alpha (U,\phi)} \tag{4}$$

The solid and broken lines in Figure 6 show the power-law regression
derived from the X-band and Ka-band data, respectively. The
correlation coefficients of these regressions are more than 0.97. It
is confirmed under the open sea condition that the spectral density in
the gravity-capillary and capillary waves increases as the surface wind
speed increases. However, considering the wind speed exponents become
small in the region below 10 GHz as mentioned above, in the short
gravity wave region, the dependence of the wave spectrum on the surface
wind becomes weak with increasing its wavelength.

Table 2 shows the values of the $\alpha(U, \phi)$ obtained for the several
cases of wind speed and azimuth observation. These $\alpha(U, \phi)$ values are

Table 2.   Comparison of the wave number exponents $\alpha(U, \phi)$ of the surface
wave spectrum obtained for each wind speed in the case of the   up-,
down-, and cross-wind observasion.

FREQUENCY (WAVE-LENGTH)	AZIMUTH DIRECTION	WIND SPEED				
		3.2 m/s SEA OF JAPAN	7.9 m/s THE PACIFIC	9.3 m/s SEA OF JAPAN	14.1 m/s THE PACIFIC	14.5 m/s SEA OF JAPAN
10.00 GHz (3.00 cm)	UP-WIND	-5.67	-5.23	-4.78	-3.59	-4.30
	DOWN-WIND	-4.92	-5.42	-4.67	-5.32	-4.30
	CROSS-WIND	-5.85	-6.29	-5.32	-5.27	-4.45
34.43 GHz (8.7 mm)	UP-WIND	-	-4.40	-3.92	-3.99	-3.58
	DOWN-WIND	-	-5.03	-4.64	-6.37	-4.57
	CROSS-WIND	-	-5.81	-4.77	-7.16	-4.42

in general larger than the value of -3.721 inferred by Valenzuela et al.
(1971) based on the data for various wind speeds.   In the table,   each
$\alpha(U, \phi)$ derived   from   the Ka-band data is somewhat   larger   than   that
derived   from the X-band data.   It probably means that the   decreasing
rate   of   the wave spectral density for wave numbers   gradually   becomes
small.   As shown in Table 2, the $\alpha(U,\phi)$ increases with the wind speed,
which   means   the slopes of the wave number spectrum in Figure 6   become
small with increasing the wind speed.   The values of the $\alpha(U,\phi)$ for the
up-wind   and   down-wind observations   derived from the X-band data   are
almost equal with each other,   which is reasonable for the approximation
formula of (3), without the cases of the wind speeds of 3.2 m/s and 14.1
m/s.   However,   in   the   Ka-band   case,   the $\alpha(U, \phi)$ for   the   up-wind
observation is larger than that for the down-wind observation,   which is
almost   equal   to those for the cross-wind observation,   the   reason   of
which cannot be recognized so well.   In the case of the wind speed   of
14.1 m/s,   the strong current flowed to the up-wind direction,   so   that
the different behavior of the $\alpha(U, \phi)$ in the case is probably because of
the influence of the current.   Generally,   the $\alpha(U, \phi)$ for the up-wind
direction is larger than that for the cross-wind direction.
    Neglecting the influence of the surface current,   the exponents for
the   surface wave frequency are -6.5 to -5 and -2.9 to -4.7 for   the   X-
band   and   Ka-band   observations,   respectively,   using   the   dispersion
relation   for sea surface waves.   Mitsuyasu and Honda (1974)   reported
the   frequency   exponent   of   the surface wave spectrum   of   -4   in   the
gravity-capillary   spectrum   region based on the wave   tank   experiment.
The   present   results   are smaller by about 1 than their   result   in   the
gravity-capillary spectrum region.

5. ACKNOWLEDGMENTS

The   authors wish to thank the Japan Meteorological Agency for   offering
the surface truth data.   They are also grateful to Showa Aviation Co.,
Ltd.,   and Mr.   H.   Okamura of Mitsubishi Electric Corporation for their
co-operation to the difficult flight experiments.

REFERENCES

Cardone, V. J., (1969); Specification of the Wind Distribution in the Marine Boundary Layer for Wave Forcasting, New York Univ. Geophys. Sci. Lab. Rep. TR69-1, NTIS No AD702490.

Fung, A. K. and K. K. Lee, (1982); A Semi-empirical Sea-spectrum Model for Scattering Coefficient Estimation, IEEE, OE-7(4), 166-176.

Jones, W. L. and L. C. Schroeder, (1978); Radar Backscatter from the Ocean: Dependence on Surface Friction Velocity, Boundary-Layer Meteorology, 13(1-4), 133-149.

Masuko, H. et al., (1984); Measurements of Microwave Back-scattering Signatures of the Ocean Surface Using X-band and Ka-band Airborne Scatterometer/Radiometer System, Digest of the 1984 International Geoscience and Remote Sensing Symposium, 321-326.

Mitsuyasu, H. and T. Honda, (1974); The High Frequency Spectrum of Wind-generated Waves, J. Oceanographical Soc. Japan, 30(4), 29-42.

Moore, R. K. and A. K. Fung, (1979); Radar Determination of Wind at Sea, Proc. IEEE, 67(11), 1504-1521.

Schroeder, L. C. et al., (1982); The Relationship between Wind Vector and Normalized Radar Cross Section Used to Derive SEASAT-A Satellite Scatterometer Winds, J. Geophys. Res., 87(C5), 3318-3336.

Valenzuela, G. R. et al., (1971); Ocean Spectra for the High-frequency Waves as Determined from Airborne Radar Measurements, J. Marine Res., 29(2), 69-84.

Valenzuela. G. R., (1978); Theories for the Interaction of Electromagnetic and Ocean Waves - A Review, Boundary-Layer Meteorology, 13(1-4), 277-293.

Wright, J. W., (1968); A New Model for Sea Clutter, IEEE, AP-16(2), 217-223.

RADIO PROBING OF OCEAN SURFACE BY OBSERVING MULTIPATH FADING SIGNAL

Yoshio Karasawa, Takayasu Shiokawa and Matsuichi Yamada
Research and Development Laboratories, KDD
2-1-23 Nakameguro, Meguro-ku, Tokyo 153 Japan

ABSTRACT. Phenomena of L-band multipath fading due to sea surface
reflection in maritime satellite communications are presented firstly.
And then, relation between multipath fading and sea parameters such as
wave height and rms slope is discussed.

1. INTRODUCTION

A global maritime satellite communication system has been operating in
the frequency band at 1.5 GHz (satellite-to-ship) with circular
polarization under the coordination of INMARSAT (International Maritime
Satellite Organization) since 1982. In the case of receiving satellite
signals on a ship, fairly large amplitude fluctuation appears,
particularly at low elevation angles. This phenomenon is called
multipath fading, and it is caused by the interference between direct
incident wave from the satellite and reflected waves coming from the
sea surface. Therefore, it is noticed that the amplitude of the
receiving signal suffering from the multipath fading includes various
information with regard to sea surface profile and ocean wave dynamics.
   In this paper, first, we will introduce multipath fading
phenomenon, the data of which were obtained by our field experiments.
Next we will make clear the relation between fading characteristics and
sea surface properties by using a rough surface scattering model,
especially, focusing our discussions on the fading depth versus the rms
slope of the sea surface, and the bandwidth of the fading spectrum
versus wave height.

2. SEA PARAMETERS EFFECTING ON MULTIPATH FADING

2.1. Multipath Fading due to Sea Surface Reflection

Multipath fading is caused by the interference between the direct wave
from the satellite and forward scattered wave at sea surfaces. In this
case, reflected waves coming from the sea surface are composed of a

Y. Toba and H. Mitsuyasu (eds.), The Ocean Surface, 329–334.
© 1985 by D. Reidel Publishing Company.

coherent component C (specular reflection component) whose phase varies
with the height of the antenna, and an incoherent component I (diffused
component) which fluctuates randomly with the motion of the sea waves.
The coherent component is predominant under calm sea conditions, whereas
the incoherent component is predominant under rough sea conditions.
      Amplitude and period of signal fluctuation mainly depend on
elevation angle of satellite, sea condition and antenna gain as well as
signal frequency and polarization. Fig. 1 shows recording examples of
multipath fading for different sea conditions. As can be seen in this
figure that the dependence of fading characteristics such as frequency
spectrum on sea condition is remarkable.

2.2. Rms Slope of the Sea Surface

Since a practical fading model applicable to L-band frequencies
concerning fading depth and spectrum was detailed by the authors
(Karasawa and Shiokawa, 1984a, 1984b), we present the brief results of
analysis in this section.
      Roughness parameter "u" indicating the roughness of the sea
surface is generally given by

$$u = \frac{4\pi}{\lambda} h_o \sin \theta_o \qquad (1)$$

where θ_o is elevation angle of satellite, λ is wavelength of radio
wave, and h_o represents the rms surface profile height of the sea.
This "h_o" approximately relates to the significant wave height "$H_{1/3}$"
by $H_{1/3} = 4h_o$.
      Considering that the ship antenna is fixed at a given height above
the mean sea level, amplitude of the coherent component is given by

$$C(u) = |R\ G_r|\ \exp(-u^2/2) \qquad (2)$$

where R is reflection coefficient of the sea and G_r is the antenna
gain toward the specular reflection point relative to that toward the
satellite. Eq. (2) indicates that the coherent component decreases
exponentially with respect to the square of the wave height. On the
other hand, the mean power of the incoherent component $<I^2>$ is
expressed by the function of both wave height "$H_{1/3}$" (or "u") and
the rms slope of the sea "β_o". Namely,

$$<I^2> = f(u,\ \beta_o) \qquad (3)$$

The rms slope "β_o" is the rms value of the slope of sea surfaces and
it is directly related to the wave steepness defined by the ratio of
wave height to wavelength of sea waves.
      When "u" grows up larger than 2, intensity of the coherent
component of the reflected waves decreases to less than -17 dB
theoretically {or -11 dB experimentally (Beard, 1961)} relative to
that of the coherent component at u=0. Therefore, effect of the
coherent component will be small for u≥2, and then dependence on "u"
for the incoherent component will be also small. In this case,

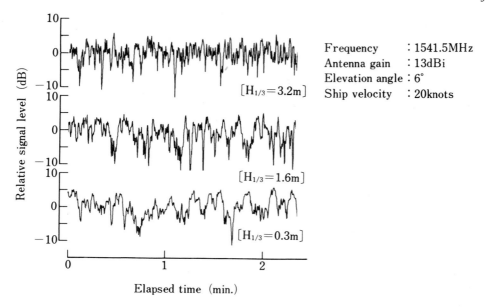

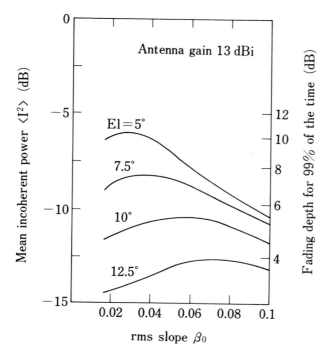

Figure 1.   Multipath fading due to sea surface reflection under three
            different sea conditions

Figure 2.   Relation between incoherent power $<I^2>$ and rms slope β_0

$$C \simeq 0, \ <I^2> = f(\beta_o) \qquad \text{for } u \gtrsim 2 \qquad (4)$$

From equations above, the significant wave height "$H_{1/3}$" corresponding to "$u \gtrsim 2$" will be higher than 1.4m (at El=5°) and 0.7m (at El=10°), respectively, at 1.5 GHz frequency.

According to the theoretical and experimental studies so far, the distribution of amplitude fluctuation follows the Nakagami–Rice distribution for the sea condition of $u \gtrsim 2$, therefore, the mean power of the incoherent component can be obtained accurately from measured fading depth defined by the signal level of the cumulative time distribution for a given percentage of the time, for example for 99 % of the time. Fig. 2 shows the relation between $<I^2>$ and β_o, based on the fading model (Karasawa and Shiokawa, 1984a), for antenna gain of 13 dBi (half power beam width : 40°) as a parameter of elevation angles. As can be seen in this figure, the lower the elevation angle becomes, the more remarkable the dependence of $<I^2>$ on β_o ranging from 0.03 to 0.1. From a view point of sea parameter measurements, it will be more appropriate by use of fading depth (or $<I^2>$) at elevation angle of 5° rather than 10°. In the case of elevation angle of 5°, $<I^2>$ decreases monotonically with increasing β_o between 0.3 to 0.1, the range of which may be probable for real ocean waves.

Fig. 3 shows a histogram of fading depth (which is defined by the signal level for 99 % of the cumulative time distribution) obtained by the on-board experiment. Fading depths obtained were ranging from 7 to 11 dB for $H_{1/3}$=1-2 m and from 6 to 9 dB for $H_{1/3}$=2-3 m, respectively. Corresponding rms slope "β_o" estimated by the fading model are also plotted in the lower abscissa. Results indicate that the rms slope for the wave height of 1-3 m, keeps the values ranging from 0.03 to 0.09. These values are not contradictory to our preliminary estimation, 0.04 – 0.07 (Karasawa and Shiokawa, 1984a), by use of ocean wave data (Hogben and Lumb, 1967). Since values of β_o of the real ocean were not measured in our experiment, it may be necessary to measure fading depth and slope parameter such as wave steepness simultaneously.

2.3. Wave Height

Dynamics of ocean waves are reflected in the frequency spectrum of signal level variation due to the multipath fading (hereafter, referred as fading spectrum). In this section, we consider the multipath fading due to sea surface reflection cause by the wind waves. We assume that the velocity "V" of each wave generated by wind is the sum of the velocity "V_o" of a stationally moving wave and the velocity "V_f" of randomly moving waves. Moreover, we assume that the power spectrum of ocean waves satisfy the "Pierson-Moskowitz power spectrum", and also each wave has the properties of the gravity wave. Under such assumptions, the fading spectrum P(f) can be expressed by using following parameters (Karasawa and Shiokawa, 1984b).

$$P(f) = P(f, \ H_{1/3}, \ \phi_w, \ V_s, \ \phi_s, \ \sigma_{roll/pitch})$$
$$\text{(for } u \gtrsim 2) \qquad (5)$$

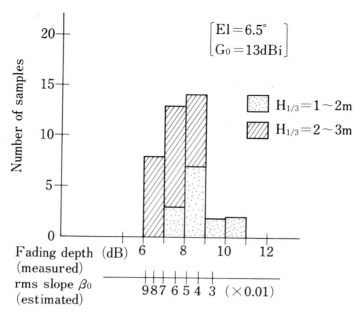

Figure 3.   Histogram of measured fading depth and estimated rms slope

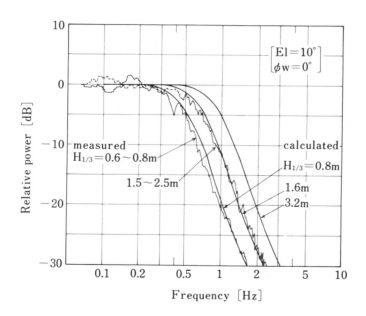

Figure 4.   Power spectra of multipath fading as a parameter of wave
     height (theoretical and experimental)

f : spectral frequency
ϕ_w : wave direction (same as wind direction)
V_s : ship speed
ϕ_s : ship heading
σroll/pitch : compensation factor for ship motion

As a matter of course, system parameters such as signal frequency and its polarization, and elevation angle are also important factors to be considered. Since "β_o" depndence on spectrum is small, this parameter is omitted in Eq. (5).

Fig. 4 shows the both calculated and experimental fading spectra as a parameter of the wave height under the condition of $\phi_w=0°$, El=10° and $V_s=0$, the case of which may correspond to the situation of experiments that the satellite signal is received at a coast, or on a fixed platform above sea. As can be seen from this figure, the bandwidth of the fading spectrum increases approximately in proportion to the square root of the wave height. Thus, it is possible to estimate the wave height from the fading spectrum.

3.   CONCLUDING REMARKS

Sea parameters effecting on L-band multipath fading due to sea surface reflection are discussed. Since the multipath fading is harmful effect in maritime satellite communications, the fading model mentioned in this paper was developed for the purpose of understanding the fading mechanism and reducing such propagation impairments. However, this type of measuring method and results may be helpful for studying the remote sensing of the ocean wave dinamics.

AKNOWLEDGEMENT.   The authors would like to express their sincere thanks to Dr. H. Kaji, Dr. K. Nosaka and Dr. A. Ogawa of KDD R & D Labs. for their contineous supports for this study.

REFERENCES

Beard, C.I., 1961: Coherent and incoherent scattering of microwaves from the ocean, IRE Trans. on Antenna and Propagation, vol. AP-9
Hogben, N. and F.E. Lumb, 1967: Ocean wave statistics, London Her Majesty's Stationally Office
Karasawa, Y. and T. Shiokawa, 1984a: Characteristics of L-band multipath fading due to sea surface reflection, IEEE, Trans. Antennas Propagat., vol. AP-32, no. 6, pp. 618-623
Karasawa, Y. and T. Shiokawa, 1984b: Spectrum of L-band multipath fading due to sea surface reflection, Trans. IECE of Japan, vol. J67-B, No.2, pp 171-178

EFFECT OF WAVE-CURRENT INTERACTION ON THE DETERMINATION OF VOLUME SCATTERING FUNCTION OF MICROWAVE AT SEA SURFACE

Y. Sugimori,[1] K. Akagi[2] and M. Ogihara[1]
1. Faculty of Marine Science & Technology, Tokai University
2. Nihon Electric Company

ABSTRACT. The microwave scatterometer onboard SEASAT satellite was used to determine the wind field over the world's oceans. The principle of measurement relies on the sensitivity of microwave radar backscatter to the capillary waves generated by surface wind. The purpose of this report is to estimate the magnitude of possible errors arising in the wind vector algorithm based on an empirical σ_0 model involving the two-dimensional wave spectrum $W(K_x, K_y)$ for the sea surface. The problems for determining the precise wind vector seem to depend on the knowledge of a realistic expression for $W(K_x, K_y)$ consistent with the mechanism of generation of wind waves. More specifically it will reguire : (1) the determination of an expression for the one-dimensional spectrum together with the availability of a Semi-Empirical or Pierson spectrum in the capillary wave range, (2) the establishment of a realistic two-dimensional wave dispersion law for the short gravity-capillary waves, (3) pertwbation of the wave pattern through wave-current interaction.

1.   INTRODUCTION

The method for determining sea surface wind with scatterometer is based principally on two kinds of physical processes (e.g. hydrodynamic and electromaguetie). The sea clutter, contributing to the radar cross section, is essentially an echo of the incident microwave due to the ocean waves generated by the sea surface wind. Hence, the concept of scatterometer to determine the surface wind is constituted by two different kinds of physical processes through the sea surface clutter. Firstly are the hydrodynamic processes responsible for the development of ocean waves. The ocean waves contributing to the microwave backscatter are in the high frequency region, in particular the capillary waves, 1-3cm wavelength, which are proportional to the momentum transfer from the atmosphere to the surface wind (stress). The empirical models for the generation on wind waves in the capillary region were established by Pierson (1976) and Fung (1982) by using the laboratory data of Mitsuyasu et al. and others, but these models were based on a one-dimensional frequency spectrum.

Y. Toba and H. Mitsuyasu (eds.), The Ocean Surface, 335–344.

Secondly are the electromagnetic scattering processes for the reflection of microwaves from the sea surface. In regard to these processes,
it should be resolved on whether their contribution to the normalized
radar cross section (NRCS) σ_0 is dependent on wind speed or not.

The microwave scattering from the sea surface is principalty contributed by two processes. One of the radar processes is specular scattering,
in which the backscatter cross section $\tilde{\sigma}_0$ for an incident angle θ is
given by the expression

$$\sigma_0(\theta) = \frac{|R(0)^2|}{S^2} \sec^4\theta \exp(-\tan^2\theta/S^2) \tag{1}$$

where S^2 is the mean-squared wave slope of the sea surface.

The other process which also plays an important role in the microwave
backscatter from the surface is Bragg resonant scattering. On the assumption of progressive surface waves of wave number Kw in the same direction
of the microwaves of wave number K, the resonance condition for Bragg
scattering is

$$Kw = 2K\sin\theta \tag{2}$$

Hence, the first order backscatter normalized radar cross section $\tilde{\sigma}_0$ can
be expressed by the equation

$$\sigma_0(\theta)_{ij} = 4\pi k^4 \cos^4\theta \left| Gij^{(1)}(\theta) \right|^2 \cdot W(Kx, Ky) \tag{3}$$

where $Gij(\theta)$ is a function of polarization, electric parameters of the
surface and angle of incidence. The difference of the above two models
can be demonstrated in Fig.1 as a function of the incident angle of the
microwaves at the surface (Valen zuela, 1978).

In order to improve the capability of scatterometers, it seems
necessary to estimate the error factors that are fundamentally involved
in the process of determination of the surface winds.
These are:
1). Error factors involved in the scattering process of microwaves at
the sea surface.
    (1) Estimation of the variability of the microwave conductivity by
the sea surface temperature.
    (2) Determination of the two-dimensional wave number spectrum $W(Kx,$
$Ky)$ in Eq.(3). It must be established on what features of $W(Kx, Ky)$ are
known to determine the wind speed. They include:
        (a) The two-dimensional dispersion of the wave spectrum.
        (b) Effect of the sea surface current on the wave spectrum in the
            capillary region.
        (c) Effect on wave-wave interaction on the wave spectrum in the
            capillary region.
2). Mechanism for the wind generated waves in the capillary region.
    (1) Temporal and fetch dependence of the wind generated waves.
    (2) Determination of the functional form of capillary waves spectrum.

2.   WAVE SPECTRUM IN THE CAPILLARY REGION

It is unfortunate that one-dimensional spectrum at the intermediate frequencies of Mitsuyasu-Honda and Pierson-Moskowitz spectra have not yet
been confirmed.

Nevertheless, recently a few spectral forms for the wave spectrum
have been proposed. Among them. Pierson (1976) has suggested

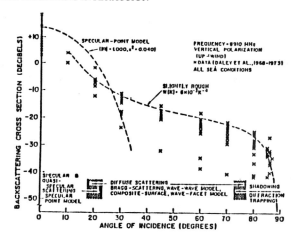

Fig.1 Illustration of mechanisms which predominate in the radar backscatter from the ocean at various angles of incidence.

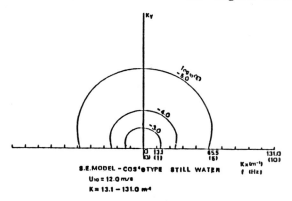

Fig.2 Two-dimensional wave spectrum on a still water containing the Semi-Empirical model and the angular spreading function $\cos^{2\ell}\theta$.

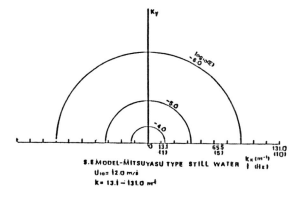

Fig.3 Two-dimensional wave spectrum on still water for the Semi-Empirical model and the angular spreading function of Mitsuyasu-type.

$$S(w) = (0.875/2\pi)(2\pi/w)^P \qquad (4)$$

where w is the radian frequency and $p = 5-\log_{10}(U_*)$. The friction velocity U_* is related to the wind speed U at altitude z above the mean sea
level by

$$U = \frac{U_*}{0.4}\ln(\frac{z}{Zo}) \qquad (5)$$

where $Zo=(0.684/U_*)+4.28\times10^{-5}U_*^2 - 0.0443$ (cm). In order to apply Eq.(4)
to wave number less than those of capillary waves, the complete dispersion relation must be used. That is $w^2 = gk(1+K^2/Km^2)$, $\qquad$ (6)

where $Km^2 = g\rho/z$ and z is the surface tension. The computed spectrum as
a function of wave number, Sc(K), is related to S(w) by the equation

$$Sc(K) = \text{Needs a factor} \frac{1}{R} \text{ for two-dimensional spectra}$$

$$= S(w)\ dw/dk = 0.4735(2\pi)^{P-1}\cdot(1+3K^2/Km^2)g^{(1-q)/2}$$

$$\cdot\left[K(1+K^2/Km^2)\right]^{-(p+1)/2} \qquad (7)$$

where K is in radians per centimeter. The empirical sea spectrum Sc(K)
proposed for radar scatter theory by Fung et al.(1982) is identical with
our Eq.(7), and is denoted as the Semi-Empirical model.

3. ANGULAR SPREADING OF THE SURFACE WAVES IN THE CAPILLARY WAVE REGION

The different models for the angular spreading of wind waves(Table 1)
seem to offer the possibility of estimating of backscatter cross section
σo through the two-dimensional wave number spectrum $W(K_x, K_y)$. There were
several investigations to derive the angular spreading function, $G(w,\theta)$,
and some of the representative models are:(1) $\cos^{2\ell}\theta$ type from Longuet-
Higgins et al.(1963) and Cartwright (1964), and (2) SWOP type by Cote et
al.(1962). Moreover, (3) Mitsuyasu et al.(1980) proposed an improved
type the measurements from the clover-leaf buoy, results which are
comparatively similar to the model of Longuet-Higgins et al., However,
it is unfortunate that no one has proposed angular spreading functions
for the capillary and short gravity waves regions or has confirmed the
applicability of the above mentioned types to waves in the high frequency
region.
        With an angular spreading function of Mitsuyasu type, the two-
dimensional spectrum, $D(f,\theta)$, can be expressed by the equation

$$D(f,\theta) = S(f)G(f,\theta) \qquad (8)$$

and the angular spreading function $G(f,\theta)$ can be represented by

$$G(f,\theta) = G'(\measuredangle)\cos^{2\measuredangle}\frac{\theta}{2} \qquad (9)$$

where $\qquad G'(\measuredangle) = \frac{1}{\pi}2^{2\measuredangle-1}\frac{\Gamma(\measuredangle+1)}{\Gamma(2\measuredangle+1)} \qquad (10)$

Further more, $D(f,\theta)$ can be expressed in the form

$$D(f,\theta)=S(f)\cos^{2\Lambda}\frac{\theta}{2}\frac{1}{\pi}2^{2\Lambda-1}\frac{\Gamma^2(\Lambda+1)}{\Gamma(2\Lambda+1)} \quad (11)$$

and Λ is a parameter to be called the concentration function. The concentration function Λ to be proposed for the high frequency region is as follows,

$$\Lambda = 0.30f^{-2.0} \quad (12)$$

In order to investigate the effect of the angular spreading factor of the ocean waves in the capillary region, a comparison of the different types of angular spreading functions $G(K,\theta)$ has been made using the two dimensional spectrum. The different angular functions, are used for the $\cos^{2\ell}\theta$ and Mitsuyasu type.

Some results of the calculation in the wave number plane (K_x, K_y) are shown in Figs.2 and 3 as a function of wind speed U_{10}. Finally, an evaluation of the backscatter cross section σ_0 for different models of angular spreading functions, is made by using Eq.(3). The result of the numerical calculation is presented in Fig. 4. It is clear that there are remarkable differences in the value of σ_0 as a function of wave number, K.

4. COUPLING WITH THE WIND WAVES AND UNDERCURRENT

Jones and Schrogder (1978) pointed out the importance of the coupling of waves with currents, resulting in energy transfer and deformation of wave shape, especially the steepness and amplitude.

In order to investigate the effect of the interaction between capillary waves and undercurrents on the Bragg backscattering of microwaves from the surface, Longuet-Higgins and Stewart's (1962) radiation stress will be used to develop a coupling model for capillary waves for numerical calculations, that can provide the backscatter cross section σ_0.
1) Single wave on a shearing current.
Waves propagating against a current with horizontal shear, develop amplitudes greater than those for still water. Longuet-Higgins and Stewart (1960) showed that the effect of shearing current on the wave amplitude is proportional to the so-called radiation stress and that the current does work on the waves at a rate $\frac{1}{2}$Sij$(\partial vi/\partial xj+\partial vj/\partial xi)$ per unit distance where the suffixes i and j (i,j=1,2) refer to rectangular coordinates arbitrarily chosen, Sij being the radiation stress.

The energy equation is generally written as

$$\frac{\partial E}{\partial t} + \nabla \cdot \left\{ E(\underline{C}g+\underline{V}) \right\} + \frac{1}{2} \cdot Sij\left(\frac{\partial Vi}{\partial xj} + \frac{\partial Vj}{\partial xi} \right) = 0. \quad (13)$$

On the assumptions that the current velocity has only y-component V, which is not variable in the y axis and depth, and that the apparent angular frequency is constant outside and inside the current, they obtained the equations

$$E\cos\theta \sin\theta = \text{const} \quad (14)$$

and

$$\frac{a}{a_0} = \left(\frac{E}{E_0}\right)^{\frac{1}{2}} = \left(\frac{\sin 2\theta_0}{\sin 2\theta}\right)^{\frac{1}{2}} , \quad (15)$$

where θ and θ_0 denote the angles between the direction of the current and waves, for current and still water respectively.

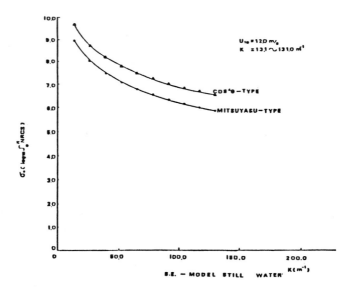

Fig.4   Difference of backscatter cross sections σ_0
        calculation by using a wave spectrum model
        containing the functions of angular spreading
        $\cos^{2\ell}\theta$ and Mitsuyasu-type.

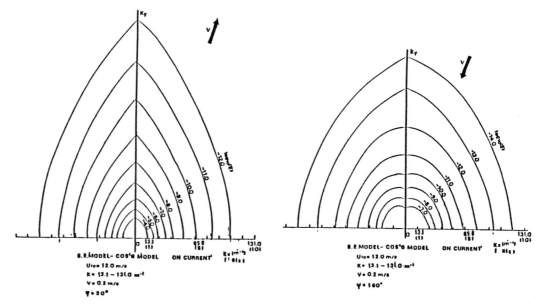

Fig.5   Two-dimensional wave spectrum
for wave in a favorable current. The
wave model is the Semi-Empirical
formula-$\cos^{2\ell}\theta$ type.

Fig.6   Two-dimensional wave spectrum
for waves in an opposing current.
The wave model is the Semi-Empirical
formula-$\cos^{2\ell}\theta$ type.

2) Random spectrum of waves on a shearing current.
When the local wave frequency is in site observed on a shearing current,
the energy ratio between the waves on a still water and on a shearing
current has been studied and developed by Sugimori (1976) as follows.

$$\frac{E(K,\beta)}{E(Ko,\beta o)} = \frac{1}{1+\sqrt{\frac{K}{g}}\ Vsin\beta} = 1-\sqrt{\frac{Ko}{g}}\ Vsin\beta o \qquad (16)$$

Equation (16) is defined as the spectral growth rate on a shearing
current. The wave energy will be reduced at this rate, when the incident
waves turn in the direction of the shearing current and will be
increased with same rate when the waves and shearing current are
orienting in opposite direction.
    A model for the directional wave spectrum for numerical calculation
is given by using the Semi-Empirical model to show the change of the
wave number spectrum in a current. First of all, the Semi-Empirical
spectrum in still water is reduces to

$$Eo(ko,\beta o) = 0.4735(2\pi)^{p-1}(1+3\frac{Ko^2}{Km^2})g^{(1-p)/2}\left[Ko(1+\frac{Ko^2}{Km^2})\right]^{-(p+1)/2}$$

$$\cdot \frac{1}{\pi}2^{2\lambda-1}\frac{\Gamma^2(\lambda+1)}{\Gamma(2\lambda+1)}\cdot \cos^{2\lambda}(\frac{\varphi}{2}-\frac{\beta o}{2}) \qquad (17)$$

The angular spreading function for the case of Mitsuyasu type is substituted in Eq.(17). After mathematical manipulations of Eq.(17) for still
water to current, the coupling model for the Semi-Empirical spectrum on
a current is given by

$$E(K,\beta) = 0.4735(2\pi)^{p-1}\left[1+3\frac{K^2}{Km^2}(1+\sqrt{\frac{K}{g}}Vsin\beta)^4\right]g^{(1-p)/2}$$

$$\cdot \left[K(1+\sqrt{\frac{K}{g}}\ Vsin\beta)^2\{1+\frac{K}{Km^2}\ (1+\sqrt{\frac{K}{g}}\ Vsin\beta)^4\}\right]^{-(p+1)/2}$$

$$\cdot \frac{1}{\pi^2}2^{2\lambda-1}\frac{\Gamma^2(\lambda+1)}{\Gamma(2\lambda+1)}\cos^{2\lambda}\left[\frac{\varphi}{2}-\frac{1}{2}\sin^{-1}\{\sin\beta\ (1+\sqrt{\frac{K}{g}}\ Vsin\beta)^{-2}\}\right]$$

$$\qquad (18)$$

The results of numerical calculation for the case of Semi-Empirical
model (Eq.(18)) combined with the angular spreading functions of
Mitsuyasu and $\cos^{2\ell}\beta$ types are shown in Figs.5,6,7. Fig.5 shows the
directional spectrum with $\cos^{2\ell}\beta$ in a favorable current and Fig.6 shows
the result for an opposing current. Fig.7 also illustrates the
directional spectrum combined with Mitsuyasu type angular spreading
function on an opposing current. One of the remarkable result is the
effect on the spectrum by favorable and opposing currents and this is
shown in Fig.8,
    At the final stage a tentative calculation to derive NRCS was made
using Eq.(3). Fig.9 shows the results of comparison of NRCS σo
dependent on various current systems and angular spreading functions of
the directional spectrum $\overline{W}(Kx,Ky)$.
    For the case of Mitsuyasu-type angular spreading function there is
a distinct σo pattern even when no current system is present. Meanwhile,
the value of NRCS σo derived with the model in combination with the
$\cos^{2\ell}\beta$ function, indicates a noticeable difference that is dependent on

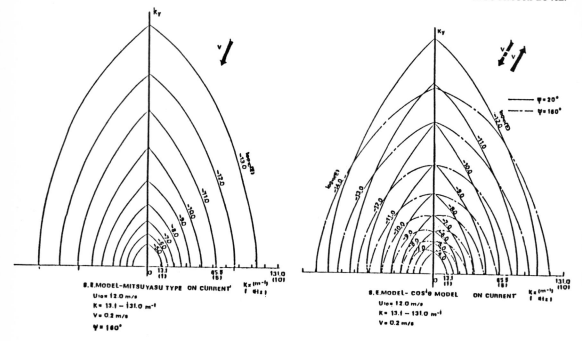

Fig.7 Two-dimensional wave spectrum
for waves in an opposing current.
The wave model is the Semi-Empirical
formula-Mitsuyasu type.

Fig.8 Summary and representative
illustration derived from the
directional wave spectrum on
two-sorts of current systems,
favorable and opposing currents.

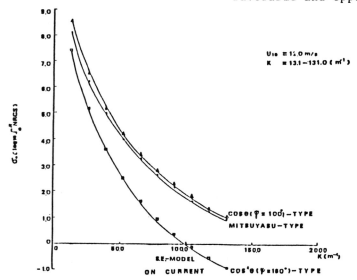

Fig.9 The backscatter cross section σ̄o calculation by using different
types of two-dimensional spectra on two-sorts of current systems,
favorable and opposing currents.

Table 1 Different types of angular spreading functions.

Case-	Type of Angular Spread	$A(\sigma)$	$S(k_1)$
1	Isotropic	$\delta(\sigma)$	$B/2\,k_1^{-3}$
2	Swop $A_0 + A_2 \cos^2\theta$	$\frac{1}{2}\delta(\sigma) + \delta(\sigma - 2)$	$\frac{5B}{8}\,k_1^{-3}$
3	$(\cos\theta)^2$	$2\delta(\sigma - 2)$	$\frac{3B}{4}\,k_1^{-3}$
4	$(\cos\theta)^4$	$\frac{8}{3}\delta(\sigma - 4)$	$\frac{5B}{6}\,k_1^{-3}$
5	$(\cos\theta)^\sigma$	$(\pi)^{1/2}\,\dfrac{\Gamma\left(\frac{S+2}{2}\right)}{\Gamma\left(\frac{S+1}{2}\right)}\delta(\sigma - S)$	$\left(\dfrac{S+1}{S+2}\right)Bk_1^{-3}$
6	Delta Function $\delta(\theta)$ Undirectional	$\displaystyle\lim_{S\to\infty}(\pi)^{1/2}\,\dfrac{\Gamma\left(\frac{S+2}{2}\right)}{\Gamma\left(\frac{S+1}{2}\right)}\delta(\sigma - S)$	$\displaystyle\lim_{S\to\infty}\dfrac{(S+1)}{(S+2)}Bk_1^{-3} = Bk_1^{-3}$

the current systems and the descrepancy of $\tilde{\sigma}o$ increases for larger wave numbers.

It is most important that the effect of undercurrent on waves should be taken into account for the determination of NRCS $\tilde{\sigma}o$, because in the deep ocean it is usual to encounter under currents larger than 20cm/sec.

5. CONCLUSION

The conclusions of this work involve several items:
(1) It is recognized that the directional spectrum for wind waves $W(K\varkappa, K_y)$ involves different types of angular spreading functions which impact greatly on the determination of NRCS $\tilde{\sigma}o$. On ther words, it is necessary to obtain a general empirical and suitable formula for the angular spreading function applying for the capillary region.
(2) The directional spectra combined with $\cos^{2\ell}\theta$ and Mitsuyasu types make some difference in spectral shape and magnitude in the large wave number region. Former spectrum shows a shape of ellipse and later give a concentric circle.
(3) The most important result of this study involves the coupling of waves and currents(especially for the large wave number region) because it provides a large effect on the directional spectrum $W(K\varkappa, K_y)$. It

means that one can not neglect the coupling of waves and currents in the determination of the NRCS σ_0. Largest changes in σ_0 is produced by opposing current systems.
(4) There are no large changes for the coupling of waves and currents when using the directional wave spectrum in combination with Mitsuyasu angular spreading function. It might be reasonable to assume that the angular spreading function of Mitsuyasu-type is not useful for the region of short gravity to capillary waves.

The present angular functions were derived from wavetanks and field experiments using buoys and wave gages that were available to measure the directional spectra only in the gravity wave range. It seems extreemly necessary to establish the angular function in the short gravity to capillary waves that may be suitable for the investigation of coupling process of microwave and sea waves.

REFERENCES

Cartwright,E., and Smith,D.(1964): Buoy Technology, Washington, M.T.S..
Cote,J.,J.Dravis, W.Marks,R.Mcgough,E.Mehr,W.Pierson,J.Ropek,g.
        Stephenson, and R.Vetter(1960): The directional spectrum of a
        wind generated sea as determined from data obtained by the
        Stereo Wave Observation Project. Meteorol. Papers. N.Y.U., Coll.
        of Eng.Vol.2, No.6 PP88.
Fung,A.K. and K.Lee(1982): A semi-empirical sea-spectrum model for
        scattering coefficient estimation, IEEE Jour. of Oceanic Engi.,
        Vol.OE-7, No.4, PP166-176.
Jones,L. and L.Schroeder(1978): Radar backscatter from the ocean;
        Dependence on surface friction velocity, Boundary Layer
        Meteorology, Vol.13, No.1, PP133-149.
Longuet-Higgins,S. and R.Stewart(1962): Radiation stress and mass
        transport in gravity waves with application to "surf-keats",
        Jour. of Fluid Mech., Vol.13, PP481-504.
Mitsuyasu,H.,F.Tasai,T.Suhara,S.Mizuno,M.Ohkusu,T.Honda and K.Rikiishi
        (1980): Observation of the power spectrum of ocean waves using a
        cloverleaf buoy, J. Phys. Oceanogr.,Vol.10,PP286-296.
Pierson,W.(1976): The theory and applications of ocean wave measuring
        systems at and below the sea surface, on the land, from aircraft,
        and from spacecraft, NASA Contract Rep., CR-2646, N76-17775.
Sugimori,Y.(1976): Two-dimensional surface wave on the shearing current,
        Report of the National Research Center for Disaster Prevension,
        No.13, PP75-86.
Valenzuela,G.(1978): Theories for the interaction of electromagnetic
        and oceanic wave-a review, Boundary-Layer Meteorology, Vol.13,
        PP61-85.

SURFACE OCEAN CIRCULATION AND VARIABILITY DETERMINED FROM SATELLITE
ALTIMETRY

Richard Coleman
Ocean Sciences Institute, Sydney University
Sydney,   N.S.W.   2006. Australia

ABSTRACT. Measurements of the elevation of the ocean surface,
obtained from radar altimeter data, are emerging as a powerful
data set available to oceanographers for enhancing their knowledge
of the oceans. Data (from the GEOS-3 and SEASAT satellites) obtained
over a period of 3.6 years from April, 1975 to November, 1978, were
used to study the spatial distribution of mesoscale sea surface
variability in the Tasman Sea. Satellite data generally agreed with
existing hydrographic measurements. Patterns of higher sea surface
variability (up to 64 cm) were shown to be associated with the East
Australian current and eddy areas. The low sea surface variability
levels in the mid-Tasman Sea suggest some sort of topographic
influence by the western edge of the Lord Howe Rise. No evidence
for a permanent surface circulation front across the Tasman Sea was
found. Altimetric eddy kinetic energy values calculated in the
region are of the same order as those obtained from cruise results.

1. BACKGROUND

The large scale hydrographic features of the Tasman Sea region are
already known. An excellent review of our existing knowledge of the
ocean circulation within this region, both observation and theory, is
given by Bennett (1983). However, to date there exists only meagre
information about the mid-Tasman circulation and its mesoscale
variability.
      The basic picture that emerges from past surveys of the ocean
circulation within the region shows the East Australian Current (EAC)
flowing southwards along the Australian coastline from about 27°S until
about 33°S. Around this position, the current separates from the coast
and turns in an easterly direction. Warm-core eddies (anti cyclonic)
are formed from the pinch-off of meanders of the EAC at the rate of
about 2-4 per year and drift generally south westwards. After leaving
the Australian coast, the EAC is thought to meander eastward to north
of New Zealand as a zonal front. The observational evidence for this
zonal front, referred to as the Tasman Front, has predominantly come

345

from ship cruises made during the winter season (August–October).
There is however, some conjecture as to the existence of this mid-
Tasman circulation (Heath, 1980).

2.   ALTIMETER DATA

The radar altimeter measures the height of the spacecraft above the
instantaneous ocean surface, of order 800 km, based on the travel time
of short pulse, microwave signals. This measurement is an average
height over the area covered by the radar footprint(about 3–12 km
depending on sea state). The orbital geometry of the altimeter
measurement relates the geoid height (N) and the height of the sea
surface (Z) above a reference ellipsoid (both N and Z of order ± 100 m),
and the dynamic sea surface height (Z_S) by the relation: $Z_S = Z - N$.
Z_S, of order ± 1 – 2 m, contains information on ocean dynamic
phenomena such as ocean currents, eddies, tides, waves, etc.
        To obtain the oceanographic signal from the altimeter measurement,
a precise (10 cm or less) knowledge is required of (i) the orbit of
the satellite and (ii) the shape of the geoid. However, dominant
error sources above the 10 cm level come from geoidal and orbital
contributions and to a lesser extent from environmental phenomena
(sea state, tides, atmospheric pressure gradients) – see Wunsch and
Gaposchkin (1980).
        The horizontal gradients of Z_S have a dominant influence on the
geostrophic circulation of the surface layer of the oceans. But no
independently determined geoid exists with sufficient resolution for
large scale, geostrophic circulation to be defined except for wavelengths greater than 8,000 – 10,000 km. Some areas, such as the
western north Atlantic, have however, regional geoids with adequate
resolution (30 – 50 cm) at mesoscale wavelengths, for relevant studies
to be made (e.g. Cheney & Marsh 1981, Kao & Cheney 1982). There are
no geoid models of similar resolution for the Tasman Sea.
        However, the time-variable component of the ocean surface
circulation can be determined without a precise knowledge of the
geoid. This is because the influence of the dominant error sources
can be effectively removed by subtracting a first degree polynomial
fit to the altimeter data of each satellite pass. But it should be
remembered that any long wavelength oceanographic signal will also be
filtered out by this detrending. Using data from both the GEOS-3
(April 1975 to November 1978) and SEASAT (June 28 to October 10, 1978)
satellites, the mesoscale variability in the Tasman Sea is determined
and compared to hydrographic measurements. In all there are some
403 altimeter passes within the area 40° to 15°S, and 150° to 175°E
covering the period April 1975 to November 1978.

3.   METHODS OF ANALYSIS

The geoid-independent analysis techniques most frequently used fall
into three basic categories. These are (i) crossover differences,

(ii) overlapping profiles, and (iii) mean ocean surface models.
Each analysis technique has advantages and disadvantages. Techniques
(i) and (ii) allow simple removal of the large and constant geoid
signal but have shortcomings since they use only a sub-set of the
available altimeter data, and are limited in their spatial resolution
and temporal sampling rate. Technique (iii) uses all available data
in the area of study but is computationally more complex. Full
details of the analysis procedures are described in Coleman (1984a)
and references quoted therein.

4.   RESULTS

Only results from Techniques (i) and (ii) are presented below. A
more detailed treatment of this study can be found in Coleman (1984a).

4.1 Crossover Differences

Fig. 1 summarises the 3.6 year spatial distribution of mesoscale
variability obtained from the 1° x 1° averaged crossover differences.
Some 11,028 crossover differences were computed from the 403 satellite
passes, with time separations between crossovers of from 1 to 1260
days. The average number of crossover differences per 1° square was
18 with the range being from 1 to 192. The noise level of the
analysis is about ±20 cm. Also indicated in Fig. 1 are sections of
the 180 m (dashed line) and 1800 m (dotted line) bathymetry contours.
These contours approximate the continental shelf edge and the general
area of the Lord Howe Rise/New Caledonia Basin respectively. Squares
containing less than 6 crossovers were not used in analysis and these
areas are indicated by dots.
     The EAC system and its eddy energetic areas, as well as the
general low variability in mid-ocean areas, is evident in the figure.
The northeast trending tongue of rms fluctuation at about 28°S, 156°E
is largely associated with EAC variability and inflow into the Coral
Sea. The 30 cm variability contour in this area and further south to
40°S closely follows the 1800 m depth contour associated with the
western edge of the Lord Howe Rise suggesting some sort of topographic
influence. No large rms values appear to the east of this contour
(162°E) until the New Zealand mainland is approached. Significant
variability is absent in the region of expected Tasman Front activity
and it is concluded that the front is not a permanent feature across
the Tasman Sea.

4.2 Overlapping Profiles

Data from the 3 day repeating groundtracks of the SEASAT satellite
were analysed. However, the number of profiles per set (an average
of 4 passes) and their distribution in both time and space is far
from optimal. The mean profile of a set is at best a 1 month mean
estimate and only significate oceanic features would contribute to
the variability estimates.

    The computed rms variability estimates were averaged into 1/2°
bins along the mean profile (about 10 data points per bin). In general
the profile displayed high variability in the meandering flow regions
in the Tasman Sea and a lower variability of 3-6 cm in areas away from
any known current activity. Figure 2 illustrates the rms variability
along an overlapping profile set as a function of position. Higher
rms values occur at the boundary regions (as expected) of the Tasman
Front defined by the cruise data of Andrews et al. (1980). Also
indicated is the corresponding eddy kinetic energy (EKE) estimates
computed using the procedures set out in Menard (1983). These EKE
values are of the same order as values computed by Wyrtki et al.
(1976) using ship drift records.

    Fig. 3 displays the EKE distribution of the Tasman Sea computed
using the reliable overlapping profile sets (> 4 passes per set) and
averaged into 5° x 5° squares. These estimates are based on data
taken over about a 1 month period. The Wyrtki et al. values, shown
in the upper left hand corner of each square, are derived from
merchant ship data collected over a 70 year period. Strictly speaking
the results give entirely different temporal estimates but they do
yield the same order of magnitude values. It is considered that more
reliable global and regional EKE estimates can be computed using
crossover differences of the presently available altimeter data (see
Coleman (1984b)).

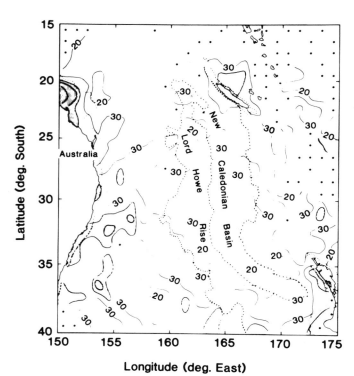

Figure 1 Mesoscale
Variability in the
Tasman Sea (1975-1978)
. Contours are in cm.
The shaded area is
variability greater
than 40 cm. The
dashed line is 180 m
bathymetry contour and
the dotted line is
1800 m bathymetry
contour. Dotted areas
refer to insufficient
data squares.

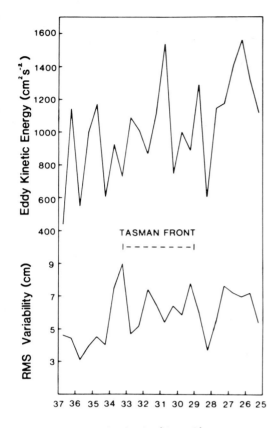

Figure 2 RMS Variability and Eddy Kinetic Energy of an overlapping Profile Set.

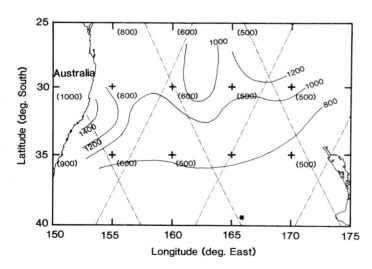

Figure 3 Altimeter Eddy Kinetic Energy Estimates. Contours are in cm² s². Values in brackets are from Wyrtki et al. (1976). The altimeter groundtracks are indicated with the profile of Fig. 2 denoted by a cross.

5. DISCUSSION

The above results demonstrate some applications of altimeter data for
oceanographic studies of the variability of current and eddy fields.
It is perhaps important to add that these remotely sensed data will
not be the panacea for oceanographers. The data will, however, provide
much needed additional and complementary information on ocean surface
phenomena over a wide range of time and space scales previously either
too slow, too expensive or impossible to measure using existing
techniques. It is hoped oceanographers will give support to future
missions such as TOPEX and ERS-1 due for launch about 1988.

REFERENCES

Andrews, J.C., M.W. Lawrence & C.S. Nilsson, 1980. Observations of the
    Tasman Front. J. Phys. Oceanogr., 10, 1859-1869.

Bennett, A.F., 1983. The South Pacific Including the East Australian
    Current. Eddies in Marine Science, Robinson, A.R., Ed. Springer-
    Verlag, Berlin, 219-244.

Cheney, R.E. & J.G. Marsh, 1981. SEASAT Altimeter Observations of
    Dynamic Topography in the Gulf Stream Region. J. Geophys. Res.,
    86, 473-483.

Coleman, R., 1984a. Investigations of the Tasman Sea Using Satellite
    Altimetry . Aust. J. Mar. Freshwater Res., 35, 619-634.

Coleman, R., 1984b. A comparison of Eddy Kinetic Energy Estimates in
    the Tasman Sea. (In preparation).

Heath, R.A., 1980. Eastwards oceanic flow past northern New Zealand.
    N.Z. J. Mar. Freshwater Res., 14, 169-182.

Kao, T.W. & R.E. Cheney, 1982. The Gulf Stream Front : A comparison
    between SEASAT altimeter observations and theory. J. Geophys.
    Res., 87, 539-545.

Menard, Y., 1983. Observations of Eddy Fields in the Northwest Atlantic
    and Northwest Pacific by SEASAT Altimeter Data. J. Geophys. Res.,
    88, 1853-1866.

Wunsch, C. & E.M. Gaposchkin, 1980. On using satellite altimetry to
    determine the general circulation of the oceans with application
    to geoid improvement. Rev. Geophys. Space Phys., 18, 725-745.

Wyrtki, K., L. Magaard, & D. Hager, 1976. Eddy Energy in the Oceans.
    J. Geophys. Res., 81, 2641-2646.

APPLICATION OF THE SEASAT ALTIMETER DATA FOR ESTIMATIONS OF SEA
SURFACE HEIGHT AND OCEAN TIDE IN THE NORTHWEST PACIFIC OCEAN

K. Sato[1], M. Ooe[1] and T. Teramoto[2]
1 International Latitude Observatory, Mizusawa 023 Japan
2 Ocean Research Institute, Univ. of Tokyo, Tokyo 164 Japan

ABSTRACT.   For the study of effects of oceanic load upon the solid
Earth, the SEASAT altimeter data were analyzed to estimate variations
of the sea surface height including the ocean tide. Instrumental error,
orbital ephemeris and atmospheric effects such as ionospheric ones and
dry and wet tropospheric ones were corrected by using the data given by
NASA. Meteorological perturbations of the sea surface height were also
corrected. Constrained least square equations were solved for analyses
of aliased tidal effects counting for major tidal constituents.
Discussed are uncertainties of the current ocean tide models, which
induce residual signals in the altimeter data, and possibilities for
improving the tidal models in active areas of the ocean.

1.   INTRODUCTION

     To observe the solid Earth tides precisely, the ocean tide must be
well determined, especially for observation sites placed near coast
such as those in Japan, because it affects upon the Earth's crust by
its load in proportion to the sea surface height (for an example Sato
et al., 1983). Besides, the ocean tide is coupled with the Earth's
rotation by changing angular momentum of the Earth and by acting
frictional force on sea bottom. Thus, according to progress in
researches of the solid Earth tides and the Earth's rotation, it became
very important to know the ocean tide more precisely and to build up
its detailed model. Past observations of the ocean tides, however,
were not sufficient for these purposes. This is due to technical
difficulties and to cost for observations in deep sea areas. Even the
best ocean tide model at present developed by Schwiderski (1980, 1981)
needs further improvement for studying details of the Earth's
deformation and rotation. Usually, for observations of the ocean tide,
tidal gauges installed along coastal lines or sea-bottom pressure
gauges set on continental shelves are used. But it is difficult to
increase observation sites rapidly because it is expensive to deploy
them in effective areas.
     On the other hand, recently, experiments were performed to measure

351

Y. Toba and H. Mitsuyasu (eds.), The Ocean Surface, 351–356.

directly global ocean tides including deep sea areas by satellite
altimetry. The altimetries with the GEOS-3 and with the SEASAT were
fluitfully accomplished and their data were analyzed by a lot of authors
(Cartwright and Alcock,1981 ; Brown and Hutchinson,1981 ; Ganeko,1982 ;
Segawa and Asaoka,1982).

     In this paper, we attempt to estimate the ocean tide, paying
attention to the variations of the sea surface height in the Northwest
Pacific Ocean including the sea of Japan by using the SEASAT altimeter
data. We compare obtained results with Schwiderski's and Nishida's
(1980) ocean tide maps.

2. METHOD OF ANALYSIS

     The SEASAT orbit is nearly circular with a period of 101 minutes
and with an inclination angle of 108 degree, covering 95% of the area of
the global ocean. We used the data at crossovers during the 'fixed
orbit' period between Sept. 13 and Oct. 9, 1978. Figure 1 shows the
SEASAT orbit around Japan during the period. All the passes shown in
the figure are overlapping several times. The numbers of revolutions
and of data at crossovers used for analysis were 139 and 374,
respectively. Examples of raw sea surface profiles projected in the
north-south direction are shown in Figure 2. These profiles cross the
Izu-Mariana trench and/or the Kurile-Kamchatka trench.

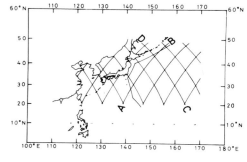

Figure 1. The SEASAT ground tracks
       around Japan during the
       'fixed orbit' period between
       Sept. 13 and Oct. 9, 1978.
       All of passes shown in the
       figure are overlapping
       several times. Coastal lines
       and trench axes are also
       illustrated schematically.

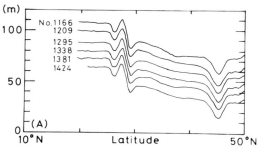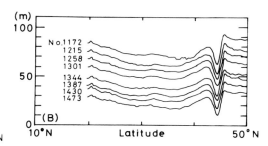

Figure 2. Raw sea surface profiles from altimetry projected in the
       direction of north-south. (A) and (B) correspond to the profiles
       along the tracks 'AB' and 'CD', respectively, shown in Figure 1,
       with artificially vertical spacing.

Instrumental error, orbital ephemeris and atmospheric effects such as ionospheric ones and dry and wet tropospheric ones were corrected by using the corresponding data given by NASA. Also, changes of the sea surface height caused by low and high atmospheric pressures and waves were corrected. Finally, corrections for the solid Earth tides and the geoid, say, the GEM10B (Lerch et al., 1981) were performed. After these corrections were applied, abnormal data were rejected with a statistical criterion.

We denote the corrected sea surface height by Z. Then, observational equations are set up as follows.

$$Z = G + S + W + O + E \qquad\qquad (1)$$

where E is unmodeled random error and the other parameters on the right are variables to be solved, namely : G, remaining geoid term ; S, steric variation of the ocean ; W, ocean tide ; and O, remaining error of orbital ephemeris. These equations were solved with several constraints that the sum of G all over the concerned region is nearly equal to zero, S and O could be modeled by low order polynomials and/or harmonics, W is smooth spatially, and O varies gradually in time. In practice, S was approximated by linear function of time at each crossover point and O was expressed by a constant for each pass, because the number of data was not large enough.

3. RESULTS OF ANALYSIS

Effect of each term in equation (1) is ensured by computing standard deviations (SD) of residuals. The SD of original data is about 7 meters. The SD is reduced to 2.7 m and to 0.8 m after solving corrections for the geoid and after solving additional corrections for the orbit of the satellite, respectively. Further, this SD is reduced to 0.3 m after solving for all parameters in equation (1).

Obtained corrections for the geoid (GEM10B) are illustrated in Figure 3. The figure shows that the altimetry geoid differs from the GEM10B by 11 m at the maximum around the Izu-Mariana trench. Error of orbital ephemeris was solved for each arc along the ground tracks. The corrections attain several meters along these tracks.

Estimation of the tidal effects is the most interesting but difficult problem in the present analysis because the period of the observation is very short and variations due to the other error sources are considered to exist in the frequency band close to the ocean tide. Moreover, aliasing periods are longer than 10 days, for instance, 16.24 days for M2 wave. Therefore, we estimated only a major tidal constituent, M2, after correcting for 7 tidal constituents which are S2, N2, K2, K1, O1, P1 and Q1 by using the Schwiderski's and/or Nishida's ocean tide maps. Obtained results for M2 are shown in Figures 4 and 5, where amplitudes and phase angles are given at each crossover point. The phase angle is defined as the Greenwich phase and a positive sign corresponds to the phase delay. For a comparison, the tidal constants of M2 in the Schwiderski's or Nishida's map are shown by contour lines.

Differences in amplitude from the Schwiderski's or Nishida's map attain 20 cm or more at some crossover points.

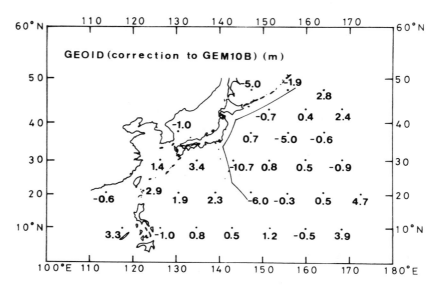

Figure 3. Altimetric geoid obtained by the analysis. The values are given as the corrections to the GEM10B at each crossover point.

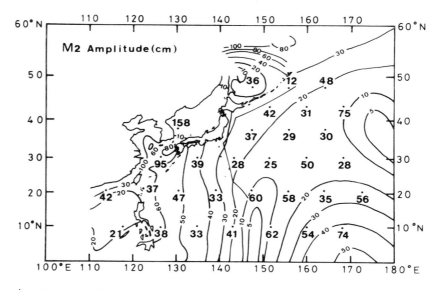

Figure 4. Altimetric M2 tide solutions. Amplitudes obtained at each crossover point are given and those in the Schwiderski's or Nishida's map are also shown by contour lines.

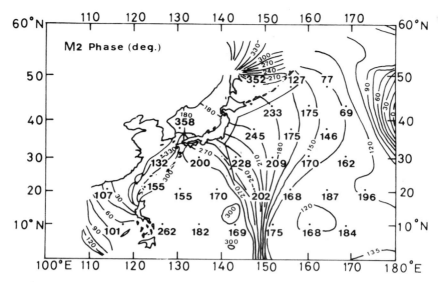

Figure 5. Altimetric M2 tide solutions. Phase angles obtained at each
    crossover point are given and those in the Schwiderski's or
    Nishida's map are also shown by contour lines, where phase angle
    is defined as the Greenwich phase and the positive sign
    corresponds to the phase delay.

4. DISCUSSION

The profile of M2 ocean tide in the Northwest Pacific Ocean was
obtained using the SEASAT altimeter data, although the period of data
used was only 27 days. The precision of estimates could not be enough
to improve the Schwiderski's ocean tide map, because the error of
estimates amounts to about 30 cm. This seems to be due to effects of
ocean currents around Japan, ambiguities in correction of other tidal
constituents and in the orbital ephemeris. The effects of above terms,
however, could be separated with enough accuracy, if data of long period
was used. In the case of short period data, optimum choice of
parameters of models and constraints for them are very important. A
Bayesian procedure will be effective for this purpose as shown by
Ishiguro et al. (1983) in analysis of Earth tides data using 'ABIC (A
Bayesian Information Criterion)'. An application of 'ABIC' to the
present subject is now under extended study.

ACKNOWLEDGEMENTS

The authors wish to express their gratitudes to the Remote Sensing
Technology Center of Japan for kind managements for providing the SEASAT
data of NASA, and to Dr. R. Coleman of Univ. of Sydney for useful advice.

REFERENCES

Brown, R. D. and M. K. Hutchinson, 1981: Ocean tide determination from satellite altimetry. *Oceanography from Space,* Plenum, New York, 897–906.

Cartwright, D. E. and G. A. Alcock, 1981: On the precision of sea surface elevations and slopes from SEASAT altimetry of the Northeast Atlantic Ocean. *Oceanography from Space,* Plenum, New York, 885–895.

Ganeko, Y., 1982: A 10' x 10' detailed gravimetric geoid around Japan. *Marine Geodesy,* 7, 291–314.

Ishiguro, M., H. Akaike, M. Ooe and S. Nakai, 1983: A Bayesian approach to the analysis of Earth tides. *Proc. of the Ninth Int. Symp. on Earth Tides,* 283–292.

Lerch, F. J., B. H. Putney, C. A. Wagner and S. M. Klosko, 1981: Goddard Earth models for oceanographic applications (GEM10B and 10C). *Marine Geodesy,* 5, 145–187.

Nishida, H., 1980: Improved tidal charts for the western part of the North Pacific Ocean. *Rep. Hydrogr. Res.,* 15, 55–69.

Sato, T., M. Ooe, and N. Sato, 1983: Tidal tilt and strain measurements and analyses at the Esashi Earth Tides Station. *Proc. of the Ninth Int. Symp on Earth Tides,* 223–237.

Schwiderski, E. W., 1980: On charting global ocean tides. *Rev. Geophys. Space Phys.,* 18, 243–268.

Schwiderski, E. W., 1981: NSWC global ocean tide data tape.

Segawa J. and T. Asaoka, 1982: Reevaluation of geoid based on the SEASAT altimeter data – Geoid around Antarctica –. *J. Geodetic Soc. Japan* , 28, 162–171.

MIROS - A MICROWAVE REMOTE SENSOR FOR THE OCEAN SURFACE

Ø. Grønlie, D.C. Brodtkorb, J.S. Wøien
A/S Informasjonskontroll
P.O. Box 265
N-1371 Asker
Norway

EXTENDED ABSTRACT [1]

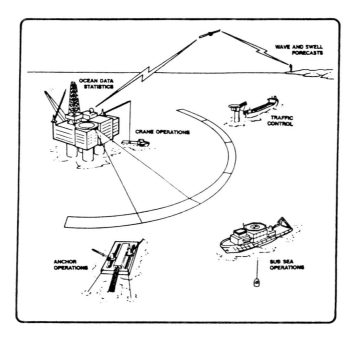

Figure 1.   Typical MIROS applications

[1] The full paper is published in the Norwegian Maritime Research No. 3,
Volume 12, 1984. The paper has also been presented at the "Symposium
on Description and Modelling of Directional Seas", June 18-20 in
Copenhagen, and at the WMO Technical Conference TECEMO", September 24-28
1984 in the Netherlands.

Y. Toba and H. Mitsuyasu (eds.), The Ocean Surface, 357–360.

INTRODUCTION

MIROS is a microwave radar system for real time directional measurements
of water particle speed, ocean waveheight frequency spectra and surface
current. This paper gives a system description, including both hardware and software. The theories behind the principles of operation are
also reviewed. MIROS has been subjected to extensive field trials, and
a selection of data collected during these trials are presented and
compared to data collected by conventional instruments. The field
trials has demonstrated that microwave remote sensing, as implemented
in MIROS is a realistic alternative to conventional techniques. MIROS
is now commercially available.

SYSTEM DESCRIPTION

MIROS is primarily designed for operation on stationary offshore
installations, see figure 1.
      The MIROS hardware consists of an outdoor mounted microwave sensor
head connected by a cable assembly to an in-door installed central
processor cabinet. The sensor head is to be installed on top of the
platform building section typically 50 to 100 metres above the sea
level.
      The MIROS software is written in extended PASCAL, with a few time-
critical operations implemented in assembly language.
      The directional properties of MIROS depends on the actual geometry
and the antenna beam-width in azimuth. The typical geometry gives 30°
directional resolution for wind waves up to 200 m wavelength. The
resolution is gradually decreasing for longer wind waves.
      During a typical measurement sequence, observations are taken in 6
directions with an angular increment of 30°. This represents a 180°
rotation of the antennas assembly, which is sufficient for a complete
directional measurement since the system observes both approaching and
receding waves.
      A typical observation time per direction is 12 minutes,
corresponding to 72 minutes for a complete directional scan.
      MIROS operates in two different modes, the pulse-doppler mode and
the dual frequency mode. Directional waveheight-frequency spectra are
measured in the pulse-doppler mode, basically obtaining water particle
velocity information. A wave-model is applied to transform the
velocity spectrum to a waveheight spectrum.
      Surface current information is obtained by measuring the phase
velocity of a gravity wave component in the dual frequency mode.

DATA COMPARISON

In the following we present a selection of MIROS data collected from
Lindesnes lighthouse, located at the most southern point of Norway,
during a field test in January - February 1983.

Nondirectional and directional wave parameters are compared with the
corresponding data from a heave, pitch and roll buoy (NORWAVE, ODAS
494).   Surface current estimates are compared with data collected by an
accoustic current meter (UCM-6).   Both instruments provided by the
Continental Shelf Institute, Norway.

A typical MIROS data output is shown in figure 2.

Figure 2.   Typical directional
frequency spectrum information
collected by MIROS.

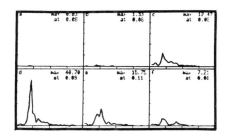

The significant waveheight (HMO) estimated by MIROS is based on
integration of the heave spectrum.   A regression line relating the HMO
for MIROS and the buoy is shown in figure 3.
      The spectral peak periods recorded by MIROS and the buoy are also
compared.   A regression line is shown in figure 4 and displays high
correlation.
      The temporal progression of the spectral peak direction observed
by MIROS and the buoy during 30 January and 5 February 1983 is shown
in figure 5.   The wind direction measured by the buoy is also shown.
      Figure 6 compares the MIROS current measurements with the component
of the current vector measured by the accoustic current meter in the
direction of the radar.   This data was collected 29 January 1983.   The
correspondance is good.   As expected, the radar indicates higer current
speed at the surface than the current meter measuring 2 meters below.

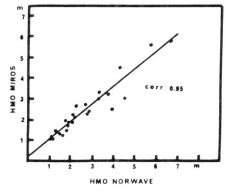

Figure 3.   Relation between
significant waveheights.

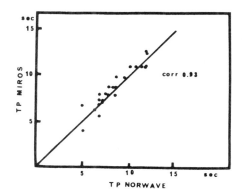

Figure 4.   Relation between
spectral peak periods.

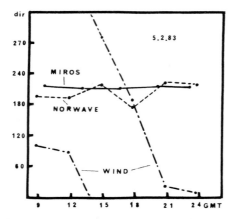 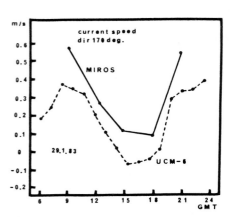

Figure 5.   Spectral peak direction.   Figure 6.   Current speed.

ACKNOWLEDGEMENT

The authors would like to acknowledge the valuable input to the MIROS
project given by Dr. D. Gjessing and his team with the Remote Sensing
Technology Programme, Royal Norwegian Council for Scientific and
Industrial Research.   The MIROS project was initiated at A/S
Informasjonskontroll on the basis of their studies.   We also would like
to thank Mr. L. Staveland, Statoil A/S, for his valuable contribution
to the system specification and design.

THE DESIGN OF SPACEBORNE MICROWAVE SCATTEROMETER

Hiroyoshi YAMADA[1], Masanobu SHIMADA[1]
Michimasa KONDO[2], Tetsuo KIRIMOTO[2]
1 National Space Development Agency of Japan Tsukuba
  Space Center, Ibaragi 305 Japan
2 Mitsubishi Electric Corp. Information Systems &
  Electronics Development Lab., Kanagawa 247 Japan

ABSTRACT. The basic design of the microwave scatterometer (SCAT)
planned in Japan is presented. The scheme has been encouraged by the
success of Seasat-A satellite scatterometer (SASS). While SCAT is basically fan beam system like SASS, characterized by 3 directional beams
and an observation cell composed of multiple unit cells. As a result of
the system design, it is clear that SCAT would be able to measure sea-
surface wind vector with high accuracy. Furthermore, it can be expected
SCAT has flexibility in measurement corresponding to various field conditions.

1. INTRODUCTION

It was almost verified a fan beam type microwave scatterometer to be a
practical global seasurface wind vector sensor by the success of
Seasat-A satellite scatterometer (SASS)[1]. There is a scheme to
develop a space-borne microwave scatterometer (SCAT) in Japan[2].
Present status of the scheme is on the way of physical evaluation using
bread board model (BBM)[3]. Some trade-off studies of SCAT in design
stage have been done. It was considered even pencil beam system with
conical scan as one of candidate systems, however it has been resulted
finally the most important thing in present stage is to improve fan beam
system followed SASS. There would be some improvement factor in fan
beam system; alias removal, reduction of measurement error etc.
    This paper describes the basic design process and its results of
SCAT. SCAT is an improved fan beam system and characterized by follows;
in order for alias removal, the third beam is added, and for reduction
of measured error, a concept of unit cell is proposed. An observation
cell is composed of multiple unit cells. It is also expected flexible
observation cell configuration due to changing combination of unit
cells.
    The least performance verification in design stage is shown by
means of computer simulation. The verification using BBM will be shown
another paper[3] in this symposium record.

Y. Toba and H. Mitsuyasu (eds.), The Ocean Surface, 361–368.
© *1985 by D. Reidel Publishing Company.*

2. BASIC CONFIGURATION OF SCAT

Basic Configuration of SCAT, which is a three directional fan beam system with HH and VV mode, was decided by the results of computer simulation on the wind measurement accuracy and feasibility study of hardware.

System requirements for SCAT are shown by Table 1, and it is same as one to SASS. However improvement of measurement accuracy under the various field condition is considered.

Tab. 1   SYSTEM REQUIREMENTS

RANGE OF WIND SPEED	4 - 25 m/s	WIND DIRECTION RANGE	0 - 360°
ACCURACY OF WIND SPEED	±2 m/s or 10 (whichever is greater)	ACCURACY OF WIND DIRECTION	±20°
		WIND SAMPLE AREA	50 x 50 km^2 (typical)

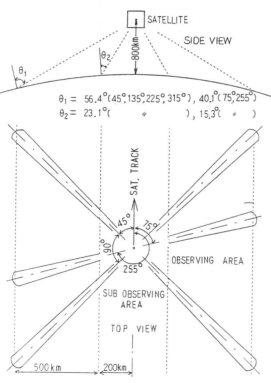

$\theta_1 = 56.4°(45°,135°,225°,315°), 40.1°(75°,255°)$
$\theta_2 = 23.1°(\quad \text{"} \quad), 15.3°(\text{"})$

Beam footprint of SCAT is shown in Fig. 1. The third beam angle (75° and 255°) is decided by computer simulation presented some examples of the results in Fig. 8, 9. (Detailed simulation method will be shown in Ref. [3]).

All of the beams can be operated in both of VV and HH polarization modes. The swath width is 500 km in each side with 200 km off-nadir distance.

Fig. 1   FOOTPRINT OF THE MICROWAVE SCATTEROMETER

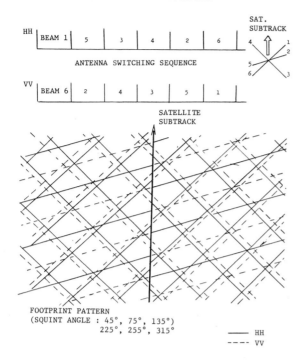

HH
| BEAM 1 | 5 | 3 | 4 | 2 | 6 |

ANTENNA SWITCHING SEQUENCE

VV
| BEAM 6 | 2 | 4 | 3 | 5 | 1 |

SAT.
SUBTRACK

SATELLITE
SUBTRACK

FOOTPRINT PATTERN
(SQUINT ANGLE : 45°, 75°, 135°)
225°, 255°, 315°

——— HH
---- VV

Fig. 2   ANTENNA SWITCHING SEQUENCE
AND FOOTPRINT PATTERN

Satellite Altitude shown in fig. 1 (800 km) is nominal value. Antenna switching sequence and the footprint pattern of SCAT is shown by Fig. 2.

In Fig. 2, antenna switching sequence is selected by considering uniform radiation to the swath. If any other sequence were selected, inequality of the beam location would be occurred in the swath, and cannot form the uniform distributed observation cell. And an observation cell can be formed by alternate patch in the footprint pattern as depicted in Fig. 2. In the footprint pattern, the interval lengths L_f of adjacent beam is given by (1).

$$L_f = n \cdot (\tau_p + \frac{2R_{cmax}}{C} + K_t \tau_p + t_m) \qquad (1)$$

where

$$K_T = \frac{\tau_N}{\tau_{S+N}}$$

n;   integral numbers of PRF   C;   light velocity
τ_p;   transmitting pulse width   tm;   signal processing time
Rcmax;   maximum slant range in the swath

In practical design, L_f differs between 45° (135°) beam and 75° (255°) beam.

The blockdiagram of SCAT hardware is shown in Fig. 3. Basically SCAT has two channels of T/R function for HH and VV modes. Both channels are independent except antenna switching sequence. A sequence switch prepared in antenna portion controls that opposite beams become HH and VV respectively in order for reducing interference between them. Basic operation of each channel is almost same as SASS. However, there are several differences in basic design parameters, which is shown in Table 2. This is caused by expectation of improvement in aliasing removal and measurement error reduction.

Tab. 2   BASIC DESIGN PARAMETERS

BEAMWIDTH	45°&135° BEAMS 0.5°x24°	TRANSMITTING FREQUENCY	13.99 GHz
	75° BEAM 0.5°x20°	TRANSMITTING POWER	100 W PEAK
NUMBERS OF DOPPLER CHANNELS		PULSEWIDTH	5 ms
	20/POL.	PULSE REPETITION FREQUENCY	40 Hz
BEAM POLARIZATION	VV&HH	ANTENNA GAIN	32 db
SATELLITE ALTITUDE	800 km NOM.	ORIENTATION OF THE BEAMS	45°, 135°,
SWATH WIDTH	500 km x 2		75°

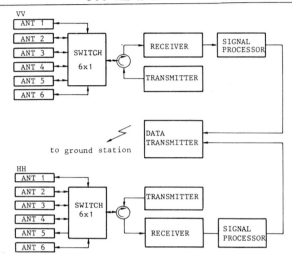

Fig. 3   BLOCK DIAGRAM OF THE MICROWAVE SCATTEROMETER

3.   RESOLUTION CELL OF SCAT

Resolution cells of SCAT are
formed by narrow beam width and
Iso-doppler line like SASS.
A basic configuration of a resolution cell is shown in Fig. 4.
As shown in Fig. 4, the received
beams (only 45° beam is depicted
in the Figure for simple expression) are divided iso-Doppler
line with Doppler filter
installed in the signal processor. The typical way of
dividing is that the cross point
of iso-Doppler line and beam LOS
line coincides with the corner
of a minimum resolution cell.

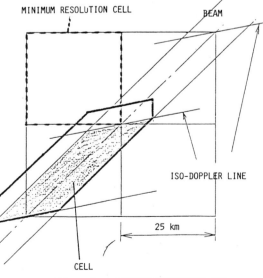

Fig. 4   CONFIGURATION OF
RESOLUTION CELL

The cell formed by this way is named "Unit Cell". Minimum resolution
cell is selected 25 km square within the swath and distributed a side to
be in parallel with satellite subtrack and means the minimum resoluable
cell by using unit cells. 50 km square formed by adjacent 4 minimum
resolution cells is a standard observation cell. Aii of the beams in an
observation cell are shown by Fig. 5.

As shown in Fig. 5, many unit cells are contained in an observation
cell. A direct observation parameter is a receiving power P_{Ru} of unit
cell. P_{Ru} of the same direction within an observation cell is summed
and estimated $\sigma°$ by the method presented in appendix. After all, minimum resolution cells are distributed in a swath as shown in Fig. 6, and
observation cells are formed by combinating 4 of them.

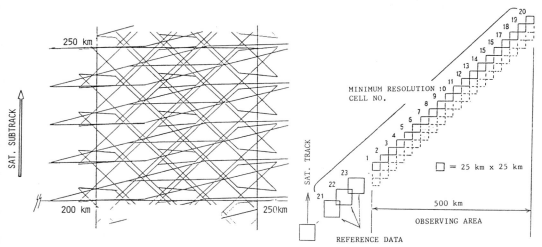

Fig. 5 AN EXAMPLE OF STANDARD
        MEASUREMENT CELL AND
        CONTAINED BEAM FORMATION

Fig. 6 MINIMUM RESOLUTION
        CELL DISTRIBUTION

4. MEASUREMENT ACCURACY OF RECEIVING POWER

Measurement accuracy of receiving power can be expressed by its standard
deviation K_p[1].

$$K_p = \frac{[Var(P_r)]^{\frac{1}{2}}}{P_r} \qquad (2)$$

In the case of the SCAT, an observation cell is formed multiple unit
cell. So, if K_{pui} is standard deviation of ith unit cell's receiving
power P_{Rui}, K_{psj} of an observation cell is shown by[4](3).

$$K_{psj} = \frac{1}{n}\{\sum_{i=j}^{n} K_{pui}^{2}\}^{\frac{1}{2}} \qquad\qquad (3)$$

where   n;   numbers of unit cells

It is assumed P_{Ru} within a considered observation cell is almost equal in (3).

Meanwhile, in the SCAT design, integration numbers and antenna LOS direction of each beam are selected under the condition which K_{pui} at the edge of the swath is as equal as possible. In the case of 45° beam, 8 unit cells are contained in an observation cell. By considering these conditions it can be calculated K_{psj} value and the results are shown in Fig. 7.

It is clear that the measurement accuracy of spatial integrated observation cell is fairly well from Fig. 7. Then SCAT would be able to be expected high accurate observation. Furthermore, observation would be feasible even by using minimum resolution cell.

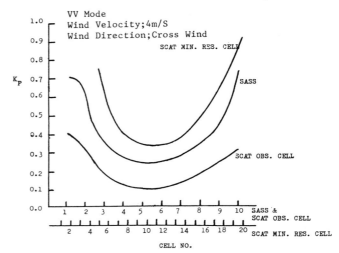

Fig. 7 K_p FOR SASS AND SCAT

5. DESIGN EVALUATION

SCAT is characterized by a fan beam system using 3 beams. In order to verify the effect of the 3 beam system, several simulation studies using computer in design stage have been done. One of the results is shown by Fig. 8. Fig. 8 is the case of the wind direction with high probability alias occurrence. Alias occurred in the 2 beam system is reduced

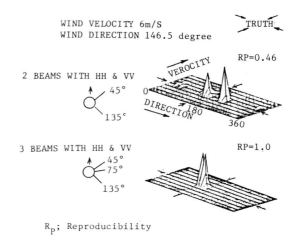

R_p; Reproducibility

Fig. 8   ACCURACY OF WIND VECTOR ESTIMATION IN 2 & 3 BEAMS MICROWAVE SCATTEROMETERS WITH VV & HH

in the 3 beam one. Another simulation result is shown by Fig. 9.
Fig. 9 presents the effects using HH & VV beams. Wind direction is the
same as Fig. 8. Alias occurred in 3 beam with HH is reduced in 3 beam
with HH & VV. By the way, to use HH & VV channels, though it will be
enlarged on board hardware, improves system reliability. More detailed
simulation results will be shown in [3].

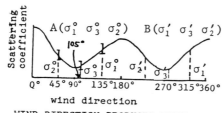

WIND DIRECTION PRODUCED LARGE MEASUREMENT ERROR

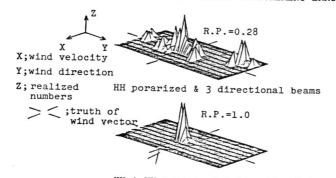

X;wind velocity

Y;wind direction

Z; realized
   numbers

$>$ $<$;truth of
       wind vector

HH porarized & 3 directional beams

R.P.=0.28

R.P.=1.0

HH & VV porarized 3 directional beams

$$R.P. = \frac{\text{trial numbers with specified accuracy}}{\text{total trial numbers}}$$

EFFECTS USING HH & VV BEAMS

Fig. 9   EFFECTS USING HH & VV BEAMS

6.   CONCLUSION

The basic design of the microwave scatterometer (SCAT) planned in Japan
was presented. SCAT is characterized by 3 directional fan beams and an
observation cell composed of multiple unit cells. The system configuration was discussed which enhanced these points. As a result of the
discussion, it was clear that SCAT would be able to measure seasurface
wind vector with high accuracy and expect flexibility in measurement
corresponding to various field conditions.

REFERENCES

W.L. Grantham et al., 1979.
M. Shimada et al.: Development of Microwave Scatterometer for Earth
    Observation Satellite in Japan. *24 Convegno Internazionale
    Scientifico Sullo Spaxio*, 1984.
M. Shimada et al.: Simulation of Wind-Vector Estimation-design estima-
    tion of microwave scatterometer- *this symposium record.*
H. Yamada, M. Kondo et al.: Microwave scatterometer. *13th ISTS in Japan*,
    1982.

APPENDIX

Fig. A shows footprint pattern and observation cells of SCAT.   An
average $\sigma°A$ ($\sigma°A$ of cell A) estimation algorithm is presented as follow;

First Estimation of $\sigma_1 A$, $\sigma_1 B$, $\sigma_1 C$, $\sigma_1 D$ ...

$$\sigma_1 A = \frac{1}{N}\sum_{i=1}^{N}\sigma_{MAi} \qquad \sigma_1 B = \frac{1}{N}\sum_{i=1}^{N}\sigma_{MBi} \qquad \sigma_1 D = \frac{1}{N}\sum_{i=1}^{N}\sigma_{MBi}$$

If $\sigma_1 A \sim \sigma_1 B$ and $\sigma_1 A \sim \sigma_1 B < \varepsilon$, $\sigma_1 A$ can be considered observation value.   If
$\sigma_1 A \sim \sigma_1 B$ and $\sigma_1 A \sim \sigma_1 B > \varepsilon$ following compensation should be done.

$$\sigma_2 A = \frac{1}{N-2}\sum_{i=1}^{N-2}\sigma_1 i + \frac{A\sigma MA_2}{S_1 + S_2\frac{\sigma_1 B}{\sigma_1 A}} + \frac{A\sigma MA_7}{S_3 + S_4\frac{\sigma_1 D}{\sigma_1 A}}$$

where

$S_1 \sim S_4$;   Shown in Fig. A
N   ;   Unit cell number
           contained in cell A
σMAi;   Measured $\sigma°$ by unit
           cell Ai
A   ;   Area of Unit cell

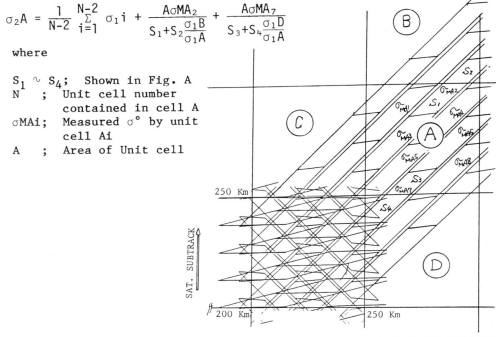

Fig. A   SCATTERING COEFFICIENT DISTRIBUTION
         IN OBSERVATION CELLS

ON-BOARD PROCESSING OF MICROWAVE ALTIMETER
- NUMERICAL SIMULATION AND REAL TIME SIMULATION -

YUJI MIYACHI[1], HIDEO KISHIDA[1], MAKOTO ISHII[2]
[1] Tsukuba Space Center,
    National Space Development Agency of Japan
    2-1-1 Sengen, Sakura-mura, Niihari-gun, Ibaraki 305 Japan
[2] Nippon Electric Co. Ltd.
    1-10 Nisshin-cho, Fuchu-shi, Tokyo 183 Japan

ABSTRACT. Microwave Altimeter is a useful sensor for the oceanic
research. The feature of Altimeter is that it estimates and tracks
altitude and other parameters on-board by using an on-board computer.
The algorithm for this tracking, which uses MLE and MMSE, is studied in
this report. As the result, the realization of the high accurate
tracking by this algorithm has been verified.

1. INTRODUCTION

Recently, remote sensing has been playing a very important role in the
oceanic research. Microwave Altimeter (Altimeter) is the one of useful
sensors for this research, which measures a distance (altitude, Ho)
between a mean sea level and a satellite, significant wave height (SWH),
and ocean backscatter coefficient (σo). Several altimeters were flown
aboard USA satellites (SKYLAB, GEOS-3, SEASAT-1). These experiments
demonstrated the usefulness of Altimeter. In Japan, the research and
development of Altimeter has been started in 1979, reflecting upon the
above results. The study of the tracking algorithm reported here was
conducted as the part of this development.
   The feature of Altimeter is that it estimates and tracks Ho, SWH,
and σo on-board by using an on-board computer. This tracking improves
the quality of the received power signals and decreases an amount of
telemetry data. This report describes the study of the normal tracking
and initial acquisition algorithm. In the normal tracking algorithm,
maximum likelihood estimation (MLE), minimum mean square error
estimation (MMSE), and the combined method of MLE and MMSE (Hybrid) –
these three methods are generally named Template Tracker – are applied,
although SEASAT-1 Altimeter used Adaptive Split Gate (ASG) Tracker.[4]
The basic theory of this algorithm was reported in the literatures.[3,6]
But in this application to Altimeter, several modifications are done,
by which the tracking accuracy is much improved and this algorithm is
able to be performed by the on-board computer. In addition to the
normal tracking, the initial acquisition algorithm is studied and the

369

Y. Toba and H. Mitsuyasu (eds.), The Ocean Surface, 369–377.
© 1985 by D. Reidel Publishing Company.

transition from the initial acquisition to the normal tracking is
verified.

2. TRACKING ALGORITHM

In this section, the normal tracking algorithm and the initial
acquisition algorithm are described. Moreover, before this description,
the received power used in the later sections is reviewed, because this
has an important effect on the tracking algorithm.

2.1 The Received Power[5,6,8]

The received power is derived to be the following form;

$$Pr(t) = \frac{Po \cdot Go^2 \cdot \lambda^2 \cdot \sigma R \cdot |R(0)|^2}{(8\pi)^{1 \cdot 5} \cdot s^2 \cdot L \cdot Ho^3 \cdot (1 + Ho/Ae)} \exp(-8 \cdot \ln 2 \cdot (\frac{\xi}{\theta bw})^2) \cdot$$

$$\cdot \exp(.5 \cdot (\sigma e \cdot \mu)^2 - \mu \cdot z) \cdot \Phi(z/\sigma e - \sigma e \cdot \mu) + No$$

$$= No \cdot \{ ao \cdot \exp(.5 \cdot (\sigma e \cdot \mu)^2 - \mu \cdot z) \cdot \Phi(z/\sigma e - \sigma e \cdot \mu) + 1 \} \quad (1)$$

$$z = c \cdot t/2 - Ho \qquad \sigma R = K \cdot c \cdot \sigma \tau/2 \qquad \sigma e = (\sigma R^2 + \sigma h^2)^{1/2}$$

$$\mu = \frac{2}{Ho \cdot (1 + Ho/Ae)} \cdot (\frac{8 \cdot \ln 2}{\theta bw^2} + \frac{(1 + Ho/Ae)^2}{s^2} - (8 \cdot \ln 2 \cdot \frac{\xi}{\theta bw})^2)$$

$$\Phi(x) = \frac{1}{\sqrt{(2\pi)}} \int_{\infty}^{x} \exp(-.5 \cdot y^2) dy$$

Po : transmitted peak power   Go : antenna gain
λ : wave length $\sigma\tau$: pulse width
K : correction factor (.362)   L : system loss
θbw : antenna beam width ξ : antenna pointing error
No : noise power
Ho : altitude   Ae : earth radius
R(0) : Fresnel reflection coefficient of sea surface
           at normal incidence
s^2 : total mean square surface slope
σh : RMS wave height ($\approx$SWH/4) c : light speed

2.2 Normal Tracking Algorithm

The algorithm applied to the normal tracking is Template Tracker, whose
basic theory depends on MLE and MMSE. First, the basic theory of MLE and
MMSE are described. Secondly, their application to Altimeter is
described. And lastly, the performance of Template Tracker is discussed.

2.2.1 The basic Theory.[3,6] It is assumed that the M parameters ($\theta =$
[$\theta 1, \cdots, \theta M$]) are estimated from the Ns data ($W = [V1, \cdots, VNs]$). If Vi is
assumed to be statistically independent and follow an exponential
distribution, the following probability density function and joint
probability density function are defined as follows;

$$pi(Vi) = (1/\bar{V}i) \cdot \exp(-Vi/\bar{V}i) \qquad P(W) = \prod_{i=1}^{Ns} pi(Vi) \qquad (2)$$

$\bar{V}i$: expected value of Vi

In MLE, the parameters are estimated to be θMLE, which maximizes the joint probability density function P(W). If the penalty function ΛMLE is defined to be $-\ln(P(W))$, θMLE minimizes ΛMLE. Therefore, θMLE can be determined from the following equation;

$$\nabla\Lambda MLE = 0 \qquad \nabla = [\frac{\partial}{\partial\theta 1}, \cdots, \frac{\partial}{\partial\theta M}]^T \qquad (3)$$

In MMSE, the mean square error ΛMMSE is equivalent to the penalty function in MLE. The parameters are estimated to minimize ΛMMSE.

$$\nabla\Lambda MMSE = 0 \qquad \Lambda MMSE = \sum_{i=1}^{Ns}(Vi - \bar{V}i)^2 \qquad (4)$$

From eq.(3) and eq.(4), the following equation can be derived;

$$\mathbb{B} \cdot \mathbb{N} \cdot (\mathbb{V} - \bar{\mathbb{V}}) = 0 \qquad (5)$$

$$\mathbb{B} = \mathbb{D} \cdot \mathbb{N} \qquad \mathbb{D} = [\nabla\bar{V}1, \cdots, \nabla\bar{V}Ns]$$

$$\mathbb{N} = \begin{bmatrix} 1/\nu i, & & \mathbb{0} \\ & \ddots & \\ \mathbb{0} & & ,1/\nu Ns \end{bmatrix} \qquad \nu i = \{\begin{matrix} 1 & \text{for MMSE} \\ \frac{1}{\bar{V}i} & \text{for MLE} \end{matrix}$$

The parameters are estimated by solving eq.(5) using νi for MLE or MMSE. In Hybrid, νi is defined to be $\bar{V}i$.[5]

2.2.2 Application of Template Tracker to Altimeter. Altimeter receives return signals (received power) reflected by a sea surface and it includes the information of parameters to be estimated (Ho, SWH, σo). So, W (in the previous section) is to be the received power. As the received power generally follows the assumption of eq.(2), it is appropriate to apply MLE to Altimeter.

The received power represented by eq.(1) becomes unique, if 5 parameters are determined − z, σe, ao, No, μ. Altitude is determined from z and SWH from σe. ao and No determine σo, but when No is measured inside Altimeter, σo is derived from ao only. μ (called decay factor below) is related mainly to antenna pointing error (ξ) and to estimate μ is not a purpose of Altimeter. But if only 3 main parameters are estimated, the estimated values have bias errors (steady state error) due to using an incorrect μ value. To reduce these errors, it may be considered to estimate 4 parameters by using MLE, MMSE, or Hybrid. But in this case, the estimation accuracies of 3 parameters become worse. Reflecting upon the above characteristics, the new configuration of Template Tracker is considered, which estimates 3 parameters normally and estimates μ at regular intervals (Tμ) by using different method, which is suitable only to μ estimation. By this configuration, one feature of Template Tracker, the high accurate estimation, can be maintained and the bias errors due to μ can be reduced. The block diagram of Template Tracker is shown in Fig.1.

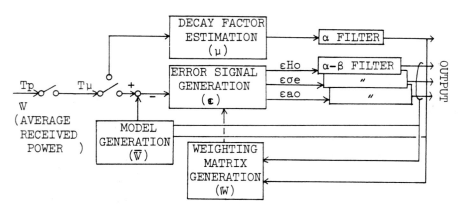

Fig.1   The Block Diagram of Template Tracker

Template Tracker is divided into 3 sections. In the first section, error signals of 3 main parameters are calculated by using MLE,etc. and in the second section, smoothed positions, predicted positions, and smoothed velocities are estimated by error signals passing α-β filter! In the third section, μ is estimated.

Error signals are determined by solving eq.(5) approximately. That is, the following equation is calculated;

$$\varepsilon k = \theta k - \theta k\text{-}1 \simeq \{\mathbb{B}(\theta s,k\text{-}Nw)\cdot\mathbb{B}^T(\theta s,k\text{-}Nw)\}^{-1}\cdot\mathbb{B}(\theta s,k\text{-}Nw)\cdot$$

$$\cdot\mathbb{N}(\theta s,k\text{-}Nw)\cdot\{\mathbb{V}k - \bar{\mathbb{V}}(\theta p,k)\}$$

$$= \mathbb{W}(\theta s,k\text{-}Nw)\cdot\{\mathbb{V}k - \bar{\mathbb{V}}(\theta p,k)\} \qquad (6)$$

$\varepsilon k = [\varepsilon Ho, \varepsilon\sigma e, \varepsilon ao]^T$: error signal at t = k·Tp
                                         (Tp : sampling period)
$\theta k = [Ho, \sigma e, ao]^T$: estimated value at t = k·Tp
$\theta s,k$: smoothed position at t = k·Tp   (eq.(7))
$\theta p,k$: predicted position at t = k·Tp   (eq.(7))
$\mathbb{V}k$: received power at t = k·Tp
$\bar{\mathbb{V}}(\theta p,k)$: received power model calculated by $\theta p,k$
$\mathbb{W}(\theta s,k\text{-}Nw)$: weighting matrix calculated by $\theta p,k\text{-}Nw$

As it needs a long time to calculate weighting matrix, it is renewed at regular intervals (Tw). Therefore, $\mathbb{W}(\theta s,k\text{-}Nw)$ is used instead of $\mathbb{W}(\theta s,k)$ in eq.(6). (Tw = Nw·Tp)
α-βfilter is used to reduce the standard deviations of parameters and to predict (track) parameters. α-βfilter is expressed by the following equations;

$$\theta p,k\text{+}1 = \theta s,k + Tp\cdot\dot{\theta}s,k \qquad \text{(predicted position)}$$

$$\theta s,k = \theta p,k + \alpha\cdot\varepsilon_k \qquad \text{(smoothed position)} \qquad (7)$$

$$\dot{\theta}s,k = \theta s,k\text{-}1 + \frac{1}{Tp}\cdot\beta\cdot\varepsilon k \qquad \text{(smoothed velocity)}$$

$$\alpha = \begin{bmatrix} \alpha Ho, & 0 \\ & \alpha\sigma e, \\ 0 & & \alpha ao \end{bmatrix} \qquad \beta = \begin{bmatrix} \beta Ho, & 0 \\ & \beta\sigma e, \\ 0 & & \beta ao \end{bmatrix}$$

Predicted positions of Ho and ao are used for receiver timing control and AGC respectively.

As z becomes large, the received power (eq.(1)) is approximated to be the following equation;

$$V(z) \simeq No\cdot\{ao\cdot exp(-\mu\cdot z) + 1\} \qquad (8)$$

V(z) derivative depends mainly on μ in this region. In other words, μ can be well estimated by using this data. As Fig.2 shows, V(z) in z = -15m ~ +15m is used normally, but for μ estimation, V(z) at z = Δzμ + Δz2 is used. μ is estimated by using this average power (Vμ) and two other average powers (V1, V2) shown in Fig.2 as follows;

$$\frac{V\mu - V1}{V2 - V1} \simeq exp(-\mu\cdot\Delta z\mu)$$

$$\mu \simeq -\frac{1}{\Delta z\mu}\cdot ln(\frac{V\mu - V1}{V2 - V1}) \qquad (9)$$

The value of μ derived from the above equation is used for calculating $\bar{V}$ and $\bar{W}$.

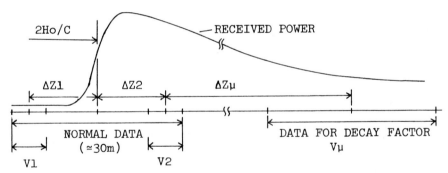

Fig.2   Received Power Range for Estimating Decay Factor (μ)

2.2.3 Performance of Template Tracker.   The system parameters used in Template Tracker are listed below;

fp (pulse repetition frequency)	1 kHz
Ns (sampled data number)	60
Δt (sampled data interval)	3.125 nsec
Np (pulse summation number)	50
Tp (sampling period = Np/fp)	50 msec
Tw (weighting matrix renewal period)	1 sec
Tμ (decay factor renewal period)	1 sec

Δzμ, Δz1, Δz2 (data position for μ estimation)
                    Δzμ = 48.52 m, Δz1 = -11.48 m, Δz2 = 11.48 m
α   (α value for 3 main parameters)   αHo = ασe = αao = .25
β   (β value for 3 main parameters)   βHo = βσe = βao = .0357
αμ (α value for decay factor)   1.

Two factors related to the estimation performance are discussed
here - the estimation accuracy (standard deviation) and the bias error
(steady state error). The estimation accuracies of 3 main parameters are
decided by MLE, MMSE, or Hybrid and α-β filter. That of decay factor
is decided by the method described in section 2.2.2 and α filter. The
results are shown in Fig.3 and 4.   (ao and σe results are not shown,
but they are less than .05 dB and 5 cm respectively in all wave height
range) In calculating the accuracies of 3 main parameters, the effect of
decay factor estimation is not included. So, these are a little better
than the real ones. In Fig.4, the bias error of decay factor is also
shown. 3 main parameters have bias errors due to this decay factor bias
error. But these errors are very little - ΔHo < 1 cm, Δσe < 1 cm,
Δao < .1 dB. α,β values used in the above evaluation are determined to
satisfy the condition on the tracker bandwidth (1 Hz).

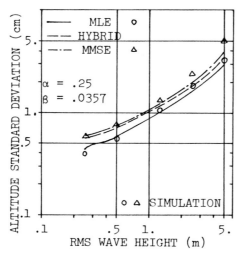

Fig.3 Altitude Estimation Accuracy   Fig.4 Estimation Accuracy and Bias
      after 1 sec average (not   Error of Decay Factor
      including μ estimation effect)

2.3 Initial Acquisition

The initial acquisition is conducted according to the following
sequence;

(1) Search Phase 1 by CW Pulse
    The received power is searched by using uncompressed CW pulse.
(2) Search Phase 2 by Chirp Pulse
    The received power is searched by using compressed pulse. First,
    the receiver timing is shifted in 1 μsec (150 m) from the point
    determined by Search Phase 1 and increased in 50 nsec (7.5 m) step.
(3) Coarse Acquisition
    After the received power enters into the receiver range, altitude
    and AGC are tracked roughly.

(4) Estimation of Initial Decay Factor
   The initial decay factor is estimated for $\bar{W}$, W. The estimation
   method described in section 2.2.2 is used.
(5) Fine Acquisition
   Altitude, SWH, and AGC are tracked for decreasing bias errors of
   3 parameters. ASG Tracker[4] is used as the tracking algorithm. In
   this stage, initial W is also calculated.
(6) Normal Tracking

In this report, the stage from (2) to (5) are studied. Altitude is
assumed to be estimated with an accuracy of ~10 m in the stage (1).

3. SIMULATION

3.1 Outline of Simulation

Two types of simulations are used for evaluating the normal tracking
and the initial acquisition. One type (Simulation 1) is the numerical
simulation performed only by the general-purpose computer. By Simulation
1, the ideal tracking performance is evaluated. Another is the real time
simulation (Simulation 2), in which a sea status and transmitter and
receiver performance of Altimeter are simulated by the general-purpose
computer like Simulation 1, but the part of the tracking algorithm is
replaced by the real on-board computer with the tracking algorithm
program. By this simulation, the performance which is related directly
to the on-board computer, for example the arithemetic accuracy, the
operation time, etc. is evaluated.

3.2 Simulation Results

3.2.1 Simulation 1. The initial acquisition performance has been
confirmed in the various conditions – various values were given to
altitude change rate ($\dot{H}o$), wave height (σh), and antenna pointing error
(ξ). The example of the initial acquisition is shown in Fig.5. All
parameters converge smoothly on the true values. The same results were
acquired in the following conditions;

   -50 m/sec $< \dot{H}o < +50$ m/sec, $.25$ m $< \sigma h < 5.$ m, $.0° < \xi < .4°$

     The received power was fluctuated according to the exponential
distribution (eq.(2)) in order to evaluate the estimation accuracies.
The results of this simulation are shown in Fig.3 and 4 together with
the theoretical values. Both results are very close. Bias errors of 3
main parameters due to antenna pointing error are very little as studied
theoretically in section 2.2.3.
     Moreover, antenna pointing error was varied according to the
following equation, in order to evaluate the perfomance of μ tracking;

$$\xi = .2 \cdot \sin(2\pi \cdot .1 \cdot t) \qquad [\text{deg}] \qquad\qquad (10)$$

The example of this simulation ($\sigma h = 1.25$ m) is shown in Fig.6. As μ is
renewed every 1 sec, the bias error are caused in both Ho and μ during
1 sec interval. But as a whole, all parameters are well tracked.

3.2.2 Simulation 2. According to the comparison of two types of
simulations, it has been confirmed that only little differences exist -
for example 1 cm in altitude estimation. As these differences can be
negligible, it can be said that Template Tracker has been verified to be
able to be performed by the on-board computer.

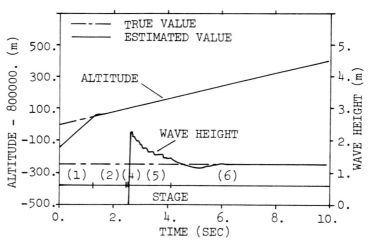

Fig.5   The Example of Simulation 1 for Evaluating Initial Acquisition

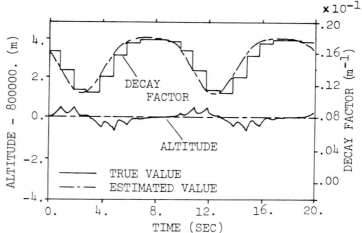

Fig.6   The Example of Simulation 1 for Evaluating Decay Factor
        Estimation

4.   CONCLUSIONS

From the results acquired by the above evaluation, it can be said that
the expected performance of the tracking, the high estimation accuracy
and reduced bias error, has been verified. But as these results were
acquired in the ideal conditions, that is, the detailed performance of

transmitter and receiver, the received power model error, and so on. are not included, the evaluation in the more practical conditions is necessary.

Although the above precise evaluation is necessary, Template Tracker has been demonstrated to have an excellent characteristic and able to be performed by the on-board computer.

REFERENCES

(1) T.R.Benedict and G.W.Border, 1962 : *Synthesis of an Optimal Set of Radar Track-While-Scan Smoothing Equations, IRE Transactions on Automatic Control*

(2) D.E.Barric, 1974 : *Wind Dependence of Quasi-Specular Microwave Sea Scatter, IEEE Transactions on Antenna and Propagation*

(3) L.W.Brooks and R.P.Dooley, 1975 : *Technical Guidance and Analytic Services in support of SEASAT-A*, NASA CR-141399

(4) J.L.MacArthur, 1976 : *Design of the SEASAT-A Radar Altimeter, MTS-IEEE Oceanic'76*

(5) G.S.Brown, 1977 : *The Average Impulse Response of a Rough Surface and Its Application, IEEE Journal of Oceanic Engineering,* Vol.OE-$\underline{2}$, No.1

(6) R.P.Dooley, L.W.Brooks, and E.N.Khoury, 1978 : *Optimization of Satellite Altimeter and Wave Height Measurements*, NASA CR-156850

(7) W.F.Townsend, 1980 : *An Initial Assesment of the Performance Achieved by the SEASAT-1 Radar Altimeter, IEEE Journal of Oceanic Engineering* Vol.OE-$\underline{5}$, No.2

(8) G.S.Hayne, 1980 : *Radar Altimeter Mean Return Waveform from Near-Normal-Incidence Ocean Surface Scattering, IEEE Transactions on Antennas and Propagation,* Vol.AP-$\underline{28}$, No.5

SIMULATION OF WIND-VECTOR ESTIMATION - DESIGN EVALUATION OF MICROWAVE
SCATTEROMETER -

MASANOBU SHIMADA[1] and MASAO SASANUMA[2]
1 Tsukuba Space Center, National Space Development Agency of
  Japan(NASDA), Ibaraki 305 JAPAN
2 Space Systems Dept., Mitsubishi Electric Corp., Kanagawa
  247 JAPAN

ABSTRACT. One of several representative features of the microwave scat-
terometer(SCAT), which has been being developed by NASDA since 1979, is
in its higher capability of wind alias removal by providing three dif-
ferently mounted antennas on each side and in each polarization. The
optimum beam configuration has been determined as 45°, 75°, and 135°,
following the results of the computer simulation with Monte Carlo technique. After completion of the bread board model's(BBM) manufacturing,
the hardware evaluation of the SCAT has been conducted with the SCAT
stimulator and measuring instruments. As a result, improved performances
of the SCAT were confirmed to be as designed, and random and bias behavior of received power by the SCAT was analyzed.

1.   INTRODUCTION

Microwave Scatterometer(SCAT) is an active remote sensor which measures
backscattering coefficients of the sea surface to infer wind-vectors
over the ocean globally. It is expected as an instrument offering significant atmospheric data, by which weather forecasting, marine meteorology will be improved effectively. There have been only two spaceborne
microwave scatterometers, Seasat-A Satellite Scatterometer(SASS), and
Skylab S-193, utilized till now. The SCAT is one of such sensors and is
planned to be installed on future marine observation satellite in Japan.
     Wind inference is based on a relationship between normalized radar
cross section(NRCS) of the sea surface and wind-vector over there. But,
due to the biharmonic and random nature of NRCS and inevitable error in
measurement, it is not easy to obtain a true wind-vector with high accuracy. In SASS, moreover, owing to the disposition of the antenna beams,
"sea chikin" has appeared as one example of such difficulties with four
or more alias solutions having nearly equal probabilities of occurences.
However, as reported in ref.1, a wind-vector extracted among several
solutions using ancillary data such as atmospheric data coincides most
closely with true wind-vector. Considering these aspects, the SCAT was
designed and manufactured so as to increase its usefulness as a remote
sensor by upgrading the capability of alias removal.

Y. Toba and H. Mitsuyasu (eds.), The Ocean Surface, 379–387.
© 1985 by D. Reidel Publishing Company.

    This paper describes the results and method of the computer simula-
tion performed to find out the optimum beam configuration, and also goes
into the hardware evaluation conducted to confirm the performances of
the BBM.

2. MEASUREMENT PRINCIPLE

Normalized Radar Cross Section(NRCS) of sea surface in microwave region
is related to wind speed(u), azimuth angle(ϕ) which is defined as the
angle between wind direction(β) and observing direction + 180°, incident
angle(θ) and polarization(ε) of radiowave, atmospheric instability, and
the underlying long waves. For the first four parameters, NRCS is usually expressed as follows,

$$\sigma°m = u^{\gamma} \cdot (a + b \cdot \cos\phi + c \cdot \cos2\phi) \tag{1}$$

where, a,b,c,and γ are coefficients depending on θ and ε of radiowave.
    In Eq.(1), $\sigma°m$ is a monotonously increasing function of u, and a
second order harmonic function of ϕ, therefore, a wind-vector(u,β) must
be obtained uniquely as a solution of nonlinear simultaneous equations
constructed by three NRCSs taken individually at different azimuths, if
there is no error in each measurement. Actually, measured NRCS is affected by various error sources, which are classified in random and bias
errors and originate in the estimation of all right side terms of radar
equation(2).

$$\sigma° = Pr \cdot (4\pi)^3 \cdot R^3 \cdot Ls \ / \ (Pt \cdot (G/G_0)^2 \cdot G_0^2 \cdot \lambda^2 \cdot \Phi \cdot L) \tag{2}$$

where, Pr, R, Ls, Pt, and G/G_0 are received power, slant range, system
loss including atmospheric losses, transmitted peak power, and relative
antenna gain at doppler cell incident angle, respectively. G_0, λ, Φ, and
L are peak antenna gain, free space wavelength, 3dB beam width in narrow
beam plane, and cell length in wide beam plane, respectively.
    Considering measurement error and scattering signature, u and β
are usually solved stochastically by using maximum likelihood technique(
  ref.2).

$$P(u,\beta|\{\sigma°i\}) = \exp[-\frac{1}{2} \sum_{i=1}^{M} \{(\sigma°i - \sigma°m)^2 \ / \ \delta i^2\}] \tag{3}$$
$$\rightarrow \text{maximum}$$

where, $P(X|Y)$, $\sigma°i$, $\{\sigma°i\}$, $\sigma°m$, δi, and M are probability of existence
of parameter X under the condition of Y, ith measured NRCS, a set of
measured NRCSs, scattering model, error occured in ith measurement, and
number of measurements, respectively.

3. IMPROVEMENT OF SYSTEM PERFORMANCE - WIND VECTOR SIMULATION

3.1. Concepts and the basic configuration

TABLE-1 SYSTEM PERFORMANCE REQUIREMENTS

WIND SPEED RANGE	4m/sec - 25m/sec
ACCURACY	±2m/sec or ±10%
WIND DIRECTION RANGE	0° - 360°
ACCURACY	±20°
SWATH WIDTH	500Km x 2
CELL SIZE	50 x 50Km

TABLE-2 HARDWARE CONFIGURATION

TRANSMITTED POWER	100W
DUTY OF TWT	20%
PULSE WIDTH	5.0msec
PRF	40Hz
ORBIT HEIGHT	800Km
ANTENNA	FAN BEAM(0.5° x 24°)
OBSERVATION	THREE BEAMS LOOKING

Utilization of more than two beams, especially three beams, is the most practical method within considerable reforms, in spite of the increment of weight. The optimum beam configuration was determined by conducting a computer simulation under actual conditions of the hardware. Table-1 is the system performance requirements settled temporally.

The SCAT is a radar system to measure the received power with square law detectors and analog integrators, and also to isolate an observing area with doppler filtering and multi fan beams scanning. Table-2 shows the basic configuration of the SCAT.

Fig.1 shows the geometry used in this simulation. The orientations of the two of three beams were fixed at 45° and 135°. Each illuminated area within an observation swath is resolved into 20 small cells, called "unit cell", by 20 doppler filters, and each unit cell is surrounded by upper and lower iso-doppler frequency lines and a 3dB beam width in narrow beam plane. As the typical deployment of unit cells included in the 45° beam and an observation cell(OC), whose dimension is 50Km x 50Km on earth, is shown in Fig.2, eight unit cells are located uniformly and densely over an OC, so, the scientific datum(NRCS) obtained averagely over an OC will provide smaller random error and show the more representative, in comparison with SASS.

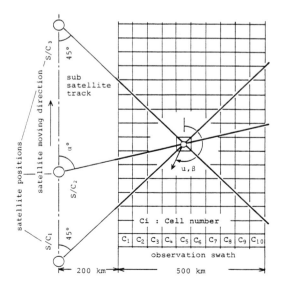

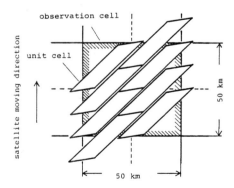

Fig.2 Typical deployment of unit cells over an observation cell

Fig.1 SCAT measurement geometry

3.2. Generation of simulated NRCS

The budget of random and bias errors, which originate in each measurement of NRCS, is estimated in Table-3, referring to the SASS data(ref.3).

TABLE-3 ERROR BUDGET FOR THE SCAT

ERROR SOURCE	BIAS (dB)	RANDOM
Φ	0.09	-
Pr	0.02	Kp
Pt	0.05	0.03
Ls	0.1	-
$(G/G_0)^2$	0.2 (TWO WAY)	-
G_0^2	0.6 (TWO WAY)	-
RMS ERROR	0.84	$\approx$ Kp

where, Kp is the normalized standard deviation of the random component
of measured Pr(ref.4). In the SCAT, because the received signal is pro-
cessed at square law detector and analog integrator, the random one is
expected to distribute normally in anti-log expression. On solving the
optimizing problem(3), however, it is far easy to obtain a solution by
dealing Pr, $\sigma°$, and all terms of Eq.(2) in dB expression rather than in
anti-log. The distribution for dB expressed Pr and its asymptotic form
holding for smaller Kp are obtained as follows,

$$P(y) = z(y)/(\tilde{a}\cdot\sqrt{2\pi}\cdot Kp\cdot\tilde{x})\cdot\exp[-\{z(y)-z(\tilde{y})\}^2/(2Kp^2\cdot\tilde{x}^2)] \qquad (4)$$

$$\approx 1/(\tilde{a}\cdot\sqrt{2\pi}\cdot Kp\cdot\tilde{x})\cdot\exp\{-(y-\tilde{y})^2/(2Kp^2\cdot\tilde{a}^2\cdot\tilde{x}^2)\} \quad (Kp<=0.5) \quad (5)$$

$$Kp = 1/\sqrt{Bc\cdot Ts}\cdot(1 + 2/SN + (1 + \tau s/\tau n)/SN^2)^{0.5} \qquad (6)$$

$$z(y) = \exp(y/\tilde{a}) \qquad (7)$$

where, x, $\tilde{x}$, and $\tilde{a}$ are Pr in anti-log, its mean value, and $10\cdot\log(e)$,
respectively. Bc, SN, τs, and τn are doppler band width, signal-to-noise
ratio, integration time for signal, and that for noise, respectively,
and y = $10\cdot\log10(x)$.
    When Kp goes smaller than about 0.5, that approximation holds in-
creasingly. Considering that NRCS of an observation cell is constructed
averagely over eight unit cells and S/N of received signal is not bad in
middle to high wind speed and all up and down wind directions, Kps used
in this simulation are less than 0.5 except a few cases of bad conditions. Consequently, Eq.(5) can be used as the distribution function for
random component of Pr in dB.
    Further, expanding this approximation to the other random component
in NRCS, and assuming normal distribution in bias component, simulated
NRCS($\sigma°$si) is made by summing nominal NRCS($\sigma°$mi) and two independently
distributed normal variables as follows,

$$\sigma°si = \sigma°mi + r_1i\cdot\xi i + r_2i\cdot0.84 \qquad (8)$$

$$\xi i = 10\cdot\log10(e)\cdot(Kpi^2 + q^2)^{0.5} \qquad (9)$$

where, ξ_i is the standard deviation of random error in ith measurement,
r_1i and r_2i are normally distributed random numbers with mean of 0.0 and
standard deviation of 1.0, and q is the quantization error.

3.3. Monte Carlo Simulation

Due to the periodic character of the reference function $\sigma°m$ in wind azimuth, some iterative method as Newton's method cannot be applied to
solve problem(3). Therefore, in the inferring routine of this simulation
, whose flowchart is shown Fig.3, next method was used to shorten the
time for obtaining a satisfactory conclusion.
  (Method): Firstly, wind speed region uA, in which $P(u \epsilon uA, \overline{\beta} | \{\sigma°i\})$ is
 comparatively high, is restricted for an arbitrary settled direction $\overline{\beta}$.
Secondly, a solution (u,β) is minutely selected as the one which makes
P( ) maximum within uA and $0° <= \beta <= 360°$.
    To evaluate the improving performance, reproducibility(Rp) defined
in Eq(10) was used.

$$Rp(ui, \beta i, ci, configuration) = Na / Nb \times 100 \qquad (10)$$

where, Na is the number of times that inferred wind-vectors satisfied
the performance requirements of Table-1, within the total trial number
Nb, and ui, βi, ci, and configuration are firstly set condition of the
Monte Carlo routine.
    As the results are shown in Fig.4, Rp increases with the orienta-
tion of the third beam till 85°, for any combination of polarization and
wind speed. The improving performance is confirmed to be about twenty or
more percentages higher than in two beams observation. But, the skew-
ness of the instantaneous field of view(IFOV) goes to bad rapidly at
more than 80°. Consequently, 75° was recommended as the orientation of
the SCAT's third beam.

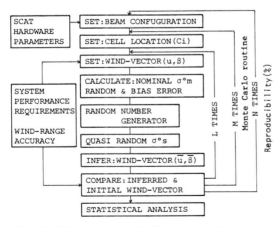

Fig.3 Flowchart of this simulation

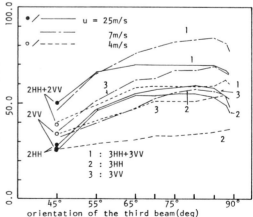

Fig.4 Simulation results. All
lines drawn in this figure are
averaged over all βs and cs.

4. HARDWARE EVALUATION

4.1. Concepts

Hardware evaluation is undergoing to confirm the fundamental performances of the BBM, which has been manufactured in Dec. 1982, and also to
construct the design base, by which the measuring accuracy of the SCAT
in flight model will be upgraded, with the accumulation and analyses of
the characteristic data taken from the BBM. All data analyzed here were
taken from following block diagram.

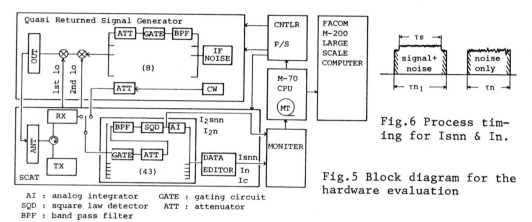

Fig.6 Process tim-
ing for Isnn & In.

Fig.5 Block diagram for the
hardware evaluation

```
AI  : analog integrator    GATE : gating circuit
SQD : square law detector   ATT : attenuator
BPF : band pass filter
```

4.2. Measurement of received power

4.2.1. Distribution of the SCAT's output. Scientific digital data sup-
plied from the SCAT includes several informations related to received
power(Pr) in question and the condition of the receiver and signal processor just at that time. They are sorted in three types as follows,
    (a) Isnn : data processed when returned signal exists
    (b) In   : data processed when returned signal does not exist
    (c) Ic   : data in calibration mode
    These data are put out after averaging over 5 – 20 msec or 65 –80
msec at analog integrator of the signal processor, following the data
acquisition sequence, but still noisy as decribes below.
    Fig.7 shows the short time variation of Isnn and In. Fig.8 shows
the correlation coefficients of the square law detector's outputs, I_2snn
and I_2n. As the correlation time(τc), which satisfies $F(\tau c) = 1/e$, of
I_2snn and I_2n are both about $30\mu sec$, and also negligibly smaller than
integration time τs and τn, Isnn and In are expected to distribute normally under the central limit theorem. The histgram resulting from these
data concludes that distribution of these two variables are similar to
normal distribution and the ratio of the standard deviation to the mean
value is nearly equal to $1/\sqrt{Bc \cdot \tau s}$(or τn), as shown in Fig.9.

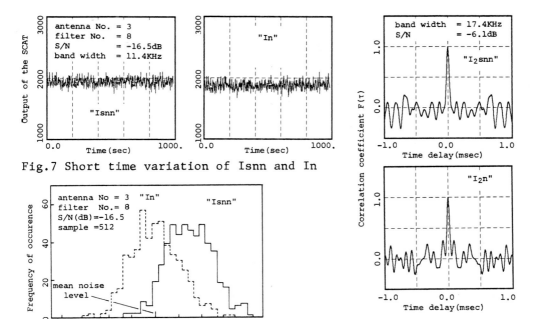

Fig.7 Short time variation of Isnn and In

Fig.9 Histgarm of Isnn and In

Fig.8 Coefficients of auto-
correlation of I_2snn and I_2n

4.3.2. Bias and random error in Pr measurement. A relationship between
the input and output of the SCAT is described theoretically and empirically as follows,

$$Pr \ = \ (\ \tau n / \tau s \cdot Psnn^k \ - \ \tau n_1 / \tau s \cdot Pn^k \ )^{1/k} \tag{11}$$

$$Psnn \ = \ Isnn^{1/k} \ / \ Grs \tag{12}$$

$$Pn \ = \ In^{1/k} \ / \ Grs \tag{13}$$

$$Grs \ = \ [\ Ic^k \ ] \ / \ Pc_0 \tag{14}$$

where, Grs, Pc_0, and k are the gain of receiver and signal processor,
calibration power, and the gradient of the input and output relation in
the signal processor, respectively. Psnn and Pn mean signal plus noise
power and noise power, and [ ] means averaging.
        Received signal is usually under receiver noise, so the dynamic
range of Psnn is almost narrower than that of Pr, which is proportional
to NRCS, and error in measured Psnn is expanded to the order of N/S,
when Pr is estimated. Among several facters degrading the accuracy,
measurement error in τn, τs, and τn_1 are the most impactable ones in the
lower S/N region of less than −15dB. As a result of Pr measurement
(translated in NRCS) performed for a range of wind-vector inferring is

shown in Fig.10, the linearlity is confirmed to be within an accuracy of less than 0.8dB, where in each point, 200 data are averaged to exclude random variation.

As for the random variation, it is evaluated statistically following the next equation.

$$Kpe = \{[Pri^2] - [Pri]^2\}^{0\cdot5} / [Pri] \tag{15}$$

Comparing with the theoretically obtained Kpt, we can confirm a good coincidence between them, as shown in Fig.11.

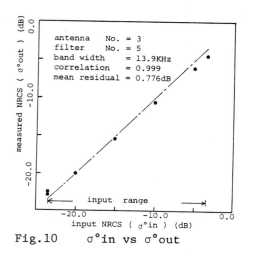

antenna   No. = 3
filter   No. = 5
band width   = 13.9KHz
correlation   = 0.999
mean residual = 0.776dB

input range

Fig.10   σ°in vs σ°out

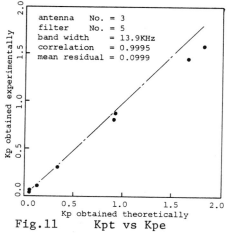

antenna   No. = 3
filter   No. = 5
band width   = 13.9KHz
correlation   = 0.9995
mean residual = 0.0999

Fig.11   Kpt vs Kpe

4.3. Realistic simulation - no bias error

The realistic simulation has been conducted using the quasi returned signal generator(QRSG) which is a computer controlled multi mission instrument having performances to generate the narrow band gaussian signals(RF & IF) for four observation cells C_1 to C_4 of Fig.1. But, because of the difficulty to make the quasi returned pulses perfectly similar to real ones, the purposes were restricted to evaluate the wind sensitivity of the BBM in ideal case of no bias error and investigate how actual random error in measured Pr behaves.

Measured Pr were firstly bias corrected and analyzed in the realistic simulation program, and following conclusions were obtained,
1) Performance is superior, 3H+3V, 3H, 3V, and two beams in that order. This is slightly different from the results in Sec.3.3, but reasonable considering that first harmonic coefficient of NRCS in wind azimuth is larger in H-pol than in V-pol.
2) If 180° ambiguity is permitted, performances of 3H+3V and 3V become fairly good with reproducibility of more than 90%.
3) In the low wind speed region approximation(5) holds.

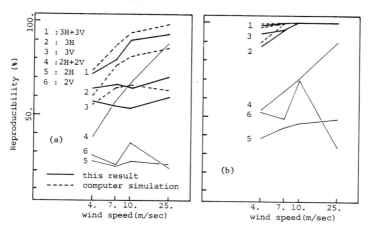

Fig.12 Results of realistic simulation a) do not permit 180° ambiguity,
b) permit 180° ambiguity, where, all measured NRCSs are bias corrected.

5. CONCLUSION

This paper describes two technical aspects related to the accuracy of
backscattered power measurement and wind-vector inference. Summarizing
all results obtained in these simulation and hardware evaluation, it can
be concluded that in doppler filtering system, the system performances
of the SCAT might be improved by three beams with dual polarizations per
each side mounted at 45°, 75°, and 135°, and random error occured in received power distributes normally as theory.
       On the contrary, bias error badly influences the inferring accuracy
especially in higher wind speed region due to the logarithmic dependence
of NRCS on wind speed. Thermal and temporal variation of signal process-
or and uncertainty of receiver gain estimate are considered as the main
source, so, understanding of these characters, including calibration
method, becomes important to upgrade the measurement accuracy.

6. ACKNOWLEDGEMENT

The authers would like to acknowledge the many contributions of Mr.S.
NIWA and the staffs of Mitsubishi Electric Co., who conceived the design
of the SCAT.

7. REFERRENCES

1) I. Halberstam,'Verification Studies of Seasat-A Satellite Scatterome-
ter(SASS) Measurement,'J.G.R. Vol.86, No.C7,pp6599-6606, July 20, 1981
2) F.J.Wentz, et al,'Algorithm for inferring Wind Stress from Seasat-A,'
J.Spacecraft, Vol.15, No.6, pp368-374, Nov-Dec, 1978
3) J.W.Johnson, et al,'Seasat-A Satellite Scatterometer Instrument Evaluation,' J.Oceanic Eng., Vol.OE-5, No.2,pp138-144, Apr.1980
4) R.E.Fisher,'Standard Deviation of Scatterometer Measurements from
Space,' Trans.IEEE, Vol.GE-10, No.2. April.19, pp106-113

WIND WAVES AND WIND-GENERATED TURBULENCE IN THE WATER

H. Mitsuyasu and T. Kusaba
Research Institute for Applied Mechanics
Kyushu University, 87
Kasuga, 816
Japan

ABSTRACT. Winds, wind-generated waves and drift currents were measured in a wind-wave tank both for pure water (ordinary tap water) and for the water containing a surfactant. For the latter no wind waves were generated within a range of the wind speed in the experiment. If the current velocity spectra were scaled with the friction velocity of the wind and the distance from the mean water surface, the low frequency part of the spectra revealed an approximate similarity irrespective of the existence of wind waves.

1. INTRODUCTION

Wind over water surface generates wind waves and drift currrent. The wind waves are random, strongly nonlinear and feed a part of their energy into the turbulence through wave breaking. The drift current co-existing with wind waves is a kind of turbulent shear flow generated by the wind stress and presumably affected by the wind waves through various processes. Knowledge of the turbulence characteristics in the water under the wind action is important for clarifying not only the structure of ocean-surface mixed layer but also the energy dissipation of wind waves. The structure of turbulence in the surface layer of the water is studied experimentally in a laboratory tank. Velocity fluctuations in the water are measured simultaneously with the wind and wind waves under the condition of steady state. Almost the same measurements are done for tap water and for the water containing a surfactant. Wind waves develop in the former water but they are surpressed in the latter. Therefore, by comparing the results of these two measurements we intend to clarify the effects of wind waves on the structure of turbulence in the water. Since the complete analysis of the data have not been finished yet, this paper is a preliminary report.

2. EXPERIMENT

Equipment and procedure The experiment was carried out in a wind-wave

389

Y. Toba and H. Mitsuyasu (eds.), The Ocean Surface, 389–394.
© *1985 by D. Reidel Publishing Company.*

tank 0.8m high, 0.6m wide and with a test-section length 14.5m. Water
depth in the tank was kept at 0.39m, and various measurements were done
mainly at the fetch F=8.9m. Wind speed in the experiment is U_r=2.5m/s,
5m/s, 7.5m/s, 10m/s, where U_r is a cross-sectional mean speed. However,
only the data for U_r=5m/s and 10m/s are analyzed and are used in the
present paper.

For velocity measurements in the air, five hot-wire I-probes and
one hot-wire X-probe were used and operated with constant temperature
hot-wire anemometer KANOMAX SYSTEM 7303. A Pitot-static tube with a
Göttingen-type manometer was also used for wind measurements and for the
calibration of the hot-wires. The drift current was measured with a
three-axis sonic currentmeter DENSHI-KOGYO DS-105. The hot-wire array
and the sonic currentmeter were traversed vertically to measure the
vertical distributions of the velocity fluctuations in the air and in
the water. In each height the measurements were done for six minutes.
Frequency response of the sonic currentmeter is approximately unity up
to 100Hz, but spatial resolution is limited by the separation distance
27mm between each pair of a transmitter and a receiver. For the
measurements of orbital velocity of wind waves, distorsions in the
spectral response is estimated to be neglisible in a frequency range 0-
4Hz which covers the dominant frequency range of wind waves in the
present experiment. Wind waves were measured with capacitance-type wave
gauges DENSHI-KOGYO VM-303. Wind, wind waves and currents were
recorded on a video-recorder TEAC PU-400, which records simultaneously a
visual signal of wave motion from a video camera and electric signals
from the instruments.

Surfactant Almost the same measurements were done both for pure water
(ordinary tap water) and for the water containing a surfactant (Sodium
lauryl sulphate $C_{12}H_{25}O$ SO_3Na: concentration 4.4×10^{-3}%). In the latter
water wind waves were surpressed up to the wind speed U_r=10m/s and only
the drift current was generated.

Analysis of the data The data of wind waves and velocity fluctuations
both in the water and in the air were digitized at a sampling frequency
200Hz. For the analysis of wind waves and current, data were skipped to
make the sampling frequency 20Hz after taking a moving average of
Δt=0.05sec. Auto and cross spectra of various quantities were computed
through FFT method using 1024 data points. Six samples of spectra were
generally obtained. Moving average of successive 21 line spectra was
applied after taking a sample mean of six samples of the spectral data.
Equivalent degrees of freedom of the measured spectra are 252. In the
present study detailed analysis of the spectrum of turbulence in the air
is not included except for the determination of the distributions of
turbulence intensities and Reynolds stress.

3. RESULTS

As usual, observed velocity profiles near the water surface showed
logarithmic distributions both for the wind and for the drift current.

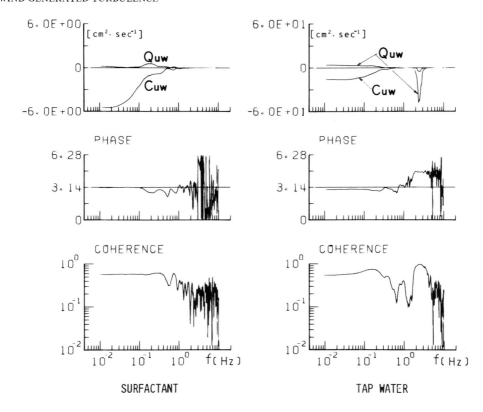

Figure 1. Cross-spectra, phase angles and coherences of the velocity fluctuations, u, w, in the drift current without wind waves (left) and with wind waves (right). U_r=10m/s, Z=-3cm, fm≑2.5Hz.

Friction velocity u_* and the shear stress τ were determined with usual profile method both for the wind and for the current. The shear stress was also determined directly as the Reynolds stress, $-\overline{\rho u w}$, where ρ is the density of the air ρ_a or the water ρ_w, and u and w are the velocity fluctuations in the air or in the water. The contribution of the wave-induced Reynolds stress $\tilde\tau_w$ to the total stress τ_w is not neglisible near the surface in the water (Figure 1). The Reynolds stress τ_w in the water was separated into the turbulent stress $\tau_w=-\rho_w\overline{u'w'}$ and the wave-induced stress $\tilde\tau_w=-\rho_w\overline{\tilde u\tilde w}$, where u' and w' are turbulent velocity fluctuations and $\tilde u$ and $\tilde w$ are the wave-induced velocity fluctuations. Practically, however, τ'_w was determined as $\int_0^{0.7 fm} C_{uw}(f)df$ and $\tilde\tau_w$ as $\int_{0.7fm}^{f_N} C_{uw}(f)df$, where f_m is the spectral peak frequency of wind waves, f_N is the Nyquist frequency and $C_{uw}(f)$ is the co-spectrum of u and w. There may be a contribution of the Reynolds stress due to the background turbulence in a frequency region $f>0.7f_m$. However, it was found from the data obtained in the water containing the surfactant that the contribution is very small. The existence of constant stress layer was

substantiated in the measured turbulent Reynold stress both for the wind
and for the drift current. The wave-induced Reynolds stress $\tilde{\tau}_w$ is
innegligible near the surface in the water, though it is still smaller
than the turbulent stress τ'_w. However, it is not clear at this time
wether it physically exists or is due to some instrumental error. It
was also found that the wind shear stress determined with the eddy-
correlation method is roughly 75% of that determined with the profile
method. The similar discrepancy has been reported by Kawamura et al
(1981), though the reason for these differences are not clear. Figure 2
shows an example of the spectra of wind waves $\eta(t)$ and velocity
fluctuations $u(t)$, $v(t)$ and $w(t)$ in the water. Although the spectra are
shown up to 10Hz, their high frequency part $f > 4$Hz are slightly affected

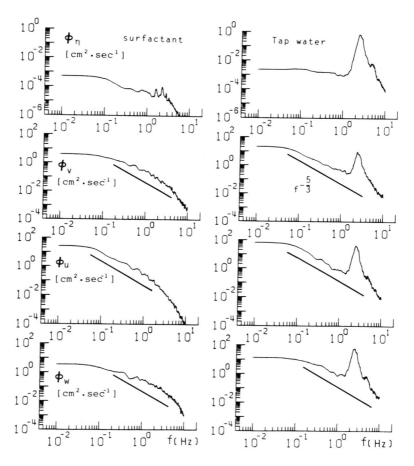

Figure 2. Frequency spectra of the surface elevation $\eta(t)$ and the
current velocity fluctuations $v(t)$, $u(t)$ and $w(t)$ under the surpression
of wind waves (left) and under the existence of wind waves (right).
U_r=10m/s, Z=-3cm. The straight line is proportional to $f^{-5/3}$.

by the spatial resolution of the sonic currentmeter. The spectra in the left side were obtained under the surpression of wind waves with the surfactant. There are no spectral peaks due to wind waves and the spectra near f=1Hz follow approximately to $f^{-5/3}$; the straight line in the figure is propotional to $f^{-5/3}$. The spectra in the right side were obtained under the existence of wind waves; the spectral peak due to wind waves is clearly seen in each spectrum, but the spectra in a frequency range $f<0.7f_m$ are quite similar to the left side spectra except for the difference in the spectral energy levels. Figure 3 (a) shows the velocity spectra which are the same spectra shown in Figure 2 but normalized with the friction velocity of the wind u_* which are measured with the eddy-correlation method, and the distance from the

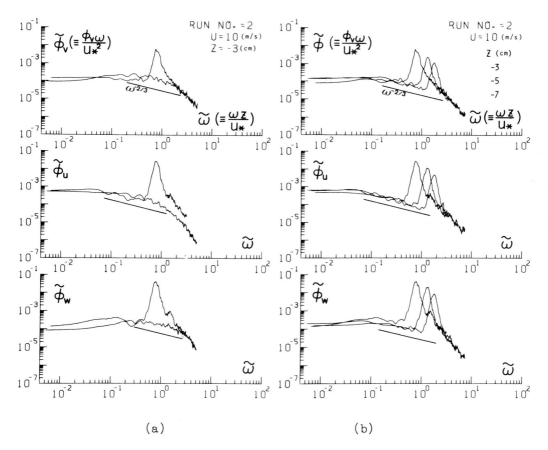

(a)   (b)

Figure 3. The normalized current velocity spectra (U_r=10m/s). (a) comparison of the spectra measured under the surpression of wind waves and those measured under the existence of wind waves (depth z=-3cm). (b) parison of the spectra measured at different depth under the existence of wind waves. The straight line is proportional to $\omega^{-2/3}$.

from the mean water surface, Z. The low frequency part of the
normalized velocity spectra observed under the existence of wind waves
agrees fairly well with that observed under the surpression of wind
waves. Figure 3 (b) show the normalized velocity spectra measurd at
different depth under the existence of wind waves. The low frequency
part of the normalized velocity spectra at different depth agree very
well with one another. The use of u_* in the current has given the
similar results, but the use of $U(z)$ instead of u_* has given more
scattered results in the present study.

4. CONCLUDING REMARKS

In a previous study (Mitsuyasu & Honda 1982) we have shown the u_*-
similarity in the wind-induced growth of water waves. The present
results show that a similar u_*-similarity is satisfied for the wind-
induced turbulence in the water. According to a recent study by
Wu(1983), the surface velocity u_s of the wind-induced current shows the
u_*-similarity. These results suggest the following simple dynamical
consequence in the air-sea interaction; The wind waves strengthen the
wind stress excerted on the water surface and then the surface velocity
and turbulence intensity of the wind-induced surface current increase.
Such a finding, in association with the fact that the low frequency
turbulence is dominant in the flow turbulence spectrum, support the view
of Jones & Kenney (1977) and Jones (1985) that inner law scaling can be
applicable to the stress carrying velocity fluctuations in the wind-
induced surface current. Discussions of the wave-induced Reynolds stress
and the energy and momentum balances at air-sea boundary are left for
the future study.

Acknowledgements. The authors are much indevted to Dr. I. S. F. Jones
of the University of Sydney for valuable comments. Our appreciations
are due to Dr. S. A. Kitaigorodskii for stimulating discussions. This
work was supported by the Grant-in-Aid for Scientific Research Project
No.57109008 by the Ministry of Education.

REFERENCES

Jones, I.S.F., 1985: Turbulence below wind waves, *Paper in this
      proceeding.*
Jones, I.S.F., and B.C. Kenney, 1977: The scaling of velocity
      fluctuations in the surface mixed layer,*J. Geophy. Res.*, **20**, 1392-
      1396.
Kawamura, H., K. Okuda, S. Kawai, and Y. Toba, 1981: Structure of
      turbulent boundary layer over wind waves in a wind wave tunnel,
      Tohoku Geophy., Journ., (Sci. Rep. Tohoku Univ., Ser.5),**28**, 69-86.
Mitsuyasu, H., and T. Honda, 1982: Wind-induced growth of water waves,
      J. Fluid Mech., **123**, 425-442.
Wu, J., 1983: Sea-surface drift currents induced by wind and waves, *J.
      Phys. Oceanogr.*, **13**, 1441-1451.

THE VERTICAL STRUCTURE OF TURBULENCE BENEATH GENTLY BREAKING WIND WAVES

E.A. Terray[1] and L.F. Bliven[2]
1 Dept. of Ocean Engineering, Woods Hole Oceanographic
   Institution, Woods Hole, MA   02543
2 Ocean Hydrodynamics, Inc., Salisbury, Maryland   21801

ABSTRACT. We present a preliminary analysis of the turbulence energy
budget beneath gently breaking laboratory wind-waves. A linear filtration procedure is employed to separate the wave-induced and turbulent motions. The data show a single Kolmogorov inertial subrange, and are consistent with observations in the outer layer of a wall-bounded shear flow.
However, the turbulent velocity is more intermittent close to the surface
than would be expected on the basis of this analogy.

1.   INTRODUCTION

When wind blows over water, it generates waves, drift currents, and turbulence. Although the dominant mechanism of turbulence production is
presumably shear layer instability driven by the wind stress, the influence of surface waves on the structure of the turbulence is not well understood. This question was recently addressed by Kitaigorodskii et al
(1983), who analyzed velocity measurements from Lake Ontario and concluded that the energetics of near surface turbulence beneath whitecapping waves departs substantially from that of a wall-bounded shear flow.
It was surmised that this deviation is the result of some wave-turbulence
interaction, either through breaking or by the turbulent diffusion of
wave energy (Kitaigorodskii and Lumley, 1983). However, the field study
did not clarify the mechanisms responsible for the conversion of wave
energy to turbulence.
    Our motivation for the present work was to examine this point
further by investigating the near-surface turbulent energy budget for
the relatively well-controlled case of laboratory generated wind waves.

2.   THE EXPERIMENT

Details of the experiment are given in Bliven et al (1983). Briefly,
the wave tank has a working section of 20 m and is 1 m wide by 1.25 m
deep, with a water depth of 75 cm. The tank is equipped with a return
pipe which was left open. A linear, programmable, hydraulic wave paddle was driven by a prerecorded spectrum of wind waves and additional
wind was drawn through a 50 cm high duct to force breaking. The friction velocity in the air was 31.7 cm/s (estimated from Pitot measurements

Y. Toba and H. Mitsuyasu (eds.), The Ocean Surface, 395–400.

of the windspeed profile). The breaking was observed to be gentle with
widespread overturning of the wave crests, but with little whitecapping
or air entrainment. A two-axis laser Doppler velocimeter (LDV) was situated 5 m downstream of the wave maker, and capacitance wave staffs were
used to measure the surface displacement at several fetches. The peak of
the wave spectrum occurred at 1.33 Hz, and the rms wave height was 1.3 cm.

3. DATA ANALYSIS

Wave and turbulent velocities were separated by linear filtration (for
references see Kitaigorodskii et al, 1983). This technique assumes a
linear deterministic relation between wave velocity and wave height, and
further requires that the wave and turbulent components are uncorrelated.
With these assumptions, the minimum mean-square error solution for the
spectral characteristic of the filter is given by $L(f) = C/S$, where C is
the cross-spectrum between the measured velocity and wave height h, and S
is the spectrum of h. There is a separate filter for each velocity component. Spectral estimates were computed by FFT, and have approximately 120
degrees of freedom. It is important to note that the filtration procedure
outlined above is uncontrolled in the sense that its results cannot be directly verified from the available data. One example of an effect not accounted for by this technique has been pointed out by Melville et al (1985
who note that the low frequency motion forced by the radiation stress gradient in a wave group is not a linear function of the surface displacement
and consequently will be misidentified as turbulence. However, we do not
see a clear indication of this component in the turbulent energy budget,
perhaps because the observed breaking is relatively gentle.

4. RESULTS

A typical turbulent energy spectrum is shown in Figure 1. Lumley and
Terray (1983) have proposed that the peak at the dominant wave frequency i
a kinematic effect resulting from the advection of a fixed spatial pattern
of inertial range turbulence by the orbital motion of the waves. Denoting
the ratio of the spectral level above to below the peak (extrapolated to
the same frequency) by R, then for Q>U this model gives

$$R = 1.31 \ (Q/U)^{2/3}$$

where $U = \langle u \rangle$ is the mean longitudinal velocity, and Q is the rms vertical wave velocity. The predicted and observed ratios are compared in
Table 1 and show good agreement.

Depth (cm)	Q/U	R (expt)	R (theor)
6.8	6.0	4.2	4.3
8.9	4.1	3.7	3.4
15.6	2.8	2.4	2.6
29.1	1.7	1.4	1.9

Table 1

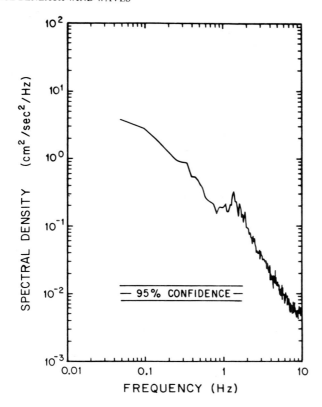

Figure 1.   Spectrum of the vertical turbulent velocity at 15.6 cm
depth, obtained by filtration.   The rms velocity is 1.1 cm/s.

      Qualitative features of the flow can be inferred from the data
shown in Figure 2.   The shape of the mean longitudinal velocity profile suggests that the flow has a three layer structure, consisting of
surface and bottom boundary layers together with a core region, presumably associated with the flow through the return pipe.   The surface
value of the drift can be estimated from the wind speed to be roughly
15 cm/s (Lin and Gad-el-Hak, 1984), so that our data evidently lie
somewhat below the layer of strong current shear.   Reynolds stress
decreases rapidly with depth, and has an extrapolated surface value
which is consistent with an upper bound determined from the windspeed
profile of 1.4 $(cm/s)^2$.   Although there is some uncertainty in the
exponents, both turbulent kinetic energy and dissipation are seen to
decay approximately as $1/z$, implying that the dissipation length scale
is only weakly dependent on depth.   Note that the dissipation rate is
determined from spectral levels outside the dominant band of wave frequencies, and hence is expected to be relatively insensitive to errors
in filtration.   Transport, due to the gradient of the turbulent kinetic energy flux, is downward, and has a depth dependence which is
similar to the Reynolds stress.

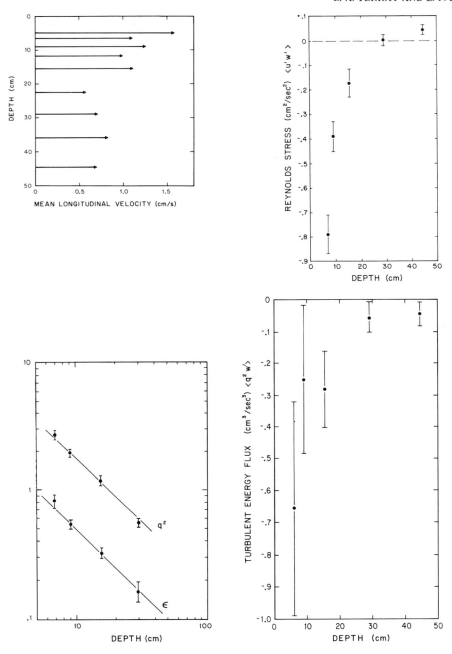

Figure 2.   Vertical profiles of the mean longitudinal velocity, Reynolds stress, turbulent kinetic energy (q^2), dissipation rate (ϵ), and vertical turbulent kinetic energy flux. The error bars denote the standard error of the block (20 s) means.

Unfortunately, because of statistical sampling error, it is not possible to determine the turbulent kinetic energy budget quantitatively; however, the dominant balance appears to be between turbulent transport and dissipation. We find that the vertical turbulent flux of wave energy (Kitaigorodskii and Lumley, 1983), is negligible compared to the observed flux of turbulent kinetic energy.

5. CONCLUSIONS

The success of a kinematic model in describing the shape of the turbulent energy spectrum implies that all of the energy input, both from wind stress and wave breaking, appears at frequencies well below the wave peak. However, the details of the turbulence energetics are complex, and although the dominant balance is between transport and dissipation, several terms appear to contribute. A major source of uncertainty is the large scale flow, which is almost certainly three dimensional, and likely consists of four counter-rotating streamwise vortices. Such a flow can contribute advection and production terms of unknown magnitude, and may be the source of long-time variability in the data.

We do not see a clear signature of breaking in either the spectrum or in the energy budget, and the data correspond, at least qualitatively, to observations in the outer layer of a wall-bounded shear flow. One reason for this may be that most of the direct energy input from breaking is dissipated above the level of our observations, which are restricted to depths below 5-6 rms wave-heights. Breaking should also enhance the momentum flux at the surface, but again, it is difficult to isolate this contribution from the wind input.

The depth dependence of the turbulent velocity kurtosis may provide an indirect signature of breaking. From Figure 3 it is seen that this quantity is large near the surface and relaxes toward the Gaussian value of 3 as the edge of the boundary layer is approached. Whether this behavior is a consequence of wave breaking remains to be seen; however, a simple additive model of shear production plus breaking is unable to account for both the observed variance and kurtosis. Multipoint velocity measurements close to the interface, together with a clear understanding of the overall flow, will undoubtedly be required to resolve this question.

ACKNOWLEDGEMENTS

This work was made possible by support from the Pew Memorial Trust (E. Terray) and NASA contract NAS6-2940 (L. Bliven). The experiment was performed by Drs. Bliven, N.E. Huang and S.R. Long at the NASA Wallops Island facility. The authors thank Mr. P.M. Dragos for his excellent assistance with all aspects of the computations.

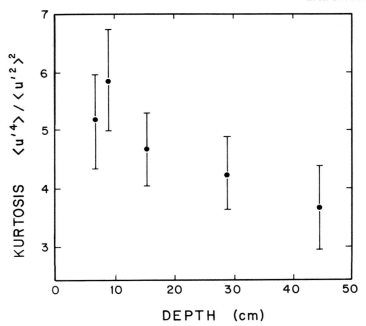

Figure 3.   Vertical profile of the kurtosis of the longitudinal
turbulent velocity.   Vertical velocity kurtosis is   similar.

REFERENCES

Bliven, L.F., N.E. Huang and S.R. Long, 1983: A Laboratory Study of
       the Velocity Field Below Surface Gravity Waves. In Gas Transfer
       at Water Surfaces, pp. 181-190, W. Brutsaert and G.H. Jirka
       (eds.), D. Reidel Publishing Co.
Kitaigorodskii, S.A. and J.L. Lumley, 1983: Wave-Turbulence Interac-
       tions in the Upper Ocean, Part I. J. Phys. Oceanogr., 13:1977-
       1987.
Kitaigorodskii, S.A., M.A. Donelan, J.L. Lumley and E.A. Terray, 1983:
       Wave-Turbulence Interactions in the Upper Ocean, Part II. J.
       Phys. Oceanogr., 13:1988-1999.
Lin, J.T. and M. Gad-el-Hak, 1984: Turbulent Current Measurements in a
       Wind-Wave Tank. J. Geophys. Res. 89:627-636.
Lumley, J.L. and E.A. Terray, 1983: Kinematics of Turbulence Con-
       vected by a Random Wave Field. J. Phys. Oceanogr., 13:2000-2007.
Melville, W.K., R.J. Rapp and E.S. Chan, 1985: Wave Breaking, Cur-
       rents and Mixing. These Proceedings.

Woods Hole Oceanographic Institution Contribution No. 5825.

WIND-INDUCED WATER TURBULENCE

Joe Wang[1] and Jin Wu[2]
1 Institute of Oceanography,
National Taiwan University, Taipei, ROC
2 College of Marine Studies,
University of Delaware, Newark, DE 19716, USA

Abstract. Systematic measurements of winds, currents and surface waves have been made in a circulating tank. Both longitudinally bounded and unbounded conditions have been simulated with a removable barricade in the tank. The experiment illustrates that the presence or absence of the return flows is significant on the vertical distributions of both the momentum flux and the turbulence energy. Spectra of currents exhibits the -5/3 slope outside the frequency range of motions induced by dominant waves, while the variances of the turbulent velocity, estimated by the coherent method, decay with depth and follow certain similarity relationships.

I. INTRODUCTION

Past studies of wind-induced aqueous surface layer flows seem to have concentrated principally on phenomena in a longitudinally bounded, small water body. Von Dorn (1953) performed experiments in a basin; Bye (1965) in a lake; Keulegan (1951), Shemdin (1972), and Wu (1968, 1975) in laboratory closed-end wind-wave tanks; and Donelon (1978) in both lake and tank. However, due to the influences of the nearby solid boundary at the downwind end, the reverse pressure gradient and its induced return flows are common features to these studies. The natural water body, like the open ocean, is not necessary to be associated with the restriction of the reverse pressure gradient. Therefore, the momentum transfer process near the air-water interface should be considered separately for the longitudinally bounded or unbounded water bodies.

An experiment has been designed and conducted in a circulating tank; by either introducing or removing a barricade, both longitudinally bounded and unbounded conditions have been simulated. The results are believed to be helpful in understanding dynamics and mechanism of wind-induced water turbulence in natural environments.

Y. Toba and H. Mitsuyasu (eds.), The Ocean Surface, 401–406.
© 1985 by D. Reidel Publishing Company.

II. THE EXPERIMENT

All experiments have been conducted in a racecourse-shape circulating wind-wave tank consisting of two straight sections and two semi-
circular sections. The tank is 7 m in overall length, 31 cm wide, and
46.5 cm in height (water was kept 25 cm deep). A variable-speed blower
was mounted on the top of the rear straight section, and all measurements including wind, waves, and currents were performed in the front
straight section. During some tests, the longitudinally bounded condition with the reverse pressure gradient and the return flows was
simulated with a barricade placing underneath the blower to block the
water flow. By removing the barricade, the flows in the tank can be
treated as longitudinally uniform, as in a large, open water body.

Winds were surveyed with a Pitot-static tube at several elevations.
Surface waves were sensed with a capacitance wave probe. The water velocity was measured with a three-beam polarized Laser Doppler Velocimeter (LDV). The surface drift currents were estimated with punched
paper discs, 0.5 cm in diameter, which were timed between two stations
1 m apart with a stop watch.

The experiments were conducted one hour after the blower had been
started. The steady state appeared to be reached according to a simple
theoretical consideration (Wang, 1983). Under this stationary condition,
waves, profiles of the wind and currents were measured simultaneously.
Since we are dealing with the wind-induced water turbulence, the wind
speed had been set at 6 m/sec, diminishing the contaminations of the
background turbulence by breaking waves. The record length of the data
was 192 seconds, and the sampling rate was chosen as 32 Hz.   Surface
drift currents were also obtained for each run.

III. VARIANCES AND SPECTRA OF VELOCITY COMPONENTS

From a dynamic point of view, the velocity of water can be expressed as the superposition of the mean motion, the wave orbital
motion, and the turbulence motion. To extract each component from overall signals is conceptually simple but practically difficult. However,
Benilov et al. (1974) and Kitaigorodskii et al. (1983) have shown that
a coherent method provides the optimum estimator of the turbulence
energy spectrum. Howe et al. (1982) and Kitaigorodskii et al. (1983)
have adopted this procedure to analyze their data. They incorporated
turbulence components correlated with waves in the calculated wave-
induced part, and non-linear wave terms in the turbulence part. This is
unavoidable, especially for those cases when the appearance of surface
waves is quite steep. Since our experiments were performed under a
condition without conspicuous breaking waves, non-linearity might not
be very strong; therefore, the coherent method has been chosen to
process our data.

Figure 1 shows the vertical variation of the wave-coherent and

wave-incoherent (or turbulent) velocity variances, and Figure 2 presents
the vertical distribution of the power spectra of the horizontal velocity. The exponentially decaying of the spectral peak associated with
dominant surface waves, and a -5/3 power slope outside this wave-
induced frequency-range imply that the superposition of background
turbulence upon wave-induced orbital motions is reasonable. A spectral
gap between 1 and 2 Hz is clearly seen in Figure 2. Since that the
separation of turbulence motions from the total fluctuations is im-
precise, and that the low-frequency part contains most of the turbulence
energy, an accumulation of energy from 0 Hz to this gap frequency, say
1.2 Hz, is helpful to illustrate the internal structure of turbulence.

For the case without the barricade, the wave-incoherent horizontal
velocity variances in this frequency band (0 - 1.2 Hz) are uniform $_{-2}$
vertically from the water surface down to 8 cm, and approach to a z
power law below 10 cm (Figure 3). Otherwise, these variances decrease
with depth monotonically and follow roughly a z^{-1} law, as illustrated
in Figure 3.

IV. THE REYNOLDS STRESS

Under a steady and longitudinally uniform condition, as in the
circulating tank, the span-wise averaged Reynolds stress, $- \langle \overline{u'w'} \rangle$, at
an arbitrary depth d, can be expressed as (Wang and Wu, 1984)

$$- \langle \overline{u'w'} \rangle = u_*^2 + \nu \frac{\partial \langle \overline{u} \rangle}{\partial z} - \text{(side-wall friction)} - \frac{1}{\rho} \int_d^o \frac{\partial \overline{p}}{\partial x} \, dz \qquad (1)$$

where $\langle \ \rangle$ indicates the laterally average process; u_* is the friction
velocity of water; ν is the kinematic viscosity; z is depth; ρ is
density of water; $\overline{p}$ is the mean pressure of water; x is the coordinate
axis along the wind direction; and $\overline{u}$ is the span-wise averaged motion
in the x direction.

For the case with the barricade, the forward momentum of wind
drift currents is confined within a narrow surface layer. Below this
layer, return flows are sluggish and the influences of the side-wall
friction on the vertical distribution of the Reynolds stress are
probably insignificant. Owing to the contribution of the reverse
pressure gradient on the momentum flux, a quasi-linear variation of the
Reynolds stress with depth is expected.

For the case without the barricade, the currents are much more
uniform vertically, the contribution of the side-wall friction on the
overall momentum balance is relatively important, and the Reynolds
stress decreases with depth monotonically (Figure 4). In the real ocean,
the surface applied wind stress is similarily supported by the Coriolis
force of the net Ekman transport within the oceanic surface layer;
therefore, the Reynolds stress within the mixed layer also decays monotonically with depth. These phenomena are interesting, with the side-
wall friction on the momentum balance in the circulating tank resemble

FIGURE 1. The normalized vertical variation of the wave-coherent and wave-incoherent velocity variances versus the dimensionless depth, kz, where k is the wave number of corresponding surface waves.

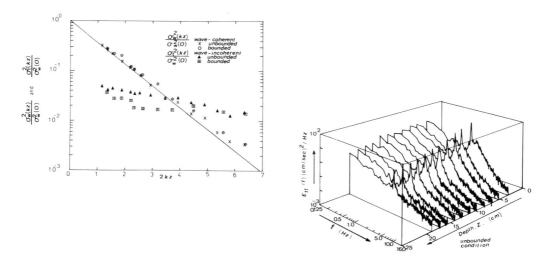

FIGURE 2. A perspective view of the vertical distribution of the power spectra of the horizontal velocity for the unbounded condition.

FIGURE 3. Vertical variation of normalized velocity variances in low-frequency band (0 - 1.2 Hz) with the dimensionless depth z/D, where D is the water depth.

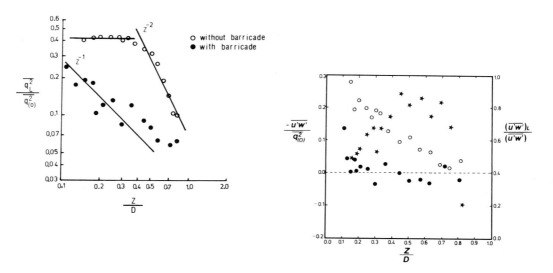

FIGURE 4. Vertical profiles of the Reynolds stress, with (●) and without (o) barricade, and (★) the contribution from low frequency band.

to the Coriolis effects in the oceanic surface layer. It is reasonable, therefore, to expect that the vertical structure of turbulence in the circulating tank is analogous to that in the oceanic mixed layer, when considering the importance of the Reynolds stress on the energy balance of a horizontally homogeneous turbulence field.

V.   DISCUSSION

For a stationary, horizontally uniform turbulence field with unidirectional mean flow, the energy balance of the fluctuating motions can be described as (Townsend, 1976)

$$\frac{\partial}{\partial z}(\frac{1}{2}\overline{q^2 w'} + \overline{p'w'}) = -\overline{u'w'}\frac{\partial \overline{u}}{\partial z} - \epsilon \tag{2}$$

where q^2 is the inner product of the fluctuating velocity vector; p' is the fluctuating pressure; ϵ is the dissipation rate of energy; and the overbar denotes the ensemble averaging.

Owing to the existence of wind waves, Kitaigorodskii and Lumley (1983) have indicated that the wave-turbulence interaction must be concerned. In our experiments, there is no conspicuous wave-breaking, and this effect may not be strong, especially for those "slow" eddies which act like a spatially inhomogeneous current field to these waves. However, fast "eddies" tend to intensify the mixing effect and extract energy from waves; therefore, the energy balance for turbulence within different frequency bands must be discussed separately.

For the longitudinally unbounded case, the nearly uniform structure of the velocity variances of the low-frequency turbulence implies an equilibrium layer below the air-water interface. The explanation is simple. Because, under a non-breaking condition, the high frequency turbulence within the wave layer is probably maintained by the energy cascading from waves. The momentum flux as well as the shear production are mainly associated with low-frequency turbulence (Figure 4), the "slow" eddies have sufficient energy supply from the current shear to compensate the locally dissipation, and the uniform structure is resulted.

Below the equilibrium layer, the z^{-2} decaying of the variances of the turbulence velocity is evident. For a shear production-free turbulence field, Toly and Hopfinger (1976) had shown $\overline{q^2} \sim z^{-2}$ from equation 2 in their experiment with a vertically oscillating grid. The z^{-2} decaying of the variances of the turbulence velocity associated with the deficiency of the Reynolds stress thus reminds us the possible correlation between our measurements and theirs. The momentum flux decreases with depth within an oceanic mixed layer; therefore, it is reasonable to speculate that the turbulence energy will decay with z^{-2} within the lower portion of a surface Ekman layer.

For the longitudinally bounded case, the downward momentum flux counterbalances with effects from the reverse pressure gradient, a linear decay of the Reynolds stress has been indicated. Associated with this distribution, the turbulent velocity variances, $\overline{q^2}$, decays with z^{-1}; which can be estimated by the equilibrium mechanism from equation 2 (Wang and Wu, 1984).

REFERENCES

Benilov, A. Yu, Kouznetsov, O. A. & Panin, G. N., 1974, On the analysis of wind-induced disturbances in the atmospheric turbulent surface layer, *Bound.-layer Met.* **6**, 269-85.

Bye, J. A. T., 1965, Wind-driven circulation in unstratified lakes, *Limnol. Oceanogr.* **10**, 451-58.

Donelan, M. A., 1978, Whitecaps and momentum transfer, in *Turbulent Fluxes Through the Sea Surface, Wave Dynamics, and Prediction*, ed. by A. Favre & K. Hasselmann, 273-87, Plenum Press.

Hopfinger, E. J. & Toly, J. -A., 1976, Spatially decaying turbulence and its relation to mixing across density interfaces, *J. Fluid Mech.* **78**, 155-75.

Howe, B. M., Chambers, A. J., Klotz, S. P., Cheung, T. K., & Street, R. L., 1982, Comparison of profiles and fluxes of heat and momentum above and below an air-water interface, *J. Heat Transfer* **104**, 34-39.

Keulegan, G. H., 1951, Wind tides in small closed channels, *J. Res. Nat. Bur. Stan.* **46**, 358-81.

Kitaigorodskii, S. A. & Lumley, J. L., 1983, Wave-turbulence interactions in the upper ocean. Part I, *J. Phys. Oceanogr.* **13**, 1977-87.

Kitaigorodskii, S. A., Donelan, M. A., Lumley, J. L., & Terray, E. A., 1983, Wave-turbulence interactions in the upper ocean. Part II, *J. Phys. Oceanogr.* **13**, 1988-99.

Shemdin, O. H., 1972, Wind generated current and phase speed of wind waves, *J. Phys. Oceanogr.* **2**, 411-19.

Townsend, A. A., 1976, *The structure of turbulent shear flow*, 2nd ed., Cambridge University Press.

Von Dorn, W. G., 1953, Wind stress of an artificial pond, *J. Mar. Res.* **12**, 249-76.

Wang, J., 1983, *Fluctuating velocity field below a wind disturbed water surface*, Ph. D. Dissertation, University of Delaware.

Wang, J. & Wu, Jin, 1984, Wind induced shear flows in lake and ocean — a laboratory study, in preparation.

Wu, Jin, 1968, Laboratory studies of wind-wave interactions, *J. Fluid Mech.* **34**, 91-111.

Wu, Jin, 1975, Wind-induced drift currents, *J. Fluid Mech.* **68**, 49-70.

THE STRUCTURE OF THE BOUNDARY LAYER UNDER WIND WAVES

Kuniaki Okuda
Tohoku Regional Fisheries Research Laboratory
Shiogama 985 Japan

ABSTRACT.   The structure of the boundary layer under wind waves is
described by reviewing our recent experimental studies using
quantitative flow visualization techniques in a wind wave tunnel.
Attention is focused on the modification of the near surface structure
by wind waves and the characteristic feature of turbulence generation.

1.   INTRODUCTION

An important feature of the surface boundary layer of the ocean is that
it forms under wind waves, and is affected by them. In certain situations wave breaking provides a significant source of turbulence energy,
and introduces factors which differentiate the boundary layer under wind
waves from that over a solid wall. Wind waves affect the drift current
and form Langmuir circulations (Leibovich and Radhakrishnan,1977).
  So far, measurements of mean velocity profiles (e.g. Wu,1975) and the
spectral forms of fluctuations (Lin and Gad-el-Hak,1984) have been
reported; as a result of these works our knowledge of the long-time
averaged structure has been greatly increased. However, the role of wind
waves in the modification and maintenance of the boundary layer has not
been clarified because of a lack of measurements which can detect both
temporal and spatial structures relating to the wind waves.
  The aim of this study is to estimate the modification of the boundary
layer by wind waves. In this paper the near surface structures observed
at individual wave crests are described, first by reviewing our recent
studies; then the characteristic features of turbulence generation under
wind waves are described in terms of new experimental results.

2.   EXPERIMENTS

The experiments were performed in a wind-wave tunnel 8.1 m long, 15 cm
wide and 70 cm high with a water depth of 53 cm. Flow velocities below
wind waves were measured by tracking "kinked" hydrogen bubble lines,
using a 16 mm cine-camera.
  This paper describes the results of two measurements. The first is a

Y. Toba and H. Mitsuyasu (eds.), The Ocean Surface, 407–412.

series of measurements of near surface structure along individual wave
profiles, made at 6 m fetch at a cross-sectional mean wind speed of 5.0
m/sec (air friction velocity u_* is 30 cm/sec); the details have been
described in Okuda(1982a). The second is a measurement of the turbulence
structure made at 6.3 m fetch at a mean wind speed of 4.7 m/sec ($u_*=$
22 cm/sec). Snapshots of the velocity and vorticity fields near the
water surface were made at 0.042 sec intervals over about 8 sec (about
40 times the wind wave period); an example is shown in Fig.1. The
vorticity field (b) is constructed from the measured velocity field (a)
by calculating circulations around small triangles formed by three
measuring points. The error in each vorticity calculation is estimated
to be less than 25%. In the figure clockwise rotation exerted by the
tangential stress of the wind is defined as plus, and high vorticity
regions are indicated by ▦ (ω (vorticity)>10 1/sec), ▥ (5<ω<10),
▨ (ω<-10) and ▦ (-10<ω<-5). Notice that the information on the
turbulent motion is clearly extracted in (b), while it is obscured in
(a) because of the predominance of the wave motion.

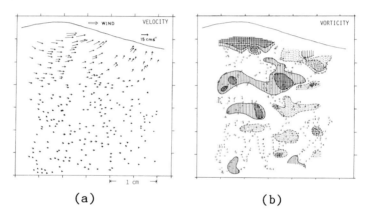

(a)   (b)

Fig.1. An example of snapshots of velocity (a) and vorticity (b).
  (b) is constructed from (a) by calculating circulations around
  small triangles formed by three measuring points. Clockwise
  rotation is defined as being plus, and high vorticity regions
  are indicated by ▦ (ω>10 1/sec), ▥ (5<ω<10), ▨ (ω<-10)
  and ▦ (-10<ω<-5).

3.   THE NEAR SURFACE STRUCTURE ALONG WAVE PROFILES

Under the conditions studied the air flow is known to separate at just
leeward of the crest(e.g. Kawai,1982), the surface velocity at dominant
crests exceeds the phase speed, although the entrainment of air bubbles
does not occur, and the surface tangential stress distribution along the
wave profile shows dominant peak near the crest(Okuda,1982a). The near
surface structure of the boundary layer under the wind waves is strongly

affected by these situations of air boundary layer and the surface
structure.

The measured vorticity distributions along the wave profiles are shown
in Fig.2. For every wave a surface vorticity layer several millimeters
thick can be distinguished. At the crest of steep waves (1) and (2), the
surface vorticity layer is noticeably thickened with particularly high
vorticity, in contrast to the vorticity distribution for (3) which is
nearly uniform along the wave profile. This thick, high vorticity region
at the crest has been found to modify the internal flow patterns and
also the dispersion relation of steep wind waves (Okuda,1982b;1983).

Considerations of the vorticity balance near the crest(Okuda,1982a)
suggest that the generation of the high vorticity region in (1) and (2)
is due to local generation of vorticity below the crest, which is
modified and intensified by the presense of both the intense surface
tangential stress and the excess flow near the crest. For wave (3) the
high vorticity region is absent in spite of the dominant peak in the
surface tangential stress, which is related to the surface velocity
smaller than the phase speed(Okuda, 1984). The combination of large
values of the surface tangential stress and the surface velocity near
the crest indicates that a significant part of the energy flux supported
by the surface tangential stress goes into the maintenance of the
phase-locked vortical motions mentioned above(Okuda,1982a).

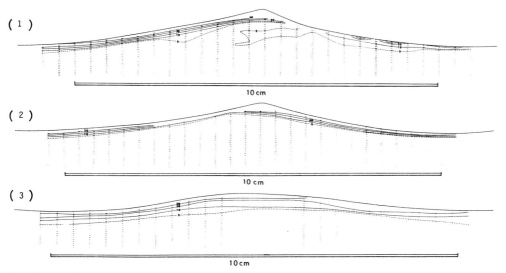

Fig 2.   Vorticity distributions along wave profiles constructed
    from velocity fields from which small scale irregular variations
    have been filtered out, from Okuda(1982a).

4.   THE ORGANIZED MOTION UNDER WIND WAVES

The wind waves studied were at the point of wave breaking; however, the

direct influence of the breaking was found to be restricted to a
shallow depth. Temporal and spatial variations of measured vorticity and
velocity patterns indicated that turbulence generation under wind waves
is governed by intense organized convective motion entailing sudden
downward intrusions of high speed near surface water with a time scale
of about ten times the wave period and a streamwise scale of about one
wavelength.

  Two examples of sequential vorticity patterns are shown in Fig.3.
The surface vorticity layer can be distinguished in every snapshot.
However, the isopleths of vorticity near the water surface suffer great
distortions, with typical spatial scales of 1 cm, although, in an
averaged version, they reproduce previous vorticity patterns(Fig.2)
constructed from the smoothed velocity field. Moreover, distinct
isolated vortical regions of both signs are found below the surface
vorticity layer, indicating strong eddy activity under wind waves. The
most noticeable result is that gross features of the vorticity patterns
and eddy activity show drastic variations with periods from 2 to 3
seconds(about 10 times the wind wave period), associated with the sudden
occurrences of downward intrusion of near surface water, which are
qualitatively similar to the bursting phenomena observed in the wall
boundary layer(Kline et al.,1967). Figure 3b is an example. The
surface vorticity layer is elongated downward, and distinct vortical
regions of clockwise rotation have become detached from the distorted
surface vorticity layer. On the other hand, in Fig.3a, which occurred at
a phase following the intrusion event, the isolated clockwise vortical
regions disappear and, on the whole, the anti-clockwise rotation
prevails. In this phase deep water is ascending.

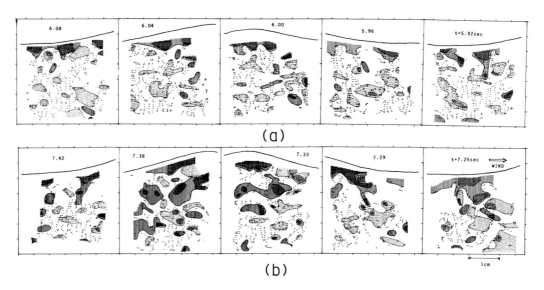

Fig.3. Typical examples of sequential patterns of vorticity.

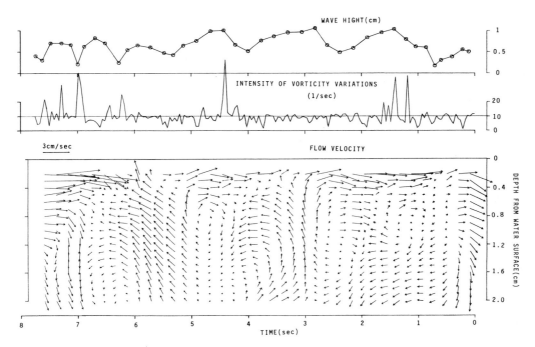

Fig.4   The velocity field of the organized convective motion under
        wind waves. Arrows indicate velocities from which the orbital
        velocity has been filtered out, and subtracted by the mean velocity
        in the measured depth range(3.03 cm/sec). In the upper panel the
        wave height and the rms of spatial variations of vorticity near
        the water surface are shown.

   The characteristic feature of organized motion can be clearly seen
from the residual velocity field after the wave orbital velocities are
filtered out(Fig.4). Arrows in the figure represent flow velocities
from which the mean velocity in the measured depth range(3.03 cm/sec)
has been subtracted. Distinct convective motions, entailing the
downward intrusion of near surface water and the ascent of deep water,
are found. The intrusion events occur at about t=0,2,4.5 and 7 sec. The
ones at t=0 and 7 sec are intense and penetrate into deep of the
boundary layer, while the ones at t=2 and 4.5 sec are shallow, and are
followed by the ascent of "high speed" deep water. This feature brings
out of the narrowness of the intrusion event in the spanwise direction
and the accompaying lateral spreading of the high speed intrusion. The
organized convective motion under wind waves seems to have a distinct
three-dimensional structure, which may be qualitatively similar to the
"horse shoe vortex" proposed for the wall boundary layer by Ofen
and Kline(1974).
   Also shown in Fig.4 are the time series of wave height and r, the
rms value of spatial variations of measured vorticity in the depth range

$0.4 < d < 0.6$ cm, which represents the intensity of the small scale
eddies near the water surface. The spikes in r appear when downward
intrusions of near surface water occur, confirming that the intrusion
event is the major mechanism of turbulence generation.

Our observations of turbulence generation under wind waves is qualitatively similar to observations in a wall boundary layer. However, the
structure of the organized motion under wind waves exhibits some
important differences. For the wall boundary layer the structure is at
an oblique angle of about $18°$ with respect to the wall(Brown and
Thomas, 1977), while under wind waves it is almost perpendicular to the
water surface(Fig.4). Distinct eddies with negative vorticity have not
been reported for the case of the wall boundary layer. The near surface
structure is strongly modified by wind waves, as described in Section 3;
they seem to be responsible for such peculiar features of the organized
motion. The clarification of this problem is a principal goal of our
future work.

REFERENCES

Brown,G. and A.Thomas(1977): Large structure in a turbulent boundary
     layer. Phys. Fluid, 20, s243–s252.
Kawai,S.(1982): Structure of air flow separation over wind wave crests.
     Boundary–Layer Met., 23, 503–521.
Kline,S.J., W.C.Reynolds, F.A.Schraub and P.W.Runstadler(1967): The
     structure of turbulent boundary layers. J. Fluid Mech., 30, 741–773.
Leibovich,S. and K.Radhakrishnan(1977): On the evolution of the system
     of wind drift currents and Langmuir circulations in the ocean. Part 2.
     Structure of the Langmuir vortices. J. Fluid Mech., 80, 481–507.
Lin,J.T. and M.Gad–el–Hak(1984): Turbulent current measurements in a
     wind wave tank. J. Geophys. Res., 89, 627–636.
Offen,G.R. and S.J.Kline(1974): Combined dye–streak and hydrogen–bubble
     visual observations of a turbulent boundary layer. J. Fluid Mech.,
     62, 223–239.
Okuda,K.(1982a): Internal flow structure of short wind waves. Part I. On
     the internal vorticity structure. J. Oceanogr. Soc. Japan, 38, 28–42.
Okuda,K.(1982b): Internal flow structure of short wind waves. Part II.
     The streamline pattern. J. Oceanogr. Soc. Japan, 38, 313–322.
Okuda,K.(1983): Internal flow structure of short wind waves. Part III.
     Pressure distributions. J. Oceanogr. Soc. Japan, 38, 331–338.
Okuda,K.(1984): Internal flow structure of short wind waves. Part 4.
     The generation of flow in excess of the phase speed. J. Oceanogr.
     Soc. Japan, 40, 46–56.
Wu,J.(1975): Wind induced drift currents. J. Fluid Mech., 68, 49–70.

WAVE BREAKING, TURBULENCE AND MIXING

W. K. Melville[1], Ronald J. Rapp[2], Eng-Soon Chan[1]
1. Department of Civil Engineering
2. Department of Ocean Engineering
   Massachusetts Institute of Technology
   Cambridge, MA 02139, USA

ABSTRACT. The evolution of programmed unstable breaking wave packets
was studied in the laboratory. Measurements of the surface elevation
were used to estimate the momentum flux lost from the wave field due to
breaking. The currents resulting from breaking were measured using
laser anemometry. (This paper is an extended abstract of work to be
published in full elsewhere.)

1.   INTRODUCTION

Wave breaking is believed to play an important role in air-sea interaction, in the dissipation of surface waves, and in the generation of
currents. With few exceptions (Banner and Melville, 1976), most of the
evidence is indirect and qualitative; a consequence of the difficulty of
unambiguously identifying and measuring breaking in the field and the
corresponding theoretical difficulties posed by the strong nonlinearity
and transition to turbulence.
       In spite of this ignorance, much progress in wave modelling has
been made in the last decade. This has been largely due to the incorporation of realistic source and nonlinear transfer terms in the spectral evolution equation. While there still exists some controversy
regarding the details of the modelling of these terms, their general
form is based on rational theory (Miles, 1957; Phillips 1957,
Hasselmann, 1962) and supporting laboratory and field measurements.
       Such is not the case for wave breaking, which is believed to play a
predominantly dissipative role in the evolution of the spectrum. There
is, at present, no rational way of predicting the evolution of a surface
beyond breaking, and there is no body of measurements that quantifies
the effects of breaking. A great deal has been learned about the instabilities that lead to breaking (Longuet-Higgins and Cokelet 1978;
Melville, 1981; McLean, 1982; Su et al., 1982). This work may ultimately lead to improved predictions of the incidence of breaking; however, it offers little or no insight into the (turbulent) breaking and
post-breaking dynamics.
       The experiments described here were undertaken with the express

413

Y. Toba and H. Mitsuyasu (eds.), The Ocean Surface, 413–418.
© *1985 by D. Reidel Publishing Company.*

purpose of measuring the dynamical consequences of breaking. These preliminary results specifically address the loss of moment flux from the
wave field and the currents generated by breaking.

2. EXPERIMENTAL DESIGN

The primary purpose of the experiments was to measure the loss of momentum flux from a wave field as a result of breaking. In order to attain
sufficient spatial resolution from a small number of instruments, the
experiments had to be repeatable (in their gross features) and predominantly two dimensional. Following Greenhow et al. (1982), we chose to
induce breaking by the superposition of sinusoidal components of
constant amplitude at a wave maker, such that all components would (according to linear theory) constructively interfere at a point down the
wave channel. Earlier experimental work by one of us (Melville, 1982),
had shown that breaking would also result from the intrinsic instabilities of initially uniform nonlinear wave fields; however, the
breaking in those experiments was not sufficiently "controlled" for the
purposes of this work. In any event, we expect that superposition is
likely to be an important cause of breaking in the ocean, and that alone
justifies the method employed here.
        Given the method of generating the breaking, there are 5 parameters
input to the experiment:
        a   - component amplitude
        f_1, f_2 - lower and upper frequency limit (or equivalently bandwidth
                         Δf and center frequency f_c)
        x_b - minimum distance to constructive interference
        n   - number of equally spaced frequency components.
        The primary variable measured was the wave amplitude η (x, t).
Thus from dimensional reasoning

$$\eta k_c = \eta k_c (x k_c, \ t f_c; \ a k_c, \ \frac{\Delta f}{f_c}, \ x_b k_c, \ n).$$

where $k_c = (f_c^2)/g$ is the linear wavenumber at the center frequency.
If n is sufficiently large (n $\gg$ 1), we expect ηk_c to be independent of
n (i.e., the spectrum is sensibly continuous). Then ηk_c will depend
on three parameters:
        $a k_c$ - amplitude parameter
        $\Delta f/f_c$ - bandwidth
        $x_b k_c$ - phase parameter.
In a random undirectional wave field $x_b k_c$ would be random and one
would specify the probability that the phases would constructively
interfere in a region $(x_b - \Delta x, \ x_b + \Delta x) k_c$. The other parameters
have direct counterparts in a random wave field.
        The processes occurring at breaking are both fast and slow, with
the fast processes being of small length scale, and the slow processes
of larger scale, as shown schematically in Figure 1. If stations upstream and downstream of the break point can be found at which the
direct effects of breaking do not influence the incident and transmitted

wave field, then a "black box" experiment can be undertaken to measure
the effect of breaking on the wave field, without directly measuring the
breaking itself. Thus, if one can measure the difference in momentum
flux passing stations 1 and 2 of Figure 1, then one has a direct measure
of the momentum flux lost from the wave field due to breaking. This
goes into generating currents. For example, from Longuet-Higgins and
Stewart (1962) and Whitham (1962), we have that conservation of mass and
momentum for a long wave group is given by

$$(\rho\bar{\eta})_t + M_x = 0 \tag{1}$$

$$M_t + gh\bar{\eta}_x = -S_x \tag{2}$$

where $\bar{\eta}$ and M are the (local) mean surface elevation and mass flux (or
momentum density) and $S = 1/2(\rho g A^2)[(2Cg/C) - 1/2]$ is the radiation
stress (or excess momentum flux), A the wave amplitude, and Cg and C the
group and phase speeds, respectively.

3. THE EXPERIMENTS

The experiments were conducted in the glass wind-wave channel at the
R. M. Parsons Laboratory, M.I.T. In these preliminary experiments,
$\Delta f/f_c$ and $x_b k_c$ were kept fixed at 0.73 and 4.6, respectively, and
three values of f_c were chosen: 0.88, 1.08, and 1.28Hz. It was found
that the number and "strength" of the breaking events varied with the
amplitude parameter ak_c. For small ak_c no breaking occurred, this
was followed by multiple spilling breakers, a single spill, multiple
spilling and plunging breakers, a single plunging breaker, and then
multiple plunging and spilling breakers, as ak_c increased.
    Measurements of wave elevation were made with five resistance wave
gauges and velocities were measured with a backscatter laser anemometer.
    Now, $S(x_1,t)$ is the excess momentum flux due to the high frequency
waves. Thus,

$$\int_{t_1}^{t_2} S(x_1,t) \, dt$$

is the excess momentum flux passing the station x_1 in the interval
(t_1, t_2). A decrease in this integral corresponds to a loss of momentum
flux from the (high frequency) wave field. Outside the breaking region,
$\partial/\partial t \simeq -Cg\,(\partial/\partial x)$ and we have that

$$\frac{\partial M}{\partial x} = -\frac{Cg}{gh - Cg^2} S_x \quad \text{(Longuet-Higgins and Stewart, 1962)} \tag{3}$$

Then changes in $\int S \, dt$ may, through equation (3) account for all the loss in momentum flux from the wave field.

4. RESULTS

Figure 2 shows photographs of spilling and plunging breakers, respectively. For fixed f_c, $\Delta f/f_c$, and $x_b k_c$, it was found that <u>single</u> breaking events occurred at two particular values of ak_c; spilling at the lesser value, and plunging at the larger. Over the range of f_c considered, it was found that these values of ak_c were constant to within $\pm$ 10%.

Figure 3 shows $\bar{\eta}^2/\bar{\eta}_0{}^2$ versus $x k_c$ for incipient breaking, spilling and plunging events with f_c = 0.88 Hz. The curves have been plotted so that $x_b k_c$ = 0 corresponds to the point of "breaking" (constructive interference) according to linear theory. The normalized change in the momentum flux, $\Delta \bar{\eta}^2/\bar{\eta}_0{}^2$, is shown as a function of ak_c in Figure 4 for f_c = 0.88, 1.08, 1.28Hz. There appears to be some influence of f_c in this figure, with the lower centre frequency showing a larger loss. The reasons for this are not clear.

Measurements of the horizontal velocity were made at a number of stations down the channel. The wave induced velocity field was subtracted from these measurements by analytical and filtering technique leaving the low frequency velocity due to breaking, an example of which is shown in Figure 5 for a position approximately 1.5 metres downstream of the break point. These measurements show that the low frequency velocities induced by the breaking may be $O(10^{-2})$ times the phase speed of the waves with a decay time of $O(10)$ times the wave period.

5. CONCLUSIONS

Our preliminary measurements of the momentum flux lost from a wave packet due to breaking and the beaking induced currents show that:

(i) The gross effects of breaking correlate with the dimensionless amplitude, bandwidth and phase parameteres.

(ii) Up to 30% of the momentum flux carried by the high frequency waves may be lost from a wave packet in a single breaking event.

(iii) The horizontal currents induced by breaking may be $O(10^{-2})$ times the phase speed and last for times of $O(10)$ times the wave period.

ACKNOWLEDGEMENTS

This work was supported by the National Science Foundation through grants 8214746-OCE and 8210649-MEA.

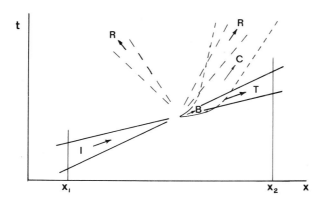

Figure 1. Schematic of fast and slow processes associated with a wave
    breaking event. I, incident waves; T, transmitted waves; R,
    radiated waves; B, fast breaking processes; C, slow processes.

Figure 2. Examples of (a), spilling and (b), plunging breakers. f_c =
    0.88Hz.

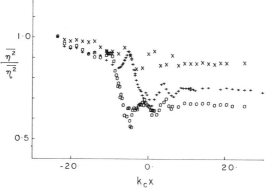

Figure 3. $\overline{\eta}^2/\overline{\eta}_o{}^2$ versus xk_c for f_c = 0.88: x, incipient breaking;
    +, spilling; O, plunging. Note the large reduction in $\overline{\eta}^2$ in the
    neigborhood of xk_c = 0.

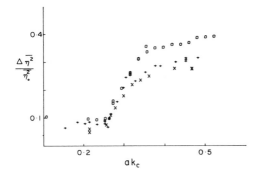

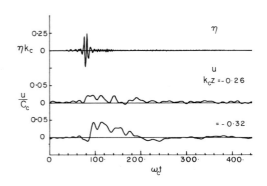

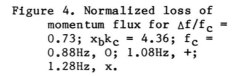

Figure 4. Normalized loss of momentum flux for $\Delta f/f_c$ = 0.73; $x_b k_c$ = 4.36; f_c = 0.88Hz, O; 1.08Hz, +; 1.28Hz, x.

Figure 5. Low frequency breaking induced currents normalized to the linear phase speed of the centre-frequency, U/C_c, downstream of a plunging wave at f_c = 0.88Hz. The time scale is $w_c t$ where w_c is the radian centre frequency. Also shown is the surface displacement ηk_c.

REFERENCES

Banner, M. L. and W. K. Melville, 1976: On the separation of air-flow over water waves. J. Fluid Mech., 77, 825–842.
Greenhow, M., P. Brevig, T. Vinje and J. Taylor, 1982: A theoretical and experimental study of the capsize of a Salter's duck in extreme waves. J. Fluid Mech., 118, 221.
Hasslemann, K., 1962: On the non-linear energy transfer in a gravity wave spectrum, Part 1. J. Fluid Mech., 12, 481–500.
Longuet-Higgins, M. S. and E. D. Cokelet, 1978: The deformation of steep surface waves on water II. Proc. T. Soc. Lond., A364, 1–28.
Longuet-Higgins, M. S. and R. W. Stewart, 1962: Radiation stress and mass transport in gravity waves, with application to surf beats. J. Fluid Mech., 13, 481–504.
Melville, W. K., 1982: The instability and breaking of deep-water waves. J. Fluid Mech., 115, 165–185.
McLean, J. W., 1982: Instabilities of finite amplitude water waves. J. Fluid Mech., 114, 315–330.
Miles, J. W., 1957: On the generation of waves by shear flows. J. Fluid Mech., 3, 185–204.
Phillips, O. M., 1957: On the generation of waves by a turbulent wind. J. Fluid Mech., 2, 417–445.
Su, M. Y., M. Bergin, P. Marler and T. Myrick, 1982: Experiments on nonlinear instabilities and evolution of steep gravity wave trains. J. Fluid Mech., 124, 45–72.
Whitham, G. B., 1962: Mass, momentum and energy flux in water waves. J. Fluid Mech., 12, 135–147.

EXPERIMENTAL STUDY ON TURBULENCE STRUCTURES UNDER SPILLING BREAKERS

Masataro Hattori and Toshio Aono
Department of Civil Engineering, Chuo University
Kasuga 1-13-27, Bunkyo-ku
Tokyo, 112
Japan

ABSTRACT. Turbulence structures under spilling breakers were examined by experimental results obtained from simultaneous measurements of the water surface elevation and fluid velocities, and from flow visualizations. Surface-generated turbulence contributes greatly to nonstationarity and intermittency in the turbulence structure, and to formation of vortex-like motions around the border zone between the outer and inner regions.

1. INTRODUCTION

Turbulence generated by breaking waves is an important factor for various processes, such as wave transformation, sediment transport and nearshore current generation, within the surf zone.

In order to examine characteristics of the turbulence under breaking waves, some experimental studies have been performed in both the field and laboratory. However, lack of adequate knowledge of the characteristics and generation mechanism of turbulence due to wave breaking is still preventing understanding results of such processes.

This paper reports experimental results on the generation mechanics and characteristics of turbulence under spilling breakers. Principal aims of the present study are (1) To develop a method for separating the wave-induced and turbulent components of the internal fluid velocity, (2) To determine the time and space dependencies of the statistical characteristics of turbulence, and (3) To examine the turbulence structure by means of flow visualizations.

2. TEST EQUIPMENT AND PROCEDURES

Experiment was performed in a glass-walled flume, 0.30m wide, 0.55m high, and 20m long. A flap-type wave maker is installed at the end of wave flume.

Most of previous laboratory experiments have been made in cases when incident waves break on a sloping bottom. Since velocity measure-

419

Y. Toba and H. Mitsuyasu (eds.), The Ocean Surface, 419–424.

ments in this study were made by an Eulerian method, stabilization of
the breaking position was essential to the acquisition of reliable data
of the velocity field under breaking waves. A flat bottom connecting to
a slope of 1/20 was used as the model beach profile. (Fig. 1).

     Properties of incident waves were determined so as that waves break
at the connecting section between the flat and sloping bottoms (Table
1). Coordinate system and notations used are also shown in Fig. 1.

     Simultaneous measurements of
the water surface elevation and
fluid velocities were made at
various locations in the breaker
zone. Water surface elevations were
measured by capacitance type wave
gages. Measurements of the
horizontal and vertical velocities,
u and w, were made using a split-
type hot film velocimeter (TSI model
1288) in a vertical plane beneath
the water surface measurement.

     Outputs of water surface eleva-
tion and fluid velocities were
recorded on a 7-channel analogue
recorder. Analogue data were
digitized by A-D converter at a
sampling frequency of 500 Hz
resulting in a Nyquist frequency of
250 Hz for convenient analyses by
computer.

Table 1   Experimental conditions.

$T(s)$	H_I (cm)	h_b (cm)	H_b/h_b
1.0	2.97	5.0	0.80

H_I: Incident wave height.
T: Wave period.
H_b & h_b: Wave height and depth
       at the breaking point.

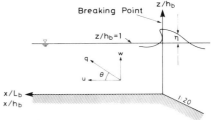

Fig. 1   Coordinate system and
         notations.

3.   SEPARATION OF WAVE-INDUCED AND TURBULENT VELOCITY COMPONENTS

In previous studies, various techniques have been employed in the
separation of the measured wave-induced and turbulent velocity
components, such as phase averaging, moving averaging, band-pass filter
and numerical filter methods.

     In the present study, the turbulent velocity is defined as the
portion of velocity field which is not correlated with the wave profile
(Thornton; 1979, Mizuguchi; 1982). Separation of wave-induced and
turbulent velocity components was made using non-recursive numerical
filters having frequency response characteristics determined by the
coherence function between measured water surface elevation and
horizontal velocity, given by Eq. (1).

$$\gamma_{\eta u}^{2}(f) = |S_{\eta u}(f)|^{2}/S_{\eta}(f)S_{u}(f) \tag{1}$$

in which $S_{\eta}(f)$ and $S_{u}(f)$; wave profile and horizontal velocity spectra,
$S_{\eta u}(f)$; cross spectrum between water surface elevation and horizontal
velocity, and f; frequency. The wave profile and velocity spectra and
cross spectrum were calculated from 70s long sample records of water
surface elevation and velocity by the FFT(N=8129, Degree of freedom=20).

$$S_{u'}(f) = S_u(f)\{1 - \gamma_{\eta u}^2(f)\}$$

and   (2)

$$S_{w'}(f) = S_w(f)\{1 - \gamma_{\eta u}^2(f)\}$$

Figure 3 and 4 are wave profile and velocity spectra. In Fig. 3, harmonic peaks due to nonlinearity of wave motions are evident in the frequency range from 1 Hz to 4 Hz. In the frequency range above the harmonic peaks, the slope of the spectra approximates a -5 power, which is consistent with the saturation spectra derived from the similarity arguments for deep water (Phillips; 1958, Thornton; 1977).

Figure 4 is a typical example of changes of the total and turbulent velocity spectra in the shoreward direction from the breaking position. The forms of the horizontal velocity spectra are very similar to those for the wave profile spectra. The turbulent velocity spectra of both the horizontal and vertical components completely coincide with the total velocity spectra in the frequency range $f > 4$ Hz. This indicates that turbulence dominates in the velocity field in the high frequency range. The slope of the velocity spectrum tails is approximated by a -3 power.

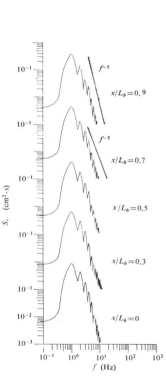

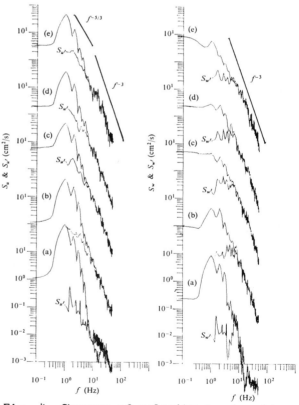

Fig. 3 Wave profile spectra.

Fig. 4 Changes of velocity spectra in the shoreward direction from the breaking position ($z/h_b = 0.4$).

4. STATISTICAL CHARACTERISTICS OF TURBULENCE

The basic properties of turbulence were examined by using four major
types of statistical functions such as probability density functions,
auto-correlation functions, power spectral density functions, and mean
square values.

4.1 Probability Structures of the Turbulence

Probability density functions (PDF) of the turbulent velocities, $p(u')$
and $p(w')$, were calculated by using the time histories of u' and w'.
    Figure 2 is an example of the time averaged PDF distributions for
u' and w'. The Gaussian forms with zero means and equal variances are
shown for the reference. Distributions of the PDF for u' and w' at a
point show the same distribution pattern. Strong deviation in the PDF
from the Gaussian means that the statistical characteristics of the
turbulence are nonstationary.

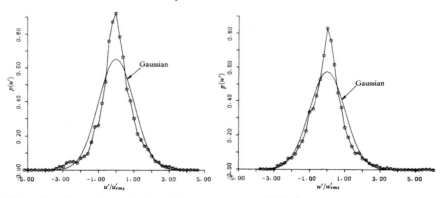

Fig. 2   Example of the time average PDF ($z/h_b = 0.4$, $x/L = 0.5$)

    According to the spatial and phase variations of the skewness and
kurtosis factors of PDF, the nonstationarity and intermittency of
turbulence structures are intensified during passage of the wave crest,
and the distribution of turbulence intensities is spotty.

4.2 Wave Profile and Velocity Spectra

Spectral analyses for the water surface elevation and for the total and
turbulent velocities were made, in order to inspect the characteristics
of turbulence structures and energy cascade processes in the frequency
domain.
    The spectra of wave profile and total velocities were calculated
from the time series data of the water surface elevation and velocities
by the FFT. Turbulent velocity spectra for u' and w' were computed by
using the following equations, which imply that the incoherent portion
of the velocity field is assumed to be turbulence.

In the low frequency range from the primary wave frequency (1 Hz) to 4 Hz, the horizontal turbulent velocity spectra exhibit a slope of about −5/3. The −3 slope of the velocity spectrum tails implies that the turbulence structure at high frequencies (small-scale) is two dimensional, whereas the turbulence at low frequencies (large-scale) is three dimensional.

4.3 Turbulence Intensity and the Reynolds Stress

The turbulence intensity was obtained by integrating the turbulent velocity spectra from the primary wave frequency to 50 Hz. The negative correlation between u' and w', −u'w', was calculated by using time series data for the horizontal and vertical turbulent velocities.

Figure 5 illustrates horizontal changes of the turbulence intensities, u'_{rms} and w'_{rms}, in the shoreward direction from the breaking position. The vertical distributions of u'_{rms} and w'_{rms} have almost the same pattern. The turbulence intensities increase greatly in the outer region (Svendsen et al.; 1978), in which rapid transition of wave profile is observed. An equilibrium region, in which the turbulence intensity is almost constant at a certain distance from the bottom, exists in the inner region (0.4 < x/L < 0.8), where the wave profile changes shape rather gradually. It is noticed from Fig. 5 that the injection of turbulence from the free surface is confined to the upper layer.

According to the spatial distribution of the Reynolds stress, the sign of −u'w' at the upper layer varies in order of minus, plus and minus in the direction of wave propagation. This suggests that vertex-like and long-lived motions are generated over the outer and inner regions.

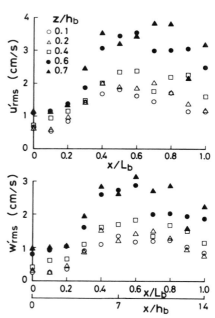

Fig. 5 Changes of u'_{rms} and w'_{rms} in the shoreward direction.

5. FLOW VISUALIZATION

Two flow visualization techniques were employed simultaneously in this experiments. Bottom-generated turbulence was visualized by using the thin-layered milk method developed by Hayashi and Ohashi (1982). Surface-generated turbulence and turbulence by internal shear were visualized by the air-bubble tracer method, similar to that by Peregrine and Svendsen (1978).

Large scale turbulence bursts intermittently near the bottom during passage of wave trough from the end of the outer region to the inner region. Such large scale turbulence or turbulent spots is stretched vertically and grows during passage of the wave crest.

Milk cloud patterns indicates that vortex-like motions are generated around the border zone between the outer and inner regions due to injection of turbulence from the free surface. Existance of such vortex-like motions are confirmed from the turbulent velocity spectra at low frequencies and from the Reynolds stress distributions at the upper layer inside the breker zone.

6. CONCLUDING REMARKS

The main findings of the present study are as follows:
(1) The probability density function and other statistical quantities of the turbulence confirm the validity of the proposed method for separating the wave-induced and turbulent velocity components.
(2) The injection of turbulence due to wave breaking gives rise to non-stationarity and intermittency in the turbulence structure, and forms vortex-like and long-lived motions around the border zone between the outer and inner regions.
(3) According to spectral analyses, turbulence at high frequencies exhibit two-dimensional characteristics, whereas the turbulence at low frequencies has three-dimensional structure.
(4) Under spilling breakers, surface-generated turbulence is confined to the upper layer. Within the inner region, there exists an equilibrium region in which the turbulence intensity is almost constant at a certain distance from the bottom.

7. REFERENCES

Hayashi, T. and M. Ohashi, 1982: A dynamical and visual study on the oscillatory turbulent boundary layer. Turbulent Shear Flows 3, 18-33.

Mizuguchi, M., 1982: Separation of turbulent component from irregular wave data. Proc. of 37th Annual Conv. of JSCE, II, 863-864 (in Japanese).

Peregrine, D.H. and I.A. Svendsen, 1978: Spilling breakers, bores and hydraulic jumps. Proc. of 16th ICCE, 540-550.

Phillips, O.M., 1957: The equilibrium range in the spectrum of wind-generated waves. JFM, 4, 426-434.

Svendsen, I.A., P.A. Madsen and J.B. Hansen, 1978: Wave characteristics in the surf zone. Proc. of 16th ICCE, 520-539.

Thornton, E.B., 1977: Rederivation of saturation range in the frequency spectrum of wind-generated gravity waves. J. of Physical Oceanography, 7, 137-140.

Thornton, E.B., 1979: Energetics of breaking waves within the surf zone. JGR, 84, C8, 4931-4938.

EXPERIMENTAL STUDY ON WIND DRIVEN CURRENT IN A WIND-WAVE TANK
- EFFECT OF RETURN FLOW ON WIND DRIVEN CURRENT -

H. Tsuruya[1], S. Nakano[1] and H. Kato[2]
1 Port and Harbour Research Institute, Ministry of Transport,
  Yokosuka 239 Japan
2 Department of Civil Engineering, Ibaraki University, Hitachi
  316 Japan

ABSTRACT. The wind driven current were investigated in a wind-wave tank
in which a false bottom was installed in order to control the return flow.
The ratio of the surface velocity to the mean wind velocity in the case
without return flow increased up to 3.8% and this is greater than that in
the case with return flow in which the ratio was 3.3%. The velocity distribution of wind driven current applicable for both cases with and without return flow was derived.

1. INTRODUCTION

When wind blows over the surface of the water, wind driven current is
induced mainly by a tangential surface stress. In studying the physics
of the air-sea system it is important to know the characteristics of the
wind driven current. Keulegan(1951) has pointed out that the ratio of
the surface velocity to the mean wind velocity in a wind tunnel was about
3.3%. It seems, however, that this ratio must be changed according to
the boundary conditions. For example, there must be no return flow in
the open sea. In the closed basin or bay, on the other hand, return flow
is induced near the bed. Therefore, it can be considered that the surface
velocities and velocity profiles differ according to such boundary conditions. In this study the drag coefficient of the water surface and the
wind driven current are investigated for both cases with and without
return flow.

2. EXPERIMENTAL PROCEDURES

2.1. Wind-Wave Tank

The wind-wave tank, shown in Fig.1, is 0.6m wide, 0.85m deep and 22m long.
Inside the tank, a false bottom was installed in order to circulate the
return flow below it and to suppress the return flow inside the flume.
In the case with return flow the entrance of the circulation path under
this false bottom was shut and the return flow was induced in the flume.

425

Y. Toba and H. Mitsuyasu (eds.), The Ocean Surface, 425–430.
© 1985 by D. Reidel Publishing Company.

As a special case detergent was added to suppress the generation of waves. The experimental conditions are listed in Table 1.

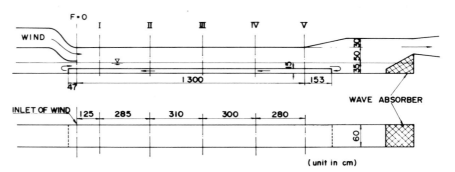

Figure 1. Wind-wave tank

Table 1. Experimental conditions

Case	Return flow	Water depth d (cm)	Water condition
A	without	20	tap water
B	with	20	tap water
C	without	15	tap water
D	with	15	tap water
CR	without	15	detergent added
DR	with	15	detergent added

2.2 Experimental Procedures

The surface velocity was measured by timing floats passing two stations at a 0.5m distance. The thin circular papers of 0.56cm diameter punched from computer cards, saturated with paraffin, were used as surface floats. Tests were conducted at least 30 times for each run. The wind driven current profile was measured by using both minute tracers of the vinyl chloride powder and a hot-film anemometer. Wind velocity profiles were measured by using a pitot static tube and a differential pressure transducer.

3. EXPERIMENTAL RESULTS

3.1 Wind Velocities and Drag Coefficients

Wind velocity profiles near the air-water interface can be represented as a logarithmic distribution

$$\frac{U(z)}{U_{*a}} = \frac{1}{\kappa} \ln \frac{z}{z_{oa}} , \tag{1}$$

where $U(z)$ is the wind velocity at an elevation z above the mean water surface, $U_{*a}(= \sqrt{\tau_a/\rho_a})$ the friction velocity of the wind, τ_a the wind shear stress at the water surface, ρ_a the density of air, κ the von Kármán's constant usually taken to be 0.4 and z_{oa} the roughness length of the water surface.

The drag coefficient C_d can be represented in terms of the wind velocity U_{10} as $C_d=(U_{*a}/U_{10})^2$, where U_{10} is the wind velocity at the height z=10m. As usual the relation between C_d and U_{10} was examined. Although the experimental data shows the large scatter, the relation could be divided roughly into the following four regions

$$
\left.
\begin{array}{ll}
\text{the logarithmic law for the hydrau-} & \\
\text{lically smooth surface,} & U_{10} < 7 \text{ (m/s)}, \\
C_d = 0.814{\times}10^{-3} \sim 1.2{\times}10^{-3}, & U_{10} = 7, \\
C_d = 1.2{\times}10^{-3}, & 7 < U_{10} \le 14, \\
C_d = 3.70{\times}10^{-6} U_{10}^{0.80}, & U_{10} > 14.
\end{array}
\right\} \quad (2)
$$

3.2 Surface Velocity

The ratios of the surface velocity u_o to the mean wind velocity $\bar{U}$ in the tunnel are plotted in Fig.2. Open symbols in Fig.2 indicate the case without return flow and solid symbols indicate the case with return flow. The figure shows that the ratio $u_o/\bar{U}$ changes corresponding to the difference of boundary conditions. In the case without return flow the ratio is about 3.8% when $u_o d/\nu_w = 7{\times}10^4$. In the case with return flow the ratio is about 3.3% as found by Keulegan (1951).

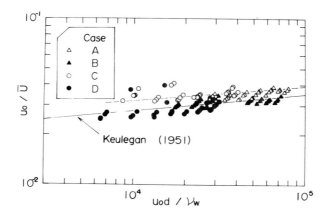

Figure 2. The relationship between
$u_o/\bar{U}$ and $u_o d/\nu_w$

4. ESTIMATION OF WIND DRIVEN CURRENT

It is recognized that the wind driven current near the water surface is well approximated by a logarithmic distribution (e.g. Kato 1974). It seems to be reasonable that the current profile near the bottom is also approximated by a logarithmic distribution. Consequently the velocity distribution of the wind driven current applicable for the entire flow region is given by

$$u(z) = \frac{u_{*s}}{\kappa} \ln \frac{z_{os}+d}{z_{os}-z} - \frac{u_{*b}}{\kappa} \ln \frac{z_{ob}+d+z}{z_{ob}} , \tag{3}$$

where u_{*s} and u_{*b} are the friction velocities in the water at the surface and the bottom respectively, z_{os} and z_{ob} the roughness length in the water at the surface and the bottom respectively and d the water depth. In deriving Eq.(3) the condition that the velocity at the bottom must vanish is applied. Integrating Eq.(3) the mean velocity v is given by

$$v = u_{*s}/\kappa - u_{*b}[\ln\{(d+z_{ob})/z_{ob}\}-1]/\kappa. \tag{4}$$

Assuming that the roughness condition at the bottom is hydraulically smooth, the roughness Reynolds number at the bottom can be represented as

$$u_{*b}z_{ob}/\nu_w = 0.111, \tag{5}$$

where ν_w is the kinematic viscosity of water. If u_{*s}, v and d are given, u_{*b} and z_{ob} are determined from Eqs.(4) and (5). In the case with return flow, the mean velocity v in Eq.(4) vanishes. Then Eq.(3) can be expressed as

$$\left.\begin{array}{l} u(z) = \frac{u_{*s}}{\kappa}(\ln\frac{z_{os}+d}{z_{os}-z} - m \ln \frac{z_{ob}+d+z}{z_{ob}}), \\[2mm] z_{ob} = 0.111\nu_w/(m u_{*s}), \\[2mm] m = u_{*b}/u_{*s} = [\ln\{(z_{ob}+d)/z_{ob}\}-1]^{-1}. \end{array}\right\} \tag{6}$$

From the first equation of (6) the surface velocity u_o is represented as

$$u_o/u_{*s} = \ln\{(d+z_{os})/z_{os} -(1+m)\}/\kappa. \tag{7}$$

Equation (7) shows that the ratio u_o/u_{*s} depends on the water depth. When the water depth is sufficiently large, the ratio m can be considered as nearly zero and Eq.(7) reduces to Spillane and Hess's result(1978). In the case without return flow the bottom shear stress in Eq.(6) is neglected. Consequently, the wind driven current profile and the surface velocity are given by

$$\left.\begin{array}{l} u(z) = \frac{u_{*s}}{\kappa} \ln \frac{z_{os}+d}{z_{os}-z} , \\[3mm] \frac{u_o}{u_{*s}} = \frac{1}{\kappa} \ln \frac{z_{os}+d}{z_{os}} , \end{array}\right\} \tag{8}$$

respectively.
    Next we assume that part of the wind momentum is transferred to the

momentum of the wind driven current. The friction velocity at the surface is, therefore,

$$u_{*s} = \sqrt{\alpha C_d (\rho_a/\rho_w)} \; U_{10},\tag{9}$$

where $\alpha(=1-\tau_w/\tau_a)$ is the ratio of the stress supported by the wind driven current to the wind shear stress, τ_w the wave drag and ρ_w the density of water. Moreover we assume that the roughness Reynolds number on and beneath the water surface coinside (Kondo et al. 1974). Then the roughness length z_{os} is determined as

$$z_{os} = \sqrt{(\rho_w/\rho_a)}(\nu_w/\nu_a)z_{oa}.\tag{10}$$

The roughness length z_{oa} is given by $z_{oa} = 10^3/\exp(\kappa/\sqrt{C_d})$. When the temperatures of both air and water are 20°C, Eq.(10) can be represented as $z_{os} = 1.9z_{oa}$. Consequently, if U_{10} and d are known, the current distribution in both cases can be calculated from Eqs.(2),(6),(9) and (10) or (2),(8),(9) and (10).

Figures 3 and 4 show the experimental data and the theoretical curves of the wind driven current in the case with and without return flow, respectively. Both figures indicate the good agreement between the experimental data and the theoretical curves.

Figure 5 shows the ratio of the surface velocity u_0 to the wind velocity U_{10} versus U_{10}. Here, we assumed $\alpha=1.0$ for $U_{10}< 7$ m/s and 0.93 for $U_{10}> 7$ m/s, $\rho_a=0.00122$g/cm^3, $\rho_w=1.00$g/cm^3, $\nu_a=0.145$cm^2/s and $\nu_w=0.01$ cm^2/s. The figure indicates that the theoretical curves of u_0/U_{10} well explain the tendency of the experimental data.

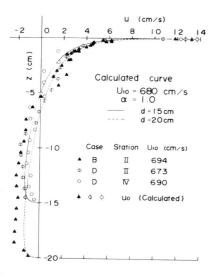

Figure 3. Wind driven current
distributions
(with return flow)

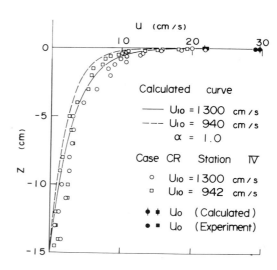

Figure 4. Wind driven current
distributions
(without return flow)

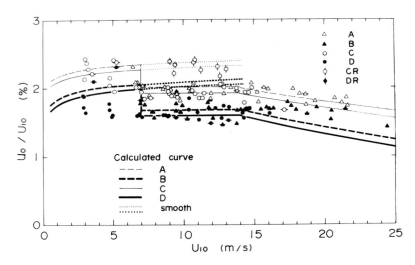

Figure 5.   The relationship between u_o/U_{10} and U_{10}

5. CONCLUSIONS

We have descrived the experiments conducted in a wind-wave tank to investigate the effect of return flow upon the characteristics of the wind
driven current. It was shown that the ratio of the surface velocity to
the mean wind velocity increases when the return flow does not exist.
The velocity distribution for the case with and without return flow was
derived. Making use of this relation, the ratio of the surface velocity
to the wind velocity could be evaluated. Close agreement between
observed and calculated values was obtained.

REFERENCES

Amorocho, J. and J. J. DeVries, 1980: A new evaluation of the wind stress
        coefficient over water surfaces. *J. Geophys. Res.*, Vol.85, No.C1,
        433-442.
Kato, H., 1974: Calculation of the wave speed for a logarithmic drift
        current. *Rept. Port and Harbour Res. Inst.*, Vol.13, No.4, 3-32.
Keulegan, G. H., 1951: Wind tides in small closed channels. *J. Res. Nat.
        Bur. Stand.*, 46, 358-381.
Kondo, J., G. Naito and Y. Fujinawa, 1974: Wind-induced current in the
        upper most layer of the ocean (*in Japanese*). *Rept. Nat. Res. Center
        for Disaster Prevention*, 10, 67-82.
Spillane, K. T. and G. D. Hess, 1978: Wind-induced drift in contained
        bodies of water. *J. Phys. Oceanogr.*, 8, 930-935.

OBSERVED STATISTICS OF BREAKING OCEAN WAVES

L.H. Holthuijsen and T.H.C. Herbers
Delft University of Technology
Department of Civil Engineering
P.O. Box 5048
2600 GA Delft
the Netherlands

ABSTRACT. Observations have been made of the occurrence of breaking waves at sea and their parameters to check the suitability of commonly used breaking wave criteria. Our analysis of the populations of breaking and non-breaking waves indicates that seemingly obvious parameters such as wave steepness or wave asymmetry cannot be used to distinguish between breaking waves and non-breaking waves.

1. INTRODUCTION

Statistical characteristics of breaking ocean waves are important to ocean engineering and air-sea interaction studies (e.g. Kjeldsen, 1983; Banner and Melville, 1976). Most of the presently available information on the statistics of breaking ocean waves is based on theoretical analyses in which the joint probability density of individual wave heights and wave periods is considered. This distribution is truncated at a certain wave steepness to obtain the marginal distribution of the wave height of breaking waves and the fraction of breaking waves (e.g. Battjes, 1971; Nath and Ramsey, 1974; Houmb and Overvik, 1976). The existence of a steepness criterion per individual wave to distinguish between breaking waves and non-breaking waves is assumed on the basis of criteria for periodic waves derived from theoretical work (e.g. Michell, 1893, quoted in Kinsman, 1965) or laboratory observations (e.g. van Dorn and Pazan, 1975).
   We performed relatively simple observations of breaking ocean waves to investigate the relevance of the above breaking criterion and also of other criteria. The statistics of some other wave parameters were also investigated. To that end we used a waverider buoy which was observed visually on two occasions.
   A brief resume of the procedures and results is presented in the following. A more comprehensive publication is in preparation.

Y. Toba and H. Mitsuyasu (eds.), The Ocean Surface, 431–436.
© *1985 by D. Reidel Publishing Company.*

2. OBSERVATIONS AND ANALYSIS

The waverider buoy was located in the southern North Sea approximately 10 km off the Dutch coast in 17.5 m deep water. This is fairly deep water for our observations for which the spectral peak frequency was always higher than 0.15 Hz. The buoy was anchored at about 100 m north of an observation tower (the Noordwijk tower) where a cup anemometer provided us with wind observations. On two occasions (May 2-3 and May 24, 1983) the buoy was observed visually from the observation tower (first occasion) and from a ship (second occasion). The wind was on-shore and its speed varied from about 8 m/s to about 12 m/s. Every time when an active white-cap passed under the buoy, the conventional wave record was electronically "flagged" by the observer. A total of about 9000 waves was thus observed of which about 1000 waves were breaking.

The analysis relates mainly to the individual waves. A wave is defined here as the sea surface profile between two successive zero-down-crossings. For each wave we determined the following primary parameters (fig. 1)
- wave height H, defined as the maximum difference in surface elevation in one wave,
- wave period T, defined as the time interval between the two zero-down-crossings of the waves,
- wave steepness s, defined as

$$s \equiv H / (\frac{g}{2\pi} T^2)$$

in which g is acceleration due to gravity.

In addition we determined the crest front steepness ε and the crest asymmetry parameters μ and λ as suggested by Kjeldsen (1983), see fig. 1.

$$\varepsilon \equiv a_m / (\frac{g}{2\pi} T_1 T)$$

$$\mu \equiv a_m / H$$

$$\lambda \equiv T_2 / T_1$$

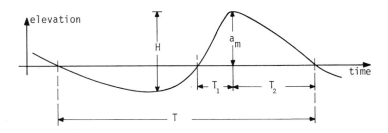

Fig. 1: Definition diagram for wave parameters.

Before the analysis of the data each record was corrected with the complex transfer function of the buoy with a high-frequency cut-off imposed at frequency f = 0.6 Hz to avoid a pronounced peak in the amplitude part of the transfer function. This cut-off frequency is three to four times the peak frequency in our observations and we therefore expect no significant influence of this low-pass filtering on the statistics of the above defined wave parameters.

Our method of estimating the steepness and asymmetry parameters does not take into account the effects of the orbital motion of the buoy. This implies that our results are biased to some extent. To illustrate this, Longuet-Higgings (1984) pointed out that our analysis would overestimate the period of a wave by a factor of 1.28 if the wave had a maximum steepness ($H/L \simeq 0.14$) and if the buoy would follow the particle motions at the surface. However such a steep wave is a very rare event in our observations, even if we would correct for this phenomenon of orbital motion. The observed steepness values (without correction) vary typically between 0.02 and 0.08 (fig. 3) so that the overestimate of the wave period is much less than 1.28.

For the wave group analysis we define a wave group as a sequence of waves each of which is higher than the H_{rms} value of the record considered.

3. RESULTS

The joint probability density function of wave height H and wave period T from all observations for both the populations of breaking and non-breaking waves is given in fig. 2. Although breaking waves are higher on the average than non-breaking waves (by about factor of 1.5), the two distributions overlap considerably so that the two populations are not well separated in the (H,T) combination. This is also illustrated with the probability density function of the steepness s (fig. 3). The breaking waves are slightly steeper on the average than the non-breaking waves but the distributions overlap too much to use steepness s as an indicator for breaking. This is also true for the crest front steepness ε and the crest asymmetry parameters μ and λ.

The wave group analysis indicated that the position of the first breaker in a group is usually slightly ahead of the centre of the group (fig. 4). This confirms the aerial observations of Donelan et al. (1972). We found that a fraction of 0.69 of the waves was breaking in a group (defined for a threshold level H_{rms}). If we exclude groups with one wave, this fraction drops to 0.53.

The fraction of breaking waves per 30 min. record as a function of wind speed (transformed to 10 m elevation wind) is given in fig. 5. The observed values are slightly lower on the average (for a given wind speed) than those of Toba et al. (1971) but the agreement is fairly good otherwise. These authors used a method of observation which is very similar to ours except that Toba et al. used a wave staff instead of a buoy. An entirely different method of observation was used by Longuet-Higgins and Smith (1983). These authors used a

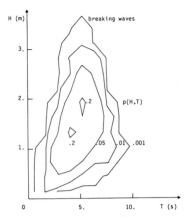

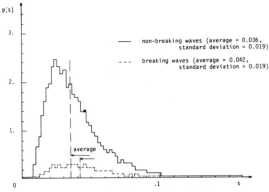

Fig. 3   Observed probability density function of wave
         steepness s, divided in breaking and non-break-
         ing waves contribution.

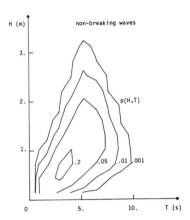

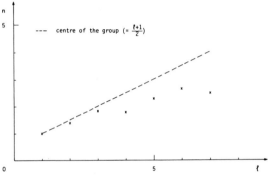

Fig. 4   The average sequence number of the first break-
         ing wave (n) in a group as function of group
         length (ℓ).

Fig. 2   Observed joint probability density
         functions of wave height H and wave
         period T of breaking and non-break-
         ing waves.

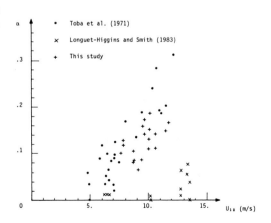

Fig. 5   Fraction of breaking waves (α) as
         function of the wind speed at 10 m
         elevation (U_{10}).

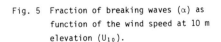

surface jump meter which gives (among other things) the number of events where the vertical speed of the surface elevation exceeds a certain threshold level. These events are interpreted by Longuet-Higgins and Smith as indicators for breaking or near-breaking. Their observations give fractions of breakers that are much smaller (by almost one order of magnitude) than found by Toba et al. and in this study.

4. CONCLUSIONS

We found that breaking waves could not be separated from non-breaking waves on the basis of wave steepness or wave asymmetry alone.

When waves break in a group, they break typically slightly ahead of the centre of the group. About two-thirds of the waves that break, break in a group (with threshold level H_{rms}).

The fraction of breaking waves in our observations was as high as 0.20 for wind speeds of about 10 m/s. Our observations agree in this respect with those of Toba et al. (1971) and not with those of Longuet-Higgins and Smith (1983).

ACKNOWLEDGEMENTS. We gratefully acknowledge the cooperation of our colleagues at the Ministry of Public Works of the Netherlands, in particular T. v.d. Vlugt, in obtaining the observations.

REFERENCES

Banner, M.L. and W.K. Melville, 1976, 'On the separation of air flow over water waves', Journal of Fluid Mechanics, Vol. 77, pp. 825-842

Battjes, J.A., 1971, 'Run-up distributions of waves breaking on slopes', Journal of Waterways, Harbors and Coastal Engineering Division, Proceedings of the ASCE, Vol. 97, No. WW1, pp. 91-114

Donelan, M., M.S. Longuet-Higgins and J.S. Turner, 1972, 'Periodicity in whitecaps', Nature, Vol. 239, pp. 449-451

Houmb, O.G. and T. Overik, 1976, 'Parameterization of wave spectra and long-term joint distribution of wave height and period', Proc. BOSS'76 Vol. 1, pp. 144-169

Kinsman, B., 1965, 'Wind waves', Prentice Hall Inc., Englewood Cliffs, New Yersey

Kjeldsen, S.P., 1983, 'Determination of severe wave conditions for ocean systems in a 3-dimensional irregular seaway', Proc. VIII Congress of the Pan-American Institute of Naval Engineering, Sept. 12-17, 1983, Washington

Longuet-Higgins, M.S. and N.D. Smith, 1983, 'Measurement of breaking waves by a surface jump meter', Journal of Geophysical Research, Vol. 88, No. C14, pp. 9823-9831

Longuet-Higgins, M.S., 1984, personal communication

Michell, J.H., 1983, 'The highest waves in water', Philosophical Magazine, Series 5, **36** (22), pp. 430-437
Nath, J.H. and F.L. Ramsey, 1974, 'Probability distributions of breaking wave heights', Proc. Symposium on Ocean Wave Measurements and Analysis, New Orleans, Sept. 9-11, 1974, ASCE, pp. 153-169
Toba, Y., H. Kunishi, K. Nishi, S. Kawai, Y. Shimada and N. Shibata 1971, 'Study on the air-sea boundary processes at the Shirahama Oceanographic Tower Station', Disaster Prevention Research Institute, Kyoto University, Annals 14B, pp. 519-531 (in Japanese with English abstract)
Van Dorn, W.G. and S.E. Pazan, 1975, 'Laboratory investigation of wave breaking, part II: Deep water waves', Scripps Institution of Oceanography, SIO RTF: No. 75-21, AOEL Report 71.

TURBULENCE BELOW WIND WAVES

Ian S.F. Jones
Marine Studies Centre
University of Sydney
SYDNEY   NSW   2006   AUSTRALIA

ABSTRACT.   Wind stress at the sea surface induces both surface waves
and turbulent shear flow.   Laboratory experiments show that 90% of the
shear stress below wind waves is supported by eddies of horizontal
wave number, k, smaller than

$$k = 3/z$$

where z is the distance from the surface.   A series of measurements of
the horizontal velocity fluctuations in the upper ocean mixed layer has
been carried out in water of 77m depth from a fixed platform.   Spherical
electromagnetic current meters were used to make the measurements and
the data analysed on three occasions where the wind velocity exceeds
$15ms^{-1}$.   These measurements support the application of inner law scaling
for the stress carrying velocity fluctuations.

1.   INTRODUCTION

Unsteady motions in the upper ocean span a large range of wave numbers
and result from processes as dissimilar as wave breaking and tidal
forcing.   In this paper we look only at those motions which we suspect
are driven by the wind and lie below the surface wave peak frequency.
Oceanic results are compared with laboratory experiments.
        In a previous paper, Jones and Kenney (1977) collected the existing
data to show that below the surface wave peak frequency, the horizontal
velocity fluctuations in the upper mixed layer of the ocean appeared to
scale on the wind stress and the distance from the surface.   In this
present paper we have carefully measured the turbulence from a fixed
platform under conditions of high wind stress and compared these new
results with wind-wave tank data that has become available since 1977.

2.   APPARATUS

The current was measured from the ESSO/BHP Kingfish B oil production

437

Y. Toba and H. Mitsuyasu (eds.), The Ocean Surface, 437–442.

platform located at $38^\circ36'S$, $148^\circ11'E$ in Bass Strait. The water depth
at this site is 77m and the nearest land is about 80 km to the north.
A spherical electromagnetic current meter was supported between two 16mm
diameter stainless steel cables tensioned to approximately 22kN to provide a very rigid support for the meter during heavy weather. This
10 cm diameter electromagnetic current meter has been calibrated in the
University tow tank, Halliday (1984). Wind speed was measured from a
radio mast, as shown in Jones and Padman (1983).
      Both wind and current were digitized twice per second and this
data used to determine the mean currents and the spectra. Each current
component spectra involved 12,288 readings extending over 102 minutes
and the resultant hanned frequency spectra defined such that

$$\int_0^\infty \Phi\ (f)\,df = \overline{u_1^2}$$

      From the mean wind speed for the 102 minutes, written U_{65}, the
friction velocity u_* in the ocean was determined as

$$u_*^2 = 1.34 \times 10^{-6}\ U_{65}^2$$

3.   COMPARISON BETWEEN FIELD AND LABORATORY RESULTS

Howe, Chambers and Street (1978) measured the horizontal velocity
spectra and the shear stress below wind waves in the Stanford wind/wave
tank. The rms wave height at the measurement station was $\zeta \cong 1.2$mm and
the measurements were made at a depth of $z/\zeta = 5$.
      Jones and Kenney (1977) suggested that most of the shear stress in
a turbulent boundary layer, be it over a rough wall or below water waves,
was carried by horizontal wave numbers k, such that

$$k < 3/z$$

      The shear stress spectra, of Howe, Chambers and Street (1978) sup-
port this contention. The above relation assumes that inner law scaling
is appropriate for the turbulence below waves as well as for solid surfaces and requires that velocities scale on the friction velocity, u_*,
and the distance from the boundary, z. A good test of this scaling is
to compare the laboratory results with the open ocean situation in
Bass Strait.
      The horizontal velocity spectra   here defined as the sum of the
two orthoginal velocity components, were converted to wave number under
the assumption that the mean velocity was much greater than the fluctuating motion. Kitaigorodskii, Donelan, Lumley and Terray (1983) discuss how for frequencies well below the surface wave peak frequency
this is reasonable. Fig. 1 shows such a representation with four
Bass Strait spectra compared with the result by Howe, Chambers and
Street (1978).
      The spectra shown extend to a frequency of 0.016 Hz and at this
depth the surface orbital motion contributes insignificant energy to the

spectra.   The same spectra when scaled by u_* and z is shown in Fig. 2 together with the friction velocity.

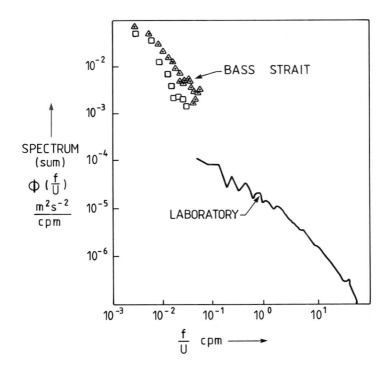

Fig. 1.   Comparison of the horizontal velocity spectra in the ocean with laboratory measurements below water waves.   See Fig. 2 for symbols.

Kitaigorodskii et al (1983) suggested the effect of surface waves may extend to frequencies below which the orbital motions (deduced from linear dynamics) are important.   The earlier data collated by Jones and Kenney (1977) showed some deviations from the inner law scaling but "mooring noise" still seems to be the most likely explanation for this early data.   For the present data the only result to clearly lie outside the 95% confidence limits about the laboratory result is the highest wind stress case.

For the spectrum with $u_* = 2$ cms^{-1} we were in a position to observe the sea surface and frequent white capping had developed before the current meter recording was made.

4.   INERTIAL SUBRANGE

Since the eddies carrying the shear stress have wave numbers less than

3/z, at wave numbers above this we expect to find an inertial subrange
where the turbulence spectrum has the form

$$\phi(k) = \alpha \varepsilon^{2/3} k^{-5/3}$$

and where α is a universal constant, say $\alpha = 0.5$. [Remember $\phi(k)$ includes both horizontal velocity components]. Now the spectra presented
in Fig. 2 extend only to kz = 3 so we will use this point to plot

$$\left[\phi(k) \ k^{5/3} \alpha^{-1}\right]^{3/2}$$

which is an estimate of the dissipation ε. If there is a balance between the generation of turbulent energy and the dissipation, and if we
accept inner law scaling,

$$= u_*^3/\kappa z$$

where κ is Karman's constant.

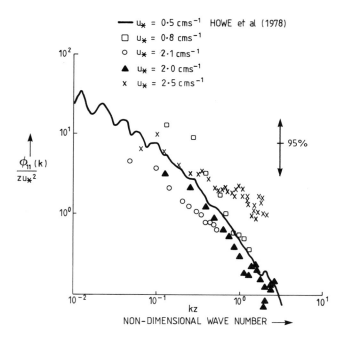

Fig. 2.   Horizontal velocity spectra scaled on inner law variables.

    Allowance in the friction velocity should be made for the age of
the waves but we have used throughout the friction velocity (in the
water) that balances the atmospheric wind stress.   In all cases
this was calculated from the wind velocity using a drag law similar to

Kitaigorodskii et al (1983) with a log law expression to reduce the wind speed to the standard height. Fig. 3 shows the present values together with dissipation estimated from Kitaigorodskii et al (1983) and Jones & Kenney (1971) in a lake where there was no "mooring noise".

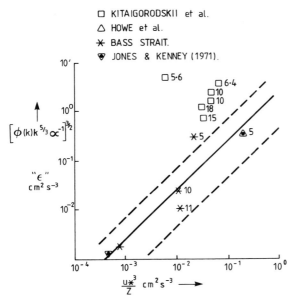

Fig. 3   An estimate of the dissipation determined from the inertial subrange for a number of measurements. Solid line represents inner law scaling. Dotted line is 95% confidence limits about the similarity relation for the Bass Strait data. Numbers associated with points are z/ξ.

Kitaigorodskii et al (1983) present results that are considerably higher than predicted by the inner law. They suggested that turbulent velocity fluctuations below the peak in the surface gravity waves may be influenced by wave turbulence interactions within a zone $z < 10\,\xi$ where ξ is rms wave height. Figure 3 does not support the idea of a zone of enhanced dissipation as deep below the surface as this. Donnelan (private communication) has suggested that rms wave height is not a good measure of the vertical length scale appropriate for breaking events. However it is hard to see a trend amongst the young laboratory waves, the older Bass Strait waves and the short fetch results of Kitaigorodskii et al (1983) which are below waves of much the same age as those in Bass Strait. Unpublished results by the author, at even smaller z/ξ, below laboratory waves where 50% of crests involved air entrainment somewhere across the tank did not show enhanced "dissipation" as defined here. Further measurements close to the sea surface are needed to determine if the results of Kitaigorodskii et al (1983) are not anomolous.

5.   CONCLUSIONS

Recent laboratory and field measurements have increased our confidence
that, at depths greater than some small number of wave heights below
the sea surface, the shear stress carrying eddies scale on the inner
law variables and the higher frequency orbital motions act as inactive
motion.   However, if one is to understand the transfer of wave energy
by breaking, further measurements close to the ocean surface at frequencies typical of breaking scales are needed.

6.   ACKNOWLEDGEMENTS

The author would like to thank ESSO-BHP for their continuing support
that allowed collection of this data.

7.   REFERENCES

Halliday, R.L. (1984)   Calibration of electromagnetic current meters.
          Uni. of Sydney, Marine Studies Centre Rep. 3/84.
Howe, B.M., A.J. Chambers & R.L. Street (1978)   Heat transfer at a
          mobile boundary, Adv. Heat Mass Transfer at Air-Water
          Interfaces, ASME, p.1-10.
Jones, I.S.F. and B.C. Kenney (1971)   Turbulence in Lake Huron, Water
          Res., 5, p.765.
Jones, I.S.F. and B.C. Kenney (1977)   The scaling of velocity fluctua-
          tions in the surface mixed layer, J. Geophys. Res., 82,
          p.1392.
Jones, I.S.F. and L. Padman (1983)   Semidiurnal internal tides in
          Eastern Bass Strait, Aust. J. Mar. & Freshw. Res., 34, p.159.
Kitaigorodskii, S.A. et al (1983)   Wave-Turbulence Interaction in the
          Upper Ocean. Part ll: Statistical Characteristics of Wave
          and Turbulent Components of the Random Velocity Field in the
          Marine Surface Layer, J. Phys. Oceanogr. 13, p.1988.

REYNOLDS STRESSES

L. Cavaleri[1] and S. Zecchetto[2]
1. Istituto Studio Dinamica Grandi Masse-CNR, 1364 San Polo
   30125 Venice, Italy
2. Telespazio Fellowship at ISDGM-CNR, Venice, Italy

ABSTRACT. We present the results of a series of wave measurements concerning surface elevation and wave kinematics taken from an offshore o-
ceanographic tower. The results oppose the theory in the sense that the
horizontal velocity component is found delayed up to 40° with respect to
the surface elevation. This corresponds to strong downward flux of horizontal momentum typically one order of magnitude larger than the wind
stress. The flux is present under active wind conditions, and it gradually disappears while sea conditions shift to swell.

1. INTRODUCTION

While studying the kinematics of the water particles under wind waves we
usually tend to forget the underlying dynamics. So, once a surface profile is given, the kinematics follows rigorously from the wave equations.
Whichever is the order of the solution, this attitude tends to bring in a
number of assumptions and consequences. In particular the phase relationships are largely maintained. Hence, when Shonting first (1964, 1970) and
then Yefimov and Khristoforov (1971 a,b) reported some odd results on the
subject, they did not receive much attention.

Due to the availability of a reliable wave measuring system in the
open sea we were capable of making an extensive series of wave measurements in quite different sea conditions. The subject of the paper is the
description of our main findings about the wave kinematics. We first describe the experimental set-up in Section 2, and then the data analysis
in Section 3. Section 4 is devoted to the presentation of the results.
Their discussion is carried out in the final Section 5.

2. EXPERIMENTAL SET-UP

The measurements have been carried out from the oceanographic tower of
ISDGM, placed in the Northern Adriatic Sea 15 km off the coast, on 16 me-

Y. Toba and H. Mitsuyasu (eds.), The Ocean Surface, 443–448.

ters of depth. Living facilities allow people to stay on board for several days, and consequently an active participation in the measurements.

The wave measuring system (Cavaleri, 1979) includes a wave gauge
(for surface elevation), two electromagnetic currentmeters at cross angles (for the 3-dimensional velocity field), and two pressure transducers
(for underwater pressure). All the subsurface instruments, rigorously on
the vertical of the wave gauge, are on a cart sliding on two vertically
tensed wires under controlled position. With a real time reading of an on
board mareograph, this allows the knowledge (and the choice) of the actual depth of measurement.

The record length varies from 30' to 60', according to the sea stability conditions. The data are stored at 4 hz scan rate on a magnetic
tape, with one out of 1000 resolution. The typical sequence is to begin
recording at the first puffs of a storm, and to leave the tower after
the last swell has disappeared.

3. DATA ANALYSIS

Each record has been analyzed with the FFT for single- and cross-spectra
at 0.01 hz interval. The energy directional distribution has been obtained both with the classical method of Longuet-Higgins et al. (1963)
and with the variational technique due to Long (1980). Actual attenuation with depth was checked against linear theory, and the relative
phase between the various couples of parameters was estimated.

Each record was graded according to the coherence γ between the signals. This was usually quite high and as a rule we limit further discussion to the cases where $\gamma^2 \geqslant 0.90$. Careful notes and pictures of the sea
conditions taken during the recording help in interpreting the results.

4. RESULTS

Only a short hint of the results is given here. For a thorough discussion
the interested reader can refer to the more extensive paper by the same
authors (1984, in press). Figure 1 shows the history of a storm which
happened on 11-12 May 1978. The numbers along the diagram of the signifi-

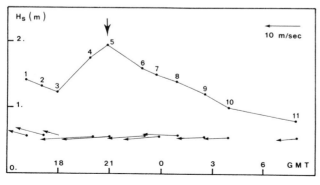

Figure 1. Wave history during 11-12 May 1978.
The numbers refer to
the available records.
Arrows show the wind
at the tower.

cant wave height H_S indicate the records during the storm. The wind present at the same time is shown by the lower arrows. The vertical arrow
at the peak of the storm (no. 5) indicates the record exploited in Figure 2. Here η is surface elevation, u and w are the longitudinal and ver-

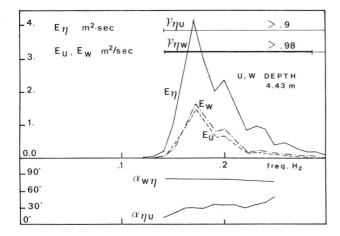

Figure 2. Spectra at the
peak of the storm. η =
surface elevation, u, w=
horizontal and vertical
velocity components. γ
is coherence. α is
phase difference.

tical velocity components. E is energy density. α_{xy} is the phase between
x and y. $\alpha > 0$ means x anticipates y. γ is coherence.
     The large phase shift of u with respect to η is immediately evident.
According to linear theory $\alpha_{\eta u}$ should equal zero; on the contrary the u-
delay reaches upper values of $40°-50°$. As w is close to its theoretical
phase relationship with η, it immediately follows that the integral

$$M = \frac{\varrho}{T} \int_0^T u\,w\,d\,t$$

has a no-zero value, i.e. there is a finite vertical flux of horizontal
momentum. Actually this turns out to be extremely large, typically (during active wind conditions) one order of magnitude larger than the surface wind stress.
     The no-orthogonality of u with respect to w is made well-evident by
the visualization of the (Lagrangian) trajectory of the water particles,
obtained by the integration of the recorded velocity components. In Figure 3a the waves are conceived as moving to the right. While the theoretical orbit should be horizontal ellipses, the u-delay moves down the nose
of the ellipses, suggesting the idea of the downward flux of a physical
quantity.
     The value of $\alpha_{\eta u}$ (or α_{uw}) is not constant throughout a storm. The
typical transition is to begin with low values at the very early stages,
$\alpha_{\eta u}$ then rapidly growing parallel to the activity of the wind. While the
storm decays $\alpha_{\eta u}$ gradually decreases, reaching practically null values
when pure swell is present. Figure 3b shows the trajectory of a water

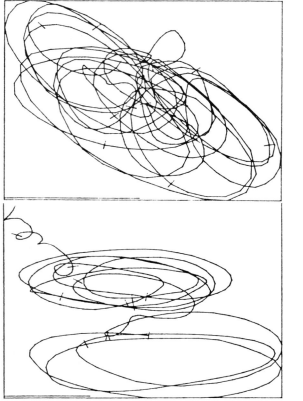

Figure 3. Orbital motion during a) active wind and
b) swell conditions.
The segment in the low-
er left-hand corner
shows 1 meter length.

a)

b)

particle during the latter case. The orbits are horizontal, indicating
the orthogonality between u and w.

It is worthwhile to point out that these diagrams are no exception,
merely representing the evidence repetitively obtained from many storms
we have recorded.

As the depth of measurement varies from record to record, to compare the different momentum fluxes we have reported all the $\overline{uw}$ values to
the surface, under the hypothesis that the attenuation with depth follows the linear theory. Figure 4 shows, in an arbitrary scale, the values of $\overline{uw}$ (for the records in Figure 1) plotted against u/c_p (c_p=phase
velocity at the spectral peak). A direct dependence between these two
quantities is an obvious suggestion.

It is very interesting to analyze a record in detail. A simple plot
of η, u and w (not shown here) leads immediately to the conclusion that
the phase shift between η and u is not constant throughout a record, but
it varies from wave to wave. Splitting the record in single waves by the
up-zero crossing method, we have estimated, again for the peak of the
storm, the M flux associated with each single wave. This has been plotted
against the corresponding single wave height in Figure 5. The line shows
the l.m.s. fit to the expression $\overline{uw} = a \cdot H^b$. b is typically close to 2.

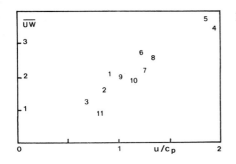

Figure 4. Reynolds stresses, in an arbi-
           trary scale, as a function of the
           ratio between wind speed and peak
           velocity. See Figure 1 for record
           number.

Figure 5. Scatter diagram of momentum
    flux and wave height for each sin-
    gle wave during record 5 (see Fig-
    ure 1).

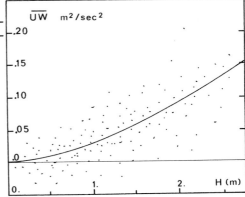

5.   DISCUSSION AND CONCLUSIONS

The results reported in the previous section oppose the actual theory, indicating downward momentum fluxes at least one order of magnitude larger
than expected. A very careful check of the measuring system and of the a-
nalysis procedure was therefore an obvious point. Following our own and
several external criticisms, and also some suggestion from the literature
(Battjes and van Heteren, 1983), we have analyzed all the steps of the
path leading from the instruments to the final output. We could not detect any fault, and we are now convinced that the data represent a physical reality.
    This of course raises the problem of explaining the findings. This
was not our aim in writing this note. We wished simply to show an experimental evidence, and to offer another problem to the scientific community.
Nevertheless a few points seem clear to us.
    Figures 2 and 4 indicate that the M flux is associated with very ac-
tive wind conditions. This could happen through the presence of breakers
or to be associated with the steepness of waves. In any case there is a
strong suggestion that the process is highly nonlinear, as shown by the
intermittancy of the process, summarized in the diagram of Figure 5.
    We suggest two different lines of research to explain the results.
One is the interaction with turbulence, the second is the convective vertical motions present at the air-sea interacting layers during very un-

stable conditions. Hopefully some future records will help in indicating
the way.

REFERENCES

Battjes, J.A. and J. van Heteren, 1983: Measurements of wind wave kine-
    matics. Rijkswaterstaat, Rep. WWKZ-G007, 33.
Cavaleri, L., 1979: An instrumental system for detailed wind wave study.
    Il Nuovo Cimento, Serie 1, 2C, 288-304.
Cavaleri, L. and S. Zecchetto, 1984: Reynolds stresses in wind waves.
    In press.
Kitaigorodskii, S.A., M.A. Donelan, J.L. Lumley and E.A. Terray, 1983:
    Wave turbulence interactions in the upper ocean. Part II. Statisti-
    cal characteristics of wave and turbulent components of the random
    velocity field in the marine surface layer. JPO, Nov.,13,1988-1999.
Longuet-Higgins, M.S., D.E. Cartwright and N.D. Smith, 1963: Observa-
    tions of the directional spectrum of sea waves using the motions of
    a floating buoy. Ocean Wave Spectra, Prentice-Hall Inc., Englewood
    Cliffs, N.J., 111-132.
Long, R.B., 1980: The statistical evaluation of directional spectrum
    estimates derived from pitch/roll buoy data. JPO, 10, 6, 944-952.
Shonting, D.H., 1964: A preliminary investigation of momentum flux in
    ocean waves. Pure and Appl. Geophys., 57, I, 149-152.
Shonting, D.H., 1970: Observations of Reynolds stresses in wind waves.
    Pure and Appl. Geophys., 81, IV, 202-210
Yefimov, V.V. and G.N. Khristoforov, 1971a: Wave-related and turbulent
    components of velocity spectrum in the top sea layer. Izv., Atmo-
    spheric and Oceanic Phys., 7, 2, 200-211.
Yefimov, V.V. and G.N. Khristoforov, 1971b: Spectra and statistical re-
    lations between the velocity fluctuations in the upper layer of the
    sea and surface waves. Izv., Atmospheric and Oceanic Phys., 7, 12,
    1290-1310.

EFFECTS OF THE ROTATION ON THE ENTRAINMENT BY GRID-GENERATED TURBULENCE
IN STRATIFIED FLUIDS

A. Masuda
Research Institute for Applied Mechanics
Kyushu University 87
Kasuga, 816
Japan

ABSTRACT. An experiment has been carried out to study the entrainment
in two-layer stratified fluids in a rotating system, where turbulence
for mixing is produced by an oscillating grid in the upper layer. Flow
visualization by dye, aluminium powder, and shadow graphs reveals
peculiar characteristics of turbulence in rotating stratified fluids:
strong spike-like turbulence above the interface and weak thin filaments
elongated downward just beneath the interface. It is found that
rotation causes even the downward (reverse) entrainment, which is never
observed in non-rotating stratified fluids.

1. INTRODUCTION

In modeling the surface mixed layer of the ocean, the most difficult
question is how to express the entrainment velocity, the rate of
increase in the depth of the mixed layer, in terms of overall parameters
such as the Richardson number. Although many researchers have
investigated this problem (Turner, 1968, or Kato and Phillips, 1969, for
example), they have neglected the rotation of the earth, another
important factor in geophysical flows; none of them have considered the
direct influence of the rotation of the earth on turbulence and
entrainment. This treatment is presumably reasonable for most ocean
cases. It is not justified, however, *a priori*. For example, the
rotation might have possible effects on the turbulence structure when
the mixed layer is sufficiently deep that the Rossby number is not
large.
    Apart from the practical application, entrainment in rotating
stratified fluids is of much theoretical interest in itself. The
rotation is expected to add a few new aspects to the usual mixing of
stratified fluids, besides the well known *indirect* effects of inertial
currents. First, the turbulence within the mixed layer will assume two-
dimensionality for sufficiently high rotation. That is, eddies will be
elongated vertically (or along the direction of the rotation vector);
turbulence intensity will decrease less in the vertical direction, which
seems favorable for mixing because entrainment is driven by turbulent

449

Y. Toba and H. Mitsuyasu (eds.), The Ocean Surface, 449–456.
© *1985 by D. Reidel Publishing Company.*

kinetic energy. The columnar eddies, however, may not be efficient in
entraining the lower fluid by scraping, because their vertical velocity
is small. Second, the rotation allows the propagation of inertial
waves. Let us note that Linden (1975) has suggested that if the water
below the mixed layer is stratified continuously, internal gravity waves
radiated from the mixed layer deprive it of considerable amount of
kinetic energy available for the entrainment and consequently retard the
deepening of the mixed layer. Inertial waves due to rotation in the
present experiments may play the role of internal gravity waves due to
continuous stratification in Linden's experiments, even for the
homogeneous lower layer. Thus we infer the enhancement of entrainment
on one ground, suppression from another reason.
    Does the rotation work positively or negatively for mixing ? The
motivation of the present paper is to answer these fundamental questions
and thus to understand the effects of rotation on the mixing, by means
of visualization experiments for rather high rotation.

2. EXPERIMENTAL ARRANGEMENTS

    The experiments were carried out in a rectangular glass tank 20.0
x 20.0 cm in cross section and 40 cm deep (see Fig.1) on a turntable,
where the typical experimental procedure is as follows. First, we set
up two homogeneous layers with an density interface. Then we begin to
revolve the turntable; acceleration follows untill the system achieves
the assigned rotation rate. After the water is spun up completely, the
grid in the upper layer begins to oscillate with the stroke a=2 cm at
the frequency fg=1.5 Hz. For detailed experimental procedure and
results, the reader can refer to Masuda (1983) and the full or extended
paper, which is to appear elsewhere.
    Three kinds of visualization were made. (1) alluminium powder (Al)
to see the whole pattern of turbulence, (2) dye in the upper layer (Dy)
to see the entrainment, and (3) shadow graphs (Sh) to see the detailed
density structure near the interface. The entrainment velocity was
determined from a sequence of photographs of (Dy), by judging the
average height of the ragged interface. Though this estimation is
rough, it suffices for our purpose of qualitative comparison.

3. RESULTS

    Before proceeding to the results, let us take a more general view
of the present experiments. That is, the subject is considered as an
extension of the simple grid turbulence in the directions of geophysical
importance; from homogeneous to stratified fluids and from inertial to
rotating system. We denote stratification by S and rotation by R; their
negations by $\bar{S}$ and $\bar{R}$. The symbol $[\bar{S},\bar{R}]$, for example, means the grid
turbulence in homogeneous non-rotating fluids.
    Among the four cases, $[\bar{S},\bar{R}]$, $[S,\bar{R}]$, $[\bar{S},R]$ and $[S,R]$, the first is
known best and the second has been studied extensively in the context of
geophysical application. On the other hand, $[\bar{S},R]$ has rarely been

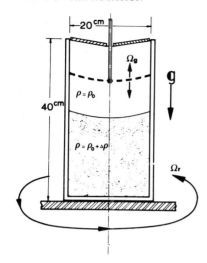

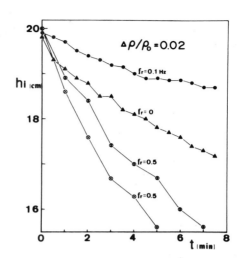

Figure 1. Schematic diagram
of experimental apparatus.

Figure 3. Deepening of the mixed layer,
where h_i is the height of the interface,
t the time measured from the onset of
the oscillation of the grid: effect of
rotation for a fixed (initial) density
difference $\Delta\rho/\rho_0$=2%.

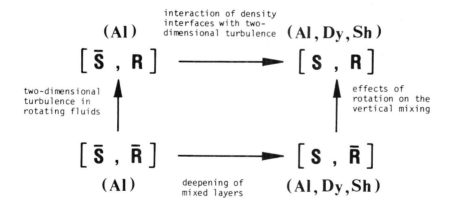

Figure. 2 Classification and interrelation of oscillated-grid
turbulence, where S denotes stratification, R Rotation and "-" their
negation.

investigated; the only published papers are those by Hopfinger and
Browand (1982), by Hopfinger, Browand and Gagne (1982) and by Dickinson
and Long (1983). As regards [S,R], the present experiment is the first
one, so far as I know.
       Fig.2 illustrates the four classes of experiments and the three
methods of visualization, together with probable motivation for each
direction of extension.

3.1   Visualization by aluminium powder for [$\overline{S},\overline{R}$], [S,$\overline{R}$], [$\overline{S}$,R] and [S,R]

Photos 1 (a) and (b) respectively show typical flow patterns of [$\overline{S},\overline{R}$]
and [$\overline{S}$,R], where f_r=0.5 Hz in the latter case. We clearly observe
three-dimensional character of turbulence in a non-rotating system,
while two-dimensional columnar vortices in a rotating system.
       For [S,$\overline{R}$], we can confirm one of the most remarkable though
familiar features that turbulence is confined to the upper layer and no
motion is found in the lower layer except the thin interfacial area.
       Photo 2 gives an example of [S,R], where f_r=0.5 Hz and $\Delta\rho/\rho_0$=2%.
In this case, the rotation allows the penetration or the propagation of
the motion into the lower layer, overcoming the resistance of density
stratification. Columnar vortices are observed both above and below the
interface.

3.2   Visualization by dye for [S,$\overline{R}$] and [S,R]

Photos 3 (a) and (b) show the height and shape of the interface for f_r=0
and 0.5 Hz respectively, where $\Delta\rho/\rho_0$=1% and the time t from the onset
of grid oscillation being 15/4 min. A notable feature in (b) is that
the upper layer deepens faster near the side boundary than around the
center. The most striking of all are thin vertical filaments, which are
descending into the lower layer. It should be remembered that in [S,$\overline{R}$]
entrainment always occurs upward and consequently the lower layer
remains transparent. To the contrary, in [S,R], thin descending
filaments mentioned above cause the reverse downward "entrainment",
which colors the lower layer gradually. Also we may consider that those
filaments take charge of a kind of turbulent "diffusion" across the
interface.
       Figure 3 gives an example of the deepening of the interface h_i. It
reveals an unexpected result that rotation restrains the deepening when
it is small (f_r=0.1 Hz), but promotes when it is large (f_r=0.5 Hz).
       We should pay some attention, however, to the case f_r=0.5 Hz, where
h_i is difficult to read by eye, partly because the interface takes the
form of a dome and partly because thin downward filaments obscure the
location of the interface. One of the two curves for f_r=0.5 Hz
corresponds to h_i near the center and the other to h_i as a horizontal
average. Furthermore, turbulent "diffusion" due to filaments does not
guarantee that the "color" interface represents the density interface
accurately. At any rate, no simple conclusion can be derived from
Figure 3, such as the positive effect of rotation on mixing or as to the
opposite.

(a)   (b)

Photo. 1   Grid-generated turbulence for homogeneous water: (a) non-rotating case when f_r=0Hz, (b) rotating case when f_r=0.5Hz.

Photo. 2   Turbulence for a strongly rotating (f_r=0.5Hz) case with the (initial) density difference $\Delta\rho/\rho_0$=2%.

(a)   (b)

Photo. 3   Visualization of the interface by dye when $\Delta\rho/\rho_0 k=1\%$: (a)
$f_r=0$Hz, (b) $f_r=0.5$Hz.   In (c), we observe the reverse entrainment due to
thin long filaments which penetrate into the lower layer, a peculiar
phenomenon unexpected from the usual entrainment.

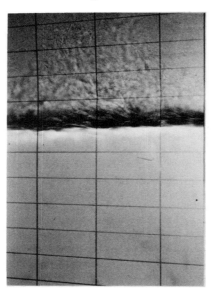 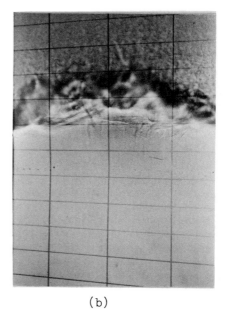

(a)   (b)

Photo.4   Shadow graphs around the interface when $\Delta\rho/\rho_0=4\%$: (a) $f_r=0$Hz,
(b) $f_r=0.5$Hz.   Turbulence intensity increases for strong rotation (6).
Note the spike-like furious structure above the interface and thin long
filaments below it.

3.3 Visualization by shadow graphs for [S,R̄] and [S,R]

Photos 4 (a) and (b) are shadow graphs for f_r=0 and 0.5 Hz when $\Delta\rho/\rho_0$=4%. They exhibit the fine structure of density near the interface. For strongly rotating fluids, we notice spike-like patterns above the interface and thin long lines below the interface, where the latters obviously correspond to the colored filaments observed in (Dy).

4. SUMMARY AND DISCUSSION

Simple flow visualization reveals unexpected characteristics of rotating stratified fluids. They are: (1) delicate dependence of entrainment velocity of the rotation rate f_r, (2) spike-like density pattern above the interace, (3) reverse entrainment or turbulent "diffusion" due to fine filaments like waterspouts, (4) doming of the interface, and (5) penetration of turbulent motion into the lower layers across a few density interfaces. Their interpretation is not provided here on account of the page limit, but the reader will find it in the full paper elsewhere.

At this point, we should make some cautionary remarks. First, since the results are presented for only a few cases under many rather artificial conditions, careless generalization or applications must be refrained from. Secondly, the rate of rotation is extremely large as compared to that for the real ocean or atmosphere, to exaggerate the effects of rotation into relief: Thirdly, the artificial method of an oscillating grid imposes external length scales to the turbulence, which is more severe and serious in rotating fluids than in non-rotating fluids because the two-dimensional columar turbulence has much longer distance of correlation vertically. Forthly, the doming of the interface will be ascribed to the effects of side walls, which are undesirable for the investigation of one-dimensional dynamics. At any rate the present experiments are intended to obtain only a primary description of the mixing in rotating stratified fluids. Further investigation should follow up to make clear the detailed dynamics which are responsible for the peculiar phenomena found in this study.

REFERENCES

Dickinson, S.C., and R.R. Long, 1983: Oscillating-grid turbulence including effects of rotation. *J. Fluid Mech.*, **126**, 315-333.
Hopfinger, E.J., and E.K. Browand, 1982: Vortex solitary waves in a rotating, turbulent flow. *Nature*, **295**, 393-395.
Hopfinger, E.J., E.K. Browand, and Y. Gagne, 1982: Turbulence and waves in a rotating tank. *J. Fluid Mech.*, **37**, 505-534.
Kato, H., and O.M. Phillips, 1969: On the penetration of a turbulent layer into a stratified fluid. *J. Fluid Mech.*, **37**, 643-655.
Linden, P.F.A, 1975: The deepening of a mixed layer in a stratified fluid. *J. Fluid Mech.*, **75**, 385-405.

Masuda, A., 1983: Entrainment by Grid-Generated Turbulence in Rotating stratified fluid. *Rep. Res. Inst. Appl. Mech., Kyushu Univ.,* **31,** 29-45.

Turner, J.S., 1968: The influence of molecular diffusivity on turbulent entrainment across a density interface. *J. Fluid Mech.,* **33,** 639-656.

DYNAMICS OF LANGMUIR CIRCULATIONS IN A STRATIFIED OCEAN

Sidney Leibovich
Sibley School of Mechanical & Aerospace Engineering
Cornell University
Ithaca, New York 14850, USA

ABSTRACT. New theoretical results concerning Langmuir circulations,
according to an extended version of the Craik-Leibovich theory, are
summarized. The issue considered is the dynamical role of buoyancy,
which can generate oscillatory and apparently aperiodic convection of
Langmuir circulation type, leading to a further level of complexity in
the signal these motions would register in current measurements and
affecting the mixing accomplished; in addition, strong density
gradients, such as those forming thermoclines, can act as a 'bottom'
for the circulation system, in the sense that the rate of mixing is
drastically reduced when the thermocline is sufficently strong.

1. INTRODUCTION

The physical mechanisms by which turbulence and consequent vertical
mixing are produced in the mixed layer of the ocean are but poorly
understood. The mechanisms identified to date may be broadly
classified into three groups: wave breaking events (of both surface and
internal waves), shear instabilities distinct from wave breaking, and
convective instabilities. An understanding of wave breaking is only
just developing; furthermore, despite the frequent attribution of mixed
layer turbulence to shear instabilities and the likelihood of their
significant contributions, direct evidence of their role seems to be
lacking. More is known about convective instabilities, driven by
temperature or salt gradients, or mechanically by the wind in the form
of Langmuir circulations. The relative importance of each of these
mixing mechanisms is unknown. Parameterizations that suppose that only
one process takes place, and base scalings upon that process, are
unlikely to be generally applicable. On the other hand, without a
preliminary assessment of the relative contributions of the various
mixing processes, more elaborate parameterizations with greater ranges
of applicability than those presently available cannot be rationally
constructed.
     One theoretical method of assessing the importance of the
contributing physical processes is to isolate each one, and to explore

457

Y. Toba and H. Mitsuyasu (eds.), The Ocean Surface, 457–464.
© 1985 by D. Reidel Publishing Company.

its characteristics in detail. To a limited extent, this has been done
for one process, the Langmuir circulations in the absence of dynamical
effects of buoyancy, by Leibovich & Paolucci[1]. In this paper, we
summarize further efforts to characterize those phenomena encompassed
by the theoretical model of Langmuir circulations proposed originally
by Craik & Leibovich[2], in which the interaction of waves and currents
controls the mean motions. This theory permits the formation of
Langmuir-like motions according to two mechanisms: the direct-drive
mechanism originally proposed, and an instability mechanism first
pointed out by Craik[3]. The latter requires fewer special assumptions,
and most attention has been given to it in theoretical work. The
present paper is devoted to effects created by the competition between
the destabilizing effect of the wave-current interaction and the
stabilizing effect of buoyancy, making use of the generalized
formulation of Leibovich[4].

       Observational evidence exists that suggests that Langmuir
circulations in some cases may extend to the thermocline (see the
survey in Leibovich[5]). Langmuir himself believed[6] that these
convective cells may be responsible for the formation of thermoclines
in lakes. A strong stable density variation, on the other hand, should
act as a barrier to the vertical motion and inhibit or prevent further
overturning by the cells, thus acting as a a more-or-less effective
'bottom' for the Langmuir circulation zone. We have done some
numerical experiments, to be described in detail elsewhere (Leibovich &
Lele[7]),to explore this question. The evolution of the mixed layer was
simulated using the full nonlinear partial differential equations of
the theory[4], starting from initial density fields that either were
linear or had piecewise constant gradients imitating pre-existing mixed
layers. An abbreviated account of this work is described in §2.

       Langmuir circulations create horizontal variability in the mixed
layer. These motions pattern the sea surface, and it is this
regularity in space and persistence in time that has drawn attention to
the phenomenon. The regularity is not perfect, of course, the driving
forces - winds and waves - are variable, and the entire circulation
system is bodily advected by larger scale currents that may have
complicated spatial and temporal features. Consequently, unless a
measurement program is specifically designed to look for Langmuir
circulations, the effects of the circulations will appear as random
fluctuations and be counted as part of the turbulence.

       It is to be anticipated that these regular motions give way (under
certain circumstances depending upon the wind and wave fields, and upon
the density stratification encountered) to more complicated ones. The
easiest kind of complications to consider theoretically are those
retaining regularity in space, but which are periodic in time or
time-dependent but nonperiodic (chaotic). More difficult to treat are
motions which lose spatial regularity. In section 4 of this paper, the
"easy" complications are addressed by consideration of a highly
simplified model problem. Full results and details of the analyses
required will be reported in due course by Moroz & Leibovich[8], and by
Leibovich et al.[9]

       The breakdown of well-ordered Langmuir circulations into more

complex convective motions is likely to render them invisible to
observers at the surface, but it may or may not diminish their
importance to vertical mixing; the delineation of various mixing
regimes is one of the long range goals of the program of which the
present paper is a part. The processes surveyed here are are related
to the development of thermal turbulence from well-ordered convection
in a fluid heated from below as the heating is increased. It is
connected even more closely to double-diffusive phenomena, but here the
instability and the turbulence that is expected to follow are directly
attributable to the wind and surface waves. We therefore suggest the
existence of a directly traceable route by which the wind and waves may
generate at least some of the turbulence in the upper ocean.

2. DEEPENING OF A MIXED LAYER BY LANGMUIR CIRCULATIONS

Numerical simulations to explore the response of mixed layers to
forcing by Langmuir circulations are reported in ref.7.   The
simulations are carried out using the instablity mechanism of the
wave-current interaction theory[2], extended to allow consideration of
convective perturbations of a water column that is not in thermal
equilibrium.   The wave-interaction theory uses the Navier-Stokes
equations filtered to remove fluctuations due to surface waves with
buoyancy is accounted for in the Boussinesq approximation.   The
residual effects of the waves are felt through a "vortex force", the
vector cross-product of the Stokes drift of the surface waves and the
filtered vorticity vector. When cast in dimensionless form[1], motion in
the form of rolls parallel to an applied surface stress or to a basic
unidirectional current depends only on three dimensionless parameters:
Ri, the "Richardson number" is a ratio of buoyant force to vortex
force; La, the Langmuir number" is a ratio of viscous force to vortex
force; Pr is the Prandtl number.
        The simulations in [7] assume an infinitely deep ocean, with a
constant temperature gradient at great depths.   At the surface, a
stress is applied at time t = 0 and held constant (unity with the
scalings used) thereafter, and either (A) the temperature is held fixed
or (B) the surface is taken to be adiabatic.   An adiabatic boundary
condition is the simplest allowing the surface temperature to vary; in
computations this leads to alternating warm and cool stripes to form on
the surface, rather like those seen in infrared scans of the surface.
The initial temperature gradient is piecewise constant, being either
(in cases A) a single constant (unity) for all depth, or (in cases B)
zero from the surface down to a depth d_1, then N from depth d_1 to a
greater depth d_2, below which the gradient is that of the deep ocean.
Cases (A) give some information on how Langmuir circulations can create
mixed layers, while cases (B) are intended to simulate the response of
pre-existing mixed layers to an imposed Langmuir circulation forcing.
In the latter cases, the depths d_1 and d_2 and are adjustable, as is N.
The region $d_2 < z < d_1$ imitates a thermocline: the temperature
gradient there is N times that of the water below.   Examples taken from
the literature values have N = 6 or so, although this number can

presumably vary over a significant range. In all of the cases
considered, the dimensionless Stokes drift taken corresponded to
monochromatic propagating surface waves.

        The computations, not surprisingly, demonstrate the anticipated
stabilizing effect of buoyancy. There are unexpected features,
however, which are illustrated by calculations for two cases of problem
A. These cases have values of Ri (measuring the ratio of the buoyancy
force, assumed to be stabilizing, to the destabilizing vortex force) of
0.1 and 0.25, respectively; La = 100 and Pr = 6.7 in both A and B. The
depth of penetration of the circulation system is smaller for Ri = 0.25
than for Ri = 0.1, but in both cases, after an initial surge of
superexponential growth of convective energy leading to formation of a
'mixed layer', further deepening of occurs at a comparable rate in the
two cases. Examination of the perturbation kinetic energy indicates
similar histories in the two cases with the main qualitative difference
being a time lag delay in the second case before superexponential
growth begins. If one takes a water friction velocity of 0.78cm/s, a
surface wavenumber of 2.4×10^{-3}cm and a wave amplitude of 30cm, the
computations give leads to a deepening rate of 0.6-0.7m/hr in the two
cases after the superexponential growth phase.

        Two of the B computed cases[7] give an indication of the extent to
which a thermocline acts as a 'bottom', preventing deepening by
Langmuir circulations. In both cases, N = 6, d_1 = 1 and d_2 = 1.5. The
only difference between the two cases was the value of Ri based on the
deep ocean temperature gradient, which was taken to be 0.1 in one
example and and 0.5 in the other. Since the uppermost layer initially
was uniform in both cases, circulations develop readily there. In the
first case, deepening of the mixed layer appeared to occur at a rate
not dissimilar to case A with Ri = 0.1 despite the existence of an
intermediate layer with six times the stratification. In the second
case, deepening was virtually completely suppressed. Other
computations showed that the stabilizing effect of the thermocline
increased with its thickness. This suggests that the overall density
variation across the thermocline may determine the extent to which it
is seen as a 'bottom' to the circulation system.

        The second B case displayed a most interesting behaviour. The
circulations, confined to the uppermost layer, were oscillatory, having
periodically alternating upwelling and downwelling motions between
neighboring cells. This led us to look at circulations in water of
finite depth, where analyses of instabilities and bifurcations are
easier to do, and to examine more systematically the effects of
stratification.

3.   ANALOGY BETWEEN LANGMUIR CIRCULATIONS AND DOUBLE-DIFFUSIVE
      PHENOMENA

The equations governing Langmuir circulations in stratified water
according to the theory of Craik & Leibovich are very similar to those
governing thermosolutal convection and for constant Stokes drift
gradient are identical to them[5]. Let the fluid be of finite depth d

(in the units used in ref.[4]), assume a constant stress in the
x-direction is applied to the mean-free water surface, and assume
isothermal boundaries and a basic state with a constant heat flux. A
possible steady equilibrium velocity field, is rectilinear motion with
x-velocity component u = U(z) where $\partial U/\partial z$ is constant throughout the
water column $-d < z < 0$. Perturbations of this basic state in the form
of rolls parallel to the x-axis are described by the following
equations

$$L_0 \nabla^2 \psi = R\ h(z)\partial u/\partial y - S\ \partial\theta/\partial y + J(\psi,\nabla^2\psi),$$
$$L_0 u = \partial\psi/\partial y + J(\psi,u),$$
$$L_1\theta = \partial\psi/\partial y + J(\psi,\theta), \qquad\qquad (1)$$

where for n = 0 or 1, $L_n \equiv \partial/\partial t + \tau^n\nabla^2$, $\tau \equiv 1/Pr$ is the inverse
Prandtl number, y is the coordinate normal to the stress (x-direction)
and the vertical (z-direction), ψ is a streamfunction in the y-z plane,
θ is temperature perturbation, u is x-velocity perturbation to the
basic state, J is the Jacobian with respect to y and z. Furthermore,
h(z) is the Stokes drift gradient, normalized by the surface value, Γ.
These equations are written[5] in variables scaled on the depth d and on
a diffusion time, but are otherwise specializations (to the assumed
basic state) of those appearing in ref. [1]. If the Stokes drift
gradient is constant (it might be sensibly be taken to be so if d <<
1), then $h \equiv 1$, and the equation set in primed variables is in the form
found in thermosolutal convection. Here R plays the role of the
thermal Rayleigh number, while S is the solute Rayleigh number, and τ
is the ratio of the salt and heat diffusivities. R and S are related
to the Richardson number and Langmuir number used in previous work[5] by
$R \equiv \Gamma d^4/La^2$, $S \equiv Ri\ d^4/La^2$. In (1), the perturbation u-current is
analogous to perturbation temperature, and the perturbation temperature
is analogous to the perturbation solute concentration.

4.   A SIMPLIFIED NONLINEAR MODEL

Evidences of oscillatory (overstable) convection were mentioned in
§2. Further numerical experiments done by Lele (private communication)
for a layer bounded below by a rigid no-slip plane show that the
periodic overstable motions may be superceded, at higher forcing
(larger R, in the notation of §3) by apparently random motion. The
object of this section is to introduce a simplified dynamical model
amenable to some analysis, that can help us to understand the kinds of
transitions (from no convective motion to periodic convection to chaos)
seen in these numerical simulations.
     Consider the system as described in eqns. (1), so that the lower
boundary is located at z = -1. Suppose water with a much larger stable
temperture gradient underlies the layer of interest. Then the lower
boundary simulates a thermocline border, and if the thermocline is deep
enough and strong enough, the studies referred to in §2 suggest that
vertical motions of and below the plane z = -1 are negligible. Then,
so far as the perturbation motion in the surface layer is concerned,
the upper boundary is stress free, since the applied stress is absorbed

by the basic flow; the lower boundary also acts much like a stress-free plane, and we therefore replace it with this idealization.   To keep the model as simple as possible, we also assume the boundaries to be isothermal.   The boundary conditions are then

$$\psi = \partial^2\psi/\partial z^2 = \theta = \partial u/\partial z = 0 \text{ at } z = 0 \text{ and at } z = -1. \qquad (2)$$

Assume the motion is periodic in the y-direction, so the equations may be satisfied with ψ developed in a Fourier sine series, and u and θ in Fourier cosine series.   When $h \equiv 1$, (2) are symmetric with respect to the mid-plane $z = -1/2$, solutions of the linearized version of (1) can be divided into functions symmetric or antisymmetric with respect to $z = -1/2$ with the symmetric and antisymmetric parts separately satisfying (1).   If h(z) is not constant, the even and odd parts are linked; this is the case as well when nonlinear terms are included.

Following Veronis[11] treatment of the thermohaline problem, we truncate the Fourier series in y after the fundamental.   Letting the wavenumber be ℓ, we introduce the truncated series $\psi = \psi_1(z,t) \sin \ell\pi y$, $u = u_0(z,t) + u_1(z,t) \cos \ell\pi y$, $\theta = \theta_0(z,t) + \theta_1(z,t) \cos \ell\pi y$ into (1). Balancing mean field distortion terms and the fundamental, and discarding second harmonics in y leads to a set of five nonlinear partial differential equations in z and t.   A possible set of Galerkin basis functions, $B_n^{(k)}(z)$ (k = 1,...,5), for the vector (ψ_1, u_1, u_0, $\theta_1, \theta_0) \equiv w \equiv (w_1,...,w_5)$ satisfying boundary conditions (2) is ($\sin n\pi z$, $-\cos (n-1)\pi z$, $-\cos n\pi z$, $\sin n\pi z$, $\sin 2n\pi z$).   We adopt this set, taking

$$w_k^{(k)} = C_k^{(k)} \sum_n A_n(t) B_n(z) \qquad , \qquad (3)$$

where $C_1 = 2\beta/\pi\ell$, $C_2 = C_4 = 4/\pi\beta$, $C_3 = C_5 = 1/\pi$, $\beta^2 = 2\pi^2(1+\ell^2)$.

Introducing (3) and these definitions into the partial differential equations proceeding in the usual Galerkin fashion and truncating at N terms leads to a set of 5N ordinary differential equations in time for the coefficients.   These we truncate at N = 1, the minimal truncation. We introduce (for neatness) rescaled time variable and system parameters $t_1 \equiv \beta^2 t/2$, $r \equiv 2K R \int h(z) \sin \pi z B_1^2(z)dz$, $s \equiv K S$, $K \equiv \ell^2/\pi^4(1+\ell^2)^3$.

With these definitions and replacing $a_1^{(k)}$, k = 1,...,5 by (a,b,c,d,e), we arrive at the set of equations analyzed in refs.[8] and [9]:

$$\dot{a} = -a + rb - sd; \quad \dot{b} = -\gamma b + a(2/\pi - 1/2 \ c);$$

$$(4)$$

$$\dot{c} = \omega(-c/4 + ab); \quad \dot{d} = -\tau d + a(1-e); \quad \dot{e} = \omega(-\tau e + ad);$$

where the dot stands for the derivative with respect to t_1, $\omega \equiv 4/(1+\ell^2)$, and $\gamma \equiv \omega\ell^2/4$.   Except for the numerical factors, these equations are the same as those discussed by DaCosta et al.[12] The stability of the unperturbed state, and the dynamics of this system may be considered[8],[9] with the hope that the qualitative

behaviour (nature of the transitions and the sequence in which they
occur) reflects the physical system that inspired them, and with the
understanding that quantitative correspondences between the two
dynamical systems ought to be regarded as fortuitous.

We briefly summarize the main features of the dynamical behaviour
exhibited[8],[9] by (8) for $\tau = 0.15$, and $\ell = 1.0$.   (Work on ref.[9] is
in progress, including attempts to establish correspondences between
the dynamics of (8) and (3).)   Fig.1 is a sketch of the linearized
stability diagram for (8) for $s > 0$.   The unperturbed system is stable
to infinitesimal disturbances in the region below the solid line, L,
and unstable above it.   This line has a break in slope at a point
$(s^D, r^D) = (0.096, 1.290)$.

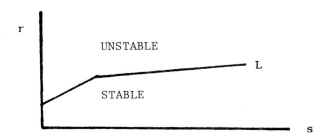

Figure 1.   Stability diagram for system (8).

The system behaviour is conveniently discussed by fixing s and
allowing r to increase.   When $s < s^D$, crossing line L by increasing r
causes instability leading to a stable steady convective state.   For s
$> s^D$, crossing L causes instability leading, just above the line, to
a stable limit cycle.   When $s > 0.0148$, the system also is
subcritically unstable to steady convective motions for values of r
below L.   Thus, above L and for $s > s^D$,   a stable finite amplitude
steady response is always possible, but finite amplitude unsteady
motions of various types may occur as well.   The response of the
periodic solution appearing for parameter values above L to increases
of r depends on s.   For $s^D < s < s_1$, the oscillation period increases
smoothly with r, reaching infinity for a value $r = r_{max}$ and for
values larger than this steady convection obtains.   This behaviour can
be explained by theory[8],[13] by a local analysis near (s^D, r^D).   For s
$> s_1$, the response is more involved.   In an interval $s_1 < s < s_2$, the
period increases smoothly as r increases from L and then the period
doubles.   Further increase of r leads to smooth variation of the
period, then period halving, and further increase leads to a smooth
increase of period to infinity, followed by steady convection.   For
values of $s > s_2$, a period doubling sequence occurs that appears to be
part of an infinite Feigenbaum[14] cascade followed by chaotic motion, as
inferred from the apparent development of a continuous curve on a
Poincare map.   Period halving cascades can be followed by subsequent
period doubling cascades, and even the introduction of new basic
frequencies (other than jumps by factors of two) appearing between

chaotic regimes. Continued increase of r for fixed s always eventually
produces steady convection.

    We have described the breakdown of ordered flow to chaos by means
of a highly simplified mathematical model. The relationship of the
phenomena predicted to consequences following from a more complete
model has yet to be demonstrated. Until such a correspondence is made,
indeed until the phenomena are actually observed, the results of this
section must be regarded merely as intriguing possibilities.
intriguing possibilities.

Acknowledgements. This work was supported by the Physical Oceanography
Program of the National Science Foundation under Grant OCE-8310624.

5. REFERENCES

[1] S. Leibovich & S. Paolucci, The Langmuir circulation instability as a
mixing mechanism in the upper ocean. J. Phys. Ocean. 10, 186 (1980).
[2] A.D.D. Craik and S. Leibovich, A rational model for Langmuir circulations. J. Fluid Mech. 73, 401 (1976).
[3] A.D.D. Craik, The generation of Langmuir circulations by an
instability mechanism. J. Fluid Mech. 81, 209 (1977).
[4] S. Leibovich, Convective instability of stably stratified water in the
ocean. J. Fluid Mech. 82, 561 (1977).
[5] S. Leibovich, The form and dynamics of Langmuir circulations. Ann.
Rev. Fluid Mech. 15, 391 (1983).
[6] I. Langmuir, Surface motion of water induced by wind. Science 87, 119
(1938).
[7] S. Leibovich & S.K. Lele, Thermocline erosion due to Langmuir
circulations. In preparation.
[8] I.Moroz & S. Leibovich, Oscillatory and competing instabilities in a
nonlinear model for Langmuir circulations. In preparation.
[9] S. Leibovich, S.K. Lele & I.M. Moroz, Ordered and disordered states of
Langmuir circulation in the ocean. In preparation.
[10] O.M. Phillips, Dynamics of the Upper Ocean, Cambridge University
Press, pp. 336 (1977).
[11] G. Veronis, On finite amplitude instability in thermohaline
convection. J. Marine Research 23, 1 (1965).
[12] L.N. DaCosta, E. Knobloch & N.O. Weiss, Oscillations in
double-diffusive convection. J. Fluid Mech. 109, 25 (1981).
[13] J. Guckenheimer and E. Knobloch, Nonlinear convection in a rotating
layer: amplitude expansions and center manifolds. Geophys Astrophs.
Fluid Dynamics 23, 247 (1983).
[14] M. Feigenbaum, Quantitative universality for a class of nonlinear
transformations. J. Statistical Phys. 19, 25 (1978).

THE INTERACTION OF CROSSED WAVES AND WIND-INDUCED CURRENTS IN A LABORATORY TANK

Shinjiro Mizuno
Research Institute for Applied Mechanics
Kyushu University
Kasuga, 816
Japan

ABSTRACT.   Subsurface current measurements were made to examine the response of wind-induced currents to the onset of mechanically generated crossed waves and to compare the results with the Craik-Leibovich (1976) theory of Langmuir circulations (LCs). To separate mean current including LCs from the wave-induced current the current data have been decomposed into two frequency ranges. In this paper we only describe the results of low-frequency data that consist of mean and turbulent motion of frequencies lower than the crossed waves.

The primary features of the experimental results are : (1) The most important effect of the wave-current interaction is the deepening of mixed layer. (2) The upwelling and downwelling currents associated with the LCs were clearly confirmed at antinodes and nodes of crossed waves, respectively, but the two components of horizontal current were largely governed by the direct wind-driven current rather than by the LCs. Nevertheless, it is found that the Reynolds stress associated with organized motion still plays an important role in the downward transport of the downwind momentum.

1. EXPERIMENTAL APPARATUS AND PROCEDURE

The experiments were made in a rectangular wave tank of 70m long, 8m wide, and 3m deep, as shown in Fig. 1. A wind blower is an open jet type, 4m wide and 40cm high in the outlet. Since the blower was 4m wide, a dividing plate was mounted over 13.5m downwind of the blower along the center line of the tank, so that the half width of the tank is a test section. According to Awaya's (1961) theory, crossed waves were generated by a lateral row of 8 equally divided wave generators at one end of the tank, as shown in Fig. 1.   Each of the 8 wave generators oscillated with the same period (1.1s) but with different strokes and phases (0° or 180°). In the present experiments the crossed waves had three nodal lines in the test section at $y/L = 0$ and ±2, where y is the lateral coordinate with the origin at the center of the test section, L the interval between a node and an adjacent antinode and equals 66.7cm, and z positive upward from the surface. Subsurface currents were measured with a three-axis sonic current meter. The interval of each

465

Y. Toba and H. Mitsuyasu (eds.), The Ocean Surface, 465–470.

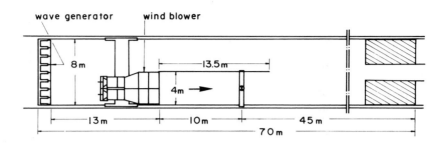

Figure 1. A schematic diagram of wind-wave tank looking from above.

pair of the sensor elements is 27mm. A 200 kHz sonar transducer was
mounted on the tank bottom to detect the depth of mixed layer.
        A wind first blew over the water surface without crossed waves.
Ten minutes later the crossed waves were generated mechanically for
another 10 min. The currents were measured in a vertical cross section
10m downwind of the blower at 64 points of + symbols in Fig. 2. We
measured 3 components of the current by fixing the current meter at each
point, and water displacement at an antinode (y/L= 1) for 20 min a point
under the same wind and wave condition. They were recorded on an analog
tape and converted into a digital data with a sampling frequency of
14.5Hz, for which a period of the crossed waves (1.1s) consists of 16
sampling points. They were allowed to pass through the low-pass filter
of 16-point simple moving average to decompose into two frequency
ranges: low-frequency data which consist of mean and turbulent motion of
frequencies lower than the crossed waves, and the high-frequency data
which consist largely of the wave-induced current and the turbulence
associated with wind-waves. In the present paper we only describe the
results of the low-frequency current data (U, V, W) and water displacement, which were divided into two main segments of 512 sampling points

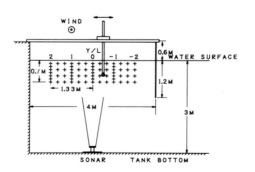

whose interval is 1.1s: One is
the data measured under the
wind alone, and the other
under the wind and crossed
wave condition. Each segment
consists of 16 non-overlapping
subsegments, each of which includes 32 subsampling points.
The symbols U, etc. are the
time averages over 10 subsegments during the quasi-steady
period, <U̅>, etc. are the
averages of currents measured
at 13 points along the lateral
direction, and (W) is the
average of currents measured
at 7 points in the vertical
direction.

Figure 2. Schematic arrangement showing
points of current measurements and sonar
position. Antinodes: y/L= ±1.

2. EXPERIMENTAL RESULTS AND DISCUSSION

Wind profiles above the surface were measured 10m downwind of the blower at 5 lateral locations. The mean wind speed was about 4.6m/s at a height of 20cm. From these profiles we estimated the water friction velocity to be approximately 0.5cm/s. Figure 3 shows the response of the wind-induced downwind mean currents to the onset of crossed waves. The time axis is adjusted in such a way that the crossed waves come to the station at time = 0. The negative time period is called 'wind-only' period, while the positive one is called 'wind-crossed wave' period. The rms wave height at the antinode during the wind-crossed wave period was approximately 4cm. In the wind-only period a high vertical shear developed near the surface and return flow at a depth of 70cm. After the crossed waves arrived to the station, a drastic change occurred in the depth profile of the currents; the currents decreased rapidly near the surface and gradually increased further away from it. The transition period after the onset of crossed waves is approximately 100 s. After that, the current profiles again became quasi-steady (cf. Fig. 3).

We now try to make an order of magnitude estimate of current $\langle U \rangle$ near the end of the transition period, which is nearly constant with depth. Conservation of the x-momentum under a constant wind stress is

$$\frac{d}{dt} h\langle U \rangle = u_*^2 , \tag{1}$$

where u_* is the water friction velocity, and h the depth of mixed layer. Equation (1) may be integrated to give

$$\langle U \rangle = u_*^2 \, t/h . \tag{2}$$

Since $u_* = 0.5$cm/s, t, which is the time from the onset of wind (-600s) to the end of the transition period (100-200s), is 700-800s, and h is approximately 1m, as will be evident later from echo images, we obtain $\langle U \rangle = 1.75$-2.0 cm/s. Thus, it is clear that $\langle U \rangle$ is mainly produced by the wind stress on the surface. Also note that $\langle U \rangle$ is not steady but slowly increasing with time after the transition period because of the continuouly applied wind stress.

Figure 4 shows the details of the vertical profiles of 3-components of the mean currents and turbulent intensities, respectively, at a node and an antinode, where the frequency range of turbulence is limited in a low frequency band between 1/35.2 and 1/1.1Hz. Note that U-profiles are almost the same at the node and antinode, and that the vertical shear varies from 0.1 down to 0.03s^{-1} by adding the crossed waves. The $\overline{V}$-profiles make little or no change by adding them. For the $\overline{W}$-profiles, downwelling current occurs at the node, while upwelling current occurs at the antinode. This will be dis-cussed later in connection with the

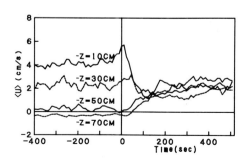

Figure 3. Time variatin of down-wind current $\langle U \rangle$ at 4 depths. Crossed waves arrive at the station at time = 0.

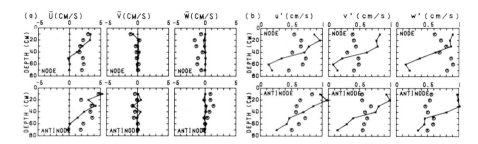

Figure 4. Vertical profiles of (a) mean currents and (b) turbulent
intensities. Solid curves: wind-only period; open circles: wind crossed
wave period.

LCs. On the other hand, the vertical profiles of the turbulent
intensities become almost uniform by adding the crossed waves, which
implies that the wave-current interaction induces strong mixing by
diffusing downward the turbulence produced by the wind-waves.

An acoustic image of the echo sounder provides us with an effective
means for visualizing the depth of mixed layer. It is known that echos
return from the bubbles produced by breaking surface waves (e.g., Thorpe
and Hall, 1983). So, Photo 1 show the time variation of the bubble
clouds within about 1.5m from the surface. The top marker denotes time
marker of 1 min-interval and time increases to the right. The figures 2
and 4 in the Photos denote the distance from the sonar transducer.
Since the tank is 3m deep the 3m horizontal line corresponds to the
water surface. In Photo 1(a) the wind sets on at the left edge. About 2
min later the layer of bubble clouds extends to 20-30 cm from the water
surface. This layer seems to correspond to the turbulent boundary layer
produced by the wind-waves. Crossed waves set on 10 min after the wind.
A few minutes after the onset of crossed waves the layer of bubble
clouds further extends downward (upward in Photo 1) to about 1m from the
surface. Photo 1(b) is another acoustic image observed mainly during the
wind-crossed wave period. When the layer of bubble clouds is assumed to
correspond to the turbulent boundary layer, the depth of mixed layer is
20-30cm during the wind-crossed wave period and approximately 1m during
the wind-crossed wave period.

It is now certain from the above-mentioned results that the wave

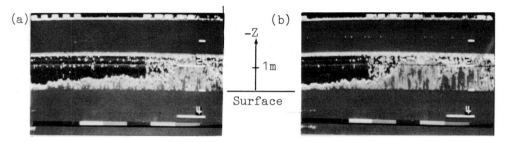

Photo 1. Acoustic images produced after onset of wind and crossed
waves.

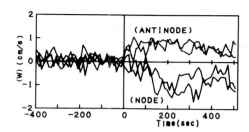

Figure 5.   Time variation of depth-mean current (W) at nodes and antinodes.

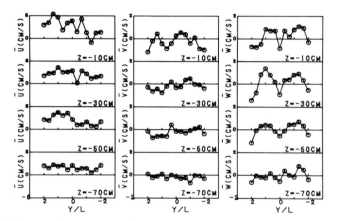

Figure 6.   Lateral profiles of mean current at 4 depthes.   Antinode: y/L= ±1; Nodes: y/L= 0,±2.

current interaction induces the deepening of mixed layer. Then we focus on the generation of Langmuir circulations. Figure 5 shows the time variation of depth-mean vertical currents at the two nodes and two anti-nodes. When the crossed waves come to the station, the down-welling current can be seen at the nodes, and the upwelling current at the antinodes. The current magnitude is variable but, on the average, about 1cm/s. Surface float experiments made previously using the same tank by Mizuno (1983) also showed that steady windrows were aligned along the nodal lines during the wind-crossed wave period. These results are in a general agreement with the prediction of the Craik-Leibovich theory.

Figure 6 shows the details of lateral profiles of 3-components of mean Eulerian currents at 4 depths during the wind-crossed wave period.

It should be noted that, for the Eulerian currents, a good agreement with the Craik-Leibovich theory is limited to the vertical component. Contrary to the expectation, the two components of horizontal current did not show any characteristic features of the LCs that were obtained numerically by Leibovich (1977) and Leibovich and Radhakrishnan (1977), but almost the same lateral profiles as those obtained during the wind-only period. One important difference of the present results from the numerical solutions is that the downwind current U had no significant current peak at the nodes or downwelling zones. It is also important to note that there is no good correspondence between the surface (Lagran-gian) and subsurface (Eulerian) lateral currents. These results imply that the horizontal components of the Eulerian current are largely governed by the direct wind stress rather than by the LCs. Table 1 presents a summary of the experimental results of Reynolds stress, current shear, and eddy viscosity. The Reynolds stress is defined as

$$\tau/\rho \;=\; -\langle\overline{UW}\rangle \;=\; -\langle\overline{U}\rangle\langle\overline{W}\rangle - \langle\overline{U_0 \cdot W_0}\rangle - \langle\overline{u'w'}\rangle \;,\qquad\qquad (3)$$

Table 1.   Results of Reynolds stress, shear and eddy viscosity.

Depth (cm)	$-\langle\overline{U}_o\cdot\overline{W}_o\rangle$ (cm^2/s^2)	$-\langle\overline{u'w'}\rangle$ (cm^2/s^2)	$d\overline{U}/dz$ (s^{-1})	ν_e (cm^2/s)
(a) wind-only period				
10	0.16	0.37		4
30	0.10	0.35	0.1	3-4
50	0.16	0.10		1
70	0.04	0.01		0
(b) wind-crossed wave period				
10	0.79	0.20		7
30	0.11	0.16	0.03	5
50	0.26	0.16		5
70	0.25	0.08		3

where $U(y,t)=\overline{U}(y)+u'(y,t)=\langle\overline{U}\rangle+\overline{U}_o(y)+u'(y,t)$. Here, $u'(y,t)$ is the low-frequency turbulent fluctuation, $\overline{U}_o(y)=\overline{U}(y)-\langle\overline{U}\rangle$, and the subscript o means organized motion. The first term of the right hand side of (3) is the product of the laterally averaged currents $\langle\overline{U}\rangle$ and $\langle\overline{W}\rangle$, the second the Reynolds stress associated with the organized motion like LCs, and the third the turbulent Reynolds stress. Table 1 indicates that in the wind-only period, the turbulent Reynolds stress is about 3 times larger near the surface than the organized Reynolds stress, while in the wind-crossed wave period, the organized Reynolds stress becomes more dominant, except for a depth of 30cm. Table 1 also shows that the sum of the two Reynolds stresses is in general greater than the wind stress (about 0.25 cm^2/s^2). The reason for this discrepancy remains unknown in this study, however, because in discussing it we must further estimate the turbulent Reynolds stress in the high-frequency data excluded from the analysis. Anyway, it is to be noted that in the wind-crossed wave period the Reynolds stress associated with the organized motion, probably due to the Langmuir circulations, plays an important role in the downward transport of U-momentum.

REFERENCES

Awaya, Y, 1961: On the generation of crossed waves in a rectangular tank (in Japanese). *Bulletin Res. Inst. Appl. Mech.*, **17**, 31-39.
Craik, A. D. D. and S. Leibovich, 1976: A rational model for Langmuir circulations. *J. Fluid Mech.*, **73**, 401-426.
Leibovich, S, 1977: On the evolution of the system of wind drift current and Langmuir circulation in the ocean. Part 1. Theory and the averaged current. *J. Fluid Mech.*, **79**, 715-743.
----, and k. Radhakrishnan, 1977: On the evolution of the system of wind drift currents and Langmuir circulations in the ocean. Part 2. structure of the Langmuir vortices. *J. Fluid Mech.*, **80**, 481-507.
Mizuno, S, 1983: Langmuir Circulations in a Laboratory tank. *Rep. Res. Inst. Appl. Mech.*, **XXXI**, 47-59.
Thorpe, S. A. and A. J. Hall, 1983: The characteristics of breaking waves, bubble clouds, and near-surface currents observed using side-scan sonar. *Continental Shelf Res.*, **1**, 353-384.

NUMERICAL MODELLING OF LANGMUIR CIRCULATION AND ITS APPLICATION

Michael R. Carnes[1,2] and Takashi Ichiye[1]
Dept. of Oceanography, Texas A&M University[1]
College Station, TX 77843
Now at U.S. Naval Oceanographic Office[2]
NSTL Station, MS 39522

ABSTRACT. Nonlinear momentum and vorticity equations incorporating the
interaction of surface gravity waves with a drift current were derived
in previous studies of Langmuir circulation (Leibovich, 1977). The
length, time, and velocity scaling of this set of equations is shown
here, for oceanic conditions, to depend only upon the wind speed when
suitable empirical formulations are used. These equations together with
the continuity equation are solved numerically by use of the semi-
spectral method (Orszag and Israeli, 1975) by representing the x-com-
ponents (parallel to the wind) of velocity (u) and vorticity and the
streamfunction (in the y-z plane) by truncated Fourier series. Also, a
no-slip condition is used at the bottom boundary to allow the development of a quasi-steady state. Numerical results are obtained for
Langmuir numbers (an inverse Reynolds number) of 0.1 and 0.01 with
scaled depths of 1, 2, and 4. Two initial conditions are imposed: one
from total rest and the other from an initial horizontally uniform
velocity parallel to the wind. The wind stress starts abruptly and
remains constant. Numerical results indicate that the stream function
shows the cellular structure and profiles of the mean velocity u show
high shear in the upper and lower layers of each cell. The salient
features of Langmuir cells are reproduced and quantitative comparisons
show good agreement between previous field measurements and model
results using the wind velocity scaling (Faller and Caponi, 1978).

1.   INTRODUCTION

Langmuir Circulation (LC) is a common process observed by many researchers since 1938 as first reported by Langmuir. Studies of this process
were reviewed recently by Pollard (1977). Its dynamics has attracted
the attention of fluid dynamicists since the early seventies and was
reviewed by Leibovich (1983). The present study focuses on modelling of
the generation of LC by a sudden input of wind stress at the sea surface.
Since the process is nonlinear, a numerical method is used to solve the
basic equations. The method consists of application of the semi-spectral
method (Orszag and Israeli, 1975) instead of the finite difference method

471

Y. Toba and H. Mitsuyasu (eds.), The Ocean Surface, 471–477.
© 1985 by D. Reidel Publishing Company.

mainly used by Leibovich and Paolucci (1980). The results are compared with laboratory and field experiments by Faller and Caponi (1978) which showed general agreement with the numerical results.

2. BASIC EQUATIONS

The present study uses the most convenient form of the governing equations; the momentum and vorticity equations in the x (or along wind) direction as in Leibovich and Paolucci (1980) except buoyancy effects are excluded and the depth is taken as finite.

The x-y- and z axis are taken along and across the wind and vertically upward with the origin at the sea surface. Momentum and continuity equations are scaled with characteristic time T_c and horizontal distance L_c. The time T_c has three scales: one for drift current, the other for cell development and the third for turbulence. The scale for cell development can be estimated from the observed values which depend on wind stress and parameters of significant waves (Carnes, 1982). Non-dimensional momentum and vorticity equations for the x-direction and the continuity equations form the basic equations of LC dynamics and are given by

$$\frac{\partial u}{\partial t} + v \frac{\partial u}{\partial y} + w \frac{\partial u}{\partial z} = L_a \nabla^2 u \tag{1}$$

$$\frac{\partial \xi}{\partial t} + v \frac{\partial \xi}{\partial y} + w \frac{\partial \xi}{\partial z} = - \frac{\partial u_s}{\partial z} \frac{\partial u}{\partial z} + L_a \nabla^2 \xi \tag{2}$$

$$\partial v/\partial y + \partial w/\partial z = 0, \quad (v = \partial \psi/\partial z, \ w = -\partial \psi/\partial y) \tag{3}$$

respectively, where ψ and ∇^2 are the stream function and the Laplacian in the y-z plane respectively, ξ is the vorticity component in the x-direction $(-\nabla^2 \psi)$ and u_s is a prescribed, non-dimensional Stokes' drift which depends on z only. All the variables are considered uniform in the x-direction.

The non-dimensional coefficient L_a, called the Langmuir number (Leibovich, 1977) is an inverse Reynolds number equaling $\nu T_c L_c^{-2}$ where ν is the eddy viscosity assumed uniform, and T_c and L_c are time and distance scales. Empirical relationships of ν in the upper ocean layer for different values of windstress lead to $L_a = 0.01$ for winds up to 15 ms^{-1}, whereas the laboratory experiments by Faller and Caponi (1971) lead to $L_a = 0.1$ (Carnes, 1982). Therefore these two sets of L_a are used for numerical calculation.

3. BOUNDARY AND INITIAL CONDITIONS

Boundary conditions at the surface (z = 0) are

$$\partial u/\partial z = 1, \quad \partial v/\partial z = 0, \quad \psi = 0, \quad w = 0 \quad (\text{at } z = 0) \tag{4}$$

The wind stress becomes unity, since the time scale T_c is taken to make the dimensional wind stress equal $(L_c/T_c)^2$. The cell depth is taken as H and boundary conditions at $z = -H$ are

$$u = v = w = \psi = 0 \qquad (at \; z = -H) \tag{5}$$

Two initial conditions are used. The first one (ICR) is

$$u = v = w = \psi = \xi = 0 \qquad (at \; t = 0) \tag{6}$$

and the second one (ICS) is

$$v = w = \psi = 0, \qquad u = H + z \qquad (at \; t = 0) \tag{7}$$

For the ICS there is an initial current which decreases linearly with depth in the direction of the wind.

Unlike Leibovich and Paolucci (1980) who used the lateral boundary conditions at $y = 0$ and πk_o^{-1} (the cell width), here only periodicity in the y-direction is imposed.

4.   SEMISPECTRAL MODEL

Periodic boundary conditions in the lateral direction lead to the semispectral method developed by Orszag and Israeli (1975). The x-directed vorticity, velocity, and stream function are represented by truncated Fourier series as

$$\xi(y,z,t) = \sum_{n=1}^{N} \{A_n(z,t)C_n(y) + B_n(z,t)S_n(y)\} \tag{8}$$

$$\psi(y,z,t) = \sum_{n=1}^{N} \{D_n(z,t)C_n(y) + E_n(z,t)S_n(y)\} \tag{9}$$

$$u(y,z,t) = F_o(z,t) + \sum_{n=1}^{N} \{F_n(z,t)C_n(y) + G_n(z,t)S_n(y)\} \tag{10}$$

where

$$C_n = \cos(nk_o y), \qquad S_n = \sin(nk_o y) \tag{11}$$

and k_o is the basic wave number.

Truncated series of (8) to (10) are substituted into (1) to (3), forming the system of partial differential equations of $A_n(z,t)$ etc. about independent variables t and z. This system includes the first and second partial derivates about t and z, respectively, such as $\partial A_n(z,t)/t$ and $\partial^2 A_n(z,t)/\partial t^2$.

A staggered vertical grid system shown in Fig. 1 is employed for the vertical variation. Values of ξ and ψ (or their spectral representa-

tions) are defined at grid points midway between those of u. Fictitious
grid points are defined above the upper boundary and below the lower
boundary to satisfy the boundary conditions (4) and (5).

Also the first and second partial derivatives about z are approximated by the centered difference operators.

5. NUMERICAL EXPERIMENTS

The computation is started employing the two initial conditions (6) and
(7). At the initial instant, the wind stress and Stokes drift are
immediately turned on and remain constant thereafter. Seven runs were
made for different parameters as listed in Table I.

Table I

Run #	La.	H	Initial Condition	Length of run	Δt	Δz	$(2\pi/k_o)$
1	.01	1	ICR	249.82	.0334	-.04167	10
2	.01	1	ICS	249.83	.0334	-.04167	10
3	.01	2	ICR	199.8	.0370	-.04082	20
4	.01	2	ICS	199.8	.0370	-.04082	20
5	.01	4	ICR	597.5	.149	-.08163	24
6	0.1	2	ICR	499.9	.039	-.15707	40
7	0.1	4	ICR	249.9	.042	-.13793	50

N: Number of Fourier Series
$2\pi/k_o$ = Field width
ICR = Initial conditions are a state of rest,
        u=v=w=0 at t=0
ICS = Initial conditions are,
        u=w=0   t=0
        u=H+z   t=0

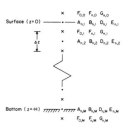

Fig. 1. The staggered vertical grid of the semispectral numerical
model.

Two cases are distinguished according to L_a. Case I (L_a = 0.01)
is more appropriate for the oceanic conditions, whereas Case II (L_a =
0.1) seems to be suitable for laboratory conditions, though both may be
applicable to the ocean under some situations. For Case I, length,

time and velocity scale for the wind speed 6 ms^{-1} are 3.7 m, 52 s and
7.6 cms^{-1}, respectively and those for the 12 ms^{-1} wind are 14.7 m, 105
s and 15.2 cms^{-1}. For Case II these scales are 2 cm, 0.65 s and 2.9
cms^{-1} for the wind speed of 4.3 ms^{-1} at 5 cm above water.

6. RESULTS OF NUMERICAL EXPERIMENTS

All cases with L_a = 0.01 are unstable and immediately begin the formation of a cellular type velocity structure. The predominant cell
spacing (cell width) is small near the beginning but grows with time
within a nondimensional time of 100 (on the order of an hour) and then
remains constant. In Fig. 2, examples of ψ in the y–z plane are shown
from run 1 of Table I.

The profiles of the mean velocity show high shear in the upper and
lower layers of each cell but are nearly vertically uniform in the mid-
depths. This may be explained by cell Reynolds stress τ_c (= $\overline{u'w}$) which
shows larger values in the mid-depth region. The large Reynolds stress
in the mid-depth transports u', the fluctuating component of u, vertically and eliminates the vertical shear of $\overline{u}$ there.

Volume transport M in the x-direction is computed versus time by
two methods: by vertical integration of the mean velocity $\overline{u}$ and by
time integration of the difference of surface and bottom stress
$(L_a|\partial\overline{u}/\partial z|_o - L_a|\partial\overline{u}/\partial z|_{-H})$. The transport computed by the two methods
are almost the same with differences increasing slightly with time.
Under the ICR the transport increases up to a nondimensional time of 50
and then reaches a steady value. With the ICS condition, the momentum
decreases rapidly between times of 20 to 50 and then decreases more
slowly.

For both ICR and ICS with L_a = 0.01, u(0), the mean downwind surface velocity, quickly rises to a peak at a time of about 25 and then
decreases. The total kinetic energy K and the kinetic energy K_n for
each cell with lateral wave number nk_o are computed. In all cases the
kinetic energy of an individual cell is from 10% to 15% at most of the
total kinetic energy. For ICR, K increases rapidly until a nondimensional time of about 25 and then settles to the steady state value.
For ICS, K remains constant initially until cells begin to develop,
after which K rapidly decreases and approaches a final steady state.
K_n may oscillate with time as energy is exchanged among cells of different modes.

7. COMPARISON WITH EXPERIMENTS

Most data available about LC are on cell widths. Faller and Caponi
(1978) combined the field experimental data with laboratory wind-wave
tank data and tried to find a general relationship between the non-
dimensional quantitites λ_c^*/H^* and λ_w^*/H^*, where λ_c^*, λ_w^* and H^* are the
dimensional cell width, surface gravity wave wavelength (or of signifi-
cant waves) and the depth (tank depth or upper mixed layer depth).
Their graph is reproduced in Fig. 3. Notice that the ordinate and

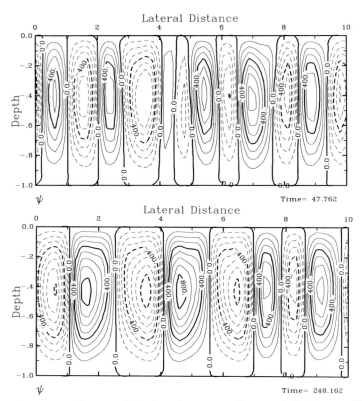

Fig. 2.   Contour plots of ψ in the y-z plane with L_a = .01 and the ICR initial conditions for the following cases: (upper) H = 1 and t = 47.762; (lower) H = 1 and t = 248.162.

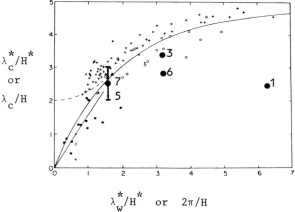

Fig. 3.   Adapted from Figure 2 of Faller and Caponi (1978), showing data obtained from laboratory experiments and oceanic data (closed circles only).   The nondimensional axis labels λ_c/H and $2\pi/H$ have been added in order to plot the numerical model results.   Each data point obtained from the numerical model results is labelled with the run number from Table 4.

abscissa can also be written in terms of the nondimensional parameters λ_c/H versus $2\pi/H$, since in the model we have used $\lambda_w^* = 2\pi\kappa_0^{-1}$ with κ_0^{-1} being the length scale. They determined the linear relationship $\lambda_c/H^* = 1.59 \lambda_w^*/H^*$ for oceanic data alone. Also they found that oceanic data always falls in the range $H^* \geq 0.5 \lambda_w^*$ or $H \geq \pi$, where H^* is the upper mixed layer depth. Presence of the minimum mixed layer depth related to λ_w^* may be due to a condition that the mixed layer is formed primarily by LC.

Results of near steady state of all numerical experiments listed in Table I are plotted in Fig. 3. Values of λ_c/H range from 2.5 to 3.35 with no clear dependence on either L_a or H. The numerical model results fit the curve of Faller and Caponi well for H = 2 and 4 ($\lambda_w^*/H^* = \pi$ and $\pi/2$), but for H = 1 (run 1), the model result is one-half of the laboratory result.

The laboratory results of Fig. 3 indicate that λ_w^*/H^* tends toward a constant 4.8 as λ_w^*/H^* increases. Such a relationship was not found for field data. This may be partly due to ambiguity about the definition of H^*. Also the bottom in the numerical model is a solid no-slip boundary, the depth of which is predetermined. Therefore the model bottom condition may be more adequate for the laboratory experiment as suggested in Fig. 3.

8. CONCLUDING REMARKS

Weak points in the present modelling are arbitrary assignment of H for the scaled thickness of the cell and the boundary conditions at the cell bottom. Models ignore the vertical stratification completely but this may be justifiable if the mixed layer is deep enough. If the vertical length scale of the LC is given in terms of some physical parameters, such as vertical eddy viscosity or surface friction velocity, the ambiguity about H may disappear.

9. REFERENCES

Carnes, M. R. (1981). Theoretical and observational study of Langmuir Circulation. Dissertation at Texas A&M University, Dept. of Oceanography, 254 pp (Extensive references are listed).
Faller, A. J. and E. A. Caponi (1978). 'Laboratory studies of wind-driven Langmuir circulations. J. Geophys. Res., 83, 3617-3633.
Leibovich, S. (1977). 'On the evolution of the system of wind drift currents and Langmuir circulations in the ocean. Part 1. Theory and averaged currents.' J. Fluid Mech., 79, 715-743.
Leibovich, S. (1983). 'The form and dynamics of Langmuir Circulation.' Ann. Rev. of Fluid Mech., 15, 391-427.
Leibovich, S. and S. Paolucci (1980). 'The Langmuir circulation as a mixing mechanism in the upper ocean.' J. Phys. Ocean., 10, 180-207.
Orszag, S. A. and M. Israeli (1974). 'Numerical simulation of viscous incompressible flows.' Ann. Rev. Fluid Mech., 6, 281-318.
Pollard, R. T. (1977). 'Observations and theories of Langmuir Circulation and their roll in near surface mixing. Deep-Sea Res., Supplement "Voyage of Discovery," 235-251.

SOME DYNAMIC FEATURES OF LANGMUIR CIRCULATION

T. Ichiye[1], James R. McGrath[2] and Matthew Howard[1]
Dept. of Oceanography, Texas A&M University[1]
College Station, TX 77843
Naval Research Laboratory[2]
Washington, D.C. 20375

ABSTRACT.   Two sets of experiments on Langmuir Circulation (LC) were carried out about 180 nm south of Nantucket Island on July 22 and 23, 1982, on board the USNS Hayes. During the experiments, dye plumes, smoke bombs and surface drifters (computer cards) were released and photographed using two airplanes equipped with special camera packages together with meteorological, wave and hydrographic measurements.   The card rows had predominent distances and angles from 2 to 23 m and between 0 and 12 degrees to the right of wind, respectively.   Edge of dye plumes perpendicular to the wind showed a wavy form.   The spectra of these waves had peaks at wave lengths between 4 to 60 m.   The wind speed was almost constant at 11 ms^{-1} and the current shear between the surface and 4 m was 0.1 s^{-1}.   Horizontal eddy diffusivity parallel and normal to the wind was determined from card row stretching and from change of row distances, respectively, with order of 10^2–10^3 cm^2s^{-1}. From the steady state non-linear, vertically integrated momentum equation scaled with wind stress and LC width along the wind direction is derived an equation for the along-wind component U of the surface current with a sinusoidal across-wind component.   U has a peak along the convergence line with about 1 to 60 cm s^{-1} corresponding to wind stress of 1 to 10 cm^2s^{-2} for LC width of 10 m.   The sharpness of the peak increases with increasing inertia effect and decreasing horizontal eddy viscosity or cell bottom stress, which becomes effective along the LC divergence lines.

1.   INTRODUCTION

Field experiments on Langmuir Circulation (LC) were carried out over the deep water about 180 n. miles south of Nantucket Island (38°30'N, 69°17'W) on board the USNS Hayes on July 22 and 23, 1982.
        The basic experiment consisted of laying down two parallel dye streaks of different colors separated horizontally by about 2–300 meters, vertically by 4 meters, and both perpendicular to the wind. Computer cards were continuously dispersed over the dye streaks. At the midpoint of each streak a smoke bomb and a vertical array of solid dye

Y. Toba and H. Mitsuyasu (eds.), The Ocean Surface, 479–486.
© *1985 by D. Reidel Publishing Company.*

cakes were deployed. The dye string had solid dye of different colors
attached at various depths to indicate the vertical variation of flow.
Photographs of the dye, cards, and markers were taken by over-flights
by two aircraft.

An ad hoc experiment consisting of laying down two dye streaks at
the same time, one over the other, was made following the end of the
regular experiment. Computer cards and smoke bombs were deployed as
before. Field data including a series of aerial photographs, wind,
wave and hydrographic data are analyzed to determine dynamic features
of LC and salient environmental conditions. Simple statistics of structures of LC are presented.

Heuristic treatment of dynamics of LC is presented to supplement a
number of models of LC which have been proposed since the sixties as
reviewed by Leibovich (1983). Our data on LC are mainly about its surface features such as surface currents and cell widths. Therefore instead of treating three dimensional dynamics of LC, attention is directed mainly to the dynamics of the surface currents induced by the
wind but strongly modified by presence of LC which is assumed to have
a simple vertical structure.

2. RESULTS OF DATA ANALYSIS

The series of film were enlarged and printed on 8x10 in. black and
white paper. The card rows which represent the convergences of the LC
were identified and digitized using a Hewlett-Packard digitizer having
a resolution of 0.01 inch. The distance between adjacent card rows
was computed using the ship for scale. Those photographs having smoke
bombs visible were used to compute the angle between the wind vector
and the cell axis. In photographs where both dye plumes were distinguishable, the distance between the dye centers was used to find the
vertical shear between the surface and 4 meters. The shear values were
almost constant and about 0.1 s^{-1}, probably due to the uniform wind
speed and direction observed during the experiment.

The cards rapidly form into rows usually in a minute after deployment. As time goes on the rows become more distinct as the area in between convergences is swept clear of cards. The frequency distribution
of the cell widths is determined for every two meters and listed in the
following table.

Cell Width (m)	2	4	6	8	10	12	14	16	18	20	24	28	32	36	40
Frequency (%)		15	12	18	13	12	4	9	7	2	8	7	4	1	3

Most of the widths fill between 2 and 12 meters. There are evidences that adjacent card rows merge and form fewer but larger separations. This is suggested by statistics of the average cell widths
against time after deployment, which indicate that the average widths
increase from about 6 m in 2 to 4 min. to 10 to 14 m in 6 to 8 min. and
to 20 to 24 m in 18 to 20 min. after deployment. Such an increase is
also predicted by numerical modelling by Carnes (1982). Pictures
of the dye streaks perpendicular to the wind show a wavy form on the

leeward edge. The amplitudes of the wavy edges are digitized along the
dye streaks from three sets of these pictures and their wave number
spectra are determined. Spectra peaks are found at about 4~5, 7~13,
20~14, and 40~60 m wave lengths, with higher peaks at larger wave
lengths. Therefore, the peaks at the lowest wave length corresponds to
the predominent card row width.
     The cell axis orientation was determined from photographs contain-
ing card rows and smoke bombs. Pollard (1977) indicated that many
experiments in the northern hemisphere showed the cell axis with angles
of 0 to 15 degrees to the right of the wind vector. Our statistics
indicate that the majority of the row direction is in agreement with
this and most measured angles fall between 0 and 12 degrees to the
right of the wind as indicated by the following table (angle: in degrees
with positive values right to the wind):

Angle	-10	-8	-6	-4	-2	0	2	4	6	8	10	12	14	16	18
Percentage		4	4	5	5	16	19	8	13	10	13	4	1	3	3

     If the cards are considered to follow the surface currents, their
distributions represent statistical features of the currents. Lengths
of card rows from the aerial photos showed increase with time, though
the rows were not straight. This is due to the fluctuations of the
current in the direction of the wind though the center of each row
moves with the mean flow. The eddy diffusivity K can be computed from
the change of the row length L with time for relation K = $(1/32)dL^2/dt$
(Carnes, 1982), yielding the mean value K = $1.2\pm0.8(\times10^3 cm^2 s^{-1})$ from 42
sets of data. The eddy diffusivity lateral to the wind can be determined from increase of variances of the cell widths with time. From
ten sets of data the lateral diffusivity is $3\pm2(\times10^2 cm^2 s^{-1})$. These estimates are based on an assumption that there is no organized motion
along or across the rows and may be meaningful only for orders of magnitudes.

3.   DYNAMICS OF THE SURFACE CURRENT

Only steady state dynamics are considered. Since the data do not give
details of the subsurface structure of the LC, the surface current is
mainly discussed. The LC is assumed to be in the y-z plane with the
x-axis in the direction of the wind stress. It is assumed that motion
is uniform with x-direction. The momentum equation in the x-direction
is given by (Ichiye and Howard, 1983)

$$\partial(uv)/\partial y + \partial(uw)/\partial z = K\partial^2 u/\partial z^2 + A\partial^2 u/\partial y^2 \qquad (1)$$

where z is the vertical coordinate positive upward with the origin at
the sea surface and K and A are vertical and horizontal eddy viscosity,
respectively. Equation (1) is scaled with characteristic speed U_0 =
$\tau_0^{1/2}$ and with a, twice the cell width for horizontal distance and with b
the cell depth for vertical distance, where τ_0 is the characteristic
wind stress, leading to the equation which replaces K and A of (1) with

κ and ν, respectively

where

$$\kappa = (Ka)(U_ob^2)^{-1} = (K/A)\nu\delta^{-2}, \quad \nu = A(U_oa)^{-1} \qquad (2),(3)$$

In (2) δ is the aspect ratio b/a. With τ_o of 1 cm²s⁻², a = 10 m, A = 10² cm²s⁻¹ from our data and other field data (pollard, 1977), ν = 0.1. Vertical eddy viscosity and aspect ratio were not conclusively determined from any field experiment. Yet a reasonable estimate is K/A~10⁻² and δ = 0.4. Then κ = 0.0625.

It is assumed that u and v in equation (2) are expressed by

$$u = Uf_1(z), \quad v = Vf_2(z) \qquad (4),(5)$$

where $f_1(z)$ and $f_2(z)$ are functions satisfying the conditions

$$f_1(0) = f_2(0) = 1, \quad f_2(-\delta) = 0 \qquad (6),(7)$$

Integrating (2) with z from $-\delta$ (the bottom) to 0 (surface) with condition w = 0 at the bottom, we have

$$md(UV)/dy = T - rU + \nu nd^2U/dy^2 \qquad (8)$$

where T is the wind stress scaled with τ_o, and m and n are constants defined by

$$m = \int_{-\delta}^{o} f_1(z)f_2(z)dz, \quad n = \int_{-\delta}^{o} f_1(z)dz \qquad (9),(10)$$

The cell bottom stress coefficient r is given by

$$r = \kappa f_1'(-\delta) \qquad (11)$$

Simplest forms are assumed for V, f_1 and f_2 as

$$V(y) = \sin 2\pi y, \quad f_1(z) = \cos\pi z(2\delta)^{-1}, \quad f_2(z) = \cos\pi z\delta^{-1}$$
$$(12),(13),(14)$$

Further it is assumed that U = 0 at y = 0 and 1, indicating there is no longitudinal flow along lines of surface divergence. Then along these lines (y = 0 and 1), equation (8) becomes

$$T + \nu d^2U/dy^2 = 0 \qquad (15)$$

This indicates that the wind stress should be balanced by the horizontal eddy viscosity term at these points, though the latter may be negligible elsewhere. On the other hand, r = 0.245 for δ = 0.4 and larger than ν but it is still smaller than unity. Therefore we neglect the r-term and retain T and ν-term on the r.h.s. of (8), leading to (15). A solution of (15) with condition that U = 0 at y = 0 and 1 is given by

$$U = qN_1^{-1}\{M_1N(y) - N_1M(y)\} \tag{16}$$

where functions $N(y)$ and $M(y)$ with $L(y)$ are defined by

$$N(y) = \int_0^y L(y)dy, \qquad M(y) = \int_0^y yL(y)dy \tag{17},(18)$$

$$L(y) = \exp\left(-S\pi \int_0^y Vdy\right) = \exp(S\sin^2\pi y) \tag{19}$$

Other constants of (16) are given by

$$q = T(\nu n)^{-1}, \qquad S = m(n\nu\pi)^{-1} \tag{20},(21)$$

$$N_1 = \int_0^1 L(y)dy, \qquad M_1 = \int_0^1 yL(y)dy \tag{22},(23)$$

The profile of U expressed by (16) is dependent on the parameter S only. When f_1 and f_2 are given by (13) and (14), respectively, we have

$$n = 2\delta\pi^{-1}, \qquad m = 2\delta(3\pi)^{-1}, \qquad S = (3\pi\nu)^{-1} \tag{24},(25),(26)$$

With the same values of physical constants as before, we have $S \approx 1.06$. The values of U/U_m for $y = 0$ to 1 is plotted for different values of S in Figure 1, where U_m is the maximum of U at $y = 1/2$. Also U_m/q is plotted versus S in Figure 2. Figure 1 indicates that the peak of U at $y = 1/2$ becomes sharper as S increases or as ν decreases. With $U_0 = 1$ cm s^{-1} and $\nu = 0.1$, $S = 1.06$, leading to $U_m/q = 0.45$. The dimensional maximum speed of U is 1.2 cm s^{-1}. When $U_0 = \sqrt{10}$ cm s^{-1} corresponding to $\tau_0 = 10$ cm^2 s^{-2}, $\nu = 0.032$, $S = 3.34$ and $U_m/q = 2.40$ and the dimensional U_m equals 61.1 cm s^{-1}. Both values of maximum speed of U are in agreement with observed data (Carnes, 1982; Pollard, 1977).

Equation (8) can be solved numerically. Also it is possible to determine behaviors of the solution without numerical integration. The ν-term may be negligible against r-term except near $y = 0$ and 1, where the effect of the former is represented implicitly by the boundary condition. Then the solution of equation (8) from $y = 0$ to $y = 1/2$ is given by

$$U = Q|\tan\pi y|^{-(r/2\pi m)}(\sin 2\pi y)^{-1}\int_0^y |\tan\pi y|^{(r/2\pi m)}dy \tag{27}$$

where $Q = Tm^{-1}$. The solution from $y = 1$ to $1/2$ is a mirror image of (27). This solution becomes infinite at $y = 1/2$, because the ν-term is omitted.

Near $y = 1/2$ boundary layer analysis is possible. For this purpose, we change the coordinate y into

$$\eta = y - 1/2, \quad (|\eta| < \varepsilon) \tag{28}$$

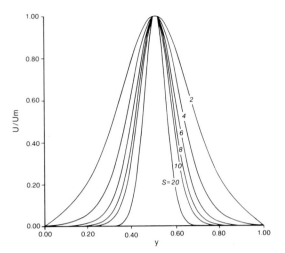

Figure 1. U/U_m versus y for different values of S.

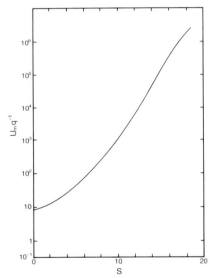

Figure 2. $U_m q^{-1}$ versus S.

where ε is small compared to unity.   Then the inertia term can be approximated by

$$md(UV)/d\eta \simeq m\{2\pi U\cos 2\pi(\eta+\tfrac{1}{2}) + (dU/d\eta)\sin 2\pi(\eta+\tfrac{1}{2})\} \simeq -2m\pi U \quad (29)$$

Thus equation (9) becomes for $|\eta| < \varepsilon$

$$\nu n d^2 U/d\eta^2 + (2\pi m - r)U + T = 0 \tag{30}$$

The solution of (30) is continuous with U given by (27) at $\eta = \pm\epsilon$. The latter is denoted as U. Also the solution should be symmetrical about η. This solution is different according to the ratio of $2\pi m$ to r or the inertia effect to the cell bottom stress.

When $2\pi m > r$, the solution can be expressed by

$$U = \frac{T}{(2\pi m - r)} (\frac{\cos\alpha\eta}{\cos\alpha\epsilon} - 1) + U_\epsilon \frac{\cos\alpha\eta}{\cos\alpha\epsilon} \tag{31}$$

where

$$\alpha^2 = |2\pi m - r|(\nu n)^{-1} \tag{32}$$

When r 2 m, the solution is given by

$$U = \frac{T}{(r - 2\pi m)} (1 - \frac{\cosh\alpha\eta}{\cosh\alpha\epsilon}) + U_\epsilon \frac{\cosh\alpha\eta}{\cosh\alpha\epsilon} \tag{33}$$

The slope of U at $x = -\epsilon$ for (31) and (33) is given by

$$dU/d\eta = \{U_\epsilon + T|r - 2\pi m|^{-1}\}\tan\alpha\epsilon \tag{34}$$

$$dU/d\eta = \{-U_\epsilon + T(r - 2\pi m)^{-1}\}\tanh\alpha\epsilon \tag{35}$$

Therefore, the slope of the solution (31) is in general steeper than that of (33), since U_ϵ is always positive. In fact the peak of U at $y = 1/2$ is caused by the convergence or the inertia effect. The bottom stress and the eddy viscosity reduce this steepness. Solutions (31) and (33) represent these two counteracting effects.

4. CONCLUDING REMARKS

The field experiments in July, 1982 provided some unique data for LC. First, the wind and waves are almost constant during two days' experimental period, with about 11 m s^{-1} of speed and significant wave heights of 0.8 to 1.4 m. By analyzing aerial photos in detail and combining wave data collected with a pitch-roll buoy, we may obtain information relating LC to the wind waves. However, a serious deficit in our data is lack of information on the vertical structure of LC. This may be remedied by simultaneous measurements with aircraft and with underwater photography. Usefulness of the latter for small scale dynamics of the upper layer ocean was demonstrated by Ichiye, McGrath and Carnes (1982). Dynamics of the surface current is treated heuristically because of lack of data on underwater structures of LC. The results seem to explain some observed facts. First, scaling of the momentum equation in the wind direction with wind stress and cell width leads to reasonable values of the speed of the surface currents of LC. Second, the treatment establishes that the along wind surface current U shows a strong peak at the convergence line of LC and that sharpness of the peak increases with decreasing horizontal eddy viscosity and cell bottom stress.   Third, the effect of horizontal eddy viscosity becomes impor- tant along the divergence lines of LC.

5.   REFERENCES

Carnes, M. R. (1981).   Theoretical and observational study of Langmuir
     Circulation.   Dissertation at Texas A&M University, Dept. of
     Oceanography, 254 pp.
Ichiye, T. and M. Howard (1983). 'Some aspects of dynamics of Langmuir
     Circulation.' Tech. Report to Naval Research Laboratory, 21 pp.
Ichiye, J. McGrath and M. Carnes (1982). 'Small scale turbulent diffu-
     sion determined with underwater flow visualization.' Pure and Appl.
     Geophysics (PAGEOPH), 120, 758-771.
Leibovich, S. (1983). 'The form and dynamics of Langmuir Circulation.'
     Ann. Rev. of Fluid Mech., 15, 391-427.
Pollard, R. T. (1977). 'Observations and theories of Langmuir Circula-
     tion and their role in near surfacing mixing.' Deep-Sea Res.,
     Supplement "Voyage of Discovery", 235-251.

MIXED LAYER AND EKMAN CURRENT RESPONSE TO SOLAR HEATING

J. D. Woods, W. Barkmann & V. Strass
Institut fuer Meereskunde an der Universitaet Kiel
Duesternbrooker Weg 20
D-2300 Kiel 1
Federal Republic of Germany

ABSTRACT. For several hours around noon on most days of the year solar heating of the ocean exceeds the rate of heat loss to the atmosphere. The surface buoyancy flux is then positive. For the rest of the day and all night the buoyancy flux is negative. This daily reversal of the surface buoyancy flux has a profound effect on the ocean mixed layer. There has been a tendency to neglect it in comparison with the effect of changes in the wind stress: witness the name "wind-mixed layer" and theories of the wind-driven current starting with Ekman (1905). This paper summarizes recent modelling studies of the diurnal variation in the upper ocean, including the development of an inertial (diurnal) jet.

1.   INTRODUCTION

A number of papers in recent years have demonstrated the power of solar heating to influence the mixed layer of the upper ocean, but there has been no systematic explanation of the physical mechanism by which the observed or modelled changes occur. The aim of this paper is to fill that gap. In order to do so I shall draw on the results of some recent publications from my research group at Kiel, the reader is referred to them for details and illustrations that were presented at the Sendai Symposium, but which must be omitted from the written version because of space limitations. The purpose of this paper is to bring those results together in a coherent account, to show how they are related to publications by other authors, and thereby to provide a firm base for including the effect of solar heating in future models of the oceanic boundary layer. The main emphasis will be on simulating mixed layer behaviour sufficiently accurately for the purposes of coupling the global circulations of the atmosphere and ocean in climate prediction models. As we shall see solar heating poses special problems that cannot be overcome by simple parameterizations.

The paper is presented in five sections. The first summarizes methods of parameterizing solar heating, convection and turbulence; the second describes a one-dimensional model of the mixed layer designed to simulate explicitly the change in convection, and consequent changes in

Y. Toba and H. Mitsuyasu (eds.), The Ocean Surface, 487–507.
© *1985 by D. Reidel Publishing Company.*

mixed layer turbulence and the depth of the turbocline. The following
section describes diurnal and seasonal variations of those variables,
given climatological mean surface fluxes and cloud cover, based on
Bunker's analysis. It is then shown that solar heating greatly compli-
cates the Ekman current response to a steady wind stress. The effects
described so far have been illustrated with a model integrated at time
steps (10 or 60 minutes) sufficiently brief to resolve the diurnal cycle
of solar heating: the next section considers ways in which longer time
steps can be taken in climate prediction models without introducting
unacceptable error. The final section describes a series of sensitivity
tests used to demonstrate the errors introduced into the seasonal cycle
by uncertainty in the surface fluxes used to drive the model, and in the
plankton concentration which determines the depth distribution of solar
heating.

2.   PARAMETERIZATIONS

2.1. Solar heating

The problem is to calculate the change of temperature as a function of
depth $\dot{T}(z)$ dt in the time step dt of the model from three sources of
information. (1) The Earth's orbital parameters (which change on
Milankovich time scales), whence the diurnal, seasonal and meridional
variation of insolation at the top of the atmosphere. (2) The Earth's
atmospheric parameters, of which the cloud cover presents the most
serious problems (see Dobson and Smith, 1985). (3) Oceanic parameters,
notably the turbidity of the seawater, which is controlled in the open
ocean by plankton concentration. Here I shall concentrate on the third
factor. Three approaches have been followed in parameterizing the effect
of plankton on $\dot{T}(z)$. In all three it has been assumed that

$$\dot{T}(z) = \left(\frac{1}{c} \right) \frac{\partial}{\partial z} I(z) \tag{1}$$

where $I(z)$ is the downwelling solar irradiance and $c = 4.2$ MJ/m^{3}K. It is
also assumed that the solar energy flux at the surface of the ocean $I(o)$
has been calculated accurately from the Earth's orbital and atmospheric
variables and the ocean albedo. The ocean surface is assumed to be flat.
The first parameterization (Woods, 1980) is based solely on the molecular
absorption of pure water (i.e. particulate and molecular scattering and
particulate absorption are neglected). The model sums the energy flux
over 27 spectral bands, using the absorption coefficients of distilled
water (the effect of dissolved salts being negligible). It is especially
useful for describing $\dot{T}(z)$ in the top metre of the ocean, where there
are no measurements of $I(z)$ on which to base an empirical model. Approximately half of the solar energy entering the ocean is absorbed in the
top metre according to this parameterization. The figure would be higher
if particulate absorption and scattering were taken into account. We
shall see later that this concentration of solar heating into the top
metre of the ocean has an important consequence for the daytime depth of
convection.

The second parameterization (Woods, Barkmann and Horch, 1984) is more useful for describing the distribution of the $1/2$ I(o) that penetrates below the first metre. It is based on Jerlov's (1976) measurements of I(z) in a wide variety of seawater turbidities which for convenience he classified into a small number of optical water types (designated J1, for the clearest, J2A, J2B, and J3 for the most turbid). He showed that these classes lie at equal intervals on a continuous "colour index", which has been used as a basis for determining seawater turbidity from satellite measurements (Højerslev, 1980). On that scale the pure water parameterization described earlier would be classified as J0. The following expression was fitted to Jerlov's data for each of the optical water types

$$I(z) = I(o) \sum_{i=o}^{3} A_i \exp(-z/\lambda_i \cos\theta) \qquad (2)$$

where θ is the zenith angle of the direct solar beam in the sea.

This empirical parameterization fits the observation to better than 4 %. For comparison, the two-exponential parameterization of Paulson and Simpson (1977) has errors exceeding 30 % in the top few metres, and the widely used single-exponential parameterization has errors exceeding 50 %. We shall see that the large errors in I(z) produced by over-simplified parameterization leads to serious error in modelling the diurnal variation of convection depth.

The third method of parameterization I(z) is more ambitious and has so far produced only preliminary results (Dörre, in preparation). It is based on explicitly modelling the growth of the phytoplankton which are responsible for seawater turbidity. This approach is too expensive in computer time to be adopted in climate prediction models, but is useful for clarifying problems arising from the inhomogeneous vertical distribution of phytoplankton (Platt et al., 1983), and for predicting optical water types in regions for which we have no optical data or under other climate regimes. The novel feature of our parameterization scheme, is that (in contrast with those described by Platt et al. (1977) in which the plankton are treated as a continuum of varying chemical concentration) we deal with photons penetrating a cloud of discrete plankters, whose life histories are known from Lagrangian integration (according to the method of Woods and Onken, 1982).

For the purposes of this paper, the first method of parameterizing $\dot{T}(z)$ has served to clarify the daytime variation of convection within the mixed layer, and the second method has permitted calculation of the temperature rise due to solar heating below the mixed layer, in the seasonal thermocline at all latitudes and below the seasonal thermocline in the tropics. A detailed climatology of solar heating, based on the first two methods, has been published in Woods, Barkmann and Horch (1984) with full details and additional illustration in Horch, Barkmann and Woods (1983). The third method has great potential for the future, but does not contribute to the results to be presented below.

2.2. Convection

The combination of evaporation, conduction and net long wave fluxes transfer energy from the ocean to the atmosphere at rate B. Heat flows up towards the surface from a depth C, the convection depth. (In this section we assume that the vertical density gradient depends only on the temperature gradient, the contribution of the salinity gradient will be considered later.) The potential energy released as heat flows upwards appears as kinetic energy of convective overturning, with energy-containing eddies of height C, which power a cascade of convectively-driven turbulence at a rate

$$\varepsilon_c = \frac{1/2 \ BC\alpha \ g}{c} \tag{3}$$

where α is the coefficient of thermal expansion, and g is the acceleration due to gravity. A small fraction m_c of that turbulent power input is consumed in entraining denser (colder) water from below the convection layer. The maximum possible value of m_c is 0.15 according to the flux Richardson number criterion of Ellison (1957), but there is experimental evidence that it may be smaller, especially when C is large (Killworth, 1983). In the limiting case where m_c = 0, the convection is said to be non-penetrative, and its depth C can be calculated simply by the method of convective adjustment (fig. 1).

Solar heating has a profound effect on the depth of convection. For several hours centred on noon each day I(o) ≥ α B with the result that more than sufficent heat is generated in the top metre of the ocean to supply the loss to the atmosphere. The depth of convective adjustment is then less than one metre, regardless of the depth of the mixed layer, which is normally much greater than one metre. At night, when there is no solar heating, the surface heat loss is supplied convectively from heat stored during the day or, in the cooling season, during earlier days. The depth of convective adjustment then nearly reaches the bottom of the mixed layer. Dalu and Purini (1982) have considered a model in which there is convective adjustment, but no turbulent entrainment. The variation of C can then be calculated from an initial temperature profile T(z), the solar heating profile $\dot{T}(z)$ and the surface heat loss B. Kraus and Rooth (1961) considered a simpler variable, the thermal compensation depth D, which depends only on $\dot{T}(z)$ and B

$$B = I(o) - I(D) \tag{4}$$

and is therefore independent of turbulent entrainment and the initial temperature profile. Fig. 2 shows that the daytime variations of C and D are remarkably similar. That is because of the concentration of solar heating close to the surface; a point overlooked by Dalu and Purini (1982), who unwisely used a single exponential to parameterize I(z). Woods (1980) showed that one can gain insight into the climatology of convection in the mixed layer of the ocean by calculating the diurnal, seasonal and meridional variations of D. This leads to a parameter P_{10}, defined as the fraction of each 24 hours in which D is shallower than 10 metres, which is close to the fraction of each 24 hours during which

C is less than the mixed layer depth H. We shall see later that P_{10} provides the key to designing a longtime step parameterization that does not suffer from the failure to resolve diurnal variation of solar heating in climate prediction models.

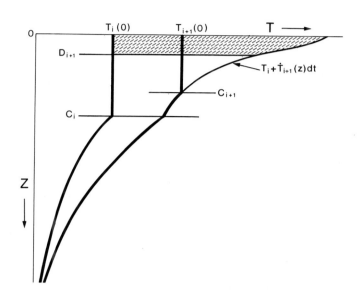

Figure 1. Schematic defining the depth of convective adjustment (C) and thermal compensation depth (D) for a one-dimensional model in which there is no turbulent entrainment. Solar heating changes the initial temperature profile from $T_i + \dot{T}_i$ dt; convective adjustment then changes it to T_{i+1}. The convection depth changes from C_i to C_{i+1}, while the thermal compensation depth is shallower at D_{i+1}. The value of D_{i+1} is independent of the temperature profile, and is therefore easily calculated from the $\dot{T}$ and B the heat flux to the atmosphere.

2.3. Turbulence

One of the major advances in the past decade has been the reliable measurement of profiles of turbulent kinetic energy in the upper ocean (Oakey and Elliott, 1982; Shay and Gregg, 1984). Broadly speaking they have confirmed the notion of a turbulent mixed layer overlying a largely laminar flow thermocline. The turbulent dissipation rate ε in the mixed layer is typically of order milliwatts per cubic metre (mW/m^3): in the seasonal thermocline ε is typically of order microwatts per cubic metre ($\mu W/m^3$). The measurements suffer from instrumental limitations and sampling errors (because the turbulence is intermittent), but even single profiles ε(z) show a clear turbocline between the mixed layer and thermocline. Now that we have such measurements the physics of the upper ocean is more clearly established when we specify a **turbocline depth** H at which ε has some standard value, say 10 $\mu W/m^3$, rather than a **mixed layer depth** or ("Deckschichttiefe") based on a criterion in the temperature profile, which is often a fossil of some earlier deep mixing event.

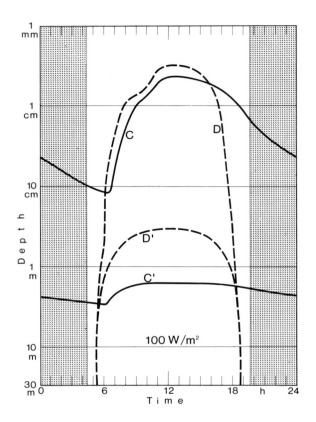

Figure 2. Diurnal variations of convective adjustment depth C and thermal compensation depth D for a model with no turbulent entrainment (as proposed by Dalu and Purini, 1982). The upper curves (C, D) are for the JO parameterization of solar heating; the lower curves (C', D') are for the much less accurate single exponential parameterization used by Dalu and Purini. It is seen that accurate parameterization of the solar heating profile is extremely important. Inaccurate parameterization underestimates the diurnal range of C and overestimates the difference between C and D. In fact the difference is negligible so long as C $\geqslant$ D, i.e. for most of the day, when I $\geqslant$ B.

The vertical distribution of ε above the turbocline is not resolved in the slab parameterization used in this paper. The emphasis will be on the power input to the turbulence per unit area (mW/m^2). The contribution from convection ε_c has already been discussed. The contribution from wind waves ε_w is complex, as we have learned from many papers

presented at the Sendai symposium. For simplicity I adopt Oakey and
Elliot's (1982) experimental result that the power supply to the turbulence from that source is 1 % of the power loss from the wind in doing
work against the wind stress.

$$\int_{o}^{H} \varepsilon_w \, dz = 0.01 \, \rho_a \, C_d \, U^3 \qquad\qquad (5)$$

where ρ_a is the air density, C_d the drag coefficient and U the wind
speed.

It is assumed that when the mixed layer is shallow the maximum
possible fraction (0.15 according to Ellison, 1957) of the power input
to the turbulence from the wind is converted to potential energy by
entrainment of denser water from the thermocline. This is equivalent
to a maximum entrainment parameter, M_w = 0.0015, similar to that found
experimentally (Denman and Miyake, 1960). The parameterization used in
our model follows that of Niiler (1977), Wells, (1979) and others (see
Woods, 1983) in exponentially decreasing the entrainment parameter as
the turbocline deepens

$$m_w = M_w \, \exp(-H/H_c) \qquad\qquad (6)$$

where H_c is the Ekman depth or a fixed depth of similar magnitude (say
100 metres). This means that, as the mixed layer deepens, a smaller
fraction of the power input to the turbulence is consumed by work
against the Archimedes force: the vertically integrated flux Richardson
number decreases as H increases. As in all slab parameterizations of the
mixed layer, the depth of the turbocline is calculated to ensure that
the Richardson number criterion is satisfied at every time step in the
integration. Thus turbulence carries no memory between time steps.

The effect of solar heating on the turbocline depth is best consid-
ered in three layers. It is neutral in the **convection layer** ($0 \leqslant z \leqslant C$),
where the density profile is adjusted by definition to be zero. It
increases the stable temperature gradient in the part of the mixed layer
below the convection layer ($C \leqslant z \leqslant H$), where there is a downward turbulent heat flux, which weakens the turbulence at the limiting Rf = 0.15.
It preconditions the density gradient in the **diurnal thermocline**
($H \leqslant z \leqslant H_{max}$) and in the **seasonal thermocline** ($H_{max} \leqslant z \leqslant D$), so that
the entrainment velocity will be slower when the mixed layer deepens as
the ocean cools in the evening and autumn respectively. (H_{max} and D are
the diurnal and annual maximum values of H, respectively.)

3.   THE ONE-DIMENSIONAL MODEL

In order to predict climate it is necessary to construct a model in
which the circulations of the atmosphere and ocean are coupled through
their respective boundary layers and the interfacial fluxes of momentum,
energy and water. One of the principal motivations for improving parameterizations of solar heating, convection and turbulence in the upper

ocean is to reduce errors in the heat and water fluxes arising from errors in the mixed layer temperature and salinity. As I have pointed out elsewhere (Woods, 1983) the errors in existing mixed layer models are often larger than the climate signals to be predicted. In this section I briefly summarize the one-dimensional model that is being used in my group at Kiel to investigate the sources of those errors and to test possible improvements.

The model is derived from that of Turner and Kraus (1957). The parameterizations described in the previous section are applied sequentially in each time step of integration. The stages in each time step are as follows:

1. Initial temperature profile $\qquad$ $T_1(Z), S_1(Z)$
   (From the last time step)

2. Solar heating $\qquad$ $T_2(Z), S_2(Z)$
   (Astronomic and atmospheric variables, seawater turbidity)

3. Convective adjustment $\qquad$ $T_3(Z), S_3(Z)$
   (Surface heat and water fluxes)

4. Turbulent entrainment $\qquad$ $T_4(Z), S_4(Z)$
   (Convective and wind stress contributions to turbulent kinetic energy)

5. Upwelling $\qquad$ $T_5(Z), S_5(Z)$
   (Wind stress curl; synoptic and meso-scale motion; internal waves)

6. Advection $\qquad$ $T_6(Z), S_6(Z)$
   (Ekman transport; climatological mean horizontal gradients of temperature and salinity)

7. Initial profiles for next time step

$$T_6{}^i(Z) = T_1{}^{i+1}(Z); \qquad S_6{}^i(Z) \equiv S_1{}^{i+1}(Z)$$

This intra-time-step sequence of changes to the temperature and salinity profiles does not include a term for the divergences of the geostrophic fluxes of heat and (fresh) water. Woods, Barkmann and Horch (1984, fig. 15) have shown that the geostrophic term dominates the seasonal changes at sites such as OWS "C" where it produces an annual heat flux divergence of 3 GJ/m²y according to Budyko (1974), and an annual water flux divergence of $1/3$ T/m²y according to Baumgartner and Reichel (1975). Two problems prevent us from including geostrophic flux divergence terms in the model. The first is that we have no reliable map of the mean circulation. The second is that we do not yet know how to parameterize the contributions of transient quasi-geostrophic eddies. Some insight into the magnitude of the problem is gained from Lagrangian integration of the above model, using observed drifter trajectories (Woods,

1985). Optimistically, the Lagrangian studies will suggest methods of economically representing the seasonally-broadening catchment of water passing through a site such as OWS "C". The variance of mixed layer temperature and salinity generated by eddy-induced deviations from the mean streamlines will limit the predictability of climate models, because too few eddies influence a water parcel during one season. Pessimistically, that loss of predictability will prevent reliable forecasting of the oceanic contribution to the thermal response of the climate system to CO_2 pollution.

For the purposes of this paper we shall brush aside those serious problems, and concentrate on Eulerian integration of the model, without a geostrophic term, at sites where the climatological mean surface heat and water fluxes are negligible, and where the mean geostrophic current and eddy kinetic energy are small. Such a site is 41° N 27° W, located in the North Eastern Atlantic sector of the anticyclonic gyre, not far from the β-spiral site of Armi and Stommel (1983). The results to be presented below refer to that site. They are based on Eulerian integration of the model with the annual cycle of surface fluxes derived from interpolation between Bunker's (1976) monthly mean values. Fluctuations due to weather events are not considered explicitly because we are concerned with developing a model for decadal climate prediction. However, we shall later consider the effect of random deviations from the climatological mean forcing, which can be thought of as representing either the uncertainty in the latter, or a parameterization of weather events in the spirit of Frankignoul and Hasselmann (1982).

4.   RESULTS OF MODEL INTEGRATIONS AT 41°N 27°W

We now demonstrate that the formulation of the model in terms of an intra-time-step sequence of adjustments to the temperature and salinity profiles leads to stable predictions in the presence of the diurnally reversing surface buoyancy flux, and the seasonally reversing net daily surface buoyancy flux. The calculations are made for the mean climatological cycle of surface fluxes derived from Bunker's climatology. The solar heating profile was calculated from astronomical variation of the solar elevation, plus a climatological mean atmosphere (including seasonally-varying cloud cover) and a constant seawater turbidity.

The model was integrated for eighteen months with a one hour time step, starting on day 270 which is the last day of the cooling season, after which the daily heat input from the sun exceeds the heat loss to the atmosphere and the daily maximum depth of the turbocline begins its vernal ascent. The initial temperature and salinity profiles were uniform to a depth of 170 metres (based on Robinson, Bauer and Schroeder, 1979), with constant gradients below, corresponding to climatology for the permanent pycnocline.

4.1. Seasonal variation

The variation of turbocline depth for the last year of the integration is shown in fig. 3. The diurnal variation, which will be described in

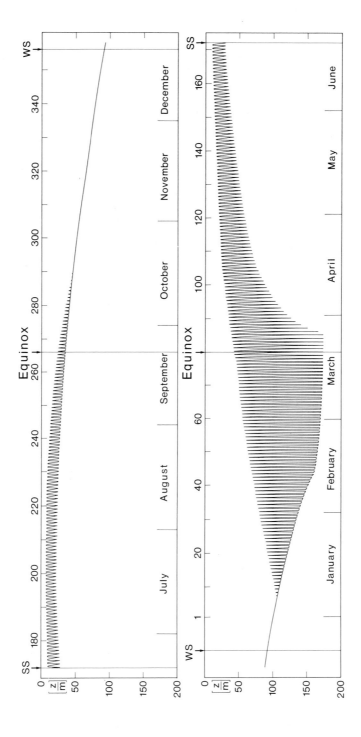

Figure 3. Variation of turbocline depth over one year at 41° N 37° W as predicted by our model for the diurnal and seasonal variation of solar elevation plus Bunker's (1976) monthly mean climatology for cloud cover, surface cooling, evaporation and wind stress, and the annual mean precipitation of Baumgartner and Reichel (1975).

more detail in the next section, reaches a maximum at the end of the
cooling season, just after the vernal equinox, and virtually disappears
from mid-October until early January, when the solar heating is weakest.
The seasonal cycle of the turbocline is best described in terms of the
envelopes of its daily maxima and minima (H_{max}, H_{min}). The former occurs
an hour or so after sunrise, when the solar heating first exceeds the
heat loss to the atmosphere. The latter occurs close to noon, when the
sun is highest (there is a slight offset owing to the seasonal variations
of cloud cover and, to a lesser extent, of surface heat loss). The temperature, salinity and density profiles every ten days are shown in
fig. 4. Note that the seasonal storage of heat occurs mainly in the top
100 metres. At locations such as OWS "C" the seasonal variation of heat
storage is more uniformly distributed between the surface and the annual
maximum depth of the turbocline D, because the geostrophic flux divergence term dominates where D exceeds 100 metres. The value of D is
determined by the net annual surface buoyancy flux and the potential
energy introduced by the geostrophic flow (Woods, 1985). A model that
does not include the latter cannot therefore predict a stable value of D
given an unchanging seasonal cycle of surface fluxes. The secular trend
in D that arises in such models decreases with the net annual buoyancy
flux which is negligible at the site chosen for the present illustration.

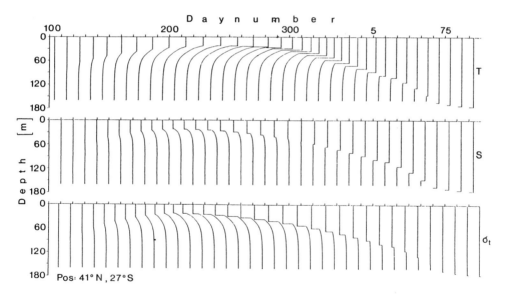

Figure 4.   Temperature, salinity and density profiles every ten days for
the annual cycle described in fig. 3.

4.2. Diurnal variation

The most rapid forcing used in the integration, arising from the diurnal
variation of solar elevation, is resolved quite well by the one hour

time step. We can therefore examine in detail the response of the mixed layer to solar heating by the diurnal variation of key variables, as they are calculated at intermediate stages within each time step. For the purposes of illustration here (more examples are given in Woods and Barkmann, 1985) fig. 5 shows the variation of:

(1) the net surface energy flux $Q_{net} = I(o) - B$,

(2) the mixed layer temperature $T_1(o)$,

(3) the depths of convective adjustment, C, and the turbocline, H and

(4) the contributions of wind stress ε_w and convection ε_c to the power supply to turbulence in the mixed layer,

for days 92, 93 and 94, at the start of the heating season. The daily range of mixed layer temperature is ten times the daily rise; if the mixed layer depth were constant at its average value the daily range would be only twice the daily rise. The convection depth rises from over 100 metres to less than 1 metre in 3 hours after sunrise, and the turbocline depth rises to a noon value of 35 metres. As soon as the surface buoyancy flux changes sign in the afternoon, the convection depth descends rapidly, meeting the turbocline at about 40 metres and thereafter descending more slowly with it. The difference between the convection and turbocline depths when they are descending together represents the contribution to the latter of turbulent entrainment: the nocturnal descent of the mixed layer is due mainly to convective adjustment. The contribution of wind stress to the turbulence is steady (apart from the slight seasonal trend), but the contribution from convection is reduced by a factor of one thousand during the day when the convection depth is shallow. At night convection contributes almost as much power to the turbulence as the wind stress. The general conclusion is that the diurnal variation of convection in the mixed layer, induced by solar heating, is the dominant factor in controlling the diurnal variation of turbocline depth, which in turn increases the diurnal range of mixed layer temperature by a factor of five.

5.   EKMAN CURRENT

The classical Ekman spiral was the solution for a steady wind stress, an infinitely deep mixed layer with constant eddy viscosity (K) independent of depth. These assumptions were perhaps appropriate for the ice drift problem posed by Nansen, but they do not evenly remotely approximate to the mixed layer as described in the previous section. We must consider a mixed layer for which the turbocline depth is not infinite, which exhibits significant diurnal variation, and in which K varies diurnally with the power supply to the turbulence and the scale of the energy containing eddies. Even if the wind stress remains constant these changes in K(z) induced by solar heating will lead to a solution quite different from Ekman's. This problem has been studied by Woods and Strass (1985),

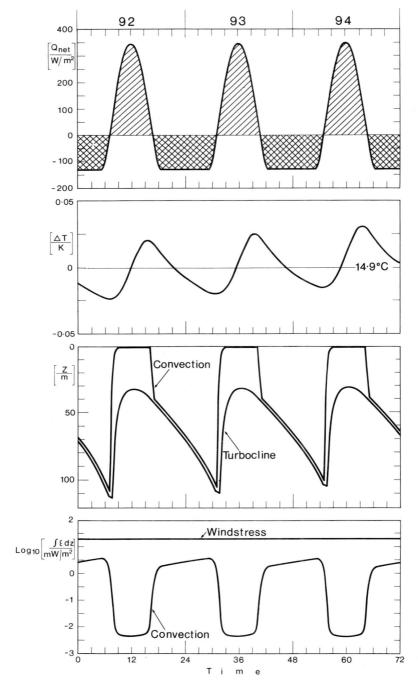

Figure 5. Diurnal variations of key variables in the mixed layer on days 92, 93 and 94 in the annual cycle described in fig. 3.

using a model in which, for simplicity, the diurnal variation of turbocline depth was prescribed as a saw tooth variation between the diurnal
extrema predicted by the mixed layer model, as described in the previous
section. A detailed description of the model, and the results of many
runs are presented in the paper cited. Here I summarize the principal
results. Firstly, we tested the model by checking that it gave

1. the classical Ekman result with a homogeneous constant K

2. Gonella's (1971) result for a mixed layer of constant depth and K,
   but shallower than the classical Ekman depth

3. the correct value for the Ekman transport in cases (1) and (2).

Next we investigated the response to a constant wind stress when solar
heating reduces the turbocline depth and K during the day. As expected,
the result is similar to that found for the atmosphere boundary layer
over land at night, when the turbocline height is reduced by the cessa-
tion of convection once the sun no longer heats the ground. Momentum
mixed down deep into the sea at night becomes isolated from the surface
during the day when the turbocline rises, and rotates as an inertial
current until the turbocline descends again next night, where-upon it is
re-entrained, albeit with a vector rotated by an angle that depends on
the number of hours that the turbocline was shallower, and the inertial
period at that latitude. The result is a complicated modulation of the
current vector, as illustrated in fig. 6.
    It is not surprising that a classical Ekman spiral has never been
detected by moored current meters (the only exception is a set of measurements made in polar seas, where day length greatly exceeds the inertial period; see (Hunkins, 1966). The absolute current (fig. 7) exhibits
a jet-like structure during the day; this is the dynamical analogue of
the atmospheric nocturnal jet, and is therefore called the daytime jet.
The progressive vector diagrams at selected depths (fig. 8) more clearly
resemble an Ekman spiral, suggesting that the best experimental test
would be to follow the motion of constant depth drifters. Saunders (1980)
tried to do that during the JASIN experiment (Charnock and Pollard,
1983). Unfortunately they tracked the drifters acoustically: a method
that worked well at night, but failed during the day when the turbocline
was shallow, permitting a diurnal thermocline to form and refract the
sound downwards (an example of the "afternoon effect", well known in
ocean acoustics). The experimental design contained the seeds of its own
destruction making it impossible to test the results of our model.
Nevertheless, given a different method of tracking the drifters, the
Lagrangian method does offer the best prospect for a test.
    The results presented here were obtained for a steady wind stress
and climatological mean variation of the mixed layer. Even with those
restrictions we can see some implications that go beyond our preliminary
study. For example, during spring, when the turbocline is becoming
progressively shallower, some of the momentum mixed down at night will
not be re-entrained the next night, but will be abandoned as an inertial
current. The leakage of momentum into the seasonal pycnocline by this

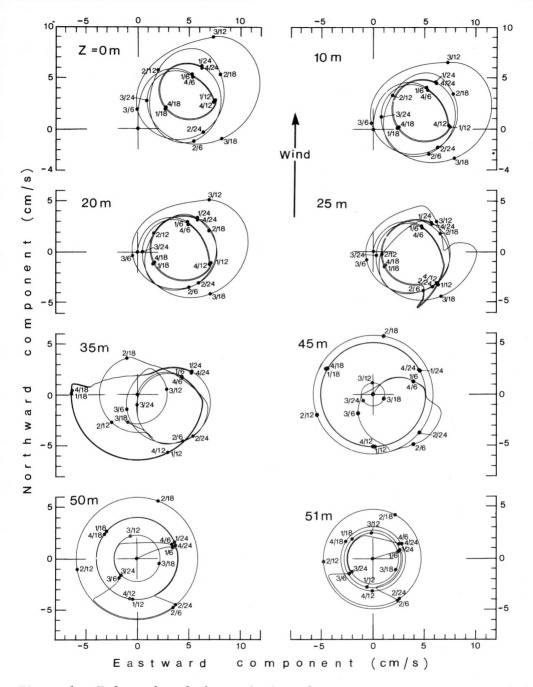

Figure 6. Hodographs of the variation of current vectors over a period of 4 days at selected depths. Results of a model with constant wind stress and diurnal variation of the turbocline depth.

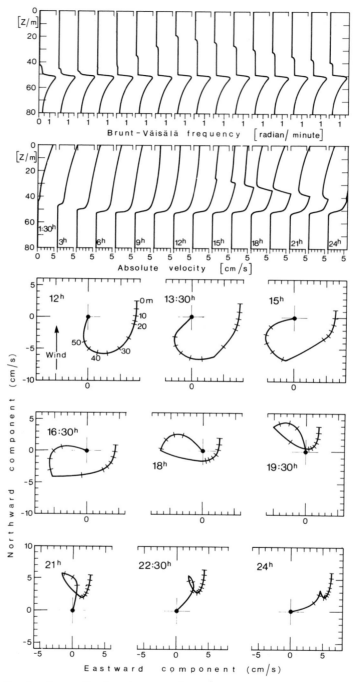

Figure 7. Variation of the Brunt-Väisälä frequency profile, the absolute current profile and the hodograph of current vector versus depth at intervals of 90 minutes showing the development of the daytime jet.

process will reduce the mean Ekman transport associated with a steady wind stress. In general, the calculation of mean Ekman transport for a wind stress that varies with the weather will be made more complicated by the variation of turbocline depth induced by solar heating and changes in the weather. And, although the shallow Lagrangian response to a steady wind stress is remarkably like the Ekman spiral solution, the deeper response, which depends on brief injections of momentum when the turbocline descends to its maximum depth in the diurnal cycle, is quite different, requiring care in calculations of the dispersion of pollution.

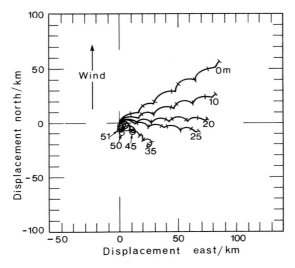

Figure 8. Progressive vector diagrams for four days at selected depths showing that the Lagrangian mean velocity profile resembles an Ekman spiral.

6. CONCLUSION

The prediction of changes in mixed layer temperature and depth has often been presented as a problem of wind mixing, with the major problem aris- ing from errors in the parameterization of turbulent entrainment, both on synoptic scales (Niiler and Kraus, 1977) and on the seasonal time scale (Niiler, 1977). This paper has considered the alternative view that the major problem arises from errors in the measurement (or atmospheric model prediction) and parameterization of the buoyancy fluxes, including solar heating and surface cooling to the atmosphere, on the diurnal, sea- sonal and longer time scales. A case has been made that nocturnal convec- tion is the dominant factor in the diurnal and seasonal cycles of turbo- cline depth variation, which greatly increase the amplitudes of diurnal and seasonal cycles of mixed layer temperature. Turbulent entrainment makes a relatively small contribution to those cycles, although work by other authors has demonstrated that it can be more important in mixed layer response to weather events (e.g. Pollard, Rhines and Thompson, 1973). The diurnal cycle of turbocline depth variation induced by solar heating interacts non-linearly with variations due to other causes and cannot therefore be ignored in studying mixed layer response to changes of wind stress. The importance of the non-linear interaction was illustrated by considering the response of the Ekman current to a steady wind stress in the presence of daily solar heating.

This paper has concentrated on describing the physical processes that occur in the upper ocean as the consequence of solar heating, in particular the diurnal variation of the convection depth, which leads in turn to diurnal variations of the turbocline depth and a complicated modulation of the Ekman flow, involving a daytime jet. The space allocated is insufficient to describe the part of my talk at Sendai that dealt with the modelling of the seasonal cycle with time steps that do not resolve the diurnal cycle, and the investigation of the sensitivity of the seasonal cycle to fluctuations in the cloud cover, seawater turbidity and surface cooling. These results will be published elsewhere.

REFERENCES

Armi, L. and H. Stommel, 1983: 'Four views of a portion of the North Atlantic Subtropical Gyre [27°N, 32°30'W]'. J. Phys. Oceanogr. **13**(5), 828-857.

Baumgartner, A. and E. Reichel, 1975: The World Water Balance. Elsevier, Amsterdam, 179 pp.

Budyko, M.I., 1974: Climate and life. Academic Press, New York, 508 pp.

Bunker, A.F, 1976: 'Computations of surface energy flux and annual air-sea interaction cycles of the North Atlantic Ocean'. Mon. Wea. Rev. **104**, 1122-1140.

Charnock, H. and R.T. Pollard, 1983: 'Results of the Royal Society Joint Air-Sea Interaction Project'. Proceedings of Royal Society Discussion Meeting held on 2 and 3 June 1982. Royal Society, London.

Dalu, G.A. and R. Purini, 1982: 'The diurnal thermocline due to buoyant convection'. Quart. J. Roy. Met. Soc. **108**, 929-935.

Denman, K.L. and M. Miyake, 1973: 'Upper layer modification at ocean station Papa: Observations and simulation'. J. Geophys. Oceanogr. **3**, 185-196.

Denman, K.L. and T. Platt, 1976: 'A theoretical variance spectrum for the patchiness of phytoplankton'. J. Mar. Res. **34**, 593-601.

Dobson, F. and S.D. Smith, 1985: 'Estimation of solar radiation at sea'.
    The Ocean Surface. H. Mitsuyasu and Y. Toba, Eds., D. Reidel,
    Dordrecht.

Dörre, F., 1985: Diploma thesis (in preparation).

Ekman, V.W., 1905: 'On the influence of the earth's rotation on ocean
    currents'.
    Arch. Math. Astron. Fysik 2, 1-53.

Ellison, T.H., 1957: 'Turbulent transport of heat and momentum from an
    infinite rough plane'.
    J. Fluid Mech. 2, 456-466.

Frankignoul, C. and K. Hasselmann, 1977: 'Stochastic climate models,
    part II. Application to sea-surface temperature anomalies and
    thermocline variability'.
    Tellus 29, 289-305.

Gonella, J., 1971: 'The drift current from observations made on the Bouée
    Laboratoire'.
    Cah. Oceanogr. 23, 19-33.

Højerslev, N.K., 1980: 'Water color and its relation to primary produc-
    tion'.
    Bound.-Layer Met. 18, 203-220.

Horch, A., W. Barkmann and J.D. Woods (1983) 'Die Erwärmung des Ozeans
    hervorgerufen durch solare Strahlungsenergie'.
    Ber. Inst. Meeresk. Kiel Nr. 120, 190 pp.

Hunkins, K., 1966: 'Ekman drift currents in the Arctic Ocean'.
    Deep-Sea Res. 13, 607-620.

Jerlov, N.G., 1976: Marine Optics.
    Elsevier, Amsterdam, 2nd edition, 231 pp.

Killworth, P.D., 1983: 'Deep convection in the world ocean'.
    Rev. of Geophys. and Space Phys. 21(1), 1-26.

Kraus, E.B. and C. Rooth, 1961: 'Temperature and steady state vertical
    heat flux in the ocean surface layers'.
    Tellus 13, 231-239.

Niiler, P.P., 1977: 'One-dimensional models of the seasonal thermocline'.
    The Sea. E.D. Goldberg, I.N. McCave, J.J. O'Brien, J.H. Steele,
    Eds., Wiley-Interscience, New York, 97-115.

Niiler, P.P. and E.B. Kraus, 1977: 'One-dimensional models of the upper
    ocean'.
    Modelling and prediction of the upper layers of the ocean.
    E.B. Kraus, Ed., Pergamon Press, 143-177.

Oakey, N.S. and J.A. Elliott, 1982: 'Dissipation within the surface
    mixed layer'.
    J. Phys. Oceanogr. **17**(2), 171-185.

Paulson, C.A. and J.J. Simpson, 1977: 'Irradiance measurements in the
    upper ocean'.
    J. Phys. Oceanogr. **7**, 952-956.

Platt, T., K.L. Denman and A.D. Jassby, 1977: 'Modeling the productivity
    of phytoplankton'.
    The Sea. E.D. Goldberg, I.N. McCave, J.J. O'Brien, J.H. Steele,
    Eds., Wiley-Interscience, New York, 807-856.

Pollard, R.T., P.B. Rhines and R.O.R.Y. Thompson, 1973: 'The deepening
    of the wind-mixed layer.
    Geophys. Fluid Dynamics **3**, 381-404.

Robinson, M.K., R.A. Bauer and E.H. Schroeder, 1979: Atlas of North
    Atlantic - Indian Ocean monthly mean temperature and mean
    salinities of the surface layer.
    U.S. Naval Oceanographic Office Ref. Pub. 18, Washington D.C.

Saunders, P.M., 1980: 'Near surface Lagrangian shear measurements'.
    JASIN News, **15**.

Shay, T.J. and M.C. Gregg, 1984: 'Turbulence in an oceanic convective
    mixed layer'.
    Nature **310**, 282-284.

Turner, J.S. and E.B. Kraus, 1967: 'A one-dimensional model of the
    seasonal thermocline. I. A laboratory experiment and its interpre-
    tation. II. The general theory and its consequences'.
    Tellus **19**, 88-97 and 98-106.

Wells, N.C., 1979: 'A coupled ocean-atmosphere experiment: The ocean
    response'.
    Quart. J. Roy. Met. Soc. **105**, 355-370.

Woods, J.D., 1980: 'Diurnal and seasonal variation of convection in the
    wind-mixed layer of the ocean'.
    Quart. J. Roy. Met. Soc. **106**, 379-394.

Woods, J.D., 1983: 'Climatology of the upper boundary layer of the
    ocean'.
    WCRP Pub. Ser. **1**, 147-179.

Woods, J.D., 1984: 'The upper ocean and air-sea interaction in global
        climate'.
        The Global Climate. J. Houghton, Ed., Cambridge University Press,
        141-187.

Woods, J.D., 1985: 'Physics of thermocline ventilation'.
        Coupled Atmosphere-Ocean Models. J.C.J. Nihoul, Ed., Elsevier
        (in press).

Woods, J.D. and W. Barkmann, 1985: 'The response of the upper ocean to
        solar heating. I: The mixed layer'.
        Quart. J. Roy. Met. Soc. (in preparation).

Woods, J.D., W. Barkmann and A. Horch, 1984: 'Solar heating of the
        oceans - diurnal, seasonal and meridional variation'.
        Quart. J. Roy. Met. Soc. 110, 679-702.

Woods, J.D. and R. Onken, 1982: 'Diurnal variation and primary produc-
        tion in the ocean - preliminary results of a Lagrangian ensemble
        model'.
        J. Plankton Research 4, 735-756.

Woods, J.D. and V. Strass, 1985: 'The response of the upper ocean to
        solar heating. II: The wind-driven current'.
        Quart. J. Roy. Met. Soc. (in preparation).

AN OCEANIC MIXED LAYER MODEL SUITABLE FOR CLIMATOLOGICAL STUDIES :
RESULTS OVER SEVERAL YEARS OF SIMULATION

Ph. Gaspar
Catholic University of Louvain
Institute of Astronomy and Geophysics G. Lemaître
B-1348   Louvain-la-Neuve
Belgium

ABSTRACT.   A simple version of the Niiler and Kraus (1977) mixed layer
model has been tested by simulating the upper ocean evolution at Ocean
Station Papa for several years. As suggested by the authors themselves,
the weakest point of this model is the parameterization of the turbulent dissipation. Consequently, the performed simulations show that the
model's wind-mixing efficiency constant (m) has to be tuned to the
observations and exhibits seasonal and interannual changes. In an
attempt to solve this problem a new parametric formulation of the dissipation is proposed. The resulting model (CMO) is calibrated independently of the observations to be simulated. When compared to the
Niiler-Kraus model on a four-year simulation of the mixed layer changes
at Station Papa, CMO presents the best overall agreement with the observations. The maximum error in the CMO-predicted monthly-mean sea surface
temperature is 0.5 K. Nevertheless CMO, like the Niiler-Kraus model,
systematically overestimates the sea surface temperature during spring
and summer and underestimates it during fall and winter. Some explana-
tions for this are suggested.

1. INTRODUCTION

A realistic prediction of the sea surface temperature (SST) is a major
objective to be reached by the models dedicated to the study of ocean-
climate interactions. As shown by Gill and Niiler (1973), the forcing of
the upper ocean by the surface heat fluxes dominates the advective
effects on the annual time scale over many parts of the world's ocean.
Therefore, one-dimensional models of the upper oceanic layers are rather
successful in predicting the SST.
     Most of these models however have been conceived and calibrated for
short-term simulations of a few days to a few weeks. Even on these short
time scales, their calibration can be difficult (e.g. Davis et al.,
1981) and their use for simulations over several years raises additional
questions concerning their skill (Woods, 1983) and ability to simulate
repeatable annual cycles without a progressive increase of the maximum
mixed layer (ML) depth (Stevenson, 1979).

Y. Toba and H. Mitsuyasu (eds.), The Ocean Surface, 509–516.
© 1985 by D. Reidel Publishing Company.

    In an attempt to cope with these problems, a new ML model has been
specifically designed for simulations on the annual time scale (Gaspar,
1984). This paper presents the basic structure and first results of this
model.

2. CLASSICAL BULK MODELS

Due to their computational efficiency, bulk models of the ML are often
preferred to turbulence closure ones for long-term integrations. In
those models, the vertical homogeneity of the ML is assumed. Therefore,
the SST is identical to the ML temperature and is directly predicted
from the ML heat budget :

$$h \frac{\partial}{\partial t} SST = \frac{1}{\rho c_p} [F_{SOL} + F_{NSOL} - F_{SOL} I(-h)] - w_e \Delta T \qquad (1)$$

where h is the ML depth, ρ and c_p the volumic mass and specific heat of
sea-water, F_{SOL} the solar irradiance absorbed at the sea surface, F_{NSOL}
the sum of all other surface heat fluxes (latent, sensible, IR), $I(z)$
the fraction of the solar irradiance that penetrates to the depth z, ΔT
the temperature jump at the base of the ML and w_e the entrainment velocity defined by :

$$w_e = \partial h/\partial t \text{ if } \partial h/\partial t > 0 \quad \text{and} \quad w_e = 0 \text{ if } \partial h/\partial t \leq 0$$

The surface fluxes being given from observations or from the coupling
with an atmospheric model, the prediction of the SST from (1) only
requires an additional equation for the determination of w_e. Since the
pioneer work of Kraus and Turner (1967), this closure equation is generally obtained from a parameterized form of the Turbulent Kinetic Energy
(TKE) budget of the ML. A widely used basic formulation of it is:

$$0.5h\Delta bw_e = m_p u_*^3 + DIT - 0.5hB - h\varepsilon_m \qquad (2)$$

where Δb is the buoyancy jump at the base of the ML, u_* the surface
friction velocity and DIT the Dynamic Instability Term first introduced
in this type of model by Pollard et al. (1973). B is equal to the total
surface buoyancy flux (positive downward) plus a term due to the absorption of the solar irradiance within the ML. ε_m is the mean value of the
TKE dissipation within the ML.
    Details about the parameterization of this equation can be obtained
from Niiler and Kraus (1977) or Zilitinkevich et al. (1979).
    Pollard et al. (1973) and de Szoeke and Rhines (1976) have shown
that DIT is a fast responding transient term, its response time being
the order of half an inertial period. As we are mainly interested by
longer timescales, this term will be neglected here. The TKE buoyant
production/destruction term (-0.5hB) is obtained analytically and the
parameterization of the mechanical production term ($m_p u_*^3$) is widely accepted. In fact, the main differences between most existing ML models
arise from the parameterization of the dissipation. Among the numerous
proposed solutions, that of Niiler and Kraus (1977) (hereafter referred

to as NK) is now widespread. It assumes that the TKE dissipation is simply a constant fraction of the production terms, i.e.,

$$h\varepsilon_m = m_d\ u_*^3 + 0.25(1-n)h\ (|B| - B) \tag{3}$$

with $0 < m_d < m_p$ and $0 < n < 1$.
Consequently, (2) reduces to

$$0.5h\Delta b\ w_e = m\ u_*^3 - 0.25h\ [(1+n)B + (1-n)|B|] \tag{4}$$

where $m = m_p - m_d$

A commonly accepted value for n is 0.2. The estimation of m is more problematical as it now seems that the results of the rotating-screen annulus experiments that were conceived to determine m, are largely dependent upon the geometry of the used experimental apparatus (Scranton and Lindberg, 1983). For our purpose it is believed that the MILE experiment provides the best available data set from which Davis et al. (1981) have inferred a value of m comprised between 0.4 and 0.5.

The long term response of the NK model has been tested by simulating the evolution of the upper ocean at the Ocean Weather Ship Papa (OWS P; 50°N, 145°W) during the years 1969 to 1972. The used surface forcing (u_*, F_{SOL}, F_{NSOL}, B) has been computed by Tricot (1984) from the three-hourly meteo-oceanographic observations. For the studied period, the ML heat content computed from the bathythermographs is generally in good agreement with that estimated from the surface fluxes, indicating that only weak advective effects are present. Several simulations were performed, all starting from the observed temperature profile on the first January of a given year and then integrating the model over one to four years with a time step of three hours. The grid for storing the vertical temperature profile has a uniform resolution of 2 m from the surface to a depth of 150 m.

The main results of these numerical experiments are the following :
i) no reasonable simulation of the SST at the OWS P is obtained with the suggested values of m. The best simulations are obtained with values ranging from 0.55 to 0.7. It is interesting to note that Davis et al. (1981) themselves have used m=0.69, that lies in our range of value, to obtain a good simulation of the MILE data set with a NK model in which DIT is neglected.
ii) the best value of m for one year is not necessarily adequate for another year.
iii) the skill of the model varies with the period of the year. An adequate change in the value of m generally increases the skill of the model over a certain period but decreases it over the rest of the year.

These changes in the value of m are probably needed to compensate for shortcomings in the parameterization of the dissipation. This is indeed the weakest point of this model, as indicated by NK themselves. Hence there is scope for further works on parameterizing dissipation as a function of the bulk variables. Such an attempt is briefly described in the next paragraph.

3. A BULK MODEL WITH A NEW PARAMETERIZATION OF THE DISSIPATION

Like in most turbulence closure models, the dissipation can be expressed as a function of σ_e, a characteristic value of the turbulent velocity, and 1, a dissipation length (Kolmogorov, 1942) :

$$\varepsilon_m = \sigma_e^3/1 \tag{5}$$

In the framework of a bulk model the best possible choice for σ_e is :

$$\sigma_e = E_m^{1/2} \tag{6}$$

where E_m is the mean value of the TKE within the ML. On the other hand, ML physics indicate that (Resnyanskiy, 1975)

$$1 = F(h,L,\lambda) \tag{7}$$

where F is a function to be determined,
    $L = u_*^3/B$, a bulk Monin-Obukhov length
    $\lambda = u_*/f$, the Ekman length scale, f being the Coriolis factor.

Using dimensional analysis (7) reduces to

$$h/1 = G(h/L,h/\lambda) \tag{8}$$

Following Therry and Lacarrère (1983) we shall require from the dimensionless function G that :
    i. in absence of rotational constraint $(h/\lambda \to 0)$, G asymptotically reaches its minimum value in the case of pure convection $(h/L \to -\infty)$.
    ii. G is an increasing function of h/L, the stability parameter.
    iii. when h exceeds the neutral height scale (0.4λ), this height becomes the dominant dissipation length, except for the convective cases, as suggested by Deardorff (1983).

A simple expression of G that meets these requirements is (Gaspar,1984)

$$G(h/L,h/\lambda) = a_1 + a_2 \text{ Max}(1, h/0.4\lambda) \exp (a_3 h/L) \tag{9}$$

where a_1, a_2, a_3 are positive constants.
However, contrary to (3), Eqs. (5), (6) and (9) do not achieve the final closure of (2) but bring it at an higher order since the explicit determination of a second-order correlation E_m is needed. Following Garwood's (1977) lead, a supplementary unknown is introduced in the model, namely W_m, the mean value of $\overline{w'^2}$ within the ML. Then (w_e, E_m, W_m) are determined from a set of three equations, i.e.,

1) Eq.(2) in which DIT is neglected and the dissipation parameterized following (5), (6) and (9).
2) the Garwood's (1977) entrainment equation :

$$h\Delta b \, w_e = m_1 E_m W_m^{1/2}.$$

where m_1 is a constant to be determined.

3) the W_m equation deduced from a parameterized depth-integrated quasi
   stationary version of the $\overline{w'^2}$ equation (Gaspar, 1984).

The whole model (hereafter referred to as CMO) is calibrated on several
laboratory measurements of turbulent flows and on results of André and
Lacarrère (1984) third-order turbulence closure model. The entrainment
rate in neutral conditions is in accordance with the MILE results, i.e.,

$$0.5 \; h\Delta b \; w_e = 0.45 \; u_*^3$$

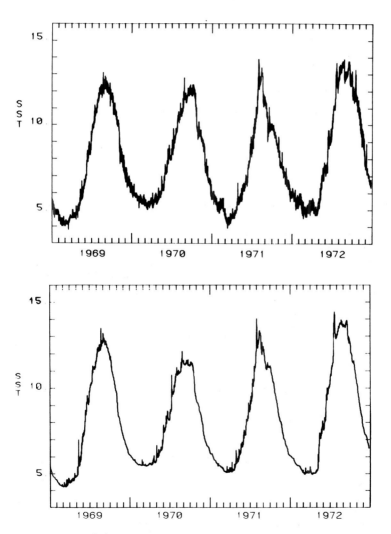

Figure 1.   Observed (a) and simulated by the CMO model (b) SST at OWS P
for years 1969 to 1972.

4. NUMERICAL RESULTS AND CONCLUSIONS

A four-year simulation of the upper ocean evolution at OWS P has been
performed with CMO using the same forcing, vertical resolution and time
step as used with the NK model. Figs. 1a and b show that CMO reproduces
the SST annual cycle with a good accuracy. Large errors are only
observed during March-April 1971 and the late summer of the same year.
In fact, the surface heat fluxes cannot account for the observed surface
cooling in March-April and hence this SST drop cannot be simulated by
the model. The abrupt decrease of the SST in August-September is due to
successive storms occuring over a shallow ML. This is typically a case
where DIT greatly contributes to entrainment. As this term is not present in CMO, the ML deepening is too slow and consequently the simulated
SST remains too high.

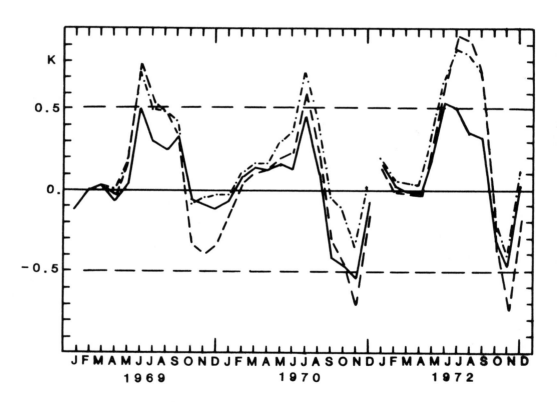

Figure 2. Difference between model-predicted and observed monthly mean
SST at OWS P for the years 1969, 1970 and 1972. _____ CMO, _ _ _ _ _ NK,
m=0.65, _._._ NK m=0.8 exp(-hf/u$_*$) (see Wells, 1979).

Finally, to facilitate the comparison between models, the simulated minus observed monthly mean SST are shown in Fig. 2, for CMO and two versions of the NK model using different estimations of m. Due to the above-mentioned problems, the year 1971 is kept out of the statistics. Without tuning to the observations, CMO presents the best overall agreement and maintains the monthly mean SST error under 0.5K. Yet all the model's errors exhibit an annual cycle. The SST is generally overestimated during spring and summer and underestimated during fall and winter. Data indicate that the fall-winter excessive cooling is due to excessive entrainment. Further work on the parameterization of the rotational constraint could help in reducing this error. The crude simulation of the diurnal cycle as performed by the models used, is probably one of the reasons for which the spring and summer SST is overestimated. The NK model is particularly sensitive to this problem.

5. ACKNOWLEDGMENTS

The support of the National Fund for Scientific Research (Belgium) is gratefully acknowledged. The data from OWS P were kindly provided by the Canadian Marine Environmental Data Service. Special thanks go to J.C. André, P. Lacarrère and Ch. Tricot for their helpful collaboration. Thanks are extended to A. Berger for advice and encouragement. It is also a pleasure to thank R. Pollard and N. Wells for their useful comments and suggestions. F. Dobson's advice on the estimation of the surface forcing was greatly appreciated.

6. REFERENCES

André, J.C., and P. Lacarrère, 1984. On the mean and turbulent structures of the oceanic surface layer as determined from one-dimensional, third-order simulations. To be published in J. Phys. Oceanogr.
Davis, R.E., R. de Szoeke and P.P. Niiler, 1981. Variability in the upper ocean during MILE. Part II : Modelling the mixed layer response. Deep-Sea Res., 28A, 1453-1475.
Deardorff, J.W., 1983. A multi-limit mixed-layer entrainment formulation. J. Phys. oceanogr., 13, 988-1002.
de Szoeke, R.A., and P.B. Rhines, 1976. Asymptotic regimes in mixed-layer deepening. J. Mar. Res., 34, 111-116.
Garwood, R.W., 1977. An oceanic mixed layer model capable of simulating cyclic states. J. Phys. Oceanogr., 7, 455-468.
Gaspar, Ph., 1984. Modelling the long term response of the upper ocean. Scientific Report 1984/4, Institute of Astronomy and Geophysics G. Lemaître, Catholic University of Louvain-la-Neuve, Belgium.
Gill, A.E., and P.P. Niiler, 1973. The theory of the seasonal variability in the ocean. Deep-Sea Res., 20, 141-177.
Kolmogorov, A.N., 1942. The equation of turbulent motion in an incompressible fluid. Izv. Akad. Nauk SSSR, 6, 56-58.

Kraus, E.B., and J.S. Turner, 1967. A one-dimensional model of the sea-
   sonal thermocline. II. The general theory and its consequences.
   Tellus, 19, 98-105.
Niiler, P.P., and E.B. Kraus, 1977. One-dimensional models of the upper
   ocean. Modelling and prediction of the upper layers of the ocean. E.B.
   Kraus (Ed.), Pergamon Press, 143-172.
Pollard, R.T., P.B. Rhines and R.O.R.Y. Thompson, 1973. The deepening of
   the wind-mixed layer. Geophys. Fluid Dyn., 3, 381-404.
Resnyanskiy, Yu.D., 1975. Parameterization of the integral turbulent
   energy dissipation in the upper quasihomogeneous layer of the ocean.
   Izv. Atmos. and Oceanic Phys., 11, 453-457.
Scranton, D.R. and W.R. Lindberg, 1983. An experimental study of entraining, stress-driven, stratified flow in an annulus. Phys. Fluids, 26,
   1198-1205.
Stevenson, J.W., 1979. On the effect of dissipation on seasonal thermocline models. J. Phys. Oceanogr., 9, 57-64.
Therry, G. and P. Lacarrère, 1983. Improving the eddy kinetic energy
   model for planetary boundary layer description. Bound. Lay. Meteor.,
   25, 63-88.
Tricot, Chr., 1984. Estimation of the surface heat fluxes at Ocean Station Papa. Scientific Report 1984/5, Institute of Astronomy and Geophysics, Catholic University of Louvain-la-Neuve, Belgium.
Wells, N.C., 1979. A coupled ocean-atmosphere experiment : the ocean
   response. Quart. J.R. Met. Soc., 105, 355-370.
Woods, J.D., 1983. Climatology of the upper boundary layer of the ocean.
   WCRP Publications Series, 1, vol. II, 147-179.
Zilitinkevich, S.S., D.V. Chalikov, and Yu.D. Resnyanskiy, 1979. Modelling the oceanic upper layer. Oceanol. Acta, 2, 219-240.

RESPONSE OF THE UPPER OCEAN TO ATMOSPHERIC FORCING

G.A. McBean[1] and M. Miyake[2]
1 Atmospheric Environment Service
2 Institute of Ocean Sciences
Sidney, B.C., Canada, V8L 4B2

ABSTRACT. During the Storm Transfer and Response Experiment (STREX) in November–December, 1980, extensive measurements, including 3 AXBT surveys, were made in the boundary layers of both the atmosphere and the ocean. The greatest cooling of the upper ocean occurred between the first two surveys. The mean temperature of the upper 60 m of the ocean decreased by 1.5 C. The deduced net effect of atmospheric surface fluxes was −1.2 C; horizontal advection was important, contributing another 0.8 C cooling. The Ekman pumping was small. During the second period the observed cooling was small and approximately balanced by the surface cooling.

1. INTRODUCTION

The upper ocean is of critical importance for the thermodynamics and dynamics of the whole of the oceanic system. Woods (this volume) has reviewed many of the aspects of the upper ocean and Gaspar (this volume) has presented results from the comparison of an upper-ocean model with long-term observations from OWS P, the same location as the data used in this paper. It has been found that a one-dimensional balance of heat is adequate for OWS P for the spring and summer months (Tabata, 1965; Denman and Miyake, 1973; Davis et al, 1983). However, for other times of the year, Tabata found that other processes were required to balance the budget.

In November–December, 1980, a joint American-Canadian experiment, the Storm Transfer and Response Exeriment, i.e., STREX (Fleagle et al, 1982), was conducted in the northeast Pacific Ocean to study the interactions of the atmospheric and oceanic boundary layers during storms. The objective of this paper is to describe the evolution of the upper ocean during a month-long period and to quantify some of the atmospheric forcings that affect the evolution.

Y. Toba and H. Mitsuyasu (eds.), The Ocean Surface, 517–524.
© *1985 by D. Reidel Publishing Company.*

2. DATA SET

Routine weather observations were made at three hourly interval on the ship. In addition, the incoming solar radiation and the net all-wave radiation were directly measured and integrated over hourly periods. The surface fluxes were calculated from:

$$Q_H = \rho c_p C_H |V| (T_s - T_a)$$

$$Q_E = \rho L \, C_E |V| (q_s - q_a)$$

$$\tau_x = \rho \, C_D |V| \, v_x$$

$$\tau_y = \rho \, C_D |V| \, v_y$$

with the C_H, C_E and C_D being based on Smith (1981) and depend on stability and wind speed.

The second source of meteorological information was the winds produced by the atmospheric boundary layer model (Brown and Liu, 1982) which used the surface pressure field to compute the geostrophic wind field. The geostrophic winds are corrected for stability, thermal advection, variable surface roughness, humidity and streamline curvature. The Ekman pumping

$$w_E = (1/\rho_w f)(\partial \tau_y / \partial x - \partial \tau_x / \partial y)$$

and the drift currents, based on the observations of drifting buoys by McNalley (1981),

$$u_M = 0.015 |V| \sin(\tan^{-1}(v_y/v_x) - \pi/6)$$

$$v_M = 0.015 |V| \cos(\tan^{-1}(v_y/v_x) - \pi/6)$$

were computed. Standard CTD measurements were from the ship Vancouver and about 28 AXBT's were deployed on each of November 14, December 7 and December 15, 1980.

3. WEATHER PATTERNS DURING THE PERIOD

For the first two weeks, there was a succession of lows with associated frontal sytems passing through the area. The winds were generally brisk with fronts passing every 2-3 days. On November 26-27, a storm passed near OWS P and deepened to 950 mb as it moved further into the Gulf of Alaska. The maximum observed wind at the ship was 26 m/s at 00Z, 27 November. The model diagnosed winds over the area up to 33.5 m/s at the same time.

After the passage of this storm the meteorological pattern changed. The upper flow structure became difluent over the region and for the period November 29-December 12, there was generally weak winds over the region as the low centres moved south of the experimental area. Only at the very end of the period, December 13, did another

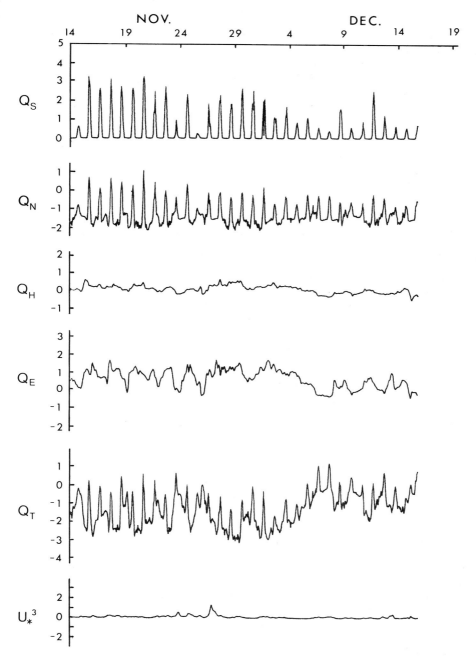

Figure 1. Time series of atmospheric forcing on the ocean surface for November–December, 1980, at OWS P.
From top: solar radiation; net radiation; sensible heat flux; latent heat flux and total heat input (all in w/m^2); and u_*^3.

substantial low pressure system move near OWS P. The results of these two different weather regimes was evident in the oceanic response, as will be discussed below.

4. ATMOSPHERE-OCEAN HEAT EXCHANGES

The time sequence of the atmospheric heat fluxes are shown in Figure 1. The daily cycle is very evident in the radiation data. The latent heat flux is generally much larger than the sensible heat flux but the two are highly correlated. The total heat loss from the ocean surface shows considerable variability. In the period up to about the beginning of December the net heat flux is large and negative, indicating a large heat loss (except for two short periods). Through most of December the net heat loss is much smaller and averages only slightly positive (heat loss). There are more numerous periods where the net flux is into the ocean.

The averages fluxes over the period of the experiment and the long-term averages for the area, based on the computations of N. Clark (pers. com.), are:

Average Fluxes (w/m^2)

	Nov 14-Dec 15	Long-term (Nov-Dec)
Solar rad.	33.2	36.3±11
Back rad.		33.9±6
Net rad.	(-41.1)*	2.4
Q_H	17.0	23.1±17
Q_E	67.5	75.9±27
Total	-125.6	-96.5±42

* This net radiation value appears unusually large relative to the long term mean. A more "normal" value would result in less radiation-induced cooling.

The computed change in the upper-ocean average temperatures is shown in Figure 2. During the period Nov 14 to Nov 25, the 0-60 m layer temperature would have decreased, due to surface heat loss, by -0.54 C. It would have decreased another 0.23 C by Nov 29. The total temperature decrease by Dec 7 is -1.22 C. For the last week of the experiment the heat loss is lower such that the temperature would have only decreased another -0.15 C. The decreases over the 0-160 m layer follow these values but are reduced by a factor 60/160.

5. OBSERVED CHANGES IN UPPER-OCEAN HEAT CONTENT

The observed upper-ocean thermal changes are shown in Figure 3. It can be seen that the depth of the mixed layer gradually increased from 60 m at the begining of the period to about 80 m by the end.

Table 1. Summary of Observed and Computed Heat Changes

Layer	Observed Change	Atmos. Sfc Flux	Oceanic Hor Adv.	Oceanic Vert Adv	Total
Nov 14 – Dec 7					
0–60 m	−1.5	−1.2	−0.8	0.0	−2.0
uncertainty	±0.2–0.4	±0.4	±0.1		
0–160 m	−0.42	−0.46	(−0.3)	−0.03	
Dec 7–15					
0–60 m	−0.3	−0.15	−0.2	0.0	−0.35
uncertainty	±0.1	±0.1			
0–160 m	0.0	−0.05	(−0.08)	−0.018	

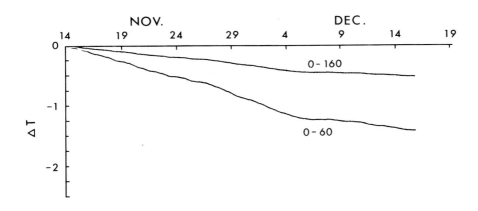

Figure 2. The calculated change, due to the net heat loss to the atmosphere, in average temperature of the 0–60 m layer and 0–160 m layer (^{o}C).

There is considerable variation due both to sampling and internal
waves.   The temperature of the mixed layer decreases from 9.3 C at the
beginning to 7.49 C at the end.   During the period when the major
storm passed through the region, the temperature decreased about 0.6 C
over four days.

To compare with the computations shown in the last section, we
have computed the mean temperatures over the depth intervals: 0-60 m
and 0-160 m.   Over the period Nov. 14 to Dec 5, the 0-60 m layer
decreased in temperature by -1.50 C; while the 0-160 m layer decreased
by -0.42 C.   For the last period of the experiment, Dec 7 to Dec. 14,
the changes were 0-60 m: -0.31 C and 0-160 m: 0.0 C.   Thus for the
last period of the experiment when the winds were generally much less,
the average temperature of the upper 160 m of the ocean did not change
significantly.   Most of the change occurred during the first two weeks
of the experimental period.

By comparing Figures 2 and 3, for the 0-60 m layer, it can be
seen that the upper ocean cooled more during the stormy period than
implied by the surface heat loss.   During the less active period, the
cooling is about equal to the surface heat loss.   This implies that
other processes must be important, at least for the stormy period.

6. WIND-INDUCED HORIZONTAL ADVECTION

We computed the following mean currents:

Period	Current	Displacement
Nov 14 - Dec 7	u_M = 0.12 m/s	240 km
	v_M = -0.04 m/s	80 km
Dec 7 - 15	u_M = 0.046 m/s	35 km
	v_M = -0.045 m/s	35 km

The horizontal gradients are subject to considerable uncertainty
because the gradients were not uniform in space. For the first period
the east-west gradient was very small while the north-south gradient
was between 0.5 and 1 C per 100 km.   Thus, the best estimate of the
temperature change over the 0-60 m layer due to horizontal advection
in the first period is -0.8 C with an uncertainty of about ±0.4 C.
For the second period the east-west gradient was larger and estimated
to be 0.2 C per 35 km.   The north-south gradient was small.

7.   VERTICAL ADVECTION

For the period Nov 14 to Dec 7, the mean Ekman pumping was
$-1.4:10^{-6}$ m/s.   For the latter period it was about $-4:10^{-6}$ m/s.
Because the Ekman pumping is negative, there is no effect on the 0-60
m layer which is all within the mixed layer. The effects on the 0-160
m layer are small.

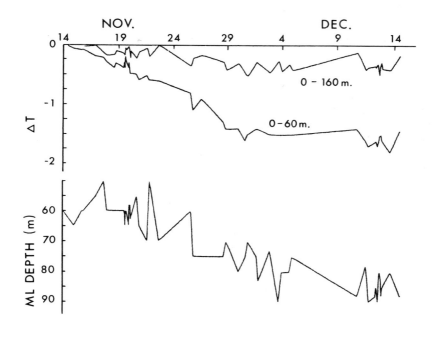

Figure 3.   The observed changes in (from top): the average temperature   of   the 0-60 m layer and 0-160 m layer (^{o}C); and the mixed layer depth (m).

8.   SUMMARY

    The observed and computed heat changes over the 0-60 m and 0-160
m are summarized in Table 1. For the first period, the observed
cooling of the 0-60 m layer is greater than the surface heat flux
induced cooling (particularly if the net radiation heat loss has been
overestimated). The effect of horizontal advection is in the right
sense but is estimated to be larger than required to make up the
difference. At the same time, the uncertainties in the temperature
changes due to all three terms are large enough to account for the
differences. For the second period, the observed and computed cooling
are in good agreement but all terms are small with large percentage
uncertainties. It should also be noted that we have not accounted for
the effects due to inertial currents and internal waves. These have
been shown to be important in the studies of D'Asaro (1984).
    To summarize, this study has demonstrated that for stormy
periods, the cooling of the upper ocean may be significantly more than
would be induced by surface heat losses alone and the dynamical
effects must be accounted for. During less stormy periods, the
surface heat losses will probably account for the observed cooling.

References:

Brown, R.A., and T. Liu, 1982: An operational large-scale marine
      planetary boundary layer model. J. Appl. Meteor., 21, 261-269.
D'Asaro, E.A., 1984: Wind forced internal waves in the North Pacific
      and Sargasso Sea. J. Phys. Ocean., 14, 781-794
Davis, R.E., R. deSzoeke, D. Halpern and P. Niiler, 1981: Variability
      in the upper ocean during MILE. Part I: The heat and momentum
      balances. Deep-Sea Res., 28, 1427-1451.
Denman, K.L., and M. Miyake, 1973: Upper ocean modifcation at Ocean
      Station Papa: Observation and simulation. J. Phys. Ocean., 3,
      185-196.
Fleagle, R.G., M. Miyake, J.F. Garrett, and G.A. McBean, 1982: Storm
      Transfer and Response Experiment, STREX. Bull. Amer. Meteor.
      Soc., 63, 6-14.
McNally, G.J., 1981: Satellite-tracked drift buoy observations of
      near-surfac flow in the eastern mid-latitude North Pacific. J.
      Geophys. Res., 86, 8022-8030.
Tabata, S., 1965: Variability of oceanographic conditions at Ocean
      Station P in the northeast Pacific Ocean. Trans. Roy. Soc.
      Canada, 3, 367-418.

ESTIMATION OF SOLAR RADIATION AT SEA

Fred W. Dobson and Stuart D. Smith
Department of Fisheries and Oceans
Bedford Institute of Oceanography
Dartmouth, Nova Scotia   B2Y 4A2
Canada

ABSTRACT. The widely-used short-wave radiation formula of Budyko
(1974) underestimates the measured solar radiation at Ocean Weather
Stations (OWS) A, I, J, K and P by as much as 30%, with the largest
errors occurring at times of high cloud amount.

Several models of hourly solar radiation as a function of solar
elevation, cloud type and cloud amount were evaluated on an hourly,
daily, monthly and long-term mean basis in comparison with a model using cloud amount but not cloud type; we find that the cloud type information does not so far improve our ability to model OWS solar
radiation.

INTRODUCTION

As a follow-up to the CAGE study (Dobson et al., 1982), sponsored by
the Committee on Climate Change in the Ocean, we are investigating the
accuracy with which existing formulae can predict, from marine meteorological observations of clouds, the short wave solar radiation incident
on the sea surface on a climatic time scale. This is part of an investigation of all the "bulk formulae" used to estimate the fluxes of
heat, water and momentum through the sea surface from marine meteorological measurements. One of our objectives is to compare fluxes computed from what we consider to be well-calibrated bulk formulae with
similar computations made by Bunker (1976). We seek thereby to confirm
Bunker's inferred meridional oceanic heat transport for the North
Atlantic Ocean, but conclude that Bunker's value may be high by as much
as 50%, or 25 W m^{-2} (Smith and Dobson, 1984). Here, we present preliminary formulae for incoming solar short wave radiation, investigate
expected errors, and indicate what remains to be done.

TEST OF BUDYKO SHORT-WAVE RADIATION FORMULA AT OWS P

The monthly incident solar radiation at OWS P calculated from the

Y. Toba and H. Mitsuyasu (eds.), The Ocean Surface, 525–533.
© 1985 by D. Reidel Publishing Company.

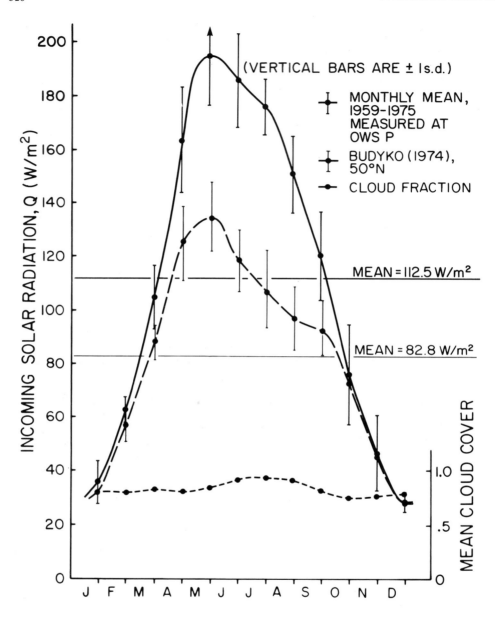

Figure 1. Annual cycle of incident shortware solar radiation at OWS P,
averaged over the years 1959–1975. Error bars are ± 1 standard
deviation of the monthly means. Solid line: measured value;
dashed line: prediction of Budyko (1974) for 50°N latitude.

formula of Budyko (1974) has been averaged over 17 years (1959-75) to obtain a seasonal cycle (Figure 1) which agrees well with the corresponding means of measured incident solar radiation during the winter months but which badly underestimates during the remainder of the year. The 17-year mean from the Budyko formula: 82.8 W m^{-2}, is 26% lower than the mean of 112.5 W m^{-2} measured by the Canadian Atmospheric Environment Service. On the other hand, for clear skies the Budyko model was found to agree with measurements at OWS P within ±2%. The errors must therefore be in Budyko's formula for attenuation by clouds, due originally to Berliand (1960)

$$Q = Q_c[1 - (a + bn)n]$$

where n is the monthly mean cloud fraction, a is an empirical function of latitude (0.40 at 50°N) and b = 0.38 is an empirical constant. This formula does not allow for variations in solar elevation, which is lower in winter than in summer. The formula was presumably calibrated from the Russian terrestial actinometric network. The coefficients given by Budyko work well during winter for OWS P but the shorter path length through clouds with the higher summer sun, coupled with differences in marine and terrestial clouds, presumably allows more radiation to pass than predicted by the Berliand formula. It would be possible to recalibrate Budyko's tabulated coefficient for 50°N to correct the mean insolation at OWS P, but then we would overestimate in winter, underestimate in summer, and consequently underestimate the seasonal variability.

Similar envelopes fitted to maximum daily solar radiation measured at Weather Stations A, I, J and K by the British Meteorological Office lie about 5% below the Budyko clear-sky values for appropriate latitudes, while Canadian data from Sable Island, N.S. are, like the OWS P values, well-predicted. Communication with Canadian and British radiation measurement experts led to the conclusion that we should expect no better than ±5% agreement among field radiation measurements, even in the long-term mean. The observed 5% discrepancy between Canadian and British measurements represents a limit on the accuracy of modelling based on data which we must accept and which, as it turns out, is our largest single source of error.

Atmospheric Transmission Factor

In this study we will estimate solar radiation on an hourly basis using a bulk formula and then do monthly or other averages, thus avoiding problems with nonlinearities and physical biases inherent in working with monthly mean data. For example, cloud cover at night cannot affect insolation, but is equally weighted with daytime clouds in Budyko's method; if there is a day-night cycle in cloudiness, errors are introduced. We prefer a method which requires only surface data, since requirements for upper air (radiosonde) information severely limit the spatial coverage of available data.

The approach of Lumb (1964) appears particularly attractive. Solar radiation is estimated for each hourly cloud observation which can be fitted into one of nine categories based on cloud amount and type. For each category i a regression of the atmospheric transmission factor

$$T \equiv Q/Q_0 S$$

on the sine S of solar elevation

$$T = A_i + B_i S \qquad\qquad\qquad\qquad (1)$$

is based on a limited period of data from OWS J. The mean solar flux Q_0 is 1368 W m^{-2} (Frölich, 1983). Lumb's method cannot be directly applied, since it does not specify what to do with cloud observations which do not fit into one of the nine categories (At OWS P only 50% of the cloud observations can be fitted into a category). Furthermore, Lumb did not investigate very thoroughly the variability of his regression coefficients from one location to another, and a partial set of coefficients which he listed for OWS A shows substantial differences from OWS J for some categories.

Lind (1981) and Lind and Katsaros (1982) have expanded Lumb's categories to include all cloud observations and have reported that Lumb's coefficients for OWS J give satisfactory results with their expanded model. Because the cloud categories are different we consider the Lind model separately from the Lumb model.

In addition to the Lind model we have generated two more models on the Lumb theme. The "Okta" model contains 9 categories (0 to 8) based only on reported total cloud amount in oktas (eighths). Coefficients for this model, fitted to the OWS P data, are presented in Table 1. The other model contains 24 categories, grouped in ranges of 2 oktas, sorting the predominant type of low cloud into generally stratus and cumulus categories, and sorting predominantly middle and high cloud into thick and thin cloud types. Certain categories include more than one lightly-populated cloud type, and fog is treated separately. These models were designed to test the effects of including or excluding information on cloud type as opposed to cloud amount only. Work along these lines is continuing.

We have adapted a computer subroutine from Davies (1981) to calculate the solar elevation as a function of time and position and also to allow by a factor of $\overline{R}^2/R^2$ for seasonal variations in the earth's radius from the sun. Since radiation data are reported as averages over the preceding hour, we calculate the solar elevation one half hour ahead of the report time.

Concurrent time series of radiation and of meteorological data for the Canadian and British stations were merged, edited to remove obvious

TABLE 1. Coefficients for linear, nonlinear fits to measured inci-
          dents solar radiation at OWSP, stratified by cloud amount.
          The boxes contain the coefficients for the hybrid model.

Cloud Amount (oktas)	Linear Model A	B	Nonlinear Model E	D
0	0.406	0.391	0.0520*	0.240*
1	0.533	0.308	0.0525	0.070
2	0.479	0.389	0.0430	-0.010
3	0.426	0.419	0.0395	0.055
4	0.385	0.473	0.0375	0.070
5	0.352	0.474	0.0345	0.090
6	0.310	0.439	0.0405	0.190
7	0.235	0.388	0.0390	0.330
8	0.117†	0.304†	0.0290†	0.625†

Notes: * Based on data from Sable Island. The remainder of the
         coefficients are based on OWS P data.
       † All data for which sky was "obscured" included in Category 8.

observation errors, and used to evaluate the various models. In every
case, we plotted transmission factor against the sine of the solar
elevation, making a separate plot for each cloud category for each
station. To avoid overestimating irradiation at very low solar elevations we use T at S = 0.1 to model irradiation for S $\leq$ 0.1.

A Nonlinear Model

Measured transmission factors for 1 to 5 oktas of cloud amount show a
distinctly nonlinear behaviour in S, while for 6 to 8 oktas Lumb's
linear formula fits well (Figure 2). We account for the nonlinearity
with a simple empirical model. The direct solar radiation is attenu-
ated exponentially through the atmosphere with optical density D_0,
along a path proportional to 1/S. For the fraction of the sky, C,
covered by cloud of category i, there is an additional attenuation
through an optical density D_i. We allow for a diffuse or scattered
component E_i which depends on cloud category but which is independent
of solar elevation. This model is expressed as

$$T = E_i/S + [\exp(-D_o/S)] \, [(1-C) + C \exp(-D_i/S)]$$

It makes the implicit assumption that all clear-sky absorption of the
diffuse component of the radiation occurs below the level of the
dominant clouds.

     The data in Figure 2 have been grouped in 16 ranges of S centred
at S = 0.125, 0.175, ..., 0.875 and D_i and E_i values (Table 1) have
been fitted iteratively for minimum rms error in T.

Tests of Solar Radiation Models by Simulating Time Series of Radiation

We have tested the ability of various models to estimate the measured radiation at OWS J on an hourly, daily, or monthly basis using average coefficients for all stations analysed. For these tests we have added two more models: a hybrid combination of the Okta model (linear in S) with coefficients A and B for categories 6 to 9 and with the non-linear model with coefficients D and E for categories 0 to 5, as used by Smith and Dobson (1984). We also test the Reed (1977) model which estimates daily insolation Q_S using the Smithsonian tables (Seckel and Beaudry, 1973) for clear-sky radiation Q_C and a simple formula for attentuation by clouds,

$$Q_S/Q_C = 1 - 0.62 \ \tilde{C} + 0.0019 \ \alpha$$

where $\tilde{C}$ is the daily mean fractional cloud cover and α the noon solar elevation.

The rms errors from all the models (Table 2) are remarkably similar; the scatter appears to be inherent in the cloud and radiation data regardless of how we choose to model them. The Lind model is marginally better at reproducing the long-term mean. The Okta model performs slightly better than the others in reproducing the hourly, daily and monthly solar radiation, while the 24 Category model is worst for daily and the Lind model is worst for hourly and monthly estimation. Our preliminary conclusion is that cloud type information is not helpful in estimating insolation, but, as mentioned, we have more work to do.

Table 2. Model error for OWS J in Wm^{-2}: coefficients used were weighted averages of coefficients for OWS A, I, J, K and P.

Model Type	Hourly rms	Daily rms	Monthly rms	Long-term mean	Monthly mean rms
24 Category	78	57	50	12	6
Lind	87	48	23	9	9
Okta	76	42	19	11	7
Non-linear	78	45	20	-18	8
Non-Lin/Okta	79	46	20	- 1	6
Reed	-	42	21	- 8	17

The Reed model underestimates the long-term mean radiation at OWS J and would underestimate more strongly at OWS P. It is the only model which we have not calibrated by using the data to determine coefficients. The seasonal cycle, i.e. the month-to-month variability of the long term monthly mean errors (last column in Table 2) is reproduced worst by the Reed model (17 W m^{-2} rms), while other models reproduce

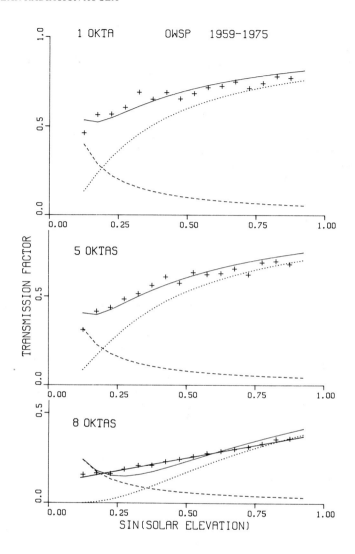

Figure 2. Atmospheric transmission factor vs sin(solar elevation) at
OWS P. Diffuse component (dashed curve) and direct component
(dotted curve) are added to obtain the fitted (solid) curve.
Linear fit used for high cloud cover is shown in lower diagram.

the seasonal cycle within 6 to 9 W m^{-2} rms. The annual cycle of long-
term mean and rms hourly errors follows a pattern which resembles the
seasonal cycle of insolation, as we would expect. The rms daily and
monthly errors are about one half and one quarter, respectively, of the
rms hourly errors.

DISCUSSION

Our coefficients apply equally well at all stations within our stated
accuracy of ±5%, but they have not been tested at low latitudes. We
would like better accuracy, since ±5 to 10 W m^{-2} accuracy in the long-
term mean sea surface heat flux budget is needed for climate prediction
modelling (Dobson et al., 1982; McBean et al., 1982). We are cautious-
ly optimistic since about 80% of the station-to-station variability in
our coefficients is due to differences in reported clear-sky radiation.
If we can track down the source of these differences then our formulae
can be expected to reproduce the measured long-term mean radiation at
any of the stations with the required accuracy.

We are still working on the problem of devising cloud categories
which sort the cloud observations into groups with similar transmission
factors. So far we have had slightly better success when we sorted by
oktas of cloud cover, finding that clouds which cover more of the sky
tend to be darker. However, we would prefer to sort by cloud type
alone to find D and E (or A and B) coefficients characteristic of
various cloud types, and we would have more confidence in extending
such a model to low latitudes, where other types of cloud predominate.
Sorting by cloud type or by total cloud amount is not independent since
some cloud types occur predominantly in high or low amounts.

Tests with new data will be necessary in applying the models to
new areas since some processes have not been included. For example,
absorption of diffuse radiation above the cloud tops, variation of
diffuse radiation with solar elevation, and a decrease in optical density with path length due to the dependence of absorption on wavelength
have been ignored in our model.

CONCLUSIONS

The Budyko (1974) bulk shortware radiation formula underestimates the
actual incoming solar radiation by about 30% at high latitudes. The
present limit on accuracy is ±5%, due mostly to radiometer calibration
differences. We cannot at present improve our model fits by including
information on cloud type; we are continuing to investigate the
problem.

Since the erroneous Budyko formula has been used in several
important studies such as Budyko (1963), Bunker (1976) and Hastenrath
(1980, for latitudes >25°), we believe that present estimates of
oceanic meridional heat transport need revision.

REFERENCES

Berliand, T.G. (1960) Methods of climatological computation of total incoming solar radiation, Meteorol. Gidrol., 6: 9-12.

Budyko, M.I. (1963) "Atlas of Heat Balance of the World". Glabnaia Geophys. Observ. Also, Guide to the Atlas of the Heat Balance of the Earth", Transl. I.A. Donehoo, U.S. Weather Bureau, WB/T-106, Washington, D.C.

Budyko, M.I. (1974) Climate and Life, Academic Press, N.Y., 508 pp.

Bunker, A.F. (1976) Computations of surface energy flux and annual air-sea interaction cycles of the North Atlantic Ocean. Mon. Weath. Rev. 104: 1122-1140.

Davies, J.A. (1981) Models for estimating incoming solar irradiance, Rep. 81-2, Canadian Climate Centre*, 120 pp.

Dobson, F.W., Bretherton, F.P., Burridge, D.M., Crease, J., Kraus, E.B. and T.H. Von der Haar (1982) The 'CAGE' Experiment: A Feasibility Study. Report WCP-22, World Meteorological Organization, Geneva, 95 pp.

Frölich, C. (1983) Solar Variability. In International Experts' Meeting on Satellite Systems to Measure the Earth's Radiation Budget Requirements, Igl, Austria, Ed. H.C. Bolle. World Meteorological Organization, Geneva.

Hastenrath, S. (1980) Heat budget of tropical ocean and atmosphere. J. Phys. Oceanogr. 10: 159-170.

Lind, R.J. (1981) Models of long and short wave irradiance with radiation estimates from JASIN 1978, M.Sc. Thesis, Dept. of Atmos. Sci., U. of Washington, Seattle, 129 pp.

Lind, R.J. and K. Katsaros (1982) Tests of radiation parameterizations for clear and cloudy conditions, JASIN News #22, 6-9.

Lumb, F.E. (1964) The influence of cloud on hourly amount of total solar radiation at the sea surface, Quart. J. R. Met. Soc., 90: 43-56.

McBean, G.A., Bennett, A.F., Burridge, D.M., Hasanuma, K., Somerville, R., Vonder Haar, T.H. and W. White (1982) Proposal for the Pacific Transport of Heat and Salt. Report WCP-51, World Meteorological Organization, Geneva, 72 pp.

Paltridge, G.W. and Platt, C.M.R. (1976) Radiative Processes in Meteorology and Climatology, Developments in Atmospheric Science, 5: Elsevier, 318 pp.

Reed, R.K. (1977) On estimating insolation over the ocean, J. Phys. Oceanogr., 7: 482-485.

Seckel, G. R. and Beaudry, F.H. (1973) The radiation from sun and sky over the North Pacific Ocean (abstract), Trans. Amer. Geophys. Union, 54: 1114.

Smith, S.D. and Dobson, F.W. (1984) The heat budget at Ocean Weather Ship Bravo, Atmosphere-Ocean, 22(1): 1-22.

---

* Available from Canadian Climate Centre, 4905 Dufferin St., Downsview, Ont., Canada M3H 5T4

CRITICAL EXAMINATION OF VARIOUS ESTIMATION METHODS OF LONG-TERM MEAN AIR-SEA HEAT AND MOMENTUM TRANSFER

K. Hanawa and Y. Toba
Department of Geophysics, Tohoku University, Sendai 980 Japan

ABSTRACT. Comparisons of flux values, estimated from data which were averaged for various times from a day to a month, were made in three ways: the sampling method (SM), the scalar averaging method (SAM) and the vector averaging method (VAM). In general, neither SAM nor VAM estimations can give the correct values as estimated by SM; their errors depend on the averaging times used. Correction procedures for both SAM and VAM estimations are proposed. (This paper is an extended abstract of work to be published in full elsewhwere.)

1. INTRODUCTION

Recently, climatic change has attracted much attention among meteoro-logists and oceanographers, and averages of air-sea heat and momentum transfers over the world ocean basins have been made by various authors. These heat and momentum fluxes are usually estimated by a bulk method, and the bulk transfer coefficients used have been improved to some extent in recent years. To study climatic changes, time series of air-sea fluxes are essential; the presently available data, however, appear insufficient for this purpose. Another problem, among the many involved in the determination of surface fluxes, is the use of long-term mean (over a month or so) input variables within the frame work of the bulk method. Three different techniques are used: the sampling method (SM), the scalar averaging method (SAM) and the vector averaging method (VAM). In the present study, we compute the differences among the three methods for various averaging times and propose correction procedures for the SAM and VAM estimates.

2. THE BULK TRANSFER FORMULATION AND THE DATA

The east-west wind-stress component τ_x, for example, can be evaluated with the following bulk aerodynamic formula:

$$\tau_x = \rho C_D |\boldsymbol{v}| u \tag{1}$$

535

Y. Toba and H. Mitsuyasu (eds.), The Ocean Surface, 535–540.
© *1985 by D. Reidel Publishing Company.*

where the symbols are the usual ones. The quantity C_D is the so-called bulk transfer coefficient for wind stress. In the present study, we employ the values proposed by Kondo (1975), which are functions of v and of the bulk atmospheric boundary layer stability, which in turn is a function of v and $T_s - T_a$, where T_s and T_a are sea surface temperature and air temperature, respectively. The air density ρ is assumed constant in the present study.

The data used were taken every three hours for 42 months from June 1, 1950 to November 25, 1953 at OWS-T, which was located at 29°N, 135°E in the sea south of Japan.

3. SAM : ESTIMATION AND CORRECTION PROCEDURE

If $\overline{\tau}_x$ denotes the mean fluxes determined by the SM ,and $\hat{\tau}_x$ represents those by the SAM, the following relation holds,

$$\overline{\tau}_x = \rho\overline{C_D}\,\overline{|v|\,\overline{u}} + \rho\overline{C_D}\,\overline{|v|\,'u'} + \rho\overline{|v|\,C_D'u'} + \rho\overline{\overline{u}C_D'|v|'} + \rho\overline{C_D'|v|\,'u'}$$

$$= \hat{\tau}_x + \rho\overline{C_D}\,\overline{|v|\,'u'} + \rho\overline{|v|\,C_D'u'} + \rho\overline{\overline{u}C_D'|v|'} + \rho\overline{C_D'|v|\,'u'} \tag{2}$$

where the overbar denotes the averaging operation over a given time period. The SAM neglects the second and third moments.

The comparison between the two methods was made for six averaging times, i.e., 1, 3, 5, 10, 15 and 30 days. For averaging times from a day to 10 days, 100 data sets were taken, and for 15 and 30 days, 84 and 42 data sets were taken. When the SM fluxes are very small, the ratios scatter widely and the comparison is meaningless. Therefore, we discarded the data sets which did not exceed the following threshold values : wind-stress components τ_x and τ_y = 0.2 dyne cm^{-2}, sensible heat flux Q_H = 5 Wm^{-2} and latent heat flux Q_E = 20 Wm^{-2}.

Figure 1 shows the dependence on the averaging time of the mean SAM flux ratios to the SM fluxes and Figure 2 shows a comparison between the SM fluxes and the SAM fluxes for 42 calendar months. The SAM wind stresses become gradually smaller than the SM values as the averaging time increases: for an averaging time of a month, the SAM wind stresses are only about 70 % of those by the SM. On the other hand, the SM sensible heat fluxes are almost identical with those by the SM irrespective of the averaging time. The latent heat fuxes are about 105 % of those by the SM for averaging times from three days to a month. The above results for sensible and latent heat fluxes are in agreement with the conclusions of Esbensen and Reynolds (1981), who analysed data from 9 OWSs.

As shown in Figure 1, the difference between the SM fluxes and the SAM fluxes is systematic, and the standard deviations of the ratios are small for longer averaging times. Therefore, it is possible to define effective SAM bulk transfer coefficients as a function of averaging time of the input data. If the effective bulk transfer coefficient is given by the product of Kondo's original bulk transfer coefficient and an adjustment factor, this adjustment factor is obviously the reciprocal of the ratio of SAM flux to SM flux. The adjustment factors so obtained

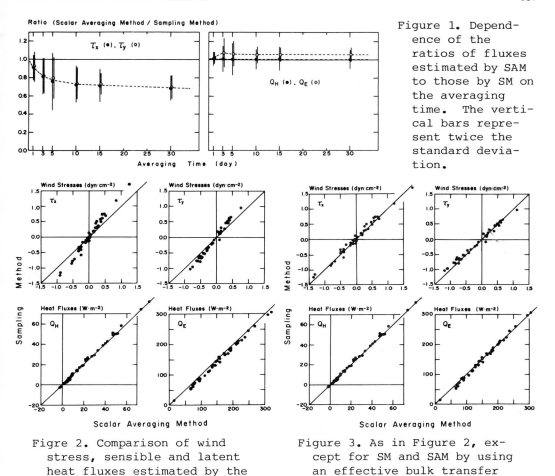

Figure 1. Dependence of the ratios of fluxes estimated by SAM to those by SM on the averaging time. The vertical bars represent twice the standard deviation.

Figre 2. Comparison of wind stress, sensible and latent heat fluxes estimated by the SM and SAM, from 42 calendar months of data.

Figure 3. As in Figure 2, except for SM and SAM by using an effective bulk transfer coefficient for the monthly averages.

Table 1. Adjustment factors for SAM estimation as a function of averaging time; to be applied to the bulk coefficients when using averaged input data.

Item	Adjustment factor	Averaging time(days) of input data					
		1	3	5	10	15	30
wind stress	m_D	1.10	1.20	1.30	1.35	1.40	1.45
sensible heat flux	m_H	1.10	1.00	1.00	1.00	1.00	1.00
latent heat flux	m_E	1.00	0.95	0.95	0.95	0.95	0.95

are shown in Table 1. Figure 3 shows a comparison between the SM and
SAM fluxes for 42 calendar months using these adjustment factors. The
good agreement between the values obtained by both methods shows the
practical usefulness of the effective bulk transfer coefficients.
Actually, the adjustment factors probably depend on the time (season),
but in practice they may be regarded as constant, judging from the
results of Kondo (1972), Esbensen and Reynolds (1981) and the present
study.

4. VAM : ESTIMATION AND CORRECTION PROCEDURE

The mean VAM flux is estimated as follows,

$$\tilde{\tau}_X = \rho C_D(|\bar{\boldsymbol{v}}|, \bar{T}_s - \bar{T}_a)|\bar{\boldsymbol{v}}|\bar{u} \tag{3}$$

where the tilde indicates that the flux was estimated by the VAM.
Figure 4 shows the dependence of the mean VAM to SM ratios on the
averaging time and Figure 5 shows a comparison between the SM and VAM
fluxes for 42 calendar months. The ratios for wind stress show a larger
decrease than did the SAM to SM ratios. For an averaging time of 30
days, the VAM fluxes can represent only about 30 % of those by the SM,
and the sensible and latent heat fluxes are only 40 %. These decreasing
tendencies are in agreement with Fissel et al. (1977), who calculated
the ratios for averaging times from 6 hours to two years from data
obtained at OWS-P.
      Similarly to the SAM estimation, in order to obtain the correct
fluxes by VAM estimation, effective bulk transfer coefficients can be
formally estimated for different averaging times. However, their use
does not produce acceptable agreement with the SM values. Therefore, we
must seek another correction method.
      Here, let us take the ratio of the VAM to the SAM flux. Since
$C_D(|\bar{\boldsymbol{v}}|) \sim C_D(\overline{|\boldsymbol{v}|})$, the ratio is given by,

$$\tilde{\tau}_X/\hat{\tau}_X = \rho C_D(|\bar{\boldsymbol{v}}|)|\bar{\boldsymbol{v}}|\bar{u}/\rho C_D(\overline{|\boldsymbol{v}|})\overline{|\boldsymbol{v}|\bar{u}} \sim |\bar{\boldsymbol{v}}|/\overline{|\boldsymbol{v}|} \tag{4}$$

That is, the ratio is approximately given by a nondimensional parameter
defined as the ratio of the vector-averaged wind speed to the scalar
averaged one for a given time interval, i.e., $S = |\bar{\boldsymbol{v}}|/\overline{|\boldsymbol{v}|}$, $0 \leq S \leq 1$.
We call this parameter the "stability of the wind field", while Roll
(1965) named it (multiplied by 100) the "constancy". Figure 6 shows the
dependence of S on the length of the averaging time. The ratio is
nearly equal to the ratio of VAM/SM to SAM/SM, i.e., $S \sim (\tilde{\tau}_X/\bar{\tau}_X)/(\hat{\tau}_X/\bar{\tau}_X)$.
The S for the 42 calendar months was calculated as shown in Figure 7.
Since the seasonal signals are relatively stable for every year, we may
be able to calculate the fluxes with the VAM using a separate value of
for each calendar month. Since the flux calculated by this method is
almost the same as that by the SAM, the calculated flux must also be
multiplied by the SAM adjustment factors. That is, the flux is calculated as follows,

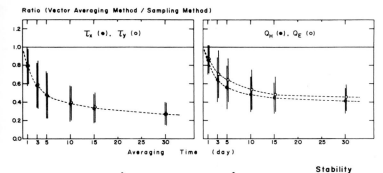

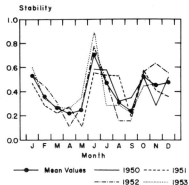

Figure 4. As in Figure 1, except for SM and VAM.

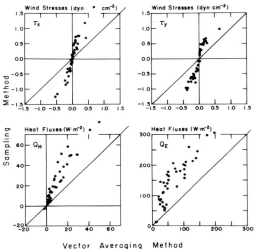

Vector Averaging Method

Figure 5. As in Figure 2, except for VAM and SM.

Figure 6. Dependence of the "stability of the wind field", S, on the averaging time.

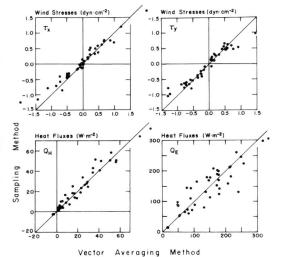

Vector Averaging Method

Figure 7. Seasonal variation of monthly S.

Figure 8. As in Figure 2, except for SM and VAM by using a separate S for each calendar month and the SAM adjustment factors for monthly averages.

$$\tilde{\tau}_x = \rho m_D C_D(|\bar{\boldsymbol{v}}|/S, \ \overline{T}_s - \overline{T}_a)(|\bar{\boldsymbol{v}}|/S)\overline{u} \tag{5}$$

A comparison between the fluxes calculated by the above equations and those by the SM for the 42 calendar months is shown in Figure 8. For the wind stress and sensible heat flux, a comparison (not shown here) of the VAM results with those using effective bulk transfer coefficients shows that a considerable improvement has been made. For the latent heat flux, however, the scattered distribution in the middle range has not changed substantially.

5. CONCLUDING REMARKS

Three methods used for the estimation of long-term mean heat and momentum fluxes, were examined using data obtained at OWS-T, and correction procedures for the SAM and VAM estimates are proposed which amount to the use of an effective bulk transfer coefficient for the SAM estimates and the use of S for VAM estimates. Although S needs to be calculated for various areas of the oceans and for each calendar month, these calculations can easily be performed. Recently, Wright and Thompson (1982) have proposed another correction method for VAM estimation and Thompson et al. (1983) have improved on this method. Marsden and Pond (1983) have also proposed a correction method. In future work, all the correction procedures will be examined and compared with the results of the present study.

REFERENCES

Esbensen, S.K. and R.W. Reynolds, 1981: Estimation of monthly averaged air-sea transfers of heat and momentum using the bulk aerodynamic method. J. Phys. Oceanogr., 11, 457-465.

Fissel, D.B., S. Pond and M. Miyake, 1977: Computation of surface fluxes from climatological and synoptic data. Mon. Wea. Rev., 105, 26-36.

Kondo, J., 1972: Applicability of micrometeorological transfer coefficient to estimate the long-period means of fluxes in air-sea interface. J. Meteor. Soc. Japan, 50, 570-576.

Kondo, J., 1975: Air-sea bulk transfer coefficients in diabatic conditions. Bound.-Layer Meteor., 9, 91-112.

Marsden, R.F. and S. Pond, 1983: Synoptic estimates of air-sea fluxes. J. Mar. Res., 41, 349-373.

Roll, H.V., 1965: Physics of the Marine Atmosphere. Academic Press Inc., New York, 426 pp.

Thompson, K.R., R.F. Marsden and D.G. Wright, 1983: Estimation of low-frequency wind stress fluctuations over the open ocean. J. Phys. Oceanogr., 13, 1003-1011.

Wright, D.G. and K.R. Thompson, 1982: Time-averaged forms of the nonlinear stress law. J. Phys. Oceanogr., 13, 341-345.

MIXED–LAYER GROWTH IN STRATIFIED FLUIDS

Harindra J. S. Fernando[1] and Robert R. Long[2]
1 Env.Eng.Sci., California Institute of Technology,
   Pasadena, CA 91125 and Mech & Aero.Eng., Arizona
   State University, Tempe, AZ 85287
2 Earth & Planetary Sci., The Johns Hopkins University,
   Baltimore, MD 21218

ABSTRACT. An experimental study of the growth of a shear-free turbulent
mixed layer in a density-stratified fluid is described. Investigations
included both linearly stratified and two-fluid systems and the results
show an entrainment law (i.e., the relationship between the entrainment
coefficient E and the Richardson number Ri) of the form $E \sim Ri^{-7/4}$.
The results also were in agreement with an earlier theoretical formulation of Long (1978a).

1. INTRODUCTION

Stratified-fluid turbulence has become of major interest in geophysical
fluid mechanics because of its importance in understanding many events
related to the atmosphere-ocean system, for example the growth of turbulent mixed layers in the upper ocean or lower atmosphere. Oceanic and
atmospheric turbulence is generated in several ways, for example from
mean shear, convective heating and cooling, wave-breaking, etc., and the
turbulence thus produced interacts with buoyancy forces to cause turbulent mixing. During oceanic wave-breaking, the turbulence is generated
within a thin layer (of the thickness of the r.m.s wave amplitude) and
is diffused downward toward the thermocline. Kitaigorodskii (1979) has
suggested that this process could be simulated in the laboratory by an
oscillating grid in a density-stratified fluid. An isolated study of a
single oceanic turbulence-generating event is an oversimplified attempt
to understand the complex events occurring in the real ocean, where many
different mechanisms interact; nevertheless, such a study may be of importance in gaining a general understanding of the mechanics of the entrainment process.

In this paper, we report some of the features of a detailed study of
turbulent entrainment both in linearly-stratified and in two-layer
fluids when the turbulence is produced by an oscillating grid. The
experimental results are compared with previous theoretical ideas by
Long (1978a).

Y. Toba and H. Mitsuyasu (eds.), The Ocean Surface, 541–546.

2. THEORETICAL DEVELOPMENTS

In this section we try to give a simplified derivation of Long's theory. The most fundamental assumption in this theory (which has been verified recently by McDougall (1979) and Piat & Hopfinger (1981)) is that the entrainment is associated with the action of quasi-isotropic eddies of the size of the wave-amplitude at the entrainment interface (δ_2) whereas eddies of the size of the horizontal integral lengthscale (ℓ) in the entrainment zone tend to flatten at the density interface. The interfacial layer is agitated by the overlying eddies and internal waves are produced, some of which may break and cause intermittent turbulent patches in the interfacial layer. The patches located at the entrainment interface are considered to be responsible for the buoyancy transfer to the mixed layer (see Figure 1).

If the vertical size of these patches, within which the turbulence is quasi-isotropic, is determined by a balance between the kinetic and potential energies of the eddies, then $w_2^2 \sim b_2 \delta_2 \sim (\Delta b/h)\delta_2^2$, where b and w are r.m.s. vertical velocity and buoyancy fluctuations, δ is the characteristic size of a patch, Δb is the buoyancy jump across the interfacial layer of thickness h and subscript 2 refers to the entrainment interface (Region R_2 in Figure 1). Within a patch the buoyancy flux $\overline{(bw)}_p$ and the dissipation (ε_p) may be written as $\overline{(bw)}_p \sim bw$ and $\varepsilon_p \sim w^3/\delta$, and taking intermittency into account, we may write the overall energy dissipation (ε) and the buoyancy flux $\overline{(bw)}$ as $\varepsilon = I\varepsilon_p$, $\overline{(bw)} = \overline{(bw)}_p I$. Notice that $\overline{(bw)}/\varepsilon \sim b\delta/w^2 \sim 0(1)$. The energy equation for the interfacial layer becomes $-\partial EF/\partial z - \overline{bw} - \varepsilon \approx 0$ where $\partial EF/\partial z (\sim \partial w^3/\partial z)$ is the energy-flux divergence. Since $\overline{bw} \sim \varepsilon$, and both are energy sinks, these terms should be of the same order as the $\partial w^3/\partial z$ term which is the only source term in the energy equation.

As mentioned previously, w_2 is determined by the velocity of quasi-isotropic eddies of the size of δ_2 and we may write $w_2 \sim \varepsilon_2^{1/3}\delta_2^{1/3}$ (Hunt & Graham 1978). Since w_2 (which is continuous across the density interface) is associated with the large mixed-layer eddies of size ℓ and r.m.s. velocity σ_u (i.e., $\varepsilon_2 \sim \sigma_u^3/\ell$), we may infer $w_2 \sim \sigma_u(\delta_2/\ell)^{1/3}$ and since $\delta_2 \sim w_2(h/\Delta b)^{1/2}$, we get $\delta_2/\ell \sim Ri^{-3/4}$, $w_2/\sigma_u \sim Ri^{-1/4}(h/\ell)^{1/4}$, where Ri is the bulk gradient Richardson number, $Ri = \Delta b\ell/\sigma_u^2$. The energy dissipation in a patch in the Region R_2 is independent of the interfacial-layer stability and we may assume that this is true within the entire interfacial layer. However, since the turbulent velocities decrease as we move away from R_2 into the interfacial layer, we may also assume that the patch dissipation is a function of the relative distance (ζ/D) where $\zeta = z - D$ and D is the mixed-layer depth. Hence $\varepsilon_p \sim w^3/\delta = (\sigma_u^3/\ell)\phi(\zeta/D)$ or $w^3 \sim \sigma_u^3 (h/\ell)^{3/4}Ri^{-3/4}\phi(\zeta/D)$ where ϕ is a functional, w is the vertical r.m.s. velocity of the turbulence and δ is the patch size at any ζ. At the outer edge of the interfacial layer ($\zeta = h$), the buoyance flux is zero, and since $\partial w^3/\partial \zeta \sim \overline{bw} = 0$, we get $\phi'(h/D) = 0$ or h/D = constant.

In view of the above finding that the interfacial layer grows linearly with the size of the mixed layer, we may infer $w_2/\sigma_u \sim Ri^{-1/4}$; $\delta_2/\ell \sim Ri^{-3/4}$. If the rate-of-change of potential energy due to interfacial mixing ($\sim u_e\Delta b$) is proportional to the kinetic-energy flux

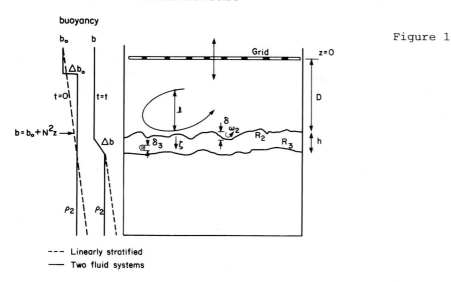

Figure 1

--- Linearly stratified
— Two fluid systems

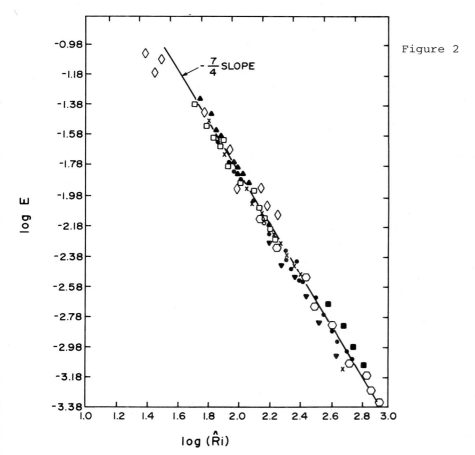

Figure 2

available to the interfacial layer (Linden, 1975), we may write $u_e\Delta b \sim w_2^3/\ell \sim Ri^{-3/4}\sigma_u^3/\ell$ or $u_e/\sigma_u \sim Ri^{-7/4}$ where u_e is the entrainment velocity. Also the intermittency I may be evaluated as $u_e\Delta b \sim \overline{bw} \sim \varepsilon \sim I\sigma_u^3/\ell$, $I \sim u_e/\sigma_u$ Ri $\sim Ri^{-3/4}$. Since experimental observations reveal that $\ell = \beta D$ where β is a constant, we may use D as the scaling parameter instead of ℓ (Thompson & Turner, 1975; Hopfinger & Toly, 1976).

3. EXPERIMENTAL RESULTS

The grid and its oscillatory motion can be parameterized by a single constant called 'action K' having the dimensions of eddy viscosity. This can be evaluated in two ways, namely by observing the mixed-layer propagation in a homogeneous fluid according to $D = (Kt)^{1/2}$ (Dickinson & Long, 1983) or $K_\ell \sim \sigma_u\ell$ (Long 1978b). A simple calculation indicates that $K = 7K_\ell$ (Fernando 1983).

Long (1978a) derived the following conservation law for an entraining stratified fluid, viz., $V_0^2 = (D + 1/2h)\Delta b - 1/2N^2(D + h)^2$ where N is the B-V frequency of the initial stratification and V_0 is a constant. For a two-layer fluid we may take $V_0^2 = (D + 1/2h)\Delta b \cong D_0\Delta b_0$ or, since $h \sim D$ (or $h = \alpha_1 D$), $V_0^2 = D_0\Delta b_0 = (1 + \alpha_1/2)D\Delta b$ where the subscript 'o' indicates values at time $t = 0$. For a linearly stratified fluid $V_0 = 0$ and we get $\Delta bD = (1 + \alpha_1)^2N^2D^2/(2 + \alpha_1)$. In our experiments, we were able to verify both of these expressions by measuring the buoyancy jump across the interfacial layer and the depth of the mixed layer with time. However, the former case showed $\alpha_1 \cong 0.2$, whereas the latter case gave $\alpha_1 \cong 0.1$.

In view of the above discussion we are able to redefine a more convenient form of Richardson number depending on the nature of the stratification. For two layer fluids, where $\ell\Delta b \sim D\Delta b \sim V_0^2$, we define $\hat{Ri} = V_0^2D^2/K^2$ and for linearly stratified fluids, where $D\Delta b \sim N^2D^2$, we define $\check{Ri} = N^2D^4/K^2$. Note that Ri $\sim \hat{Ri}$ or $\check{Ri}$. Figures 2 and 3 show the variation of the entrainment coefficient E (= u_eD/K) with $\hat{Ri}$ (= $V_0^2D^2/K^2$) for the two-layer case and E (= u_eD/K_ℓ) with $\check{Ri}$ (= N^2D^4/K_ℓ^2) for the linearly stratified fluids, respectively. Results are in good agreement with the $Ri^{-7/4}$ behavior.

Notice that the E $\sim Ri^{-7/4} \sim (\hat{Ri}^{-7/4}$ or $\check{Ri}^{-7/4})$ behavior implies a deepening law of the form D $\sim V_0^{-7/11}K^{9/11}t^{2/11}$ for two-fluid systems and D $\sim N^{1/2}K^{-7/18}t^{1/9}$ for linearly stratified fluids. We measured the variation of D with t, V_0, K and N and verified these expressions experimentally. Folse et al. (1981) have also obtained similar results for the linearly stratified case.

Since $w_2/\sigma_u \sim w_2D/K \sim Ri^{-1/4}$ and $\delta/\ell \sim \delta/D \sim Ri^{-3/4}$, we may expect that the interfacial-layer wave frequencies (ω_i) follow the law $\omega_iD^2/K \sim Ri^{-1/2}$. In our experiments we measured the wave frequencies and amplitudes at the entrainment interface and the results are in agreement with the abovementioned formulae. Figure 4 shows the variation of estimated buoyancy flux (evaluated as $Dd\Delta b/dt$) with the dissipation $\varepsilon(0.61\sigma_u^3/\ell$ - Hopfinger & Toly 1976) just above the entrainment interface. The results indicate that the ratio $\overline{bw}/\varepsilon$ is independent of the Richardson number at the entrainment interface.

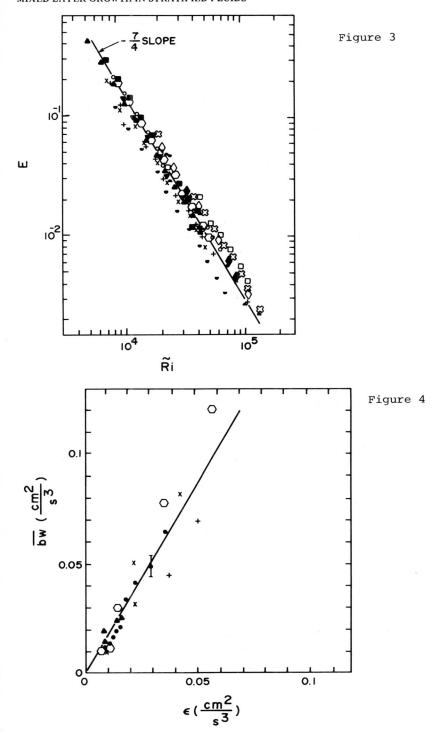

Figure 3

Figure 4

4. CONCLUSIONS

We have presented some results of an experimental study carried out to investigate the mixed-layer growth in two-fluid and linearly stratified systems where the turbulence is induced by an oscillating grid. The results showed very good agreement with an earlier theory by one of us (Long 1978a).

5. ACKNOWLEDGEMENTS

This work was supported by NSF Grant Nos. ATM-7907025, ATM-8210498, CEE-7272A1 and the Office of Naval Research (Fluid Dynamics Division) Contract No. N00014-76-C-0184.

REFERENCES

1.   Dickinson, S.C. & Long, R. R. 1983. Oscillating grid turbulence including the effects of rotation. *J. Fluid Mech.*, 126, 315.
2.   Fernando, H. J. S. 1983. Studies on turbulent mixing in stably stratified fluids. *Ph.D. Thesis*. The Johns Hopkins University.
3.   Folse, R. F., Cox, T. P., & Schexnayder, K. R. 1981. Measurements of the growth of a turbulently mixed layer in a linearly stratified fluid. *Phys. Fluids*, 24, 396.
4.   Hopfinger, E. J. & Toly, J. A. 1976. Spatially decaying turbulence and its relation to mixing across density interfaces. *J. Fluid Mech.*, 125, 505.
5.   Hunt, J. C. R. & Graham, J. M. R. 1978. Free stream turbulence near plane boundaries. *J. Fluid Mech.*, 84, 209.
6.   Kitaigorodskii, S. A. 1979. Review of the theories of wind mixed-layer deepening. *Marine Forecasting, Elsevier Oceanog. Series*, 25, 1.
7.   Linden, P. F. 1975. The deepening of a mixed layer in a linearly stratified fluid. *J. Fluid Mech.*, 71, 385.
8.   Long, R. R. 1978a. A theory of mixing in a stably stratified fluid. *J. Fluid Mech.*, 84, 113.
9.   Long, R. R. 1978b. Theory of turbulence in a homogeneous fluid induced by an oscillating grid. *Phys. Fluids*, 21, 1887.
10. McDougall, T. J. 1979. Measurements of turbulence in a zero-mean-shear mixed layer. *J. Fluid Mech.*, 94, 409.
11. Piat, J. F. & Hopfinger, E. J. 1981. A boundary layer topped by a density interface. *J. Fluid Mech.*, 113, 411.
12. Thompson, S. M. & Turner, J. S. 1975. Mixing across an interface due to turbulence generated by an oscillating grid. *J. Fluid Mech.*, 67, 349.

THE USE AND TESTING OF A MODEL FOR UPPER OCEAN DYNAMICS

L.H. Kantha[1], A.F. Blumberg[1], H.J. Herring[1] and G.R. Stegen[2]
[1]Dynalysis of Princeton, Princeton, New Jersey 08540 U.S.A.
[2]Science Applications International Corporation, Bellevue,
Washington 98005 U.S.A.

ABSTRACT. A one-dimensional turbulent Ekman layer model is used to
simulate the upper-ocean dynamics in Santa Barbara Channel. Comparison
of simulated and measured currents showed excellent agreement below
subtidal frequencies.

1. INTRODUCTION

An integrated program of field observations and numerical modeling has
been undertaken to provide a better understanding of circulation in the
Santa Barbara Channel. The Channel is a semi-enclosed basin with a deep
sill at the western end and a shallow one at the eastern end (Figure 1).
Extremely active dynamically, the Channel is characterized by highly
variable currents generated locally and remotely via synoptic-scale
atmospheric disturbances and by atmospheric fronts associated with flow
around Point Conception. Offshore ocean circulation, primarily the
California Current, is also a major source of variability in this
region. The present program is designed to provide an understanding of
the circulation dynamics on time scales longer than the subtidal period.
    A two-month pilot study was conducted in April–June 1983 to
estimate the relative importance of different processes prior to final
decisions on the observational sampling network for an intensive
twelve-month study to be conducted in 1984–85. Data from 35 subsurface
and two nearsurface current meters deployed on 15 moorings, from two
meteorological buoys and a wave buoy, and from one hydrographic survey
were used to quantify volume transports at the western and eastern ends
of the Channel and to estimate the short-time small-scale variability
within the Channel (see Blumberg et al., 1984).
    This paper addresses the comparison between observed upper-layer
structure and that predicted using a numerical model.

2. PREDICTION OF UPPER OCEAN CURRENTS

A three-dimensional prognostic shelf circulation model is used to
describe the flow interior to the Channel. A second-moment turbulence

Y. Toba and H. Mitsuyasu (eds.), The Ocean Surface, 547–552.
© 1985 by D. Reidel Publishing Company.

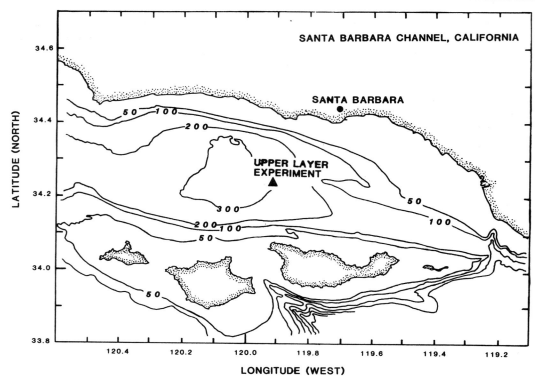

Figure 1.   Santa Barbara Channel showing bathymetry and location of
    upper-layer measurements (depths in fathoms).

closure technique is used to provide a realistic parameterization of
vertical mixing processes. Model variables are the three components of
velocity, and temperature, salinity, turbulent kinetic energy, and
turbulent macroscale. The model is driven by the observed fluxes
through the surface and at the open boundaries (east and west entrances
to the Channel).

     Although the Channel was heavily instrumented with current meters,
questions of instrument survivability in the heavily trafficked Channel
precluded the general use of current meters at less than 30 m. A good
estimate of the currents in the mixed layer can be obtained from a
knowledge of the atmospheric forcing, the density structure in the upper
layer, the subsurface currents, and the use of a turbulent mixing model.
Wind-driven (Ekman) transport in the upper layers can be an appreciable
portion of the total transport in the water column; hence, a validated
procedure for estimating the currents in the upper 30 m is essential. A
one-dimensional version of the circulation model has been constructed to
fulfill this need. This one-dimensional turbulent Ekman layer model

simulates the upper-ocean dynamics. Based on a second-moment turbulence
closure scheme, the Ekman model is used to distribute wind stress
through the water column in a dynamically consistent manner. The
governing equations are

$$\frac{\partial u}{\partial t} - fv = -\frac{1}{\rho_o}\frac{\partial P}{\partial x} + \frac{\partial}{\partial z} K_M \frac{\partial u}{\partial z} \tag{1}$$

$$\frac{\partial v}{\partial t} + fu = -\frac{1}{\rho_o}\frac{\partial P}{\partial y} + \frac{\partial}{\partial z} K_M \frac{\partial v}{\partial z} \tag{2}$$

with

$$K_M \left(\frac{\partial u}{\partial z}, \frac{\partial v}{\partial z}\right) = (\tau_{o_x}, \tau_{o_y}) \quad \text{at } z = 0$$

$$u = u_{data}, \quad v = v_{data} \quad \text{at } z = z_o$$

and

$$\frac{\partial T}{\partial t} = \frac{\partial}{\partial z} \left(K_H \frac{\partial T}{\partial z}\right) \tag{3}$$

$$\frac{\partial S}{\partial t} = \frac{\partial}{\partial z} \left(K_H \frac{\partial S}{\partial z}\right) \tag{4}$$

with

$$K_H \left(\frac{\partial T}{\partial z}, \frac{\partial S}{\partial z}\right) = (\overline{w'T'}, \overline{w'S'}) \quad \text{at } z = 0$$

$$T = T_{data}, \quad S = S_{data} \quad \text{at } z = z_o$$

Here (u,v) are the velocity components in the east-west (x) and
north-south (y) directions, f is the Coriolis parameter, P is the
pressure which includes barotropic and baroclinic effects, T is
temperature, S is salinity, and τ_o is the surface wind stress. z_o is
the shallowest depth at which currents, temperature and salinity are
available as a function of time. The turbulent mixing coefficients K_M
and K_H are obtained from the turbulent closure scheme of Mellor and
Yamada (1982). The pressure gradient terms in Equations (1) and (2) may
be computed using current data from the uppermost current meter. The
meter is assumed to be below the mixed layer so that local wind-induced
mixing is small. Under these conditions, a good approximation for the
pressure gradients at z_o is

$$-\frac{1}{\rho_o}\frac{\partial P}{\partial y}\bigg|_{z_o} = \frac{\partial u_{z_o}}{\partial t} - fv_{z_o} \tag{5}$$

and

$$-\frac{1}{\rho_o}\frac{\partial P}{\partial y}\bigg|_{z_o} = \frac{\partial v_{z_o}}{\partial t} - fu_{z_o} \qquad (6)$$

These pressure gradients are dominated by their barotropic component, especially in the upper layer of the column, so that they can be taken as approximately constant in the vertical. Equations (1) to (6) can then be used to compute the currents in the upper ocean between $z = 0$ and $z = z_o$.

To explore the precision possible in this approach, the Ekman model was driven with data from a heavily instrumented mooring near the middle of the Channel (Figure 1). Current speed and direction, salinity and temperature were measured at 30 m (z_o) with an Aanderaa RCM-4 current meter. Wind speed and direction were measured at the same location with an anemometer mounted on a spar buoy. To validate this model, a vector-averaging current meter (EG&G VMCM 630) was suspended from the spar buoy at a depth of 8 m.

The Ekman model was run prognostically with the initial temperature and salinity distributions prescribed using CTD data taken near the mooring site at the time of deployment. Wind stress τ_o was derived from the anemometer readings at the mooring. The buoyancy fluxes $\overline{w'T'}$ and $\overline{w'S'}$ were assumed to be small compared to the wind-energy flux, and have been set equal to zero in equations (3) and (4).

The Ekman model was used to simulate a 23-day period beginning 28 April 1983. The results for the final 21 days of integration yield wind stress and unfiltered and filtered velocities at 8 m and 30 m (Figure 2). The calculated unfiltered velocities at 8 m show large-amplitude inertial oscillations which are inherent in a time-dependent one-dimensional calculation. The measured unfiltered velocities, on the other hand, display a tidally modulated flow. Thus, although the measured and calculated 8-m currents appear to roughly agree in phase, a proper comparison requires filtering out the inertial components in the model velocities and tidal components in the measured data. Figure 2 also shows the calculated and measured currents, 40-hour averaged to remove inertial and tidal components. The agreement between the model and data is quite good, especially in view of the fact that surface buoyancy fluxes have been ignored. There appears to be high correlation between the measured and calculated velocity. Thus, the Ekman model seems to provide a good estimate of currents above the topmost sub-surface current meter below subtidal frequencies.

3. CONCLUSIONS

A one-dimensional Ekman model has been used to estimate flow in the upper ocean in a dynamically consistent manner. By driving the model with observed surface stress and 30-m currents, an estimate of the currents above 30 m was obtained. Comparison between model calculations and measured currents showed excellent qualitative agreement below subtidal frequencies.

4.   ACKNOWLEDGEMENTS

This work was supported by the Minerals Management Service under contracts AA851-CT2-63 and 14-12-0001-29123 (Science Applications International Corporation Contribution SAI/NW-84-944-131).

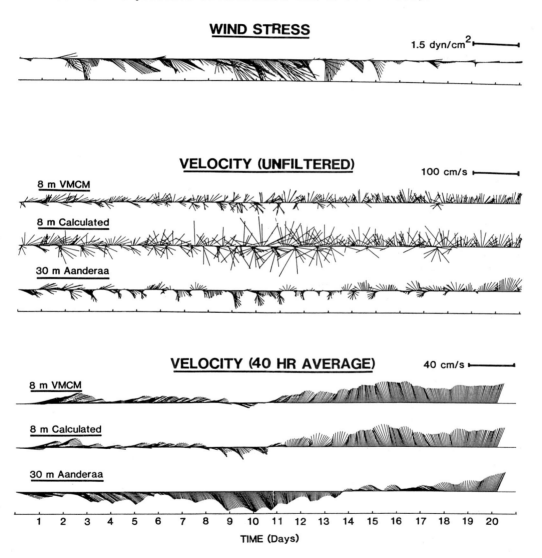

Figure 2.   Comparison of upper-level velocities at 8 m calculated with the Ekman model and measured with a VMCM.

REFERENCES

Blumberg, A.F., D.E. Cover, J.T. Gunn, P. Hamilton, H.J. Herring, L.H.
    Kantha, G.L. Mellor, R.D. Muench, L.E. Piper, G.R. Stegen and E.
    Waddell, 1984: Santa Barbara Channel circulation model and field
    study. Dynalysis of Princeton Report #86 (available from authors).
Mellor, G.L., and T. Yamada, 1982: Development of a turbulence closure
    model for geophysical fluid problems. Rev. of Geophys. and Space
    Phys., 20, 851-875.

VERTICAL MIXING ON THE BERING SEA SHELF

G. R. Stegen[1], P.J. Hendricks[1] and R. D. Muench[1]
[1]Science Applications International Corporation, Bellevue,
Washington 98005 U.S.A.

ABSTRACT.   Recent studies of the Bering Sea marginal ice zone have
focussed on the heat balance at the ice edge. The role of double-
diffusive convection in supplying heat input to the ice is evaluated.

1.   INTRODUCTION

The Bering Sea plays a major role in global ocean circulation because it
is the connection between the North Pacific and the Arctic oceans
(Stigebrandt, 1984). Water transiting the Bering can be significantly
modified by air-sea interaction processes during its passage over the
broad, shallow Bering Sea shelf. The air-sea interaction characteristics of the region change substantially throughout the year due to the
varying ice cover. From May through November the region is ice free.
Beginning in November, ice is formed in the northern Bering Sea by the
action of cold, northeasterly winds. The ice is advected southward by
the influence of the northeast winds, and replaced in the north by
newly formed ice. The ice cover continues to expand until sometime in
February-March when the ice edge has reached its most southerly extent
near the shelf break (Figure 1). The ice edge remains in this location
for 1-2 months, with ice input from the north being balanced by ice
melting at the edge (Muench and Ahlnäs, 1976). Increasing air temperature and insolation combine to initiate recession of the ice edge
sometime in April.
    During mid-winter, when the ice edge is quasi-stationary, the heat
input necessary to melt the ice must be provided by the ocean and
atmosphere. This paper investigates the role of vertical oceanic mixing
in providing heat input to the ice.

2.   DISCUSSION

Oceanographic data were obtained from the central Bering Sea marginal
ice zone (MIZ) during winter 1983 when the sea ice was at its maximum
southward extent. Temperature and salinity data, obtained along
transects across the ice edge, indicate that a well-developed frontal
structure underlay the MIZ (Figure 2); this structure has been described

Y. Toba and H. Mitsuyasu (eds.), The Ocean Surface, 553–557.

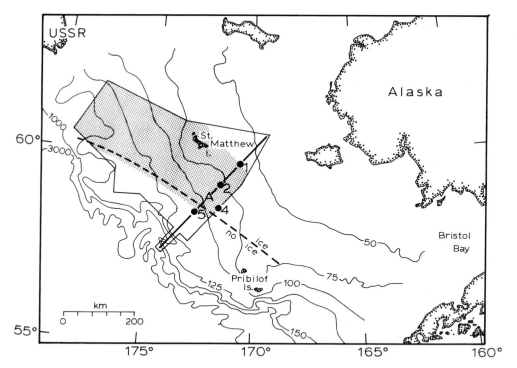

Figure 1. Geographical location of the study area. Region encompassed
    by CTD survey is indicated by polygon, and solid line labeled
    "A" shows location of CTD transect illustrated on Figure 2.
    Numbered dots indicate current moorings. Heavy dashed line shows
    approximate ice-edge location, and stippled area is approximate
    extent of two-layered, double-diffusive structure. Depths are in
    meters.

in detail by Muench and Schumacher (1985). Similar features were
observed during the winters of 1980 and 1981 (Muench, 1983), suggesting
that the observed structure is a regular feature of the Bering Sea
winter ice-edge region. The ice-edge frontal structure consists of
surface and bottom, upper- and lower-layer fronts which are separated by
a wide (>100 km) region having two layers. These two layers are well
defined on the basis of temperature, salinity and density as the lower
layer is warmer, more saline and denser than the upper. Temperature,
salinity and density each increase rapidly across the interface between
upper and lower layers. An along-front flow occurs, and monthly mean
upper-layer speeds toward the northwest are nearly 15 cm/s. Regional
tidal currents are predominantly M2, and tidal speeds of 15-20 cm/s are
typical. Both the areal extent of the frontal structure (75,000 km^2)
and the fact that the ice edge occurs over this structure suggest that

heat fluxes associated with the front are locally important to the heat
balance.

   The uniformity of the upper and lower frontal layers (Figure 2)
suggests strong mixing in those regions. Heat must be supplied in part
from the lower layer, and this discussion focusses on the processes
responsible for transfer of heat across the interface. Two such
mechanisms are considered: shear-induced turbulence and double-
diffusive convection.

   During the 1983 field program current meters were moored in the
uniform layers above and below the interface (Figures 1 and 2).
The large vertical separation of the meters (20-40 m) and their relative
accuracies preclude direct calculation of interfacial shear. However,
if we assume that speeds vary slowly in the layers, consistent with
geostrophic flow calculations, then we can estimate shear across the
layer using the speed differences between the meters and the thickness
of the layer. This yields a shear of approximately 10^{-2} sec^{-1}. The
density difference across the interface is about 0.2 g/cm^3, which for
the estimated shear gives a Richardson number $R_i \sim 4$, too large for
sustained shear-layer instability. It therefore appears unlikely that

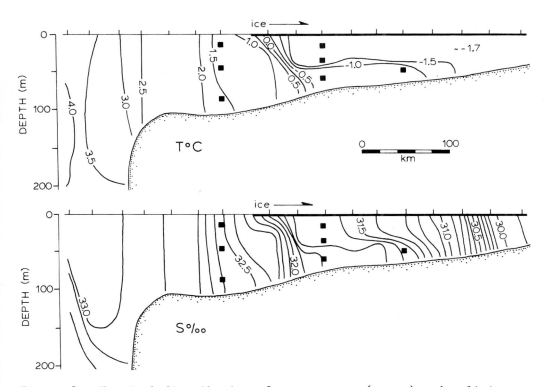

Figure 2. Vertical distribution of temperature (upper) and salinity
           (lower) along transect A (see Figure 1) on 5-9 February 1983.
           Solid squares show locations of current meters which were in line
           with transect.

shear-induced turbulence contributes appreciably to vertical mixing across the interface.

A second mechanism for heat transfer across the interface is double-diffusive convection. The vertical stratification across the interface--cool fresh water over warmer saltier water--is unstable to double-diffusive convection. Depending upon the relative gradients, this mode of convection can result in significant vertical fluxes of heat and salt.

The convective heat flux between parallel plates separated by a distance d has been calculated by Kraichnan (1962) as $Nu = 0.089\ Ra^{1/3}$, where the Nusselt (Nu) and Rayleigh (Ra) numbers have their usual definitions. For very large Rayleigh numbers ($>10^8$) typical of the interface, convective heat flux is two orders of magnitude higher than the conductive heat flux. However, the presence of the stabilizing salinity gradient inhibits overturning and hence precludes Rayleigh convection. Double-diffusive convection (DDC) can have, however, a magnitude comparable to that for Rayleigh convection. Following Marmorino and Caldwell (1976) the DDC heat flux is

$$F_T = F_T* \cdot 0.101\ \exp\ \{4.6\ \exp\ [-0.54(Tu - 1)]\}$$

where

$$F_T* = 0.089\ K_T\ (g\tfrac{\alpha}{\nu\kappa})(\Delta T)^{4/3}$$

is the heat flux for an equivalent Rayleigh convection, and the Turner number ($Tu = \beta\Delta S/\alpha\Delta T$) accounts for the stabilizing salinity influence. Using typical values across the interface of $\Delta T = 2°C$, $\beta\Delta S/\alpha\Delta T = 4$ and evaluating fluid properties at $0°C$ gives $F_T = 0.09$ kcal/m^2·s. The front is typically 100 km or more wide, so that the integrated double-diffusive heat flux per unit length along the front would be at least $\dot{q} \simeq 0.9 \cdot 10^3$ kcal/m·s.

During the winter 1983 experiments the ice was observed to move southward with a typical drift speed $U = 0.25$ m/s and a nominal thickness $t = 0.5$ m. The amount of heat $\dot{q}_i$ per unit length edge required to melt the ice is given by $q_i = \rho LUt$, where ρ is the ice density and L is the latent heat of fusion. Using values of ρ and L appropriate for the ice conditions, we find that $\dot{q} \simeq 5.3 \cdot 10^3$ kcal/m·s. Comparison of this heat requirement with the heat available by double-diffusive convection shows that about 20 percent of the required heat can be supplied by this source.

3. SUMMARY

Examination of the vertical mixing processes in the Bering Sea has shown that double-diffusive convection can supply a significant fraction--as much as 20 percent--of the heat required to maintain the ice edge at a fixed location. The results of this analysis are of particular significance because regions where cool low-salinity water overlies warm higher-salinity water occur elsewhere in the Arctic. The Bering Sea shelf is typical in many ways of other arctic shallow-shelf regions; therefore, information on mixing processes there can lead to parameterizations which will be of use in modeling other arctic shelf regions.

4. ACKNOWLEDGEMENT

This work was supported by the Office of Arctic Programs, Office of Naval Research under contract N00014-82C-0064 with Science Applications International Corporation (SAIC contribution SAI/NW-84-995-41).

REFERENCES

Kraichnan, R.H., 1962: Turbulent thermal convection at arbitrary Prandtl number. Phys. Fluids, 5, 1374-1389.

Marmorino, G.O., and D.R. Caldwell, 1976: Heat and salt transport through a diffusive thermocline interface. Deep-Sea Res., 23, 59-68.

Muench, R.D., 1983: Mesoscale oceanographic features associated with the central Bering Sea ice edge. J. Geophys. Res., 88, 2715-2722.

Muench, R.D., and K. Ahlnäs, 1976: Ice movement and distribution in the Bering Sea from March to June 1974. J. Geophys. Res., 81, 4467-4476.

Muench, R.D., and J.D. Schumacher, 1985: On the Bering Sea ice edge front. J. Geophys. Res., 90, in press.

Stigebrandt, A., 1984: The North Pacific: A global-scale estuary. J. Phys. Oceanog., 14, 464-470.

SURFACE MIXED LAYER OBSERVATION USING A METEO–OCEANOGRAPHIC SPAR BUOY, XTGP AND SEMVP SYSTEM

Sei-ichi Kanari[1] and Momoki Koga[1]
1 Department of Geophysics, Hokkaido University,
   Sapporo 060 Japan

ABSTRACT. Surface mixed layer observation using a meteo–oceanographic spar buoy, XTGP and SEMVP systems are now in progress at a site 6 km off Yoichi coast of Hokkaido.
   This paper describes general features of these systems and presents some examples of measurements by XTGP and SEMVP made independently at different sites. It is shown that these systems provide important information on the evolution and turbulent structures in the surface mixed layer.

1. Introduction

Observations of mixed layer evolution is now underways at Oshoro coast of Hokkaido using three new measurement systems. The main system consists of a Spar Buoy for meteorological observation near sea surface and a 11-layer thermister chain attached to the lower end of the immersed spar, which was moored off the coast at 50 m depth during July 20 to Sept. 20, 1984. The system can measure five meteorological parameters and one near surface water temperature successively at 30 minutes intervals. Besides, the underwater thermister chain hung below the spar can measure water temperature in the layer from 20 m to 40 m at 30 minute intervals.

XTGP(Expendable Temperature Gradient Profiler) is one of the three systems, which was reconstructed from a XBT system. Another one is a free-fall EMVP(Electro-Magnetic Velocity Profiler) for shallow sea, which is called the "SEMVP".

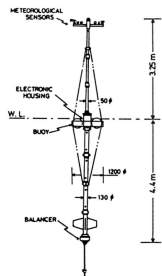

Figure 1. The meteo-oceanographic spar buoy configurations. The thermister chain is hung at the end of a line below the underwater spar.

Y. Toba and H. Mitsuyasu (eds.), The Ocean Surface, 559–564.
© *1985 by D. Reidel Publishing Company.*

As the observation with the above three systems are now in progress, an outline will be given of the three systems and also an example of an observation and its brief analysis using XTGP and SEMVP.

2. The outline of the mixed layer observation system.

2.1. Spar buoy with thermister chain.
Figure 1 shows the newly constructed Spar Buoy System.
The meteorological sensor board is on the top of the mast 3.5 m above
the sea surface. The recording unit and the other electronic compart-
ment is in the center of the buoyancy section at the base of the mast.
The buoyancy section is followed by an underwater spar of 10 cm dia
and 4.4 m long duralumin tube and a balance weight which stably holds
the spar in vertical. The meteorological mast and the under water
spar are both easily decomposed to 1 m long pieces by hand and easily
removable from one place to another.
The meteorological sensors can measure (1) wind speed (2) wind
direction (3) air temperature (4) atmospheric pressure (5) solar radiation and (6) sea surface temperature at every 30 minutes intervals.
An 11-sensor thermister chain, of 2 m intervals, is hung below
the end of the underwater spar. The chain can measure mixed layer
evolution at 30 minutes intervals in a layer depth from 20 m to 40 m.
The spar buoy system was moored 6 km off shore of Oshoro.
2.2. SEMVP
The free-fall current profiling system(Kanari, 1983) was developed for shallow water observation. The principle of measurement is
same as Sanford's EMVP(Sanford, et al.,1974), however our system use a
single velocity sensor. As well an unique data processing method has
been adopted. The free-fall unit is made from Benthos glass sphere.
The measured data were recorded on 32 KB IC-memory· RAM and the recorded
data were read out through a photo-electric coupler attached on the

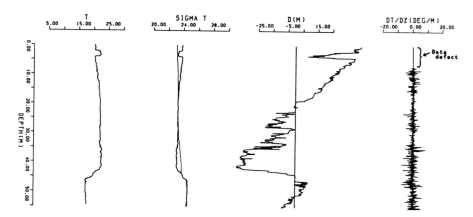

Figure 2. The profiles of temperature(T), water density
(SIGMA T), Thorpe displacement(D in meter-unit), and
temperature gradient(DT/DZ in Degree/meter).

outside of the sphere and fed to a micro computer on board. The system can measure horizontal velocity vectors about every 50 cm intervals.
2.3. XTGP
     This instrument was designed for microstructure measurements in
shallow water. The XTGP consisted of the standard XBT launcher connected to a specially designed signal conditioning circuits capable to
record both fine temperature and temperature-gradient profile. The
probe, with three drag-wings attached, descends with fall rate of 20
cm/s and can resolve temperature steps up to 2.0 cm. Measured profiles
are recorded on the analog tape recorder.

3. Examples of observed data and analysis.

     An example of
measurement by the
XTGP is shown in Fig-
ure 2 measured at the
site off Yoichi in
Sept 29, 1983. As it
is in the cooling
season, the tempera-
ture profile shows a
slight inversion in
the mixed layer from
the surface to about
40 m depth. A ther-
mocline has been
formed at the depth
from 40 m to 45 m.
The density profile
shown in Figure 2
was estimated by T-S
relations formerly
obtained in this
area. This estimate
includes an error of
±0.5 in σ_t value.
But the density in-
version seen in Fig-

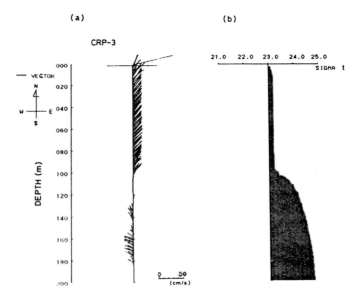

Figure 3. An example of velocity vector
profiles obtained by SEMVP(a), and mean
density profile obtained by CTD cast (b).

ure 2 is not so rare in this area in that season and is often seen in
the annual oceanographic report(Hokkaido, 1983). The estimated Thorpe
displacement D, was much greater than those in other seas (e.g. Dillon,
1982), and may be due to the density inversion trend. Correlation between r.m.s. D moving-averaged at every 1 m interval and r.m.s. temperature gradient within the mixed layer shows positive correlation
with a correlation coefficient of 0.706. Unfortunately, the data had
to be based on the pen-recorder because of troubles with the analog
data loger. Consequently, we could not estimate the fine and microscale turbulent parameters from the data described above. We will be

able to estimate these quantities from the data obtained in the observations in progress now.

Velocity profilings by SEMVP was carried out in November 16, 1983
near the weather buoy station at 29°N, 135°E of North Western Pacific
Ocean, as one of the preliminary experimental observations of OMLET(
Ocean Mixed Layer Experiment).

An example of velocity profiles obtained at every 5 hours interval
casts in shown in Figure 3 together with mean density profile measured
by CTD.   The two velocity components u and v were moving-averaged with
20 m intervals, and separated into averaged profiles and anomaly profiles respectively.   Using the averaged and anomaly profiles, vertical
shear profiles and Richardson number profiles were calculated.

In a statistically steady state with non-divergent flux, the turbulent energy balance equation(Osborn, 1980) is given by

$$-\overline{u'w'}(\partial\overline{u}/\partial Z)-\overline{v'w'}(\partial\overline{v}/\partial Z)=\epsilon-\alpha\overline{gw'T'} \qquad (1)$$

where $\tau_x=-\rho\overline{u'w'}$, $\tau_y=-\rho\overline{v'w'}$ and $B=-\alpha\overline{gw'T'}$ are the Reynolds stress and
rate of buoyancy flux per unit volume respectively, ρ is the water den-
sity, ϵ is rate of energy dissipation which is writtern in terms of
microstructure shear, and α is the specific volume of water.

Since the measureble scale range of SEMVP are confined within or
greater than the fine structure, it will not be applicable to estimate
dissipation rate by the microstructure shear.   However, if we consider
finescale turbulence in mixed layer, the dissipation term in eq.(1) can
be expressed in terms of fine scale velocity fluctuations measurable by
SEMVP.   From dimensional consideration, the fine scale dissipation ϵ_f
is given by $\epsilon_f=C\ N(Z)[\overline{(u')^2+(v')^2}]$, where $N(Z)$ is the stability frequency, C is unknown non-dimentional constand (Imberger and Hamblin,
1982).

The left-hand side of Eq.(1) may be rewritten in terms of mean
shear.   And from (1), we obtain the expression for K_Z as

$$K_z = R_i /(1-R_f)N\ C\ [\overline{(u')^2+(v')^2}] \qquad (2)$$

where, R_i is the gradient Richardson number and $R_f=B/K_z[(\partial u/\partial Z)^2+(\partial v/\partial Z)^2]$
is the flux Richardson number respectively.

On the other hand, the buoyancy flux term may be written as

$$B=-\alpha\overline{gw'T'}=\alpha gK_z (\partial\overline{T}/\partial Z) \qquad (3)$$

where K_H is the vertical eddy diffusivity of heat, and $\partial\overline{T}/\partial Z$ is mean
temperature gradient.

Using the difinition of R_f, we have

$$K_z R_f /K_H =\alpha g(\partial\overline{T}/\partial Z)/[(\partial\overline{u}/\partial Z)^2+(\partial\overline{v}/\partial Z)^2] \qquad (4)$$

The right-hand side of Eq.(4) was estimated using CTD data and
mean shear by SEMVP.   Mean value during this observation is 0.08 within
the mixed layer and 0.67 below the mixed layer base respectively.   The
ratio of K_Z and K_H cannot be specified in the present case, however,

Osborn(1980) has estimated K_z and K_H in the Equatorial Under Currents.
Osborn's estimate shows that the mean ratio of K_z and K_H is 1.5 within
the Under Current core and 9.13 in the shear layer below the core.
Then, if we take the value of 1.5 for the mixed layer and 10 for shear
layer below the mixed layer, then the flux Richardson number R_f for the
both layers becomes 0.05 for mixed layer and 0.067 for the layer below
the mixed layer. This estimate suggests that the value of R_f exhibits
nearly constant value of about 0.06 through whole layer.
        The unknown parameter C involved in Eq.(2) should to be determined
by field experiments. We also have no knowledge about it, so we search-
ed more suitable value of C so that ε_f and K_z take more realistic values.
        Figure 4 shows the estimated vertical profiles of ε_f (Fig. 4(a))
and K_z (Fig. 4(b)) for C=0.01

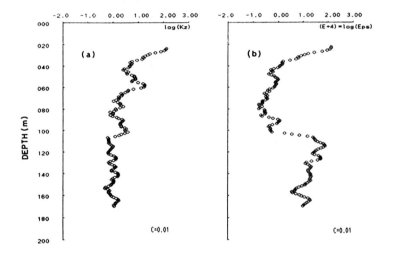

Figure 4. Mean vertical profiles of momentum
eddy diffusion coefficient(a) and finescale
dissipation rate(b) in log-scales calculated
from the Eq.(2).

4. What results can we expect from the observation in progress now ?

        As described above, variation of meteorological parameters such as
wind speed and wind direction, air temperature, air pressure, solar
radiation and near surface water temperature are measured successively
at evely 30 minutes intervals. As well the thermister chain hung below
the end of underwater spar records 11-layers water temperature variations at 30 minutes intervals. Consequently, correlation between the
heat flux through sea surface and mixed layer depth over the period of
about two months may be obtained from this observation.
        Some complementary observations using XTGP and SEMVP are also being
planned. One of such complementary observation is a concentrated survey

in which six hours interval casts will be carried out for two days.
The concentrated survey will allow us to recognize the effect of tides.
The others are temporal surveys which will be carried out before and
after windy cindition such as passing low pressure or atomospheric front.
        From such observations, we can estimate the variations of the var-
ious turbulent parameters as a function of meteorological parameters.
This will make clear the thermodynamic and hydrodynamic condition in a
coastal mixed layer and below the mixed layer base.

ACKNOWLEDGEMENTS

        The preliminary experimental observation which is a part of this
study was made by the Bosei-maru II of Tokai University. For the deployment and recovery of the SEMVP, we were aided by the staff of the
Bosei-maru II, which is gratefully acknowledged. We also thank the
research staff of Tohoku University for providing the CTD data used in
this study. This work was partly supported financially by Funds for
Scientific Research from the Ministry of Education, Japan.

REFFERENCES

Dillon, T.M., (1982): Vertical overturns: A comparison of Thorpe and
        Ozmidov length scales, J. Geophys. Res., 87, 9601-9613.
Hokkaido, (1983): Report on the marine ecological investigation near
        Ishikari Bay New Port (in Japanese), pp. 237.
Kanari, S., (1983): A desigh and construction of free-fall electro-
        magnetic velocity profiler for shallow water use. Geophys.
        Bull. Hokkaido Univ., 42, 215-228.
Osborn, T.R., (1980): Estimates of the local rate of vertical diffusion
        from dissipation measurements, J. Phys. Oceanogr., 10, 83-89.
Sanford, T.B., Drever, R.G. and Dunlap, J.H., (1974): The Design and
        performance of a free-fall electro-magnetic velocity
        profiler(EMVP), Technical Rept., Woods Hole Oceanographic Inst.,
        144
Imberger, J., and P.F. Hamblin, (1982): Dynamics of lakes reservoirs,
        and cooling ponds, Ann. Rev. Fluid Mech., 14, 153-187.

DETAILED STRUCTURE OF THE SURFACE LAYER IN THE FRONTAL ZONE BETWEEN THE KUROSHIO AND OYASHIO WATER

J. Yoshida[1], Y. Michida[2] and Y. Nagata[3]
1 Tokyo University of Fisheries, Tokyo 108 Japan
2 Hydrographic Department, Maritime Safety Agency, Tokyo 104 Japan
3 Geographical Inst., Univ. of Tokyo, Tokyo 113 Japan

ABSTRACT. From the observation of frontal region between the Kuroshio and Oyashio Water, the active phase of horizontal and vertical mixing processes are captured. The microstructure activity is found to be very high inside of the Kuroshio Front.

1. INTRODUCTION

One dimensional modelling of the surface mixed layer usually fails to simulate detailed evolution of the mixed layer structure even in the open ocean where oceanic current is weak. The effects of advection should be essential to the dynamics of the surface mixed layer in frontal regions such as the mixed water region between the Kuroshio and Oyashio Waters, because of the strong horizontal contrast in distributions of temperature, salinity and so on. The frontal regions occupy considerable area in the world ocean, and so the nature of the surface mixed layer could not be ignored in the world wide consideration on air-sea interaction. The frontal regions are also important as active water mass conversions are occurring there. Besides , it is well known that the sea temperature anomaly in the surface layer of the western North Pacific has a relation to the abnormal weather such as cold summer in the northern part of Japan. We made detailed oceanographic observations of the surface layer in the frontal region between the Kuroshio and Oyashio Fronts in July 1983. The active phase of the horizontal water mass exchange were captured in this observation, and the results suggest that the water can be lifted from deeper portion beneath the Kuroshio to the surface mixed layer in the warm eddy detached from the Kuroshio Front. The investigation of the more detailed structure of the front would be needed for the modelling of the surface layer in the mixed water regions.

2. DOUBLE STRUCTURE OF THE KUROSHIO FRONT AND THE WATER CHARACTERISTICS IN DETACHED WARM EDDY

565

Y. Toba and H. Mitsuyasu (eds.), The Ocean Surface, 565–570.
© 1985 by D. Reidel Publishing Company.

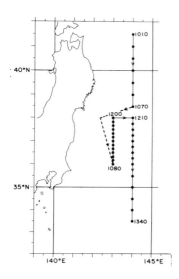

Figure 1. Observation lines in July 1983.
Solid circles (●) indicate CTD(+DO sensor)
observations.   Broken line indicate XBT
observations.

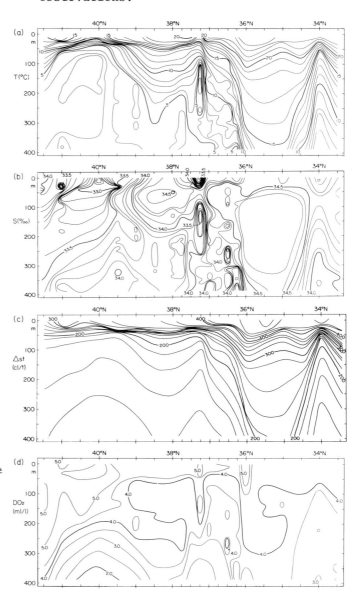

The observation lines
in July 1983 are shown
in Fig. 1.   Fine
vertical structures of
water temperature,
salinity and dissolved
oxygen were measured
and analyzed in the
layer shallower than
400m.   Temperature,
salinity, thermosteric
anomaly and dissolved
sections along 144°E
line are shown in
Fig. 2.   The Kuroshio
Front is located
between 36° and 37°N,
and has a fine structure
; two narrow zones of
large horizontal
temperature gradient or
fronts found near the
northern and southern
edges of the front, and
the relatively wide and
small gradient zone
between them(inside
zone).   Though the
temperature gradient at
the northern front is
stronger than the

Figure 2.   Temperature(a), Salinity(b), Thermosteric anomaly(c) and Dissolved oxygen(d)
sections along 144°E line.

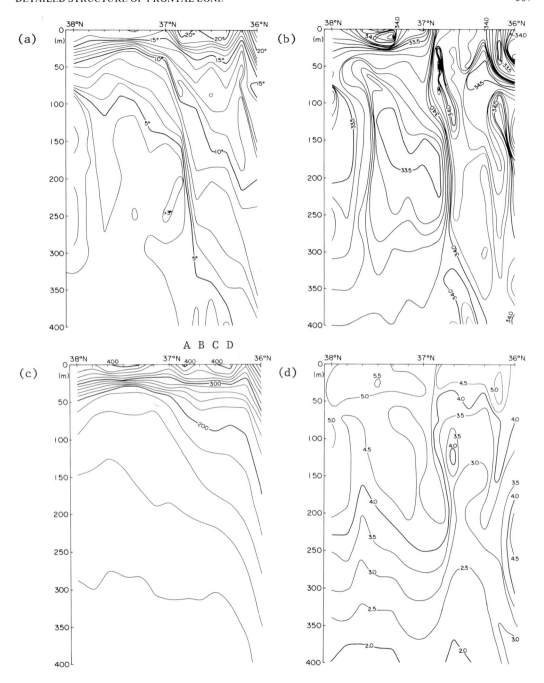

Figure 3. Temperature(a), Salinity(b), Thermosteric anomaly (c) and Dissolved oxygen (d) sections along 143°E line.

southern one, the effect of the salinty gradient on density gradient
cancels that of the temperature gradient there, and the density gradient
at the northern front is much weaker. So, we may call the northern
front as "temperature front" and the southern front as "density front".
This double structure of the Kuroshio Front is also observed along 143°
E line(Fig. 3). The detached warm eddy does not appear in the section
along 143°E, and the Kuroshio Front faces a broad water mass of the
typical mixed water between the Kuroshio and Oyashio Fronts. The narrow
low salinity and high dissolved oxygen domain just to the north of the
temperature front in the section along 144°E should be considered as a
narrow zone of this mixed water which is sandwitched between the
Kuroshio Front and the detached warm eddy.

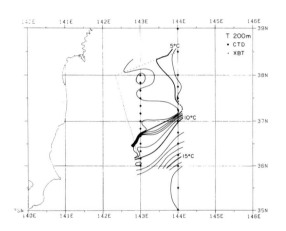

Figure 4. Horiz-
ontal temperature
distribution at
the depth of 200m.

The horizontal temperature distribution at the depth of 200m is given
in Fig. 4. These figures suggest that the double structure exists over
long distance along the Kuroshio Front. Such a double structure is
often observed in this region and reported by several investigators(e.g.
Nagata 1967,1970 and Kawai 1972).

    Warm water domain seen in the surface layer to the north of the
Kuroshio Front(between 37°30'N and 39°30'N in Fig. 2) seems to be a
warm eddy detached from the Kuroshio. The water inside of the warm eddy
has a low dissolved oxygen values. Such low dissolved oxygen value can
be found only in layers deeper than 400m in the Kuroshio region. It
should be noted, however, that the low dissolved oxygen water can be
found at relatively shallow depths in the inside zone of the Kuroshio
Front. It is clear from the configuration of iso-thermosteric anomaly
lines that the low dissolved oxygen water in the inside zone cannot be
generated by isopycnal mixing in the cross section, and that the water
should be genarated or lifted from deeper portion of somewhere upstream
frontal region. To illustrate the characteristics of this water more
clearly, the distribution of dissolved oxygen in this section is shown
in Fig. 5 taking thermosteric anomaly in the ordinate. It should be
worth-while to note here that the water having the same low dissolved
oxygen values as in the warm eddy can be found near the northern

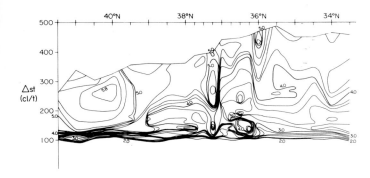

Figure 5. Dissolved
oxygen section along
144°E, taking thermo-
steric anomaly in the
ordinate.

boundary of the inside zone at the same thermosteric levels. The water
corresponding to the upper portion of the warm eddy(say, shallower than
50m or thermosteric values larger than 260cl/ton) cannot be found in
the regions of the Kuroshio Front or the Kuroshio, and this suggests
that the shallow water has been modified by some horizontal or vertical
mixings.
    The results indicate that the deeper Kuroshio Waters may be sucked
into the surface mixed layer in the detaching process of the warm eddies
, and that the detailed structure of the Kuroshio Front may play an
important role in this mechanism.

3.   Fine and micro-structure of the Kuroshio Front

Configurations of the contour lines
in Figs. 2, 3 and 5 are very
complicated especially in the inside
zone of the Kuroshio Front, and seem
to indicate that tha active mixing
processes are existing both isopycn-
ally and diapycnally there. In the
surface layer, a narrow high dissolved oxygen domain seen just to the
south of the density front(Fig. 5).
corresponding high dissolved oxygen
domain can also be seen in the
section along 143°E(Fig. 6). This
high dissolved oxygen water would
consist of a narrow stripe or patch
elongated current direction. This
may indicate the importance of the
advective effects in the surface
layer near frontal regions.
    Below the 200cl/ton surface
(the lower portion of the thermocline), many low and high dissolved
oxygen patches can be seen in the
inside zone(Fig. 5 and 6).

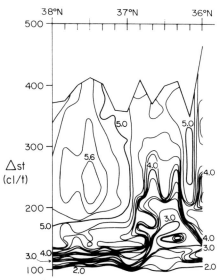

Figure 6. Dissolved oxygen
section along 143°E, taking
thermosteric anomaly in the
ordinate

It is hard to find correlation of the distributions of these patches
between two sections. These patches might be elongated oblique to the
current direction, and might indicate cross-sectional intrusion
processes in the frontal region. The vertical salinity distributions
in the inside zone along 143°E are
shown in Fig. 7. The location of
the observation points are shown in
Fig. 3(a)(the profiles A,B,C,D are
aligned from north to south and the
spacing of succesive observation
points is about 10 miles). The
profile taken just to the north of
the density front(D) has several
relatively thick homogeneous
layers as indicated by a through d
in the figure. Such homogeneous
layers tend to split or decrease
their thickness and become
ambiguous as the observation part
moves northwards or approaching to
the temperature front.

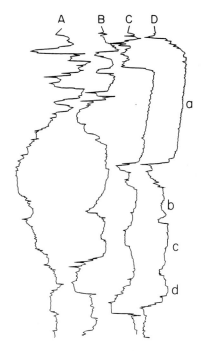

Figure 7. Change of the vertical
salinity profile along 143°E. The
location of the observation points
are shown in Figure 3(a)(A,B,C,D).
Homogeneous layers are shown as a,
b,c,d in the figure

REFERENCES

Kawai, H, 1972: Hydrography of the Kuroshio extension. In, *Kuroshio-
    Its Physical Aspects*, ed.by H.Stommel and K.Yoshida, Univ. of
    Tokyo Press, Tokyo, 235-352.
Nagata, Y, 1967: On the structure of the shallow temperature inversions
    · *J.Oceanogr.Soc. Japan*, 23(5), 221-230
Nagata, Y, 1970: Detailed temperature cross section of the cold-water
    belt along the northern edge of the Kuroshio. *J.Mar.Res.*, 28(1),
    1-14

HORIZONTAL PROCESSES INVOLVED IN THE FORMATION OF SEA SURFACE
TEMPERATURE NEAR A WESTERN BOUNDARY CURRENT

Y. Toba[1], K. Hanawa[1], H. Kawamura[1], Y. Yano[1] and Y. Kurasawa[2]
1 Department of Geophysics, Tohoku University, Sendai 980 Japan
2 Japan National Oil Corporation, Chiyoda-ku, Tokyo 100 Japan

ABSTRACT. We demonstrate, from a series of recent studies, the
importance of horizontal processes in the formation of sea surface
temperature distribution adjacent to the Pacific western boundary
current in the vicinity of Japan.

1. INTRODUCTION

Physical processes which dominate the formation of the sea surface
temperature (SST) distribution vary for different marine areas. For
example, in the area of Ocean Station Papa, where the Mixed Layer
Experiment (MILE) was carried out, and the area north to Ireland, where
the Joint Air-Sea Interaction Project (JASIN) was performed, one-
dimensional processes dominate the heat balance of the mixed layer. In
the area south and southeast of Japan, on the other hand, the
redistribution of heat by horizontal processes related to the Kuroshio
appears to dominate. We summarize a series of our recent studies to
demonstrate this situation.

2. OCEAN WEATHER STATION T

The former Ocean Weather Station T (Figure 1: hereafter call OWS-T) was
located south of Japan (29°N and 135°E). From June 1950 to November
1953, intensive marine meteorological observations (every three hours)
were made by the Japan Meteorological Agency (JMA) at OWS-T. The
oceanographic data are composed of serial observations down to 1000m
taken about 500 times during this period.
    The time series of the heat flux components through the sea surface
and the local time change $\partial/\partial t$ in the heat content H of a water column
of 200m depth were estimated from these data, and the time series of the
oceanic heat convergence F caused by horizontal advection and/or
horizontal mixing, was obtained as $F = \partial H/\partial t - Q$, where Q is the net
heat flux through the sea surface. The detailed procedure can be found
in Kurasawa et al. (1983).

Y. Toba and H. Mitsuyasu (eds.), The Ocean Surface, 571–576.
© 1985 by D. Reidel Publishing Company.

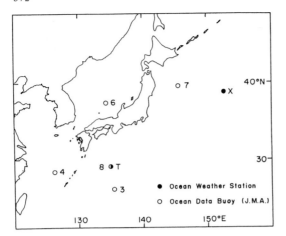

Figure 1. The former
        Japanese Ocean
        Weather Stations T
        and X, and the
        locations of the
        Ocean Data Buoy
        Stations being
        managed by the JMA.

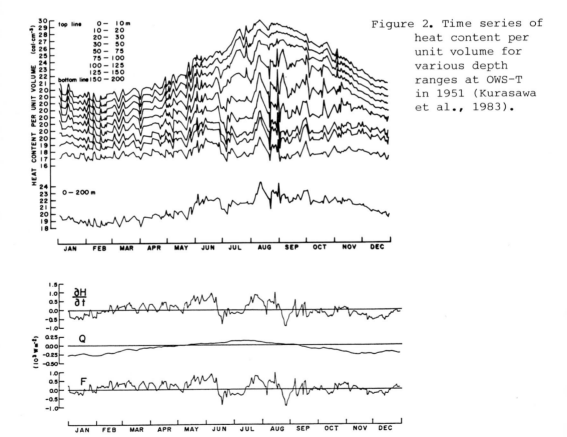

Figure 2. Time series of
        heat content per
        unit volume for
        various depth
        ranges at OWS-T
        in 1951 (Kurasawa
        et al., 1983).

Figure 3. Thirty-one day moving averaged variations of heat content
        increment of the upper 200m water column, net surface heat flux and
        of heat convergence in the sea at OWS-T in 1951 (ibid.).

The striking feature (Figure 2) is that the temperature at various depths, at least for the year 1951, varies in a highly coherent way. The range of the short-term variations is of the order of $1^{\circ}C$, with a time scale of two or three days.

For this time scale, the temperature variation of the water column is, of course, so large that the $\partial H/\partial t$ cannot be balanced by Q alone. However, even using a one month moving average, $\partial H/\partial t$ is still controlled by F (Figure 3: note that the scale of Q is two times magnified).

The monthly mean of F has a characteristic seasonal variation (Figure 4): it is positive in March, April, May and August, the maximum value reaching 500 Wm^{-2}. An overall comparison among the average monthly values of $\partial H/\partial t$, Q and F is shown in Table 1. From December to February, $\partial H/\partial t$ is determined by Q, in other words by one-dimensional processes, while from March to May and in August it is determined by F, that is, by three dimensional processes.

In a study of the velocity variation of the Kuroshio, Sekine and Toba (1981) demonstrated that abrupt increases in current velocity occur in spring, and in some years August also, causing the formation of a small meander south of the Kyushu, shown in Figure 5. Although the observation years are different, we can conjecture that the relative importance of F in the heat balance of this area is closely related to variability in the Kuroshio intensity.

Figure 6 shows time series of daily mean values of SST and the cube of the air friction velocity u_*^3 at OWS-T in 1952. The decrease in SST does not necessarily correspond to the event of strong wind. This also supports the idea of the importance to the oceanic heat balance of the passage of water masses.

The direct physical cause for the coherent variation of water column temperatures seems to be the passage of water masses with sharp fronts. From the time scale of the variability together with the typical magnitudes of the water velocity in the area, the dimensions of the water masses were 20 km or so: less than that of synoptic scale eddies.

3. JMA OCEAN DATA BUOYS

The Japan Meteorological Agency (JMA) has been operating Ocean Data Buoys (ODB) in the ocean around Japan since 1972. The present locations of the ODB's are also shown in Figure 1: OWS-T has been replaced by ODB No. 8. Data from the ODB's have been examined in order to compare the above description with the OWS-T data (Yano et al., 1984).

Figure 7 shows data at ODB No. 3, a little south of the former OWS-T, as an example of the time series of wind speed (WS) and water temperatures for 1976-1977. Although there is evidence of the development of the seasonal thermocline, depth-coherent temperature variations are evident which cannot be explained as the result of air-sea heat exchange and/or vertical wind mixing. The small water masses causing these temperature variations appear to have a variety of vertical scales.

Table 1. Seasonal variation of heat content increment of 200m water
column ∂H/∂t, the air-sea heat flux Q and the heat convergence in
the sea F at OWS-T for 1950-1953 (Kurasawa et al., 1983)

(Wm^{-2})

Month	1	2	3	4	5	6	7	8	9	10	11	12
∂H/∂t	-229	-152	105	347	228	19	143	426	-4	-166	-412	-213
Q	-271	-189	-103	-37	8	99	93	73	12	-126	-184	-262
F	42	37	208	384	220	-80	50	353	-16	-40	-228	49

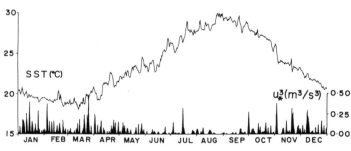

Figure 4. Monthly mean heat convegence F for the upper 200m water column
at OWS-T.   Shaded areas indicate large values in March-May (ibid.).

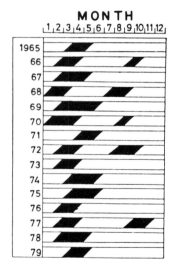

Figure 5. Periods of small meander existence off
southern Kyushu (Sekine and Toba, 1981).

Figure 6. Time series of daily mean values
of SST (upper solid line) and u_*^3 at
OWS-T in 1952.

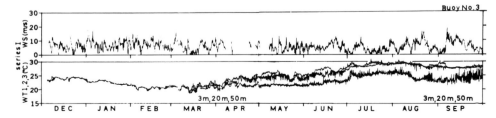

Figure 7. Time series of wind speed (WS) and water temperatures at ODB
No. 3,   17 Oct. 1976 - 13 Oct. 1977 (Yano et al., 1985).

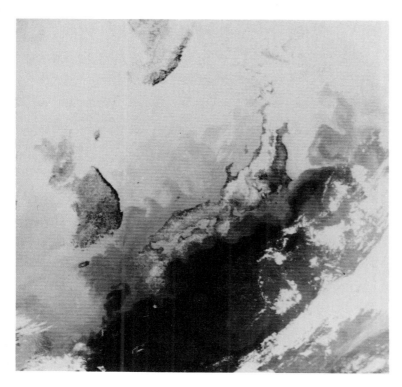

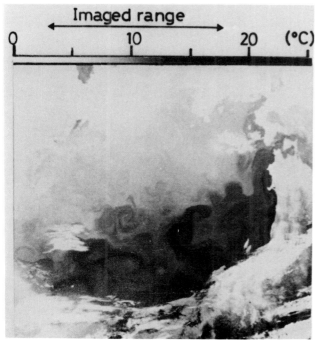

Figure 8. SST image of
      the marine area
      around Japan from
      AVHRR channel 4
      data of NOAA-6 on
      23 April 1981
      (Kawamura et al.,
      1984).

Figure 9. SST image of
      the Japan Sea from
      AVHRR channel 4 data
      of NOAA-7 on 18 May
      1982 (Toba et al.,
      1984).

4. NOAA SATELLITE INFRARED SST IMAGES

Figure 8 is an SST image derived from NOAA satellite AVHRR data for the marine area around Japan (Kawamura, et al., 1984). Fine structures in SST adjacent to the Kuroshio are seen, reminding us of the existence of small water masses passing through the area.

Figure 9 shows a similar SST-image of the Japan Sea (Toba et al., 1984). This shows many synoptic scale eddies, including two dimensional turbulence in the Tsushima Warm Current. SST structures like the bands of a spiral nebula can be seen in the eddies. These structures appear in the SST wave-number spectra as a secondary peak at a scale of 15-20 km, corresponding to that of small water masses.

5. CONCLUDING REMARKS

We point out the significance to upper-ocean heat transport in the vicinity of western boundary currents of the advection of small-scale water masses or bands torn from the boundary currents themselves. Numerical models for the prediction of SST for such an area must include these horizontal processes of heat transport in parametric form, since these small scale phenomena cannot in general be resolved by the large scale models.

REFERENCES

Kawamura, H., K. Hanawa and Y. Toba, 1984: On the characteristic structure of horizontal interleaving at the northern edge of the Kuroshio and the Kuroshio extension. *Ocean Hydrodynamics of the Japan and East China Seas*. T. Ichiye, Ed., Elsevier Sci. Pub., Amsterdam, 333-346.

Kurasawa, Y., K. Hanawa and Y. Toba, 1983: Heat balance of the surface layer of the sea at Ocean Weather Station T. *J. Oceanogr. Soc. Japan*, **39**, 192-202.

Sekine, Y. and Y. Toba, 1981: Velocity variation of the Kuroshio during formation of the small meander south of Kyushu. *J. Oceanogr. Soc. Japan*, **37**, 87-93.

Toba, Y., H. Kawamura, F. Yamashita and K. Hanawa, 1984: Structure of horizontal turbulence in the Japan Sea. *Ocean Hydrodynamics of the Japan and East China Seas*. T. Ichiye, Ed., Elsevier Sci. Pub., Amsterdam, 317-332.

Yano, Y., K. Hanawa and Y. Toba, 1985: Characteristics of the upper ocean thermal structure with its variations around Japan--From records of Ocean Data Buoys obtained by J.M.A.--. To be published in *La mer* (In Japanese with English abstract and figure captions).

AUTHOR INDEX

SUBJECT INDEX